3D and Circuit Integration of MEMS

3D and Circuit Integration of MEMS

Edited by

Masayoshi Esashi

Editor

Prof. Masayoshi Esashi
Tohoku University
Micro System Integration Center
519-1176 Aramaki-Aza-Aoba
Aoba-ku
980-0845 Sendai
Japan

■ All books published by **Wiley-VCH** are carefully produced. Nevertheless, authors, editors, and publisher do not warrant the information contained in these books, including this book, to be free of errors. Readers are advised to keep in mind that statements, data, illustrations, procedural details or other items may inadvertently be inaccurate.

Library of Congress Card No.:
applied for

British Library Cataloguing-in-Publication Data
A catalogue record for this book is available from the British Library.

Bibliographic information published by the Deutsche Nationalbibliothek
The Deutsche Nationalbibliothek lists this publication in the Deutsche Nationalbibliografie; detailed bibliographic data are available on the Internet at <http://dnb.d-nb.de>.

© 2021 WILEY-VCH, Boschstr. 12, 69469 Weinheim, Germany

All rights reserved (including those of translation into other languages). No part of this book may be reproduced in any form – by photoprinting, microfilm, or any other means – nor transmitted or translated into a machine language without written permission from the publishers. Registered names, trademarks, etc. used in this book, even when not specifically marked as such, are not to be considered unprotected by law.

Print ISBN: 978-3-527-34647-9
ePDF ISBN: 978-3-527-82325-3
ePub ISBN: 978-3-527-82324-6
oBook ISBN: 978-3-527-82323-9

Typesetting SPi Global, Chennai, India
Printing   CPI Group (UK) Ltd, Croydon CR0 4YY

Printed on acid-free paper

C095603_120321

Contents

Part I Introduction *1*

1 Overview *3*
Masayoshi Esashi
References *10*

Part II System on Chip *13*

2 Bulk Micromachining *15*
Xinxin Li and Heng Yang
2.1 Process Basis of Bulk Micromachining Technologies *16*
2.2 Bulk Micromachining Based on Wafer Bonding *20*
2.2.1 SOI MEMS *20*
2.2.2 Cavity SOI Technology *27*
2.2.3 Silicon on Glass Processes: Dissolved Wafer Process (DWP) *29*
2.3 Single-Wafer Single-Side Processes *34*
2.3.1 Single-Crystal Reactive Etching and Metallization Process (SCREAM) *34*
2.3.2 Sacrificial Bulk Micromachining (SBM) *38*
2.3.3 Silicon on Nothing (SON) *40*
References *45*

3 Enhanced Bulk Micromachining Based on MIS Process *49*
Xinxin Li and Heng Yang
3.1 Repeating MIS Cycle for Multilayer 3D structures or Multi-sensor Integration *49*
3.1.1 Pressure Sensors with PS^3 Structure *49*
3.1.2 P+G Integrated Sensors *52*
3.2 Pressure Sensor Fabrication – From MIS Updated to TUB *54*
3.3 Extension of MIS Process for Various Advanced MEMS Devices *58*
References *58*

4	**Epitaxial Poly Si Surface Micromachining** *61*	
	Masayoshi Esashi	
4.1	Process Condition of Epi-poly Si *61*	
4.2	MEMS Devices Using Epi-poly Si *61*	
	References *67*	
5	**Poly-SiGe Surface Micromachining** *69*	
	Carrie W. Low, Sergio F. Almeida, Emmanuel P. Quévy, and Roger T. Howe	
5.1	Introduction *69*	
5.1.1	SiGe Applications in IC and MEMS *70*	
5.1.2	Desired SiGe Properties for MEMS *70*	
5.2	SiGe Deposition *70*	
5.2.1	Deposition Methods *70*	
5.2.2	Material Properties Comparison *71*	
5.2.3	Cost Analysis *72*	
5.3	LPCVD Polycrystalline SiGe *73*	
5.3.1	Vertical Furnace *73*	
5.3.2	Particle Control *75*	
5.3.3	Process Monitoring and Maintenance *75*	
5.3.4	In-line Metrology for Film Thickness and Ge Content *76*	
5.3.5	Process Space Mapping *77*	
5.4	CMEMS® Process *78*	
5.4.1	CMOS Interface Challenges *79*	
5.4.2	CMEMS Process Flow *80*	
5.4.2.1	Top Metal Module *80*	
5.4.2.2	Plug Module *84*	
5.4.2.3	Structural SiGe Module *85*	
5.4.2.4	Slit Module *85*	
5.4.2.5	Structure Module *85*	
5.4.2.6	Spacer Module *85*	
5.4.2.7	Electrode Module *85*	
5.4.2.8	Pad Module *86*	
5.4.3	Release *86*	
5.4.4	Al–Ge Bonding for Microcaps *87*	
5.5	Poly-SiGe Applications *88*	
5.5.1	Resonator for Electronic Timing *88*	
5.5.2	Nano-electro-mechanical Switches *92*	
	References *94*	
6	**Metal Surface Micromachining** *99*	
	Minoru Sasaki	
6.1	Background of Surface Micromachining *99*	
6.2	Static Device *100*	
6.3	Static Structure Fixed after the Single Movement *101*	
6.4	Dynamic Device *103*	

6.4.1	MEMS Switch	*103*
6.4.2	Digital Micromirror Device	*104*
6.5	Summary	*111*
	References	*111*

7 Heterogeneously Integrated Aluminum Nitride MEMS Resonators and Filters *113*
Enes Calayir, Srinivas Merugu, Jaewung Lee, Navab Singh, and Gianluca Piazza

7.1	Overview of Integrated Aluminum Nitride MEMS	*113*
7.2	Heterogeneous Integration of Aluminum Nitride MEMS Resonators with CMOS Circuits	*114*
7.2.1	Aluminum Nitride MEMS Process Flow	*115*
7.2.2	Encapsulation of Aluminum Nitride MEMS Resonators and Filters	*116*
7.2.3	Redistribution Layers on Top of Encapsulated Aluminum Nitride MEMS	*118*
7.2.4	Selected Individual Resonator and Filter Frequency Responses	*119*
7.2.5	Flip-chip Bonding of Aluminum Nitride MEMS with CMOS	*121*
7.3	Heterogeneously Integrated Self-Healing Filters	*123*
7.3.1	Application of Statistical Element Selection (SES) to AlN MEMS Filters with CMOS Circuits	*123*
7.3.2	Measurement of 3D Hybrid Integrated Chip Stack	*124*
	References	*127*

8 MEMS Using CMOS Wafer *131*
Weileun Fang, Sheng-Shian Li, Yi Chiu, and Ming-Huang Li

8.1	Introduction: CMOS MEMS Architectures and Advantages	*131*
8.2	Process Modules for CMOS MEMS	*139*
8.2.1	Process Modules for Thin Films	*140*
8.2.1.1	Metal Sacrificial	*140*
8.2.1.2	Oxide Sacrificial	*142*
8.2.1.3	TiN-composite (TiN-C)	*143*
8.2.2	Process Modules for the Substrate	*145*
8.2.2.1	SF_6 and XeF_2 (Dry Isotropic)	*145*
8.2.2.2	KOH and TMAH (Wet Anisotropic)	*146*
8.2.2.3	RIE and DRIE (Front-side RIE, Backside DRIE)	*146*
8.3	The 2P4M CMOS Platform (0.35 µm)	*148*
8.3.1	Accelerometer	*148*
8.3.2	Pressure Sensor	*149*
8.3.3	Resonators	*150*
8.3.4	Others	*152*
8.4	The 1P6M CMOS Platform (0.18 µm)	*154*
8.4.1	Tactile Sensors	*154*
8.4.2	IR Sensor	*156*
8.4.3	Resonators	*158*

8.4.4	Others	*160*
8.5	CMOS MEMS with Add-on Materials	*164*
8.5.1	Gas and Humidity Sensors	*164*
8.5.1.1	Metal Oxide	*164*
8.5.1.2	Polymer	*170*
8.5.2	Biochemical Sensors	*173*
8.5.3	Pressure and Acoustic Sensors	*175*
8.5.3.1	Microfluidic Structures	*178*
8.6	Monolithic Integration of Circuits and Sensors	*180*
8.6.1	Multi-sensor Integration	*180*
8.6.1.1	Gas Sensors	*180*
8.6.1.2	Physical Sensors	*181*
8.6.2	Readout Circuit Integration	*183*
8.6.2.1	Resistive Sensors	*183*
8.6.2.2	Capacitive Sensors	*184*
8.6.2.3	Inductive Sensors	*188*
8.6.2.4	Resonant Sensors	*190*
8.7	Issues and Concerns	*191*
8.7.1	Residual Stresses, CTE Mismatch, and Creep of Thin Films	*192*
8.7.1.1	Initial Deformation – Residual Stress	*192*
8.7.1.2	Thermal Deformation – Thermal Expansion Coefficient Mismatch	*195*
8.7.1.3	Long-time Stability – Creep	*197*
8.7.2	Quality Factor, Materials Loss, and Temperature Stability	*199*
8.7.2.1	Anchor Loss	*201*
8.7.2.2	Thermoelastic Damping (TED)	*201*
8.7.2.3	Material and Interface Loss	*201*
8.7.3	Dielectric Charging	*203*
8.7.4	Nonlinearity and Phase Noise in Oscillators	*204*
8.8	Concluding Remarks	*205*
	References	*207*
9	**Wafer Transfer** *221*	
	Masayoshi Esashi	
9.1	Introduction	*221*
9.2	Film Transfer	*223*
9.3	Device Transfer (via-last)	*228*
9.4	Device Transfer (Via-First)	*231*
9.5	Chip Level Transfer	*236*
	References	*241*
10	**Piezoelectric MEMS** *243*	
	T Takeshi Kobayashi (AIST)	
10.1	Introduction	*243*
10.1.1	Fundamental	*243*
10.1.2	PZT Thin Films Property as an Actuator	*244*
10.1.3	PZT Thin Film Composition and Orientation	*246*

10.2	PZT Thin Film Deposition	*246*
10.2.1	Sputtering	*246*
10.2.2	Sol–Gel	*248*
10.2.2.1	Orientation Control	*248*
10.2.2.2	Thick Film Deposition	*249*
10.2.3	Electrode Materials and Lifetime of PZT Thin Films	*250*
10.3	PZT–MEMS Fabrication Process	*251*
10.3.1	Cantilever and Microscanner	*251*
10.3.2	Poling	*254*
	References	*255*

Part III Bonding, Sealing and Interconnection *257*

11 Anodic Bonding *259*
Masayoshi Esashi

11.1	Principle	*259*
11.2	Distortion	*262*
11.3	Influence of Anodic Bonding to Circuits	*263*
11.4	Anodic Bonding with Various Materials, Structures and Conditions	*265*
11.4.1	Various Combinations	*265*
11.4.2	Anodic Bonding with Intermediate Thin Films	*269*
11.4.3	Variation of Anodic Bonding	*271*
11.4.4	Glass Reflow Process	*274*
	References	*276*

12 Direct Bonding *279*
Hideki Takagi

12.1	Wafer Direct Bonding	*279*
12.2	Hydrophilic Wafer Bonding	*279*
12.3	Surface Activated Bonding at Room Temperature	*283*
	References	*286*

13 Metal Bonding *289*
Joerg Froemel

13.1	Solid Liquid Interdiffusion Bonding (SLID)	*290*
13.1.1	Au/In and Cu/In	*291*
13.1.2	Au/Ga and Cu/Ga	*294*
13.1.3	Au/Sn and Cu/Sn	*297*
13.1.4	Void Formation	*297*
13.2	Metal Thermocompression Bonding	*298*
13.2.1.1	Interface Formation	*299*
13.2.1.2	Grain Reorientation	*299*
13.2.1.3	Grain Growth	*300*
13.3	Eutectic Bonding	*301*

13.3.1	Au/Si	*302*
13.3.2	Al/Ge	*302*
13.3.3	Au/Sn	*304*
	References	*304*

14 Reactive Bonding *309*
Klaus Vogel, Silvia Hertel, Christian Hofmann, Mathias Weiser, Maik Wiemer, Thomas Otto, and Harald Kuhn

14.1	Motivation	*309*
14.2	Fundamentals of Reactive Bonding	*309*
14.3	Material Systems	*311*
14.4	State of the Art	*312*
14.5	Deposition Concepts of Reactive Material Systems	*313*
14.5.1	Physical Vapor Deposition	*313*
14.5.1.1	Conclusion Physical Vapor Deposition and Patterning	*315*
14.5.2	Electrochemical Deposition of Reactive Material Systems	*315*
14.5.2.1	Dual Bath Technology	*316*
14.5.2.2	Single Bath Technology	*318*
14.5.2.3	Conclusion DBT and SBT	*319*
14.5.3	Vertical Reactive Material Systems With 1D Periodicity	*319*
14.5.3.1	Dimensioning	*320*
14.5.3.2	Fabrication	*321*
14.5.3.3	Conclusion	*323*
14.6	Bonding With RMS	*323*
14.7	Conclusion	*326*
	References	*326*

15 Polymer Bonding *331*
Xiaojing Wang and Frank Niklaus

15.1	Introduction	*331*
15.2	Materials for Polymer Wafer Bonding	*332*
15.2.1	Polymer Adhesion Mechanisms	*332*
15.2.2	Properties of Polymers for Wafer Bonding	*335*
15.2.3	Polymers Used in Wafer Bonding	*337*
15.3	Polymer Wafer Bonding Technology	*341*
15.3.1	Process Parameters in Polymer Wafer Bonding	*341*
15.3.2	Localized Polymer Wafer Bonding	*348*
15.4	Precise Wafer-to-Wafer Alignment in Polymer Wafer Bonding	*350*
15.5	Practical Examples of Polymer Wafer Bonding Processes	*351*
15.6	Summary and Conclusions	*354*
	References	*354*

16 Soldering by Local Heating *361*
Yu-Ting Cheng and Liwei Lin

16.1	Soldering in MEMS Packaging	*361*
16.2	Laser Soldering	*362*

16.3	Resistive Heating and Soldering	*365*
16.4	Inductive Heating and Soldering	*368*
16.5	Other Localized Soldering Processes	*370*
16.5.1	Self-propagative Reaction Heating	*370*
16.5.2	Ultrasonic Frictional Heating	*371*
	References *374*	

17 Packaging, Sealing, and Interconnection *377*
Masayoshi Esashi

17.1	Wafer Level Packaging	*377*
17.2	Sealing	*378*
17.2.1	Reaction Sealing	*378*
17.2.2	Deposition Sealing (Shell Packaging)	*380*
17.2.3	Metal Compression Sealing	*385*
17.3	Interconnection	*388*
17.3.1	Vertical Feedthrough Interconnection	*388*
17.3.1.1	Through Glass via (TGV) Interconnection	*388*
17.3.1.2	Through Si via (TSiV) Interconnection	*393*
17.3.2	Lateral Feedthrough Interconnection	*395*
17.3.3	Interconnection by Electroplating	*401*
	References *404*	

18 Vacuum Packaging *409*
Masayoshi Esashi

18.1	Problems of Vacuum Packaging	*409*
18.2	Vacuum Packaging by Anodic Bonding	*409*
18.3	Packaging by Anodic Bonding with Controlled Cavity Pressure	*414*
18.4	Vacuum Packaging by Metal Bonding	*416*
18.5	Vacuum Packaging by Deposition	*417*
18.6	Hermeticity Testing	*417*
	References *420*	

19 Buried Channels in Monolithic Si *423*
Kazusuke Maenaka

19.1	Buried Channel/Cavity in LSI and MEMS	*423*
19.2	Monolithic SON Technology and Related Technologies	*425*
19.3	Applications of SON	*435*
	References *439*	

20 Through-substrate Vias *443*
Zhyao Wang

20.1	Configurations of TSVs	*444*
20.1.1	Solid TSVs	*444*
20.1.2	Hollow TSVs	*445*
20.1.3	Air-gap TSVs	*445*
20.2	TSV Applications in MEMS	*445*

20.2.1	Signal Conduction to the Wafer Backside	446
20.2.2	CMOS-MEMS 3D Integration	446
20.2.3	MEMS and CMOS 2.5D Integration	447
20.2.4	Wafer-level Vacuum Packaging	448
20.2.5	Other Applications	450
20.3	Considerations for TSV in MEMS	450
20.4	Fundamental TSV Fabrication Technologies	450
20.4.1	Deep Hole Etching	451
20.4.1.1	Deep Reactive Ion Etching	451
20.4.1.2	Laser Ablation	452
20.4.2	Insulator Formation	454
20.4.2.1	Silicon Dioxide Insulators	454
20.4.2.2	Polymer Insulators	455
20.4.2.3	Air-gaps	455
20.4.3	Conductor Formation	455
20.4.3.1	Polysilicon	456
20.4.3.2	Single Crystalline Silicon	456
20.4.3.3	Tungsten	457
20.4.3.4	Copper	457
20.4.3.5	Other Conductor Materials	459
20.5	Polysilicon TSVs	460
20.5.1	Solid Polysilicon TSVs	460
20.5.2	Air-gap Polysilicon TSVs	463
20.6	Silicon TSVs	464
20.6.1	Solid Silicon TSVs	465
20.6.2	Air-gap Silicon TSVs	467
20.7	Metal TSVs	469
20.7.1	Solid Metal TSVs	470
20.7.2	Hollow Metal TSVs	474
20.7.3	Air-gap Metal TSVs	480
	References	481

Index *493*

Part I

Introduction

1

Overview

Masayoshi Esashi

Tohoku University, Micro System Integration Center (μSIC), 519-1176 Aramaki-Aza-Aoba, Aoba-ku, Sendai 980-0845, Japan

Micro-electro mechanical systems (MEMS) have been used for versatile components as sensors and are called microsystems. This technology is based on advanced arts of microfabrication developed for integrated circuit (IC) on a Si (silicon) wafer. The microfabrication uses patterning with photolithography by which many patterns on a photomask are transferred on the surface of the Si wafer. This batch transfer enables to fabricate 10 billion transistors on a chip by using the latest technology, which corresponds to one trillion transistors on a 12 in. (300 mm) diameter Si wafer. We can fabricate thick (2.5 dimensional) structures by extending the microfabrication with etching and deposition. The extended technologies of microfabrication is called micromachining. The MEMS that have versatile components and circuits on a chip play important roles in advanced systems for user interface, wireless communication, Internet of things (IoT), and so on. This book *3D and Circuit Integration of MEMS* deals with various configurations of MEMS as shown in Figure 1.1. There are two kinds of MEMS. One is monolithic type called system on chip (SoC) MEMS and the other is hybrid type called system in package (SiP) MEMS. The former SoC MEMS means that MEMS and circuits can be fabricated on the same chip. This will be explained from Chapter 2 to Chapter 10. The SoC MEMS reduces interconnection complexity and enables arrayed active matrix MEMS; on the other hand it reduces the freedom of MEMS processes because of restrictions such as thermal budget limited by the circuits. The latter SiP MEMS means that MEMS and circuits are fabricated on different chips and assembled in a package as will be discussed later using Figure 1.5.

The bulk micromachining in Figure 1.1 will be described in Chapter 2, which uses etching of the bulk Si wafer to make MEMS structures. The surface micromachining in Figure 1.1 is shown in Figure 1.2a. Sacrificial layer is deposited and patterned followed by the deposition and patterning of structural layer. The sacrificial layer is selectively etched out, and the remained structural layer is used for the components of MEMS.

3D and Circuit Integration of MEMS, First Edition. Edited by Masayoshi Esashi.
© 2021 WILEY-VCH GmbH. Published 2021 by WILEY-VCH GmbH.

1 Overview

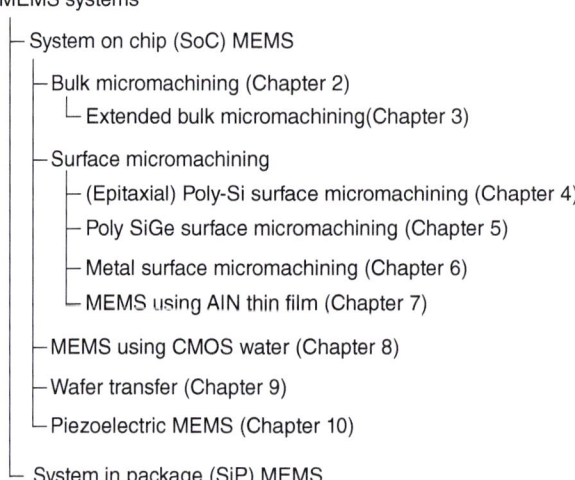

MEMS systems
- System on chip (SoC) MEMS
  - Bulk micromachining (Chapter 2)
    - Extended bulk micromachining (Chapter 3)
  - Surface micromachining
    - (Epitaxial) Poly-Si surface micromachining (Chapter 4)
    - Poly SiGe surface micromachining (Chapter 5)
    - Metal surface micromachining (Chapter 6)
    - MEMS using AlN thin film (Chapter 7)
  - MEMS using CMOS water (Chapter 8)
  - Wafer transfer (Chapter 9)
  - Piezoelectric MEMS (Chapter 10)
- System in package (SiP) MEMS

Figure 1.1 Various configurations of MEMS.

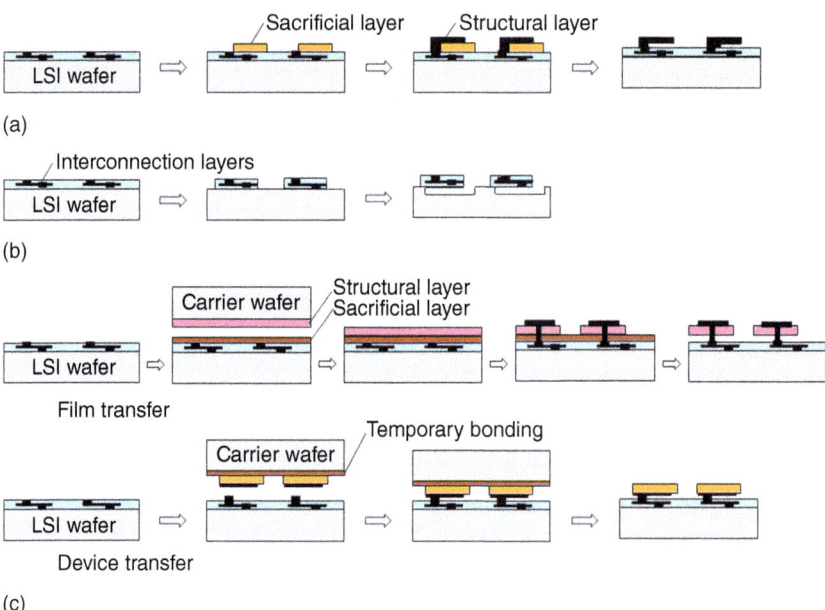

Figure 1.2 Various SoC MEMS by LSI first approach (a) Surface micromachining (deposition based) (chap. 4, 5, 6 and 7), (b) MEMS using CMOS wafer (chap. 8), (c) Wafer transfer (chap. 9 and 10).

MEMS by the surface micromachining integrated with n-channel MOS (NMOS) FET and complementary metal oxide semiconductor (CMOS) circuits are pioneered by Prof. R. T. Howe and Prof. R. S. Muller in the University of California, Berkeley, United States. The fabrication process of these will be explained later in Figure 1.3 and Figure 1.4, respectively. In recent years large-scale integration (LSI) uses standardized processes in foundries and the process is not flexible. For

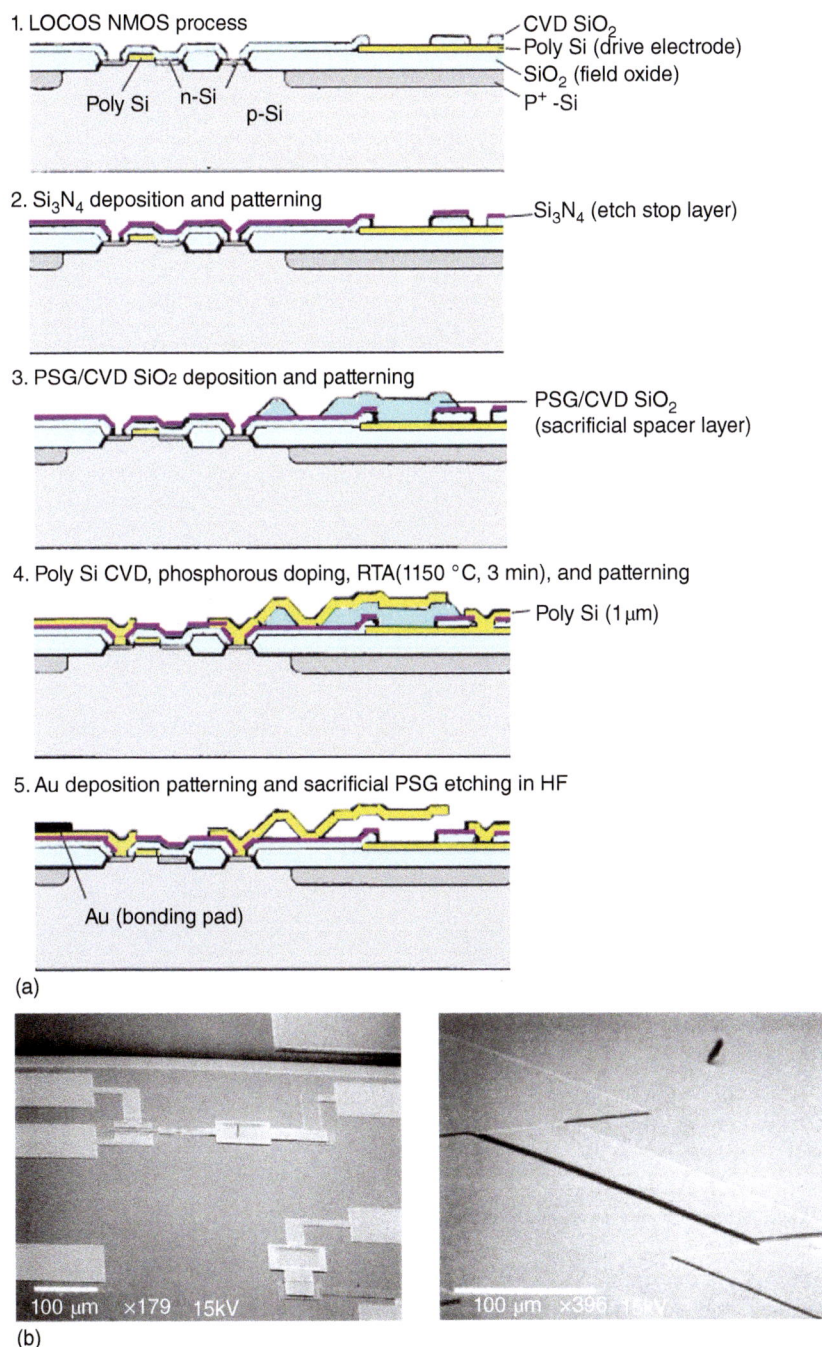

Figure 1.3 NMOS integrated poly Si resonant microstructure (SoC MEMS). (a) Fabrication process (b) Photographs of microstructure (c) Circuit with NMOSFET and resonator. Source: Putty et al. [1]. © 1989, Elsevier.

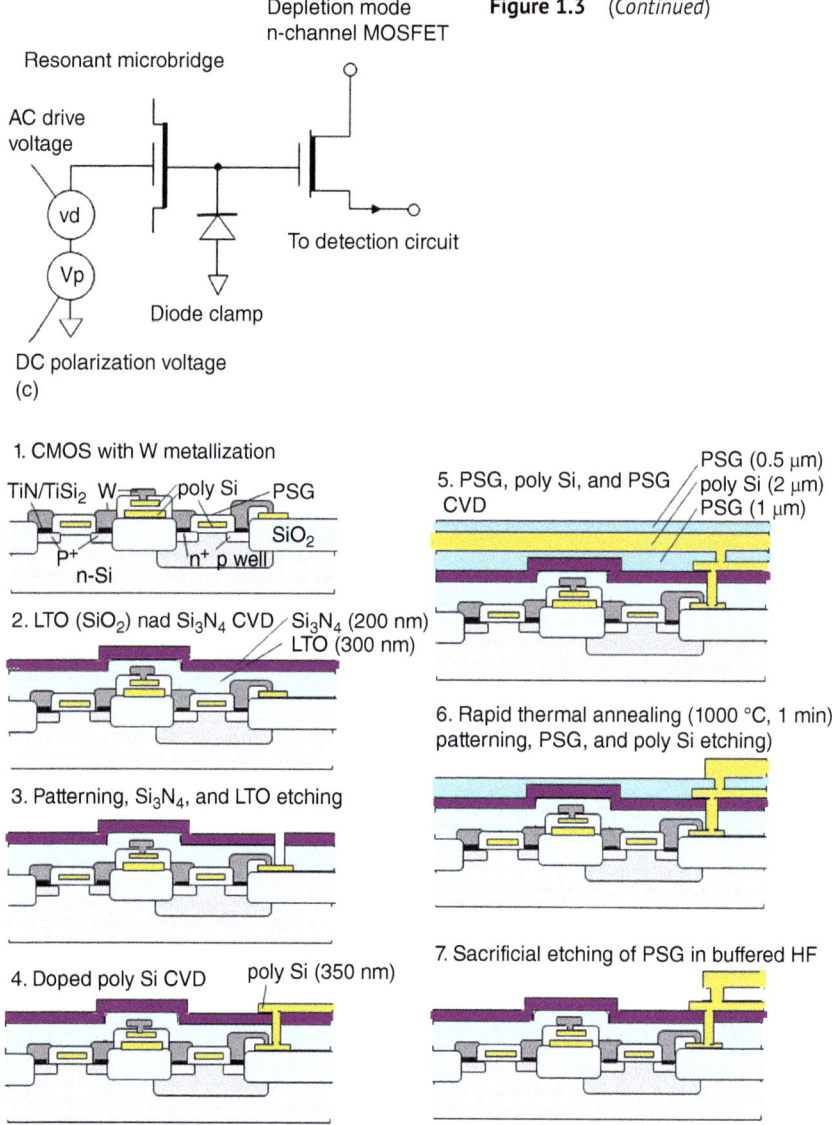

Figure 1.4 Fabrication process of surface-micromachined poly Si microstructures integrated with CMOS circuit. Source: Based on Yun et al. [2]; Bustillo et al. [3].

this reason, LSI first approach is commonly used as shown in Figure 1.2. This is required not to contaminate the LSI production line as well.

Polycrystalline Si (poly Si) surface micromachining uses poly Si as the structural layer. The surface micromachining using epitaxial poly Si (epi-poly Si) will be explained in Chapter 3. The poly Si has a good mechanical property as a material for spring, moving element, etc. and can be used for sensors or actuators as resonators. High temperature around 600 °C is required to deposit the poly Si by chemical vapor deposition (CVD). The co-fabricated IC has to stand this temperature. Resonant microstructures made of poly Si were integrated with depletion type NMOS field effect transistor (FET) [4]. The fabrication process is shown in Figure 1.3a [1, 4]. The NMOS FETs that have poly Si gate are fabricated using conventional local oxidation of Si (LOCOS) technology. The poly Si layer for the gate of the FET is also used as a drive electrode for the resonator (1 in (a)). Si_3N_4 is deposited and patterned (etched after photoresist patterning) for the purpose of etch stop layer needed for the etching of the sacrificial layer (2 in (a)). Phosphosilicate glass (PSG) and SiO_2 are deposited by the CVD for the purpose of sacrificial spacer layer (3 in (a)). Poly Si is deposited at 600 °C and doped with phosphorous by ion implantation. It is patterned after rapid thermal annealing (RTA) at 1150 °C for three minutes. The RTA is needed to control the stress of the poly Si (4 in (a)). Au (gold) is deposited and patterned for bonding pads, and finally the PSG sacrificial layer is etched out in HF (hydrofluoric acid) to make poly Si microcantilever for the resonator (5 in (a)). The Au is used on behalf of conventional Al (aluminum) because the Au is not etched in HF. Photographs of the chip surface and the self-suspended poly Si cantilever fabricated are shown in Figure 1.3b. Circuit with the NMOS FET and the resonator (resonant microbridge) are illustrated in Figure 1.3c. The circuit for capacitance detection has to be monolithically integrated with the MEMS microstructure in order to minimize the influence of stray capacitance.

The surface-micromachined poly Si microstructures were integrated with CMOS circuit as well, being called modular integration of CMOS with microstructures (MICS) [2, 3]. The fabrication process is shown in Figure 1.4. The poly Si surface micromachining requires 600 °C for deposition, and the thermal budget requires metallization different from conventional Al. W (tungsten) and $TiN/TiSi_2$ were used for the metallization and diffusion barrier at the metal/silicon contacts, respectively, to stand the high temperature process (1 in the figure). Low-temperature oxide (LTO) (SiO_2) and Si_3N_4 are deposited by CVD (2 in the figure), and the Si_3N_4 and the LTO are patterned (3 in the figure). Doped poly Si is deposited and patterned (4 in the figure). PSG, poly Si, and 2nd PSG are deposited by CVD (5 in the figure). RTA is made and the PSG and the poly Si are patterned (6 in the figure). Movable poly Si microstructures are formed by sacrificial etching of the PSG in buffered HF (7 in the figure).

The thickness of the poly Si is limited to less than 2 μm because of its stress, which causes a bending of the wafer. On the other hand, low-stress epi-poly Si that makes a layer thickness of 20 μm or more possible was developed in Fraunhofer Institute for Silicon Technologies (ISIT) in Germany and Uppsala University in Sweden [5]. Surface micromachining using the epi-poly Si has been used for capacitive sensors as accelerometer and gyroscope. The lateral capacitance of micromechanical structure can be increased because of the thick epi-poly Si layer. The capacitance detection circuits need not to be monolithically integrated for the SoC MEMS. The epi-poly Si surface micromachining will be explained in Chapter 4.

The poly SiGe surface micromachining categorized in Figure 1.1 was developed to achieve low deposition temperature (410 °C) [6]. This will be explained in Chapter 5. Ge (germanium) is used as the sacrificial layer because it can be etched out selectively in H_2O_2 (hydrogen peroxide).

Metal surface micromachining was developed for the purpose of mirror array fabricated on a CMOS LSI. Metals can be deposited by sputtering or evaporation. Since it doesn't require high temperature, thermal damage to the CMOS LSI is not a problem and photoresist can be used as a sacrificial layer. The durability was improved by using amorphous metal. The mirror array has been used successfully for video projectors and other systems as will be explained in Chapter 6.

The other surface micromachining is a MEMS using AlN (aluminum nitride). The AlN is a piezoelectric material and can be deposited at low temperature by reactive sputtering. This enables SoC MEMS that have surface micromachined structures having piezoelectric material on a circuit. This will be explained in Chapter 7.

The MEMS using CMOS LSI wafer in Figure 1.1 is the SOC MEMS and schematically shown in Figure 1.2b. The multilayers for interconnection on a CMOS wafer and the bulk Si are used as the MEMS structure, and the Si under the MEMS structure can be under-etched if necessary. This will be explained in Chapter 8.

Wafer transfer methods shown in Figure 1.2c will be explained in Chapter 9. There are two ways. One is film transfer in which structural layer on the carrier wafer is transferred on an LSI wafer, and MEMS are fabricated on the LSI wafer. The other is device transfer in which MEMS are fabricated on the carrier wafer and transferred on the LSI wafer by using bump or other methods. The MEMS are left on the LSI wafer by etching out the carrier wafer or the temporary bonding layer, otherwise by debonding (laser lift-off) the MEMS from the carrier wafer. The advantage of the wafer transfer method is that the structural layer or MEMS are not fabricated on the LSI wafer but on the carrier wafer, and hence the fabrication process has flexibility. High temperature process can be applied on the carrier wafer.

Some piezoelectric material such as PZT (lead zirconate titanate) requires high temperature (700 °C) for deposition by sputtering. The wafer transfer method can be applied for the fabrication of the SoC MEMS with the PZT. The piezoelectric MEMS will be described in Chapter 10.

Monolithic SoC MEMS have been discussed earlier. The other approach is a hybrid type called SiP MEMS. MEMS chips and LSI chips can be connected with each other as shown in Figure 1.5. MEMS that have to be exposed as microphone use side-by-side configuration with LSI as shown in in Figure 1.5a. The side-by-side configuration of MEMS chip and LSI chip (Figure 1.5b) and the stacked configuration of MEMS chip on LSI chip (Figure 1.5c) are placed in a can or ceramic package. The side-by-side configuration of packaged MEMS chip and LSI chip (Figure 1.5d) and the stacked configuration of LSI chip on packaged MEMS chip (Figure 1.5e) can be molded with polymer. An advantage of the hybrid approach is that each MEMS and LSIs can be fabricated separately using optimized processes.

Unpackaged MEMS chips can't be molded with resin because these have moving elements on them. Packaging is needed for MEMS and especially a wafer level packaging plays important roles in MEMS [7]. Bonding, sealing, and interconnection are required for the MEMS process and the packaging. The packaged MEMS chip shown in Figure 1.5d,e uses glass frit (solder glass and low-melting-point glass) for the bonding and sealing. This will be explained using Figure 4.6 (Chapter 4) and Figure 17.35 (Chapter 17). Such elementary technologies for the MEMS packaging

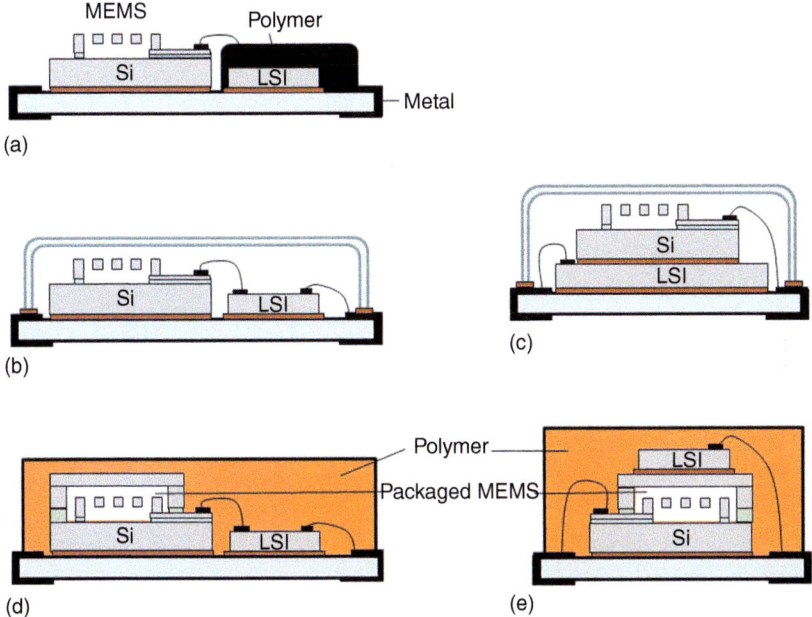

Figure 1.5 SiP MEMS (a) Side-by-side hybrid configuration of exposed MEMS chip and polymer coated LSI chip, (b) Side-by-side hybrid configuration of MEMS chip and LSI chip in can or ceramic package, (c) Stacked hybrid configuration of MEMS chip on LSI chip in can or ceramic package, (d) Side-by side hybrid configuration of packaged MEMS chip and LSI chip with polymer molding, (e) Stacked configuration of LSI chip on packaged MEMS chip with polymer molding.

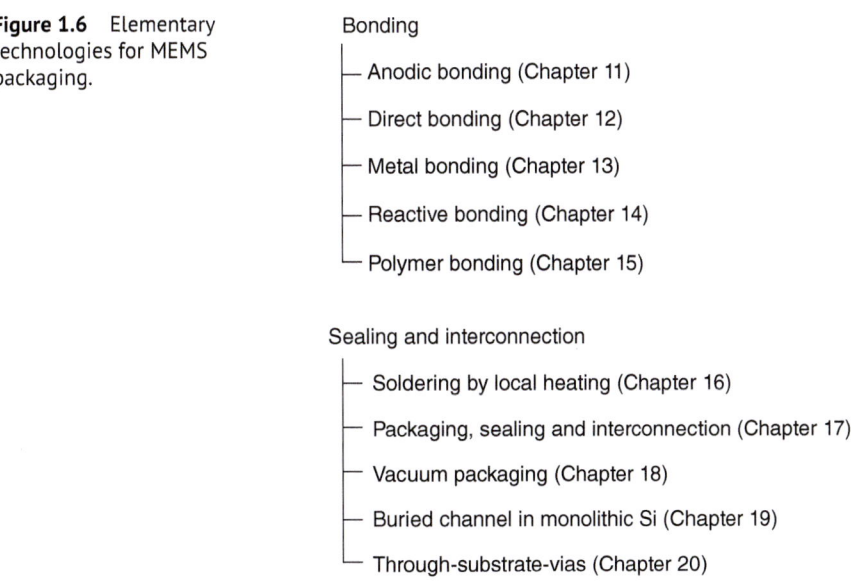

Figure 1.6 Elementary technologies for MEMS packaging.

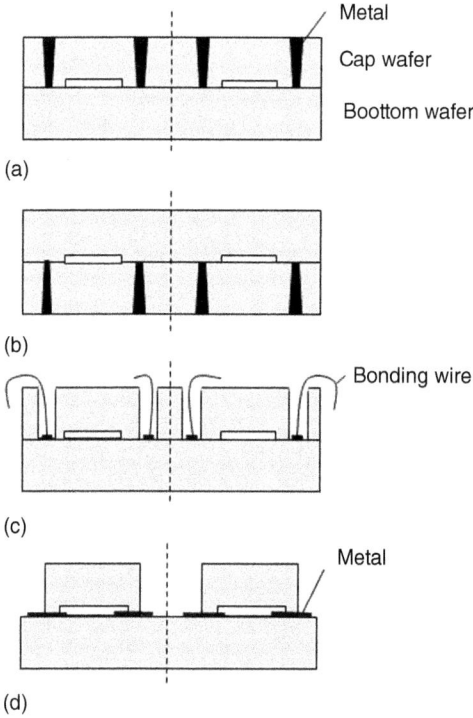

Figure 1.7 Interconnection methods from the packaged MEMS (a) Through-substrate-vias (cap), (b) Through-substrate-vias (bottom), (c) Wire-bonded through-Si vias, (d) Lateral feedthrough.

listed in Figure 1.6 will be explained from Chapter 11 to Chapter 20 of this book. Five bonding methods will be explained as follows. These are anodic bonding (Chapter 10), direct bonding (Chapter 11), metal bonding (Chapter 13), reactive bonding (Chapter 14), and polymer bonding (Chapter 15). Sealing and interconnection methods explained are soldering by local heating (Chapter 16), packaging, sealing, and interconnection (Chapter 17), vacuum packaging (Chapter 18), buried channel in monolithic Si (Chapter 19), and through-substrate vias (TSV) (Chapter 20). Interconnection methods from the packaged MEMS are categorized in Figure 1.7. The TSV in Figure 1.7a,b will be explained in Section 17.3.1 and Chapter 20. The technology called wire-bonded through-Si vias in (c) will be shown in Section 20.4.3.5 [8]. The lateral feedthrough interconnection in (d) will be discussed in Section 17.3.2.

References

1 Putty, M.W., Chang, S.-C., Howe, R.T. et al. (1989). Process integration for active polysilicon resonant microstructures. *Sens. Actuators* 20: 143–151.
2 Yun, W., Howe, R.T., and Gray, P.R. (1992). Surface micromachined, digitally force-balanced accelerometer with integrated CMOS detection circuitry. IEEE Solid-State Sensor and Actuator Workshop, Hilton Head Island, USA (22–15 June 1992), 126–131.

3 Bustillo, J.M., Fedder, G.K., Nguyen, C.T.-C., and Howe, R.T. (1994). Process technology for the modular integration of CMOS and polysilicon microstructures. *Microsyst. Technol.* 1 (1): 30–41.

4 Howe, R.T. and Muller, R.S. (1984). Integrated resonant-microbridge vapor sensor. IEEE IEDM 84, San Francisco, USA (9–12 December 1984), 213–217.

5 Kirsten, M., Wenk, B., Ericson, F., and Schweitz, J.A. (1995). Deposition of thick doped polysilicon films with low stress in an epitaxial reactor for surface micromachining applications. *Thin Solid Films* 259 (2): 181–187.

6 Takeuchi, H., Quévy, E., Bhave, S.A. et al. (2004). Ge-blade damascene process for post-CMOS integration of nano-mechanical resonators. *IEEE Electron Device Lett.* 25 (8): 529–531.

7 Esashi, M. (2008). Wafer level packaging of MEMS. *J. Micromech. Microeng.* 18 (7): 073001(13pp).

8 Fischer, A.C., Grange, M., Roxhed, N. et al. (2011). Wire-bonded through-silicon vias with low capacitive substrate coupling. *J. Micromech. Microeng.* 21: 085035 (8pp).

Part II

System on Chip

2

Bulk Micromachining

Xinxin Li and Heng Yang

State Key Lab of Transducer Technology, Shanghai Institute of Microsystem and Information Technology, Chinese Academy of Sciences, 865 Changning Road, Shanghai 200050, China

Bulk micromachining is the technology using selective etching of silicon substrates and bonding of multiple etched and/or unetched wafers to fabricate micro-electro mechanical systems (MEMS). In view of 3D micromachining capability, bulk micromachining is advantageous in fabricating large step and/or thick silicon structure, compared with its silicon MEMS process counterpart – surface micromachining techniques. When large seismic mass is needed for inertial sensor or high depth/width aspect, deep trench is needed for large electrostatic force comb-drive actuators; bulk micromachining techniques is, more often than not, preferred to be utilized. On the other hand of integrated circuit (IC) compatibility, surface micromachining techniques have advantage in process nature, where more IC-compatible thin-film (e.g. poly-silicon, SiN_x, SiO_2, or metal layer) processes are employed as structural or sacrificial layer, as well as the structural step is not that deep. Most importantly, bulk micromachining generally features high structural reliability compared to the surface micromachining counterpart where the sacrificial layer release often suffers stiction issue. Therefore, there have been many successful products of bulk-micromachined MEMS devices already applied in market. Let alone, some advanced bulk-micromachining techniques have been able to imitate surface micromachining processes to fabricate comb-drive and sacrificial-layer structures with even stronger structural robustness. And some bulk-micromachined MEMS structures have sound integration compatibility with IC. In fact, with advanced system in package (SiP) techniques like through-silicon via (TSV) for stacked MEMS chip packaging, CMOS-MEMS monolithic integration has been not the only selection towards high-density MEMS devices or microsystems. How to batch-fabricate MEMS wafers by using standard IC foundry manufacturing facilities becomes a more meaningful topic in practical MEMS industry field. The relevant bulk micromachining technical solutions will be sequentially addressed in the following parts of this chapter.

Figure 2.1 is a schematic view of a typical bulk micromachined structure – a piezoresistive accelerometer [1]. The silicon wafer is anisotropically etched to form

3D and Circuit Integration of MEMS, First Edition. Edited by Masayoshi Esashi.
© 2021 WILEY-VCH GmbH. Published 2021 by WILEY-VCH GmbH.

2 Bulk Micromachining

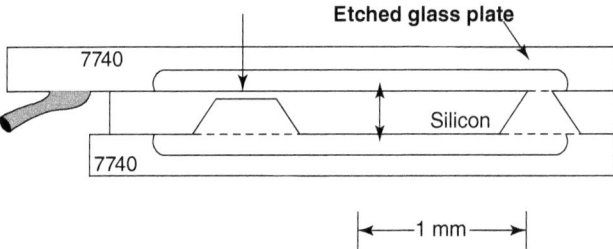

Figure 2.1 Schematic view of a bulk micromachined accelerometer. Source: Petersen [1]. © 1982, IEEE.

the cantilever beam-mass structure to serve as sensing structure of acceleration. Piezoresistors are fabricated on the surface of silicon cantilever before anisotropic etching to serve as transducers. The silicon wafer with cantilever structures is then bonded to the top and bottom glass wafers, in which gaps are isotropically etched and metal wires are fabricated and connected to the metal wires on silicon wafer.

The electromechanical structures in bulk micromachining are usually much thicker than those in surface micromachining, which are appreciated in many MEMS devices. For example, the bulk inertial sensors can attain higher performances due to their large proof mass [2]. The bulk micromachined structures are usually made by single-crystal silicon, which features excellent fatigue resistance and low residue stress level. On the other hand, improving the machining accuracies in the out-of-plane direction is a challenge in bulk micromachining because many etching processes are time controlled. Some important bulk micromachining techniques are not complementary metal oxide semiconductor (CMOS) compatible, such as anisotropic wet etching, anodic bonding, and high-temperature silicon fusion bonding. Electrical wiring through the bonded wafers is also an important issue.

The process basis of bulk micromachining is discussed in Section 2.1. Some typical process flows are discussed in Sections 2.2 and 2.3.

2.1 Process Basis of Bulk Micromachining Technologies

The etching of single-crystal silicon is one of the most important bulk micromachining techniques. As a matter of fact, bulk micromachining of silicon sometimes refers to the bulk etching of single-crystal silicon [3]. The etching techniques of Si include isotropic wet etching, anisotropic wet etching, plasma etching, and vapor phase etching.

The isotropic wet etchants etch the silicon in all crystallographic directions at the same rate. The most common isotropic wet etchant is $HF/HNO_3/CH_3COOH$ (HNA). The isotropic wet etching of silicon is not a popular method for bulk micromachining because the machining accuracy is quite low due to the isotropy. As shown in Figure 2.2, the etchant undercuts the silicon under the mask layer significantly after

Figure 2.2 Isotropic wet etching.

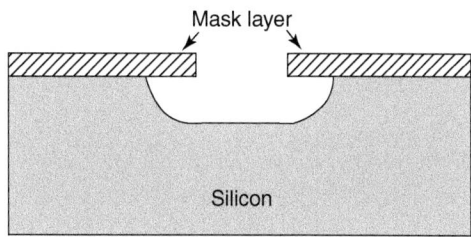

Figure 2.3 Anisotropic wet etching in KOH.

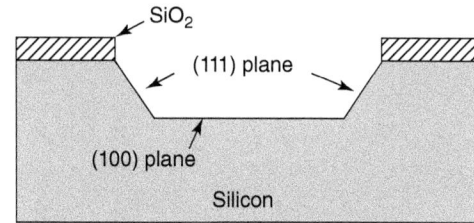

isotropic etching. The width of the trench is not determined by photolithography but by time-controlled etching. Because the etch rates are also agitation sensitive, the uniformities across wafers and consistencies among batches are poor.

The anisotropic wet etching is widely used in bulk micromachining because of the significant improvement on machining accuracy over isotropic wet etching. Most of the anisotropic wet etchants are aqueous alkali, such as KOH, NaOH, and TMAH (tetramethylammonium hydroxide). The etch rates of the anisotropic wet etching are orientation dependent. For example, the etch rate of Si in <100> orientation is about 100–400 times higher than that in <111> orientation in KOH solution. When carried out properly, anisotropic wet etching results in geometric shapes bounded by well-defined crystallographic planes. As shown in Figure 2.3, when the (100) silicon wafer covered with a layer of SiO_2 is etched in KOH solution and the edges of the rectangle window in SiO_2 layer is along <110> orientations, the etched cavity is bounded by (111) side walls. Because the etch rate of (111) planes can be neglected, the length and width of the cavity are determined by the photolithography/etching of the etch windows in the SiO_2 layer. In addition, the depth of the cavity can also be controlled in micrometer scale because the anisotropic etch rates are extremely reproducible in the etchants, such as EPW (ethylenediamine pyrocatechol water), KOH, and TMAH, when the concentration of the etchant, etchant lifetimes, and temperature are well defined. The etch rates are agitation insensitive since they are mainly determined by surface reaction rates. The anisotropic etch rates of silicon with different etching parameters have been measured extensively [4–6]. As an example, Figure 2.4 shows the etch rate distributions in $(01\bar{1})$ plane and $(0\bar{1}1)$ plane in 50 °C 40% KOH [6].

The anisotropic etching is versatile. Multiple MEMS devices and structures have been fabricated. Figure 2.5 shows two examples of multilevel structures fabricated by KOH etching [7].

The machining accuracy of anisotropic etching can be further improved by etch-stop techniques. Boron etch stop can be achieved because the heavily boron-doped silicon has a very low etch rate in EPW. A classic example of this

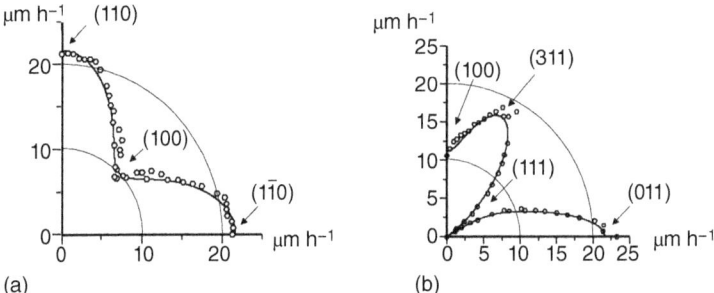

Figure 2.4 Polar plots of some two-dimensional etch rate distributions of Si in KOH. Source: Yang et al. [6]. © 2000, Elsevier.

Figure 2.5 Multilevel structures fabricated by KOH etching.

type of etch-stop process is the dissolved wafer process (DWP) developed by Gianchandani and Najafi [8], which will be discussed in the next section.

Electrochemical etch stop is another important technique, which relies on the anodic passivation characteristics of both p- and n-type silicon in anisotropic wet etchant when the bias voltage between the silicon and the solution is made more positive than the open circuit potential (OCP). Complex MEMS structures can be fabricated with the three electrodes or four electrodes etch-stop configurations shown in Figure 2.6 [9], in which n-type silicon is biased more positive than OCP in KOH and p-type silicon is biased less due to the inversely biased diode. After the p-type silicon is selectively etched, the etch stops due to the anodic passivation of the n-type silicon.

Electrochemical etch stop may also be achieved with the galvanic cells [10], instead of the external bias. The configuration of galvanic etch stop is shown in Figure 2.7. The Au, Si, and anisotropic etchant form a galvanic cell. When the Au:Si area ratio is larger than a threshold value, the anodic current is large enough to selectively passivate the n-type Si.

One of the main drawbacks of the anisotropic wet etching is the large chip area occupied by the tilted (111) sidewalls. Because the angle between (111) sidewall and

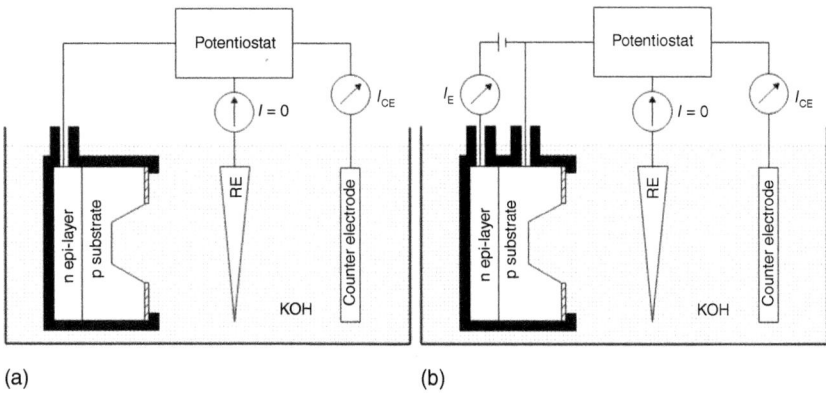

Figure 2.6 (a) Three electrodes and (b) four electrodes electrochemical etch-stop configurations. Source: Kloeck et al. [9]. © 1989, IEEE.

Figure 2.7 Configuration of galvanic etch stop.

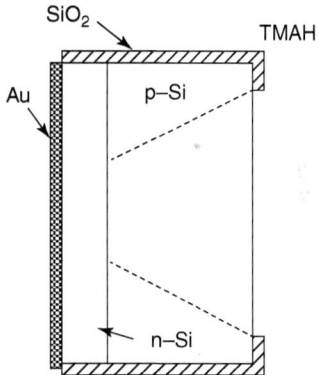

(100) surface is about 54.74°, the projection width of a (111) sidewall is about $\sqrt{2}H$, H the thickness of the wafer. The occupied chip area is unacceptable for the wafers larger than 100 mm. Another drawback is that the anisotropic wet etching is usually CMOS incompatible.

Nowadays, the anisotropic wet etching has been substituted by deep reactive ion etching (DRIE) in many applications. The Bosch process [11] is the most well-known DRIE technique, which achieves high-aspect-ratio plasma etching of Si by switching between passivation and isotropic fluoride-based plasma etching, as shown in Figure 2.8. High etch rates are achieved with inductive coupled plasma (ICP) source. Excellent machining accuracy on the wafer surface and vertical sidewalls are achieved by Bosch process. However the machining accuracy in the out-of-plane direction is worse than that of the anisotropic wet etching.

In addition to plasma-based etching, vapor phase etching is another type of dry etching. XeF_2 etching is a typical vapor phase etching, which features a high selectivity among silicon, SiO_2, and metal. The drawback is that the etching is

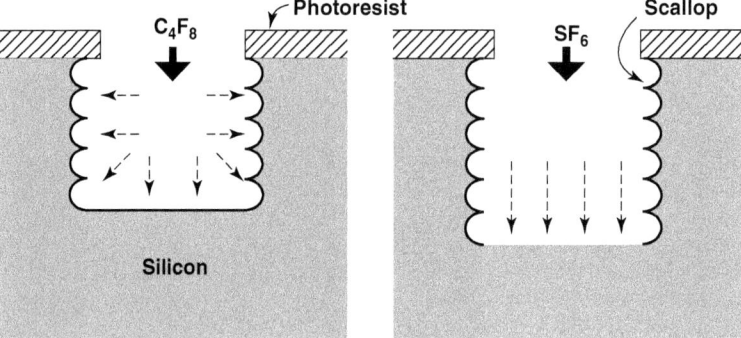

Figure 2.8 Bosch process.

isotropic and driven by diffusion. The machining accuracy and uniformity is not good.

Wafer bonding is the other key bulk micromachining technology [12]. There are roughly three types of wafer bonding: direct wafer bonding, anodic bonding, and intermediate-layer bonding.

Silicon fusion bonding is a widely used direct wafer bonding technique [13]. Good bonding quality can be obtained when the wafer surface keeps on being clean and flat. Hermetic and vacuum sealing can also be achieved. However, the technique is sensitive to the contaminations and the surface profiles. High temperature annealing is also required in many cases.

Anodic bonding is a reliable technique. High bonding strength can be achieved, and the technique is less sensitive to the contaminations and the surface profiles. However, the bonding voltage as high as 1000 V is not compatible with many MEMS structures. The electrical wiring through the glass wafer is a challenge because glass wafers are not easy to machine. The residue gas induced by bonding process also imposes an issue.

Intermediate-layer bonding has been widely used in packaging level applications [14].

2.2 Bulk Micromachining Based on Wafer Bonding

2.2.1 SOI MEMS

The MEMS technologies originate from two-dimensional integrated-circuit processes. One of the most important issues of MEMS processes is the control of the dimensions and tolerances in the third dimension, the out-of-plane direction of wafers. Silicon-on-insulator (SOI) MEMS technology provides the best control of thickness and its tolerances of the structure layer in single-crystal silicon bulk micromachining technologies, although the number of layers is limited.

An SOI wafer is a sandwich structure, with a layer of SiO_2 between a single-crystalline silicon device layer and a single-crystalline silicon substrate.

2.2 Bulk Micromachining Based on Wafer Bonding

The machining accuracies of all three layers are excellent. The thickness tolerances of the device layers are lower than 1 μm when the device layers are as thick as 100 μm. The thickness tolerances of the buried oxide layers are as low as ±50 nm. The thickness tolerances of the substrates are also micron scale. Fabrication accuracy can be achieved with SOI MEMS.

In SOI MEMS, freestanding structures are usually fabricated in the device layers. The structural layers are more than an order of magnitude thicker than those in surface micromachining. As there is a positive correlation between the thicknesses of the structures and the performances in many MEMS devices, such as inertial sensors and optical devices, superior devices can be developed using SOI MEMS processes than those using surface micromachining.

The SOI MEMS processes are versatile, and the buried oxide layers are excellent isolation layers. Not only are the device layer and the substrate of an SOI wafer isolated but also different parts on the device layer can also be isolated using trenches. Therefore, capacitive sensing and actuating structures, such as comb drives, can be fabricated easily. Piezoresistive sensing can also be easily employed in SOI MEMS devices.

The buried oxide layers are built-in sacrificial layers or provide excellent etch stops. Process simplicity can be achieved with SOI MEMS. Additionally, the top and bottom surfaces of the device layers are sufficiently smooth for optical application.

There are roughly three strategies for fabricating MEMS structures with SOI wafers: a surface-micromachining-like process with buried oxides as sacrificial layers, a surface-micromachining-like process with substrates as sacrificial layers, and the traditional bulk micromachining process with buried oxides as etch-stop layers.

(1) Surface-micromachining-like process with buried oxides as sacrificial layers

The surface-micromachining-like process with buried oxides as sacrificial layers is quite straightforward, as shown in Figure 2.9:

(a) After the device layer of the SOI wafer is doped, the metal wires are deposited and patterned by photolithography and etching. Fine wires can only be lithographed before DRIE. The device layer is patterned by photolithography and etched by DRIE to shape the structure. Release holes should be etched in large structures, such as proof masses. The following release etching is an

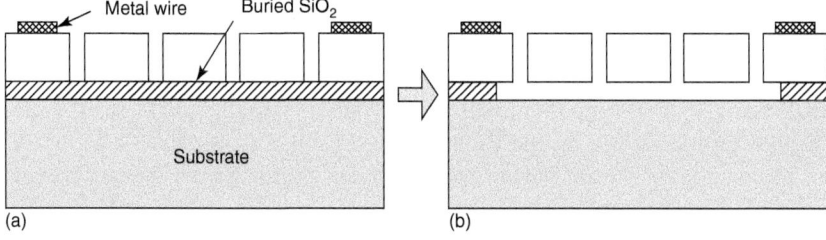

Figure 2.9 After the top layer of the SOI wafer is dry etched into desired structure such as comb fingers, the buried SiO_2 layer beneath is removed by wet or vapor HF.

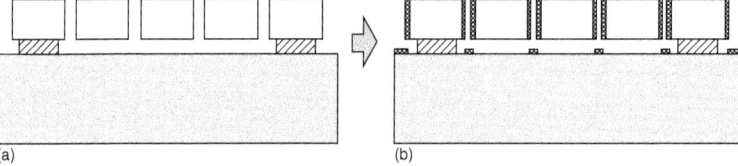

Figure 2.10 Self-aligned metallization after release etching: (a) The free-standing structures are fabricated. (b) A layer of metal is sputtered, which is disconnected by the undercut of the buried oxide layer.

isotropic time-controlled process. The time and extent of the etching can be controlled by the design of release holes.

(b) The structure is released by etching the buried oxide layer underneath and dried. The etching and drying processes in surface micromachining can be employed, such as vapor hydrofluoric acid (HF) etching, highly concentrated HF etching, freeze drying, and supercritical drying.

Depending on the release etchant, different metals are employed, similar to the case of surface micromachining. The metal wires can also be fabricated after DRIE or after release etching. However, the control of line widths may decline. It is a common choice to fabricate metal wires using a lift-off method instead of a lithography/etching method after DRIE [15], because metals and photoresist may be left over at the sidewalls of the high topography structures in the normal lithography/etching method. The released structures can also be metalized with a self-aligned process[16], which is shown in Figure 2.10.

The release etching is one of the critical steps. There are some matters needing attention. First, nearly all the SOI wafers used in the process are bond and etched back SOI (BESOI), and the quality of the bonding is critical. When the quality of the bonding is low, the etch rate of the buried oxide along the bonding surface may be tens or even hundreds of times faster than those in other directions. The SOI wafers should be qualified by an HF etching test of the buried oxide layer. Second, the etching of buried oxide is isotropic and based on time control. It is difficult to control the etched cavity precisely. The trenches refilled by low-stress SiN_x can be used as the etch-stop structures to define the releasing region, as shown in Figure 2.11.

Various devices can be fabricated with the aforementioned SOI MEMS process – inertial sensors [15], optical MEMS [16], accelerometers, and optical switches. The SOI MEMS structures may be flexible in plane or out of plane, which can be detected using capacitive sensing or piezoresistive sensing. The structures may

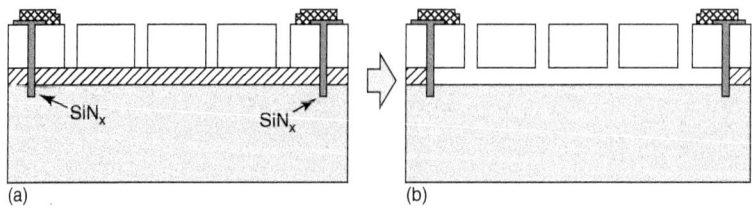

Figure 2.11 Trenches are refilled by low-stress SiNx as the etch stop structures, then, the buried SiO_2 layer is etched to free the MEMS structure.

2.2 Bulk Micromachining Based on Wafer Bonding

Figure 2.12 Capacitive accelerometer. Source: Matsumoto et al. [15]. © 1996, Elsevier.

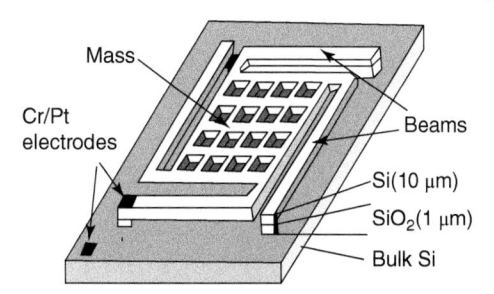

Figure 2.13 Optical switch actuated by comb drive. Source: Noell et al. [16].

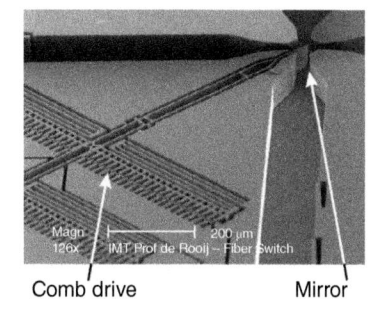

also be actuated to deform in plane or out of plane by comb drive or bottom electrode, respectively, although the displacements out of plane are limited by the thickness of the buried oxide layers. The beam-mass structure of the capacitive accelerometer shown in Figure 2.12[15] is flexible out of plane, which is detected by the capacitance between the mass and the substrate. The optical switch shown in Figure 2.13 is actuated by a comb drive to move in plane [16].

(2) Surface-micromachining-like process with silicon substrates as sacrificial layers

In the SOI MEMS process with buried oxide layers as sacrificial layers, the gaps between the freestanding structures and the substrates are equal to the thicknesses of the buried layers, which are limited to several microns. The small gaps impose limitations on the maximum sizes of the freestanding structures [17], because SOI wafers usually have an inherently high stress level, and the deformations induced by the residue stress and stress gradient may be comparable to the gap size when the freestanding structures are big [15, 16]. The maximum out-of-plane displacements of the freestanding structures are also limited by the small gaps.

The limitations induced by the small gaps are relieved when the freestanding structures are released by etching the silicon substrate underneath, and the buried oxide layers are used as etch-stop layers. The basic process flow is shown in Figure 2.14 The process flow is similar to that shown in Figure 2.13, except that the freestanding structure is released by DRIE from the back side.

The well-known SOI multiuser MEMS process (SOIMUMPs)[18] employs an SOI process with silicon substrates as the sacrificial layer. In SOIMUMPs, two different metal wires are provided. The first metal layer is deposited and patterned before DRIE of the structures to allow fine metal features, as shown in Figure 2.14a. The fine metal wires are limited to areas not etched in the DRIE process. The second

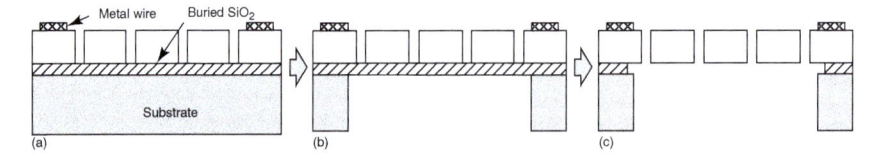

Figure 2.14 (a) After the metal wires are fabricated, the silicon device layer is patterned by photolithography and etched by DRIE to shape the structure, (b) the silicon substrate is etched by DRIE from the back side, and (c) the freestanding structure is released after the buried oxide is etched.

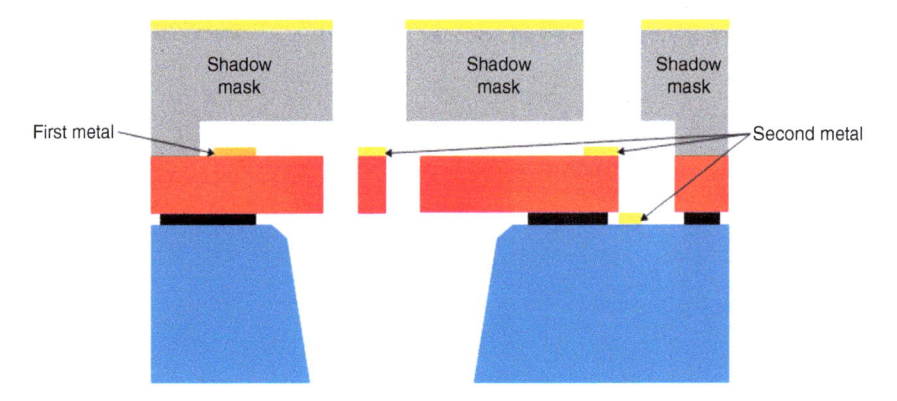

Figure 2.15 Second metal layer defined by a shadow mask, which is temporarily bonded to the wafer and removed after metal deposition. Source: Cowen et al. [18]. © 2005, MEMSCAP.

metal layer is deposited after the freestanding structures are released and defined by a shadow mask, which is temporarily bonded to the wafer and released after metal deposition, as shown in Figure 2.15 [18].

One consideration about the process in Figure 2.14 is how to protect the fragile structure during the DRIE from the backside. During the DRIE, the wafer is cooled by helium flow from the unetched side. The pressure drop across the wafer is not zero and may destroy the fragile structures, especially after the substrate is etched away and only the buried oxide layer is left. The structures may be strengthened by protection layers or temporarily bonded handle wafers.

The substrates can also be etched from the front side to release the freestanding structures. The basic process flow is shown in Figure 2.16. The process flow is similar to that in Figure 2.10, except that the device layers must be protected by mask layers during the etching of the silicon substrate, and the mask layers and buried oxide layers under the freestanding structures must be stripped after releasing etching. The silicon substrates may be etched by isotropic dry etchants, such as XeF_2, or wet etchants, such as KOH. The buried oxide layers under the freestanding structures must be stripped, because there is high compressive stress in thermal oxide layers, which induces large deformation of the freestanding structures.

Because the silicon substrates under the freestanding structures are etched, the bottom electrodes cannot be utilized. Comb drives are usually employed for the capacitive sensing and electrostatic driving; therefore, structures flexible in plane are

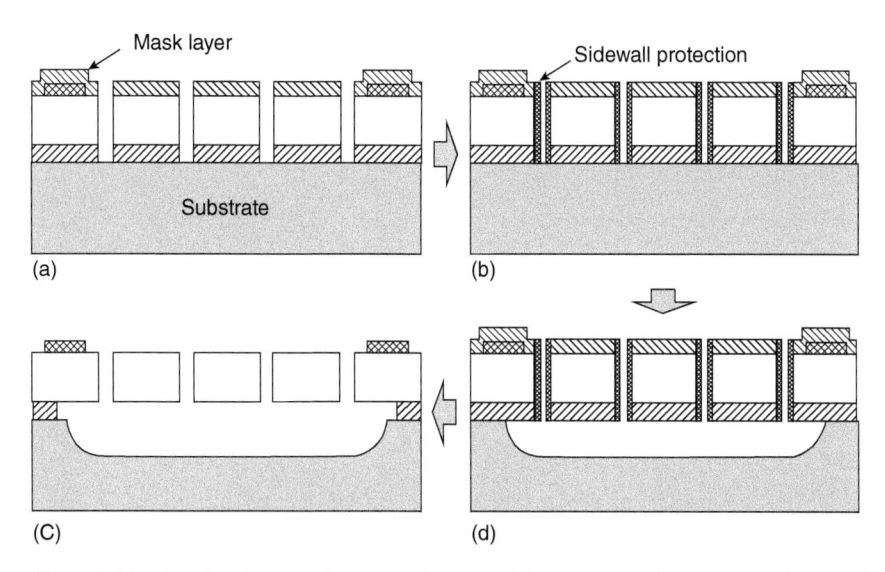

Figure 2.16 (a) After the metal wires and the mask layers are fabricated, the silicon device layer is patterned by photolithography and etched by DRIE to shape the structure, and the buried oxide layer is etched with the same photoresist, (b) a mask layer is deposited and etched masklessly to protect the sidewalls of the structures, (c) the silicon substrate is etched to release the freestanding structures from the front side, and (d) the mask layer, sidewall protection, and buried oxide layers under the freestanding structures are stripped.

widely employed for capacitive sensing and electrostatic driving. However, because the in-plane deformation is not easy to detect using piezoresistive sensing, it is a challenge to fabricate devices with piezoresistive sensing and electrostatic driving using this technology. A process presented in literature [19] provides a solution to the aforementioned problem. As shown in Figure 2.17, the piezoresistors on the sidewalls of the beams were implanted by tilting the wafers to approximately 31° from the vertical direction.

(3) Traditional bulk micromachining process with buried oxides as etch-stop layers

Although there are quite a few etch-stop techniques in a wet-etching processes, such as electrochemical etch stop and heavily boron etch stop in anisotropic wet etching, there are limitations on the applications of the techniques, because the wet etchants etch the usual metals and passivation materials in MEMS technology. Even worse, there is no similar etch-stop technique available for dry-etching processes, which are time controlled or stopped by different materials. The buried oxide layers in SOI wafers are effective etch-stop layers in the dry- and wet-etching processes. Traditional bulk structures can be fabricated with much higher machining accuracy using the SOI wafers.

Figure 2.18 shows the process flow of a piezoresistive three-degree-of-freedom (3-DOF) accelerometer using an SOI wafer [20], which is described briefly as follows:

(a) The oxide layer is thermally grown. The piezoresistors and the metal interconnection are fabricated in the device layer.

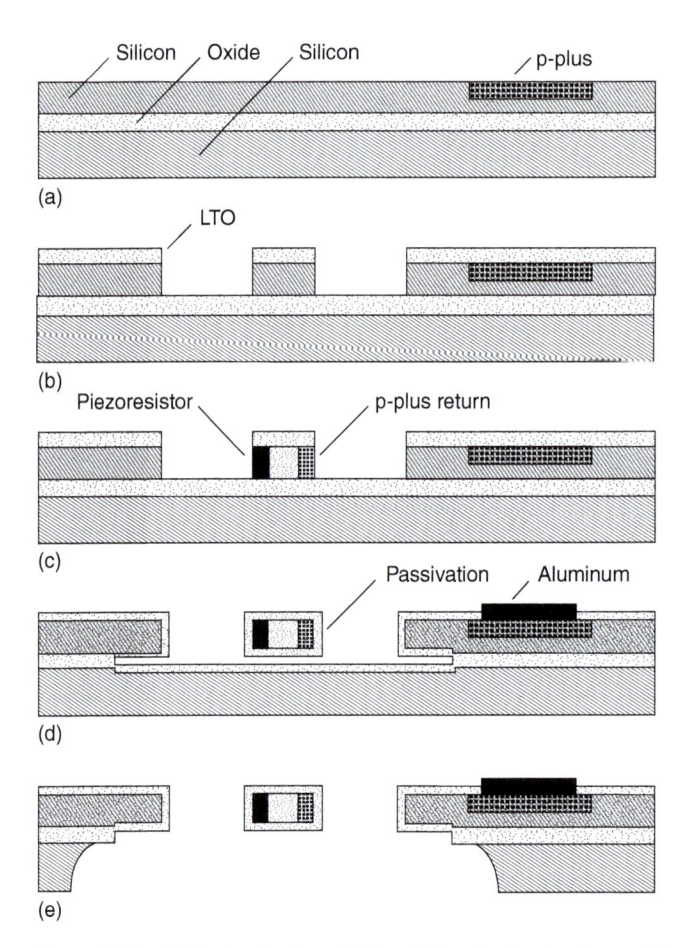

Figure 2.17 (a) P-plus doping. (b) Dry etching for the top layer of the SOI wafer. (c) P-type and p-plus doping by ion implantation at the vertical sidewalls of the beam for the formation of piezoresistors and return current path. (d) Passivation layer deposition. (e) Backside etching to free the accelerometer structure. Source: Partridge et al. [19]. © 2000, IEEE.

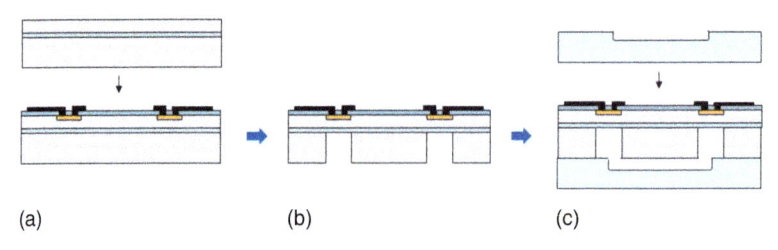

Figure 2.18 (a) The piezoresistors and the metal interconnection are processes from the front side of the SOI wafer. (b) Backside etching is processed to define the mass. (c) Then, the wafer is backside bonded with a glass wafer. Source: Dao et al. [20]. © 2010, ANSNN.

(b) After the SOI wafer is etched from the back side using DRIE to define the mass, it is etched from the front side using DRIE to define the beams.
(c) After the gaps are etched in the glass wafer, the SOI wafer is anodically bonded to the glass.

2.2.2 Cavity SOI Technology

In traditional bulk micromachining, cavities under freestanding structures are formed either during or after the formation of these structures. Thus, the protection of the freestanding structures during cleaning and bonding processes is imperative. Moreover, the formation of the freestanding structures induces bows in the wafers and roughness on the wafer surfaces, which in turn would lead to deterioration in quality and yielding of the following wafer-bonding processes.

In cavity SOI technology [12, 21–25], the cavities are formed first. Because shallow cavities are usually employed, the impacts on the bows and surface roughness of the wafers can be neglected.

Cavity SOI technology likely dates back to the 1980s, although it was not known by this term at that time. The fundamental process flow is shown in Figure 2.19 and described as follows:

(1) The cavities are etched on the front side of the substrate wafer using wet or dry etching methods, while alignment marks are prepared on the rear side.
(2) The cap wafer is directly bonded to the substrate wafer.
(3) The cap wafer is processed using grinding, polishing, or etching to obtain the desired design thickness.
(4) The mechanical structures are fabricated by etching the silicon diaphragm over the cavity.

The advantages of cavity SOI technology are as follows:

(1) The technology is CMOS compatible because the wafers with pre-etched cavities are transparent to CMOS processes.
(2) The dimensional accuracies are higher. The cavity dimensions are well defined by photolithography and shallow etching. The thickness of the silicon diaphragm over the cavity can also be controlled quite accurately when the SOI layer thickness and cavity size are within the safe region. The alignment accuracy exhibited is also good. The inaccuracy originates first from double-sided lithography of the handle wafer and then from the alignment mark transfer to the front surface of the bonded stack.
(3) The pressure inside the cavity can be adjusted by using a suitable bonding ambient.

One of the main drawbacks of cavity SOI technology is the thinning process of the cap wafers. When the cavity dimensions are large, thin diaphragms may not be able to withstand the thinning process. For the process of thinning of cap wafers by grinding and polishing [25], the safe region of the cavity sizes and diaphragm thicknesses for rectangular cavities are as shown in Figure 2.20.

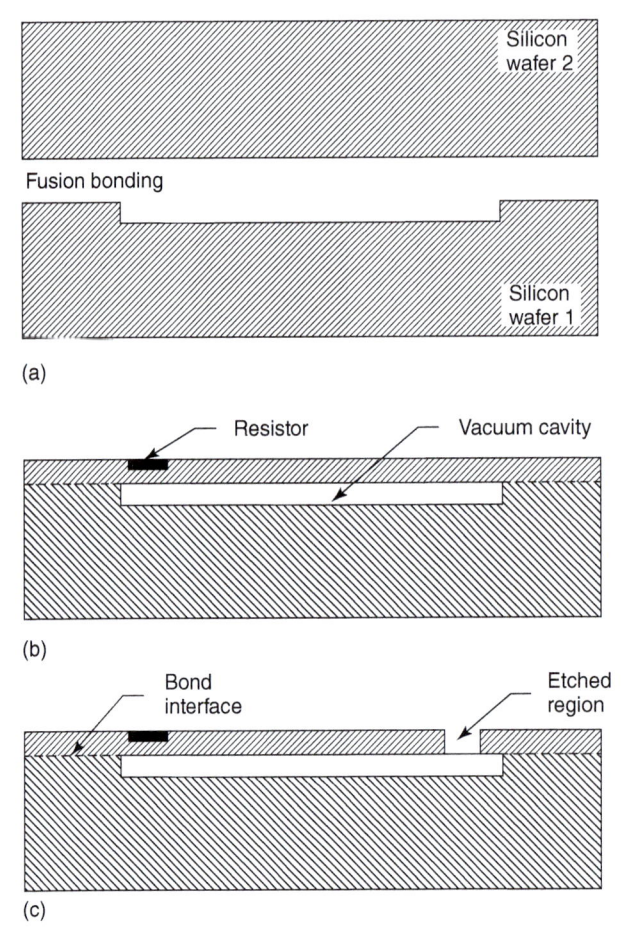

Figure 2.19 (a) The bottom wafer with previously etched cavities is directly bonded with another flat wafer. (b) The top wafer is thinned to a desired thickness for the sensing structure, and the piezoresistors are fabricated. (c) The sensing structure is defined by dry etching from the front side. Source: Petersen et al. [22]. © 1991, IEEE.

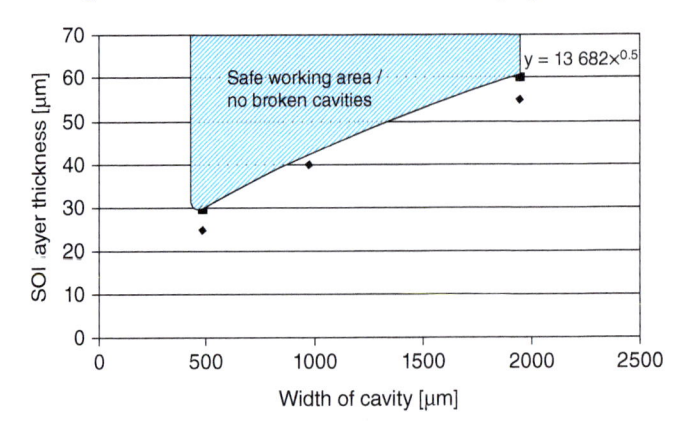

Figure 2.20 Safe region of the cavity sizes and diaphragm thicknesses for rectangular cavities. Source: Luoto et al. [25]. © 2007, Elsevier.

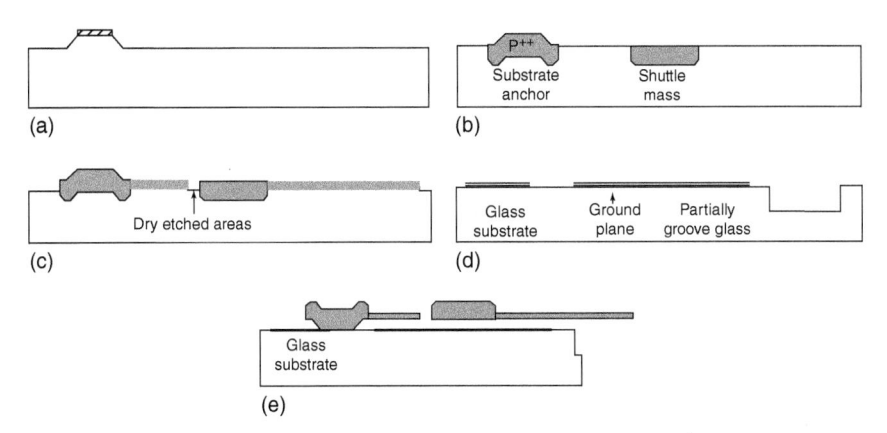

Figure 2.21 (a) Selective shallow etch at the front side. (b) Heavy boron doping at the anchors and the masses. (c) Shallow dry etching to define the structure. (d) Another glass wafer is micromachined and metalized for the following bonding. (e) The silicon wafer is anodic-bonded with the glass wafer. Then, the silicon, except for the heavy boron-doped parts, is wet etched off. Source: Gianchandani and Najafi [8]. © 1992, IEEE.

2.2.3 Silicon on Glass Processes: Dissolved Wafer Process (DWP)

Dissolved wafer fabrication is a technique to fabricate thick single-crystal silicon freestanding structures on glass with thickness-to-width ratios in excess of 10 : 1. The process is schematically shown in Figure 2.21 and described as follows:

(a) The wafer is first etched with KOH to form the anchors and gaps. Although the gaps may be etched using reactive ion etching (RIE), anisotropic wet etching is recommended in Ref. [8] because the anisotropic wet etch yields smoother trench bottoms.
(b) The second step is optional when structures of varying thicknesses are designed. The wafer is thermally oxidized and patterned. The thickest region is exposed and heavily doped with boron. The typical depth is approximately 15 μm [8]. The thermal diffusion process is time consuming, where approximately 16 h is required to diffuse to 15 μm at a temperature as high as 1175 °C [26] and where diffusion to even deeper levels is impractical.
(c) All masking layers are stripped and an unmasked boron diffusion is performed to define the thicknesses of planar structures. Boron diffusion thickness is typically in the range of 2–20 μm. Deep trenches are dry etched to cut through the boron-diffused layer to define the high-aspect-ratio structures.
(d) A glass wafer is prepared with the metal interconnects and partial grooves for dicing.
(e) After all masking layers are stripped, the wafer is anodically bonded to the glass wafer face to face. The interconnects and boron-doped anchors overlap partly to ensure the interconnects and anchors are in close contact to obtain good ohmic contacts. The resistance for a 13 000-μm² contact area is approximately 12 Ω. The metal interconnects must survive the anodic bonding and the subsequent anisotropic wet etching. Therefore, noble metal interconnects

should be employed. The wafer is finally etched in anisotropic wet etchant, such as EPW. The etch rate of the heavily boron-doped silicon is much lower than the undoped silicon and can be neglected, which is referred to as heavily boron etch stop. After the undoped silicon substrate is dissolve away, the heavily boron-doped structures on glass is released.

Although the heavily doped silicon layer has the characteristic of high stress due to the nature of boron diffusion, the measured residue stress of the structures after being released was quite low [27]. The residue stress was measured using the pull-in voltage technique, resonant frequency technique, and bent-beam strain sensors, yielding tensile stresses of 18.3 MPa[27], 18.5 ± 4 MPa[8], and 15–40 MPa[28], respectively. No obvious out-of-plane deformations of the released structures were observed, indicating that the stress gradients could be neglected [28, 29]. The Young's modulus of the p^{++} structures was measured to be approximately 175 ± 40 GPa.

Structures with different thicknesses can be fabricated using the DWP by multiple instances of heavy boron diffusion. Because the depths of boron diffusion can be controlled quite accurately, good control of the thicknesses and thickness tolerances of the structures can be achieved using the DWP. Because both the comb fingers and bottom electrodes on the glass can be fabricated, both the in-plane and out-of-plane movements of the structures can be detected by capacitive sensing or actuated by electrostatic driving. Numerous MEMS devices have been developed using DWP technology, including accelerometers, gyroscopes, thermal profilers, Pirani gauges, and microactuators. A gyroscope fabricated with DWP is shown in Figure 2.22 [30, 31].

One issue is raised when fabricating absolute pressure sensors using DWP technology. The p^{++} diaphragm is bonded to the glass substrate in vacuum to form the vacuum reference chamber of the absolute pressure sensor. The displacement of the diaphragm is capacitively sensed using the bottom electrode in the bottom chamber. Producing the external leads of the bottom electrode in the vacuum chamber without breaking the vacuum does pose a problem. Figure 2.23 shows the solution in which the glass electrodes and external leads are connected with poly-silicon [32].

In the standard DWP technology, only the bottom electrodes on the glass are integrated. Top electrodes can also be integrated by the double-bonding and dissolving processes. As shown in Figure 2.24, a tunneling-based accelerometer with a top electrode was developed [26]. After the sensing structure wafer was bonded to the glass wafer and dissolved, the top electrode wafer was bonded and dissolved again to form the top electrodes.

To integrate CMOS circuits with the p^{++} MEMS in one chip, an active DWP (ADWP) was developed in Ref. [33]. As shown in Figure 2.25, the MEMS devices were located in the p^{++} region, whereas CMOS circuits were located in the n-type region. The n-type region was anodically biased to achieve an electrochemical etch stop during the anisotropic dissolution step. Because the holes through the p^{++} protection rings were required for the leads transfer of the CMOS circuits, it was inevitable that some etchant penetrated the p^{++} protection rings to attack the

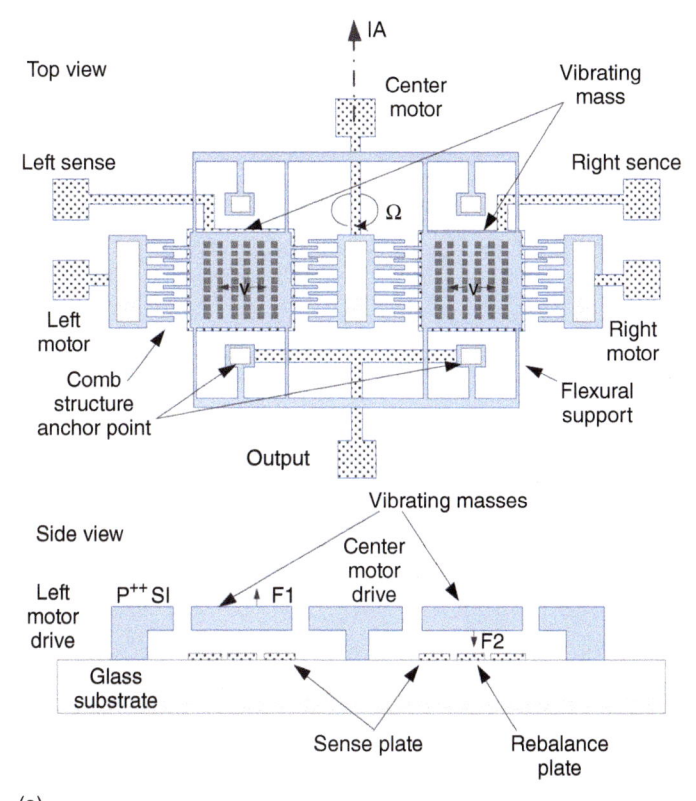

Figure 2.22 (a) Schematic view and (b) SEM image of a gyroscope fabricated by DWP. Source: (b) Kourepenis et al. [30] © 1998, IEEE.

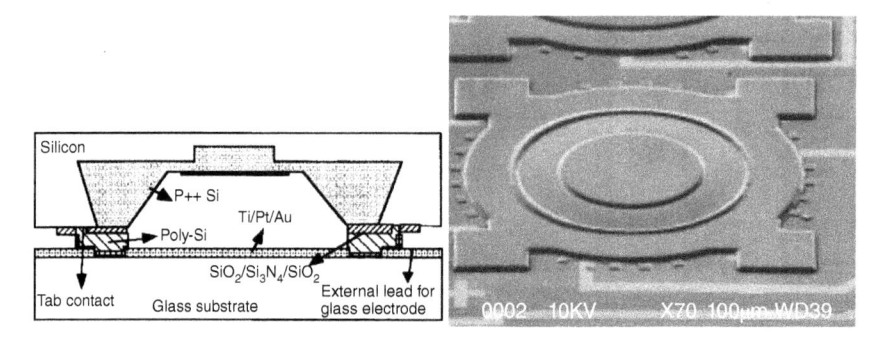

Figure 2.23 Schematic view and SEM image of a gyroscope fabricated by DWP. Source: Mason et al. [32]. © 1998, IEEE.

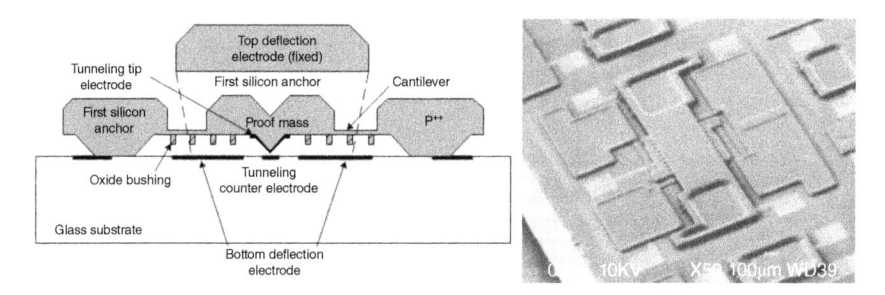

Figure 2.24 Schematic view and SEM image of an accelerometer fabricated by DWP. Source: Yeh and Najafi [26]. © 1997, IEEE.

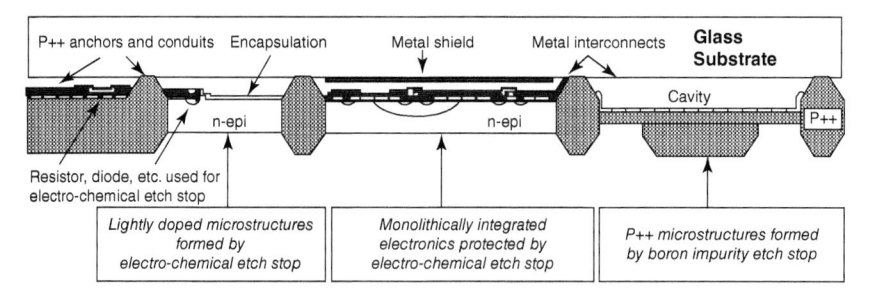

Figure 2.25 Active DWP. Source: Gianchandani et al. [33]. © 1995, IEEE.

CMOS region somewhat. To protect the CMOS region from this attack, the whole CMOS region was covered by a layer of Low-Temperature Oxide (LTO) deposited after the metallization [33].

In the DWP, the thicknesses of the p^{++} layers are limited by the time-consuming process of boron diffusion. The deep-etch shallow-diffusion process [34] may be considered as an improvement. Structures as thick as 40 μm with 2-μm wide gaps can be

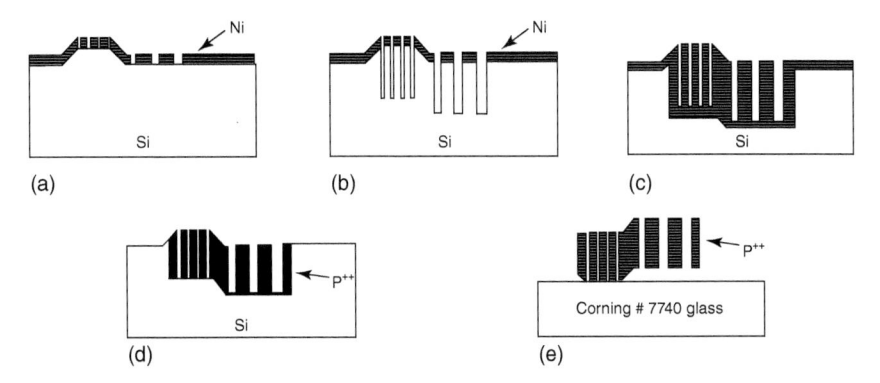

Figure 2.26 (a) The wafer is etched to form the anchors and gaps. (b) Deep trench etch is made to divide the structure region into narrow gridded structures, where electroplated Ni is used as the masking layer. (c) Shallow boron diffusion is conducted to convert the gridded structure into heavy boron-doping one. (d) Deep etching is performed to etch through the p^{++} layer at the bottom. (e) The wafer is anodically bonded to a glass wafer. The silicon substrate is dissolved away, while the heavy boron structure remains. Source: Juan and Pang [34]. © 1996, IEEE.

fabricated with a shallow boron diffusion. The process flow is shown in Figure 2.26 and described as follows:

(a) The wafer is first etched to form the anchors and gaps.
(b) After a thick masking layer is deposited and patterned, deep trenches are etched using DRIE, which divides the structure region into narrow gridded structures. The trenches are etched using the Electron Cyclotron Resonance (ECR) source in Ref. [34], and electroplated Ni is used as the masking layer. DRIE represents a better alternative approach to the standard RIE.
(c) Shallow boron diffusion is conducted to convert the entire thickness of the gridded structure into a p^{++} layer.
(d) Another iteration of DRIE is performed to etch through the p^{++} layer at the bottom.
(e) After all masking layers are stripped, the wafer is anodically bonded to a glass wafer with metal interconnects. The p^{++} freestanding structure on glass is obtained after the undoped silicon substrate is dissolved away in anisotropic wet etchant.

Very thick structures can be fabricated using the deep-etch shallow-diffusion process, which is limited by the aspect ratio of dry etching and the gap width of the design. In Ref. [35], arrays of high-aspect-ratio vertical Si mirrors were fabricated by this technology for optical switching applications, as shown in Figures 2.27 and 2.28. A 50-µm thick vertical mirror was actuated by the comb drive to enable switching between two light paths. Depending on the mirror position, input light either can be reflected to another stationary fiber or it can continue its path, as shown in Figure 2.28. The movement of the mirror can be as great as 34 µm with a DC voltage of only 30 V because of the high-aspect-ratio structure.

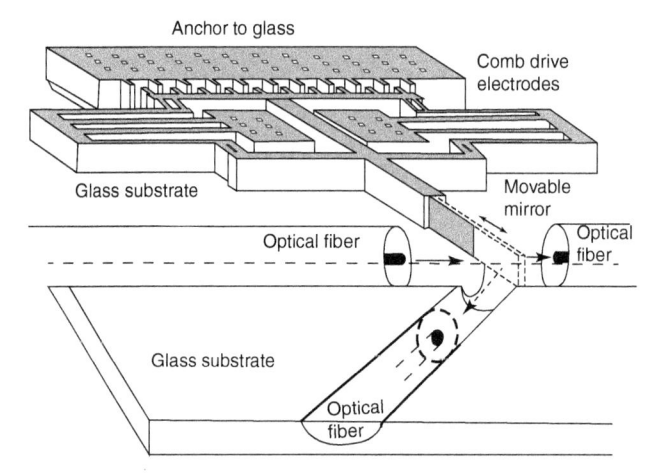

Figure 2.27 Schematic of the optical switch. Source: Juan and Pang [35]. © 1998, IEEE.

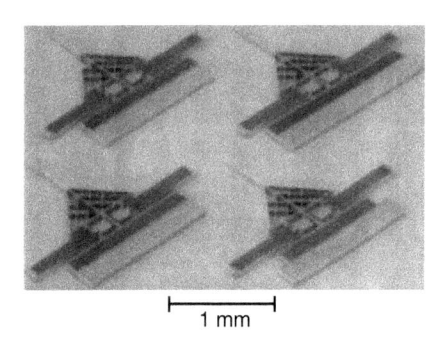

Figure 2.28 SEM image of a 2 × 2 array of optical switches. Source: Juan and Pang [35]. © 1998, IEEE.

2.3 Single-Wafer Single-Side Processes

2.3.1 Single-Crystal Reactive Etching and Metallization Process (SCREAM)

The SCREAM process [36, 37] is one of the earliest known single-sided bulk micromachining processes used to fabricate freestanding structures with high aspect ratio in single-crystal silicon substrates. Its process flow is extremely simple. Only one round of photolithography is required to fabricate the freestanding devices. The isolated metallization is achieved by self-aligned sputtering; no photolithography is required. The process is independent of crystal orientation, doping level, and doping type. The process flow is shown in Figure 2.29 [36] and described as follows:

(a) A layer of SiO_2 is deposited and patterned to serve as the mask throughout the process sequence. The thickness of the layer is critical because the layer must be able to withstand the RIE – performed in Step (d) – and protect the surfaces and upper edges of the structures during isotropic etching, which is performed in Step (f). The quality of the layer is less critical. PECVD SiO_2 may be used, although high quality SiO_2 layers are preferred.

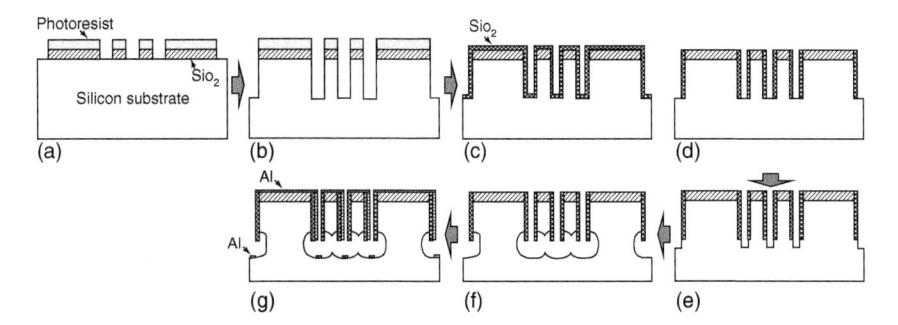

Figure 2.29 (a) Oxide layer at the top of the wafer is patterned. (b) Deep trench etching is performed. (c) Oxide layer is deposited or thermally grown. (d) The oxide layer at the bottom of the trench is removed by reactive ion etch, while the oxide at the vertical sidewalls is retained. (e) Deep silicon etching is performed to etch down to another 3–5 μm below the lower edge of the sidewall oxide. (f) Isotropic etch is performed to remove the silicon under the structures, leaving them suspended over the substrate. (g) The released structures are metalized by self-aligned aluminum sputtering. Source: Modified from Shaw et al. [36].

(b) A DRIE is performed, wherein the etch depth, D_1, is larger than the structure height, H. After this, the photoresist is stripped.
(c) A thin layer of SiO_2 is deposited conformally.
(d) The thin layer of SiO_2 at the bottom of the trenches is removed by RIE. It is important that the sidewall oxide is left intact.
(e) A deep silicon RIE is performed to etch down to another 3–5 μm below the lower edge of the sidewall oxide.
(f) An isotropic etching method is performed to remove the silicon under the structures, leaving them suspended over the substrate. The in-plane etch distance, D_3, must be larger than half of the maximum width of the freestanding structures. When the etching is fully isotropic, the height of the freestanding structures is approximately equal to D_1-D_3, although the bottom of the structures is not flat, as shown in Figure 2.29.
(g) The released structures are finally metallized by self-aligned aluminum sputtering. All the aforementioned steps may be carried out at low temperature (<400 °C) when PECVD SiO_2 is employed in Steps (a) and (c).

As discussed in Ref. [36], some design rules need to be followed:

(1) The first SiO_2 layer must be much thicker than the layer deposited in Figure 2.29 (c). Extra attention must be paid to the upper edges of the structures and their protection. The consumption of the SiO_2 layer at the upper edges may be greater than that on the surface, which results in the unwanted etching at the upper edges during the release etching in Figure 2.29 (f), as shown in Figure 2.30 [38]. The extra consumption of SiO_2 at the corners may be due to the high DC bias in Cl_2 RIE [36] or the tapered profile in the first mask layer. In some cases, a primary SiO_2 layer as thick as 2.5 μm is adequate to protect the upper edges [36]. The methods of protecting the upper edges are widely discussed in Ref. [38].
(2) For the self-aligned metallization, all the isolated set of electrical interconnects, contact pads, and capacitive plates must be surrounded by trenches to ensure

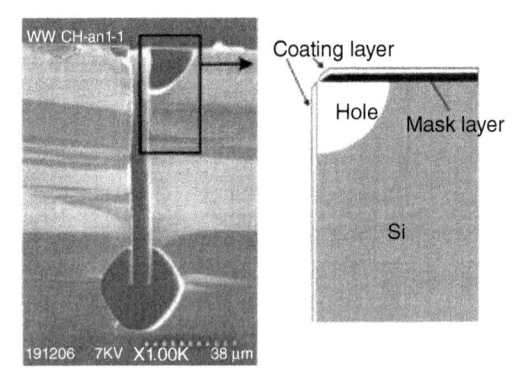

Figure 2.30 Unwanted etching at the upper edges during the release etching. Source: de Boer et al. [38]. © 2000, IEEE.

electrical isolation, as shown in Figure 2.31a. The cross section of an interconnect is shown in Figure 2.31b. The processes are simplified by the self-aligned metallization. On the other hand, however, the MEMS structures cannot be connected to IC by CMOS metallization.
(3) Released beams in a device should be of the same width and much narrower than the anchors.

Figure 2.32 shows a comb-drive microactuator fabricated by the SCREAM process [36]. The grid-shaped movable plate D is supported by springs B and actuated by comb-drive A. The microactuator moves in plane, but the movable structure is anchored to the substrate through C. The undercut under the anchors C and the fixed plates E can be observed clearly.

The SCREAM process is simple and versatile. Many surface micromachining-like structures have been fabricated using this process, and their structure layers are much thicker than those of surface micromachining structures. Very low residue stress can be achieved, and no fatigue is found due to the single-crystal silicon structures.

However, there are some limitations as well. First, the thicknesses of the freestanding structures and the gaps between the freestanding structures and the substrates are determined by isotropic etching, the machining accuracies of which are much lower than those in the in-plane direction, which are determined by photolithography and DRIE. The bottoms of the freestanding structures are not flat due to isotropic etching. Second, as the entire structures are coated with metal by self-aligned metallization, only electrostatic actuation and capacitive sensing can be employed. Third, the out-of-plane movements of the freestanding structures cannot be actuated or sensed because the silicon parts of the freestanding structures are connected to the substrate electrically, which also serve as shielding for the metal layers. Almost all of the SCREAM structures are actuated or sensed in the in-plane directions. Fourth, the freestanding structures are grid shaped, which results in the loss of mass and strength. Finally, once again because of the use of self-aligned metallization, different structures cannot be wired by planar processes.

Being an early single-wafer single-side process, the SCREAM process inspired many additional novel processes and improvements, some of which are discussed in the next section.

Figure 2.31 Self-aligned metallization. Source: Shaw et al. [36]. © 1994, Elsevier.

(a)

(b)

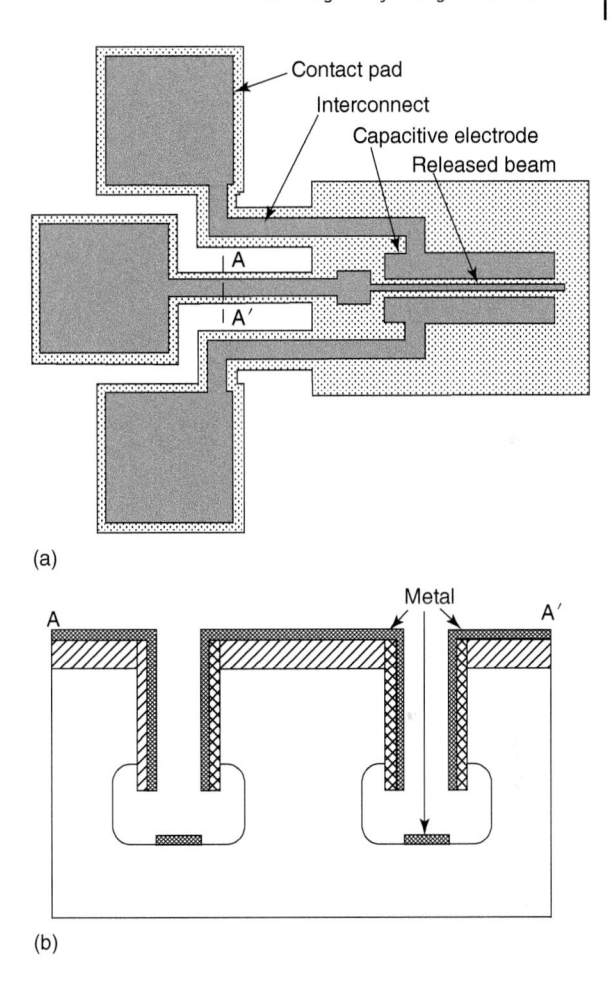

Figure 2.32 A comb-drive microactuator fabricated using the SCREAM process. Source: MacDonald [37]. © 1996, Elsevier.

2.3.2 Sacrificial Bulk Micromachining (SBM)

In the SCREAM process, the width of released structures is limited by isotropic release, and the structure under surfaces are unevenly etched. The sacrificial bulk micromachining (SBM) [39] is another surface-micromachining-like bulk process, which can be used to fabricate wide structures with flat undersurfaces.

The SBM process is based on the anisotropic etching and the self-etch-stop effect of (111)-oriented silicon wafer in anisotropic wet etchants, say KOH. There are eight {111} planes in (111) wafer, six of which are shown in Figure 2.33a [39], the other two being the top and bottom. Three of the six {111} planes are tilted at 19.47° from

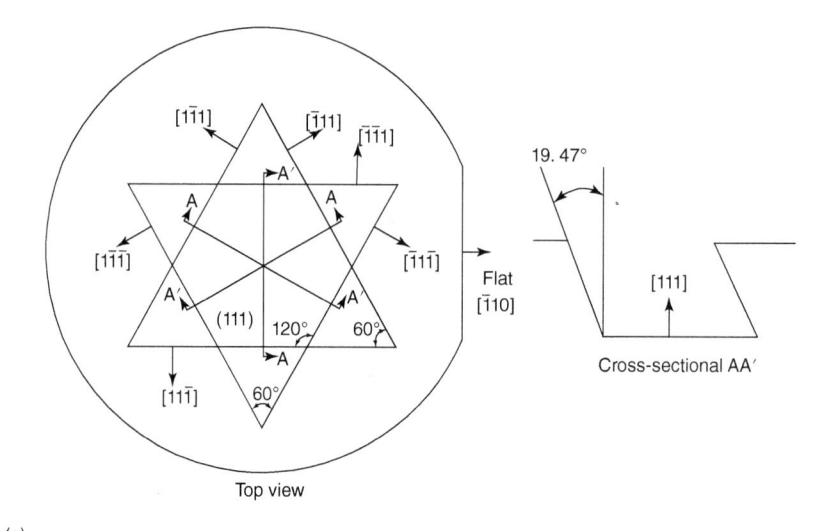

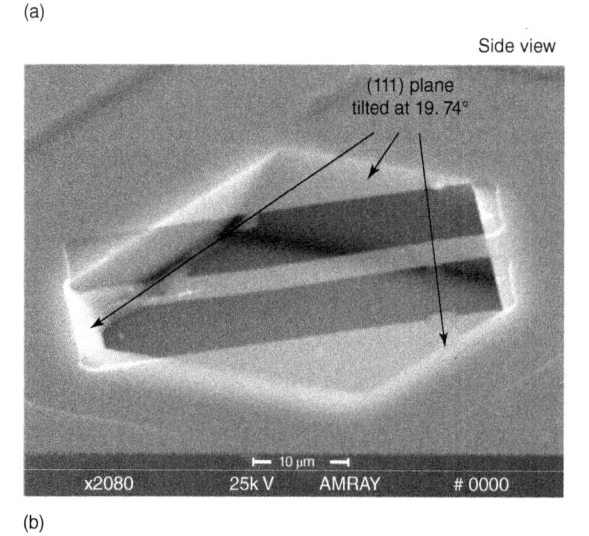

Figure 2.33 (a) All the [111]-group crystal orientations in a {111} wafer are denoted. (b) SEM image shows the exposed {111} plane exposed in an etched cavity in a {111} wafer. Source: Lee et al. [39]. © 1999, IEEE.

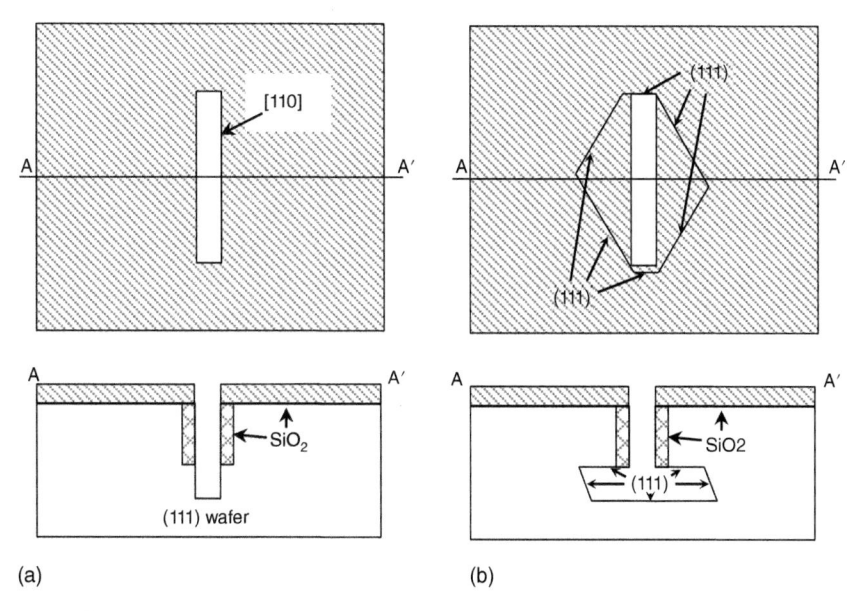

Figure 2.34 (a) Top view of a trench-etching pattern is shown at the top and an etched trench with the sidewall partly covered by oxide layer is shown at the bottom. (b) Top view showing the automatically etching-stopped hexagonal-shaped cavity at the top. The corresponding cross-sectional view is shown at the bottom.

the vertical, while the opposite three are tilted at −19.47°, as shown in Figure 2.33b. The etch rates of {111} planes are much lower than those of the other planes in KOH, which can be neglected. When the structure shown in Figure 2.34a is etched in KOH, the silicon is etched in lateral direction, because the top and bottom surfaces are {111} planes. After sufficiently long etching, the cavity is bounded by all eight {111} planes, as shown in Figure 2.34b. The self-etch stop is achieved.

The process flow of the SBM is shown in Figure 2.35 and described as follows:

(a) A layer of SiO_2 is deposited and patterned. Because freestanding structures are released by long anisotropic wet etching, both the thickness and the quality of the SiO_2 layer is important, which is different from that in the SCREAM I. The high-temperature SiO_2 is preferred, such as thermally grown SiO_2 and low-pressure chemical vapor deposition (LPCVD) tetraethoxysilane (TEOS). A DRIE is used. The etch depth is a bit larger than the structure height.
(b) A thin layer of SiO_2 is deposited conformally. The high-temperature SiO_2 is preferred.
(c) The thin layer of SiO_2 at the bottom of the trenches is removed by RIE. It is important that the sidewall oxide must be left intact.
(d) After a deep silicon RIE is used to etch down another 3–5 μm below the lower edge of the sidewall oxide, the structures are released by an anisotropic wet etching.

Though the self-aligned metallization in the SCREAM can be duplicated in the SBM as claimed in Ref. [39], the design of the anchors is complex because

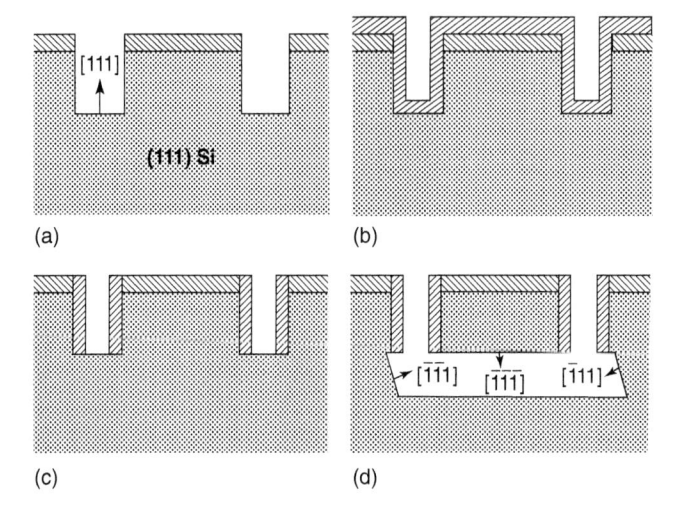

Figure 2.35 (a) After SiO$_2$ was deposited and patterned. Deep silicon etching is processed to let the etch depth be a bit larger than the structure height. (b) A thin layer of SiO$_2$ is deposited. (c) The thin layer of SiO$_2$ at the bottom of the trenches is removed by reactive ion etching. (d) After a deep silicon reactive ion etching is used to etch down another 3-5 μm below the lower edge of the sidewall oxide, the structures are released by anisotropic wet etching. Source: Lee et al. [39]. © 1999, IEEE.

the anchors of the self-aligned metallized structures are undercut very quickly in anisotropic wet etchant. The p–n junction isolation method is a better choice.

2.3.3 Silicon on Nothing (SON)

The silicon on nothing (SON) structures are single-crystal structures with vacant regions formed inside. The technology was initially proposed for low-power and high-speed metal oxide semiconductor (MOS) devices. It is now being used for the mass production of pressure sensors.

Sato developed the empty-space-in-silicon (ESS) process [40, 41] to fabricate SON structures with self-organizing recrystallization. As shown in Figure 2.36a, when a pure single-crystal silicon sample with a high-aspect-ratio trench is annealed in H$_2$ at about 1100 °C, surface migration occurs to seal the trench to minimize the surface energy. A deep trench is transformed to a spherical ESS, whose diameter is larger than that of the trench. When the silicon samples with closely arranged trenches are recrystallized, pipe-shaped ESS and plate-shaped ESS can be obtained, as shown in Figure 2.36b,c.

Some rules must be followed in ESS technology according to Ref. [41]. First, the surface migration must be made to occur only at the surface without the oxide layer and in deoxidizing ambient. Second, the initial aspect ratio and diameter of the trenches play important roles. When the aspect ratio of the trenches is lower than 3, the trenches disappear without ESS after recrystallization. When the aspect ratio is larger than 9.5, multiple ESSs are formed vertically. The diameter of the initial trenches must be smaller than a critical value, which is determined by the diffusion

(a)

(b)

(c)

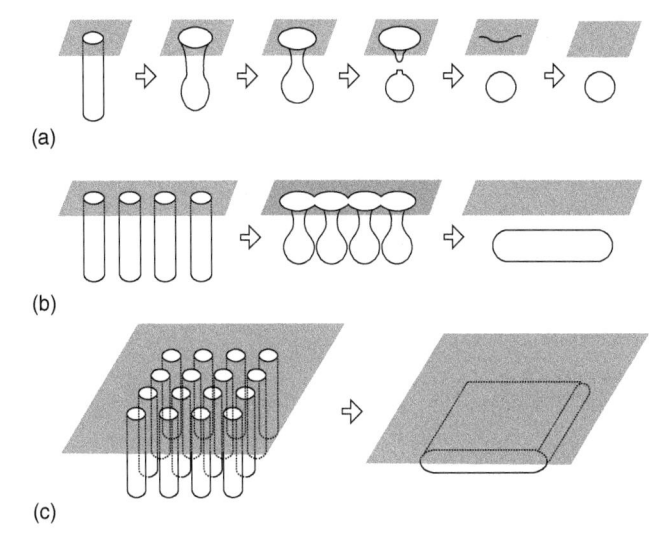

Figure 2.36 ESS process. Source: Sato et al. [41]. Copyright (2004) The Japan Society of Applied Physics.

Figure 2.37 Diffusion coefficient of silicon atoms on the surface. Source: Sato et al. [41]. Copyright (2004) The Japan Society of Applied Physics.

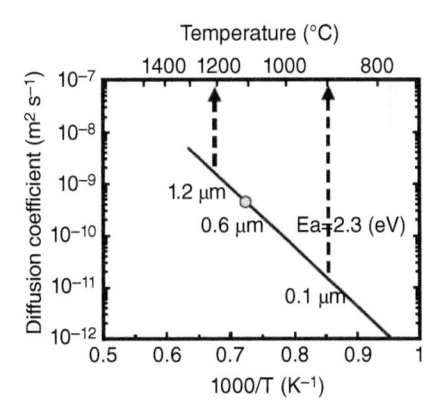

coefficient of silicon atoms. The diffusion coefficient of silicon atoms is a function of the annealing temperature, as shown in Figure 2.37 [41]. The trenches with 0.6 μm diameter can be transformed to ESS by the annealing at 1100 °C for 10 min. When the annealing temperature increases to 1200 °C, the trenches with 1.2 μm diameter can be transformed. In order to form the plate-shaped ESS, the pitches between trenches must be smaller than the diameters of spherical ESS, which are approximately twice those of the initial trenches. Figure 2.38 shows the plate-shaped ESS [41].

In order to let the top silicon layer reach desired thickness for the fabricated MEMS devices like pressure sensors, silicon epitaxy process is normally needed to thicken the original top silicon layer. The Venice process for sensors (VENSENS) MEMS fabrication technology developed by STMicroelectronics for fabricating barometer pressure sensors is just a modified ESS process where epitaxial top layer serves as the pressure sensitive diaphragm.

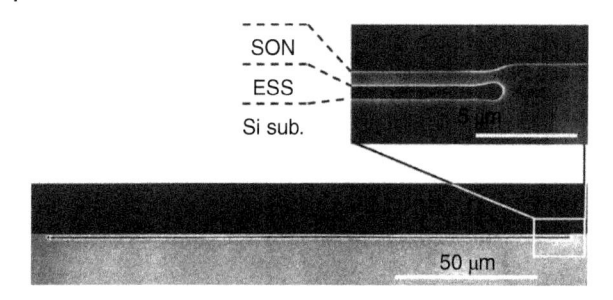

Figure 2.38 Cross-sectional view of a plate-shaped ESS structure. Source: Sato et al. [41] (Copyright (2004) The Japan Society of Applied Physics).

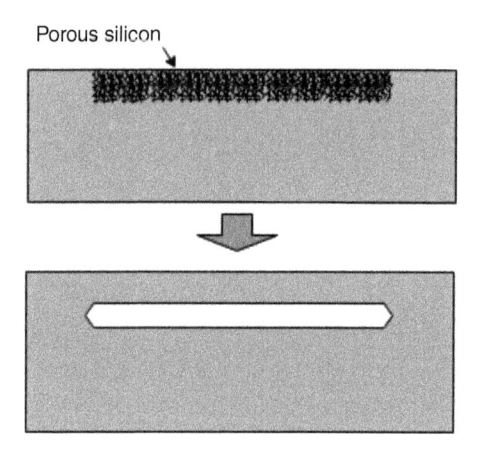

Figure 2.39 APSM process.

Another SON technology termed advanced porous silicon membrane (APSM) was developed at Robert Bosch GmbH [42]. The process flow is shown in Figure 2.39. After the porous silicon is locally etched in the single-crystal silicon wafer, an epitaxial layer is grown in a non-oxidizing condition. Because the atoms inside the pore walls retain the original monocrystalline order, a single-crystal epitaxial layer can be obtained. During the epitaxy process, cavities are formed in the porous area due to the recrystallization of the porous silicon to minimize the surface energy. When the epitaxy is based on the decomposition of SiH_4, the residue gas in the cavities is H_2, which can penetrate through the epitaxial layer at the elevated temperature. High vacuum can be obtained in the cavities using high-temperature annealing in N_2. Compared to the ESS process, thick silicon layers above the cavities can be obtained, the thickness of which is mainly determined by the epitaxy. Thus, it is possible to fabricate large diaphragms. A series of pressure sensors have been manufactured using this process. Figure 2.40 shows the close-up of a pressure sensor fabricated using the APSM process.

The thicknesses of diaphragms in the APSM process are only limited by the pressure difference across the diaphragms, while those in the cavity SOI technology are also limited by the grinding and polishing processes.

A recently developed micro-holes interetch and sealing (MIS) process is deserved to be introduced where no epitaxy is used in its single-wafer single-sided bulk-micromachining fabrication [44, 45]. Similar to the SBM process described in the Subsection of 2.3.2, the MIS process is also performed in (111) wafer and

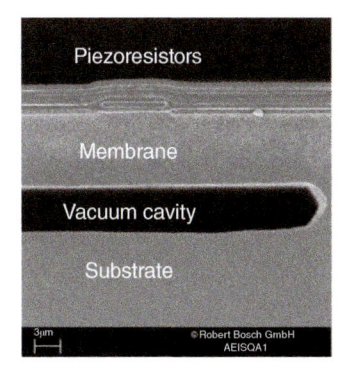

Figure 2.40 Close-up SEM of a pressure sensor. Source: Melzer [43].

the unique anisotropic wet etching properties are utilized [46]. However, herein the MIS process in (111) wafer does not for comb structure of inertial sensors or electrostatic actuators but for pressure sensitive diaphragms and cantilever beams, under which a cavity is formed to build the SON-like structure. Xinxin Li's group developed the process, and they categorized the MIS process together with other SON processes as the 3rd generation process; in contrast, the fabrication techniques for the silicon-glass-bonded pressure sensing structures and the cavity-SOI ones belong to the 1st and the 2nd generations, respectively. Apparently the 3rd generation process is single-wafer and single-sides performed, which is compatible with normal fabrication for IC. Besides, the fabricated devices are even smaller and has low cost.

Figure 2.41 shows the MIS process fabricated piezoresistive pressure sensors, where the process is only conducted from the front side of the (111) wafer. The two rows of micro-holes (trenches) with microns of diameter are opened for lateral inter-etch (under-cutting etch with TMAH) to controllably form the cavity and diaphragm (with desired thickness). Finally the micro-holes are sealed with LPCVD poly-silicon to form the vacuum cavity.

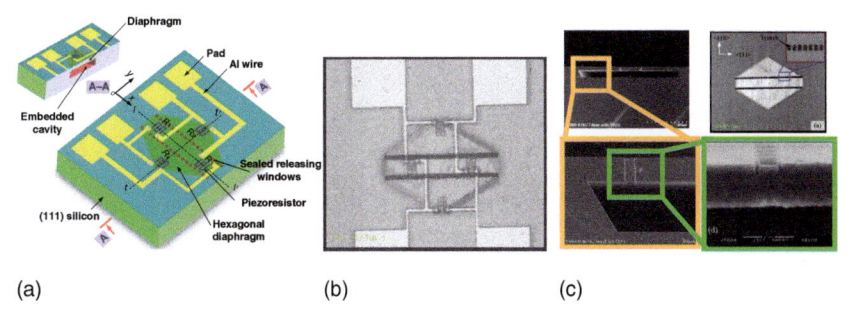

Figure 2.41 MIS process for piezoresistive pressure sensors in (111) wafers. (a) Schematic of the designed single-wafer-based single-side processed pressure sensor. (b) Infrared top-view image showing the hexagonal-shaped diaphragm/cavity structure of a fabricated piezoresistive pressure sensor. (c) Magnified infrared top-view image shows the two rows of micro-holes opened for anisotropic inter-etching to form the hexagonal diaphragm and the vacuum cavity at beneath. The sequentially magnified cross sections of the diaphragm/cavity structure show the sealed micro-holes (the trenches) with low-stress poly-silicon.

2 Bulk Micromachining

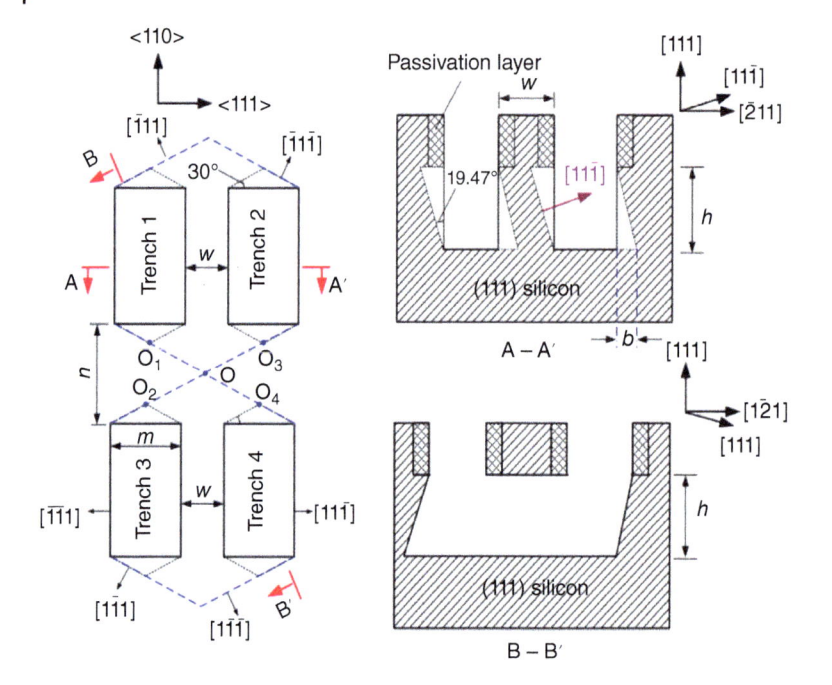

Figure 2.42 Hexagonal-shaped silicon diaphragm formation scheme in (111) wafer. The lateral under-etch along <110> orientation is shown at the top view. And the lateral under-etch along <211> orientation is sketched along the cross-sectional A-A'.

Based on anisotropic etching properties, the design rules for laying out the adjacent micro-trenches is schematically shown in Figure 2.42, where the following structural relationships should be satisfied [47]:

$$b = \tan(19.47°)h \geq w$$

$$n \leq [i\,m + (i-1)\,w]\tan(30°)$$

where i is the number of trenches in <111> direction.

Being really like the medical MIS, the MIS process is operated not for human being but for silicon wafer. Moreover, as is demonstrated in Figure 2.43, all the fabrication techniques used for the MIS process steps have been ready in standard IC foundries, where neither wafer bonding nor double-sided alignment/photolithography is needed.

Compared with the earlier described single-wafer single-sided silicon-on-nothing processes of ESS and APSM, which have both been used for fabrication of piezoresistive pressure sensors, the MIS process does not use special processes like the porous silicon treatment in APSM and the silicon epitaxy (used in the both process counterparts). More importantly, the MIS process can be multiply cycled or modified for more complex 3D structures. The following part of Chapter 3 will address these interesting technical progresses.

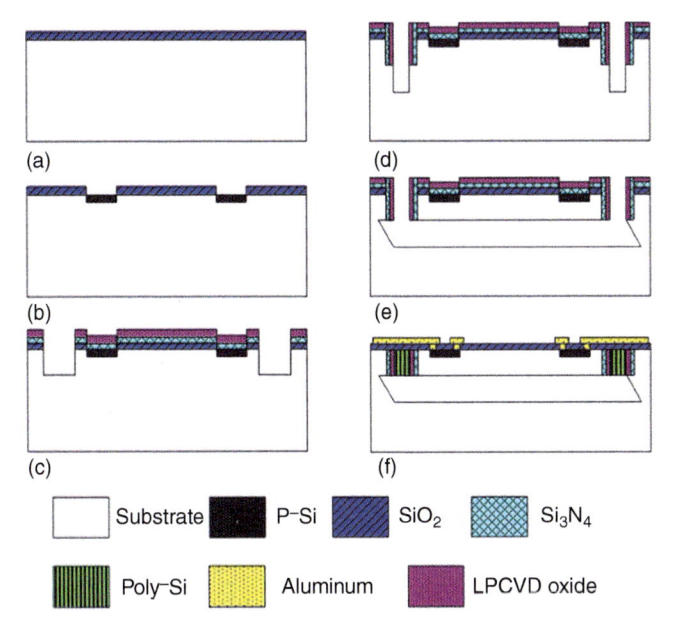

Figure 2.43 Main steps of MIS process for pressure sensors. (a) Oxidization. (b) Piezoresistor patterning and boron ion implantation. (c) LPCVD Si_3N_4 and SiO_2 depositing and releasing trench patterning and dry etching. (d) LPCVD Si_3N_4 and SiO_2 sequential depositing, passivation layers at the bottom of the releasing trench stripping, and depth of the pressure-cavity definition. (e) Aqueous TMAH lateral under-etching for hexagonal pressure-sensitive diaphragm. (f) The openings sealing with LPCVD poly-silicon and the reference-pressure cavity formation. Contact holes opening and aluminum sputtering and patterning.

References

1 Petersen, K. (1982). Silicon as a mechanical material. *Proc. IEEE* 70: 420–457.
2 Yazdi, N., Ayazi, F., and Najafi, K. (1998). Micromachined inertial sensors. *Proc. IEEE* 86: 1640–1659.
3 Kovacs, G.T.A., Maluf, N.I., and Petersen, K.E. (1998). Bulk micromachining of silicon. *Proc. IEEE* 86: 1536–1551.
4 Sato, K., Shikida, M., Matsushima, Y. et al. (1998). Characterization of orientation-dependent etching properties of single-crystal silicon: effects of KOH concentration. *Sens. Actuators A* 64: 87–93.
5 Shikida, M., Sato, K., Tokoro, K., and Uchikawa, D. (2000). Differences in anisotropic etching properties of KOH and TMAH solutions. *Sens. Actuators A* 80: 179–188.
6 Yang, H., Bao, M., Shen, S. et al. (2000). A novel technique for measuring etch rate distribution of Si. *Sens. Actuators A* 79: 136–140.
7 Li, X., Bao, M., and Shen, S. (1996). Maskless etching of three-dimensional silicon structures in KOH. *Sens. Actuators A* 57: 47–52.

8 Gianchandani, Y.B. and Najafi, K. (1992). A bulk silicon dissolved wafer process for microelectromechanical devices. *J. Microelectromech. Syst.* 1 (2): 77–85.

9 Kloeck, B., Collins, S.D., de Rooij, N.F., and Smith, R. (1989). Study of electrochemical etch-stop for high-precision thickness control of silicon membranes. *IEEE Trans. Electron Devices* 36: 663–669.

10 Connolly, E.J., French, P.J., Xia, X.H., and Kelly, J.J. (2004). Galvanic etch stop for Si in KOH. *J. Micromech. Microeng.* 14: 1215–1219.

11 Laermer, F., Schilp, A., Funk, K., and Offenberg, M. (1999). Bosch deep silicon etching: improving uniformity and etch rate for advanced MEMS applications. Technical Digest. IEEE International MEMS 99 Conference. Twelfth IEEE International Conference on Micro Electro Mechanical Systems (Cat. No.99CH36291), Orlando, FL, USA, 211–216.

12 Schmidt, M.A. (1998). Wafer-to-wafer bonding for microstructure formation. *Proc. IEEE* 86 (8): 1575–1585.

13 Miki, N., Zhang, X., Khannaa, R. et al. (2003). Multi-stack silicon-direct wafer bonding for 3D MEMS manufacturing. *Sens. Actuators A* 103: 194–201.

14 Chen, K. (2005). *Copper Wafer Bonding in Three-Dimensional Integration*. Massachusetts Institute of Technology.

15 Matsumoto, Y., Iwakiri, M., Tanaka, H. et al. (1996). A capacitive accelerometer using SDB-SOI structure. *Sens. Actuators A* 53: 267–272.

16 Noell, W., Clerc, P., Dellmann, L. et al. (2002). Applications of SOI-based optical MEMS. *IEEE J. Sel. Top. Quantum Electron.* 8 (1): 148–154.

17 Sari, I., Zeimpekis, I., and Kraft, M. (2012). A dicing free SOI process for MEMS devices. *Microelectron. Eng.* 95: 121–129.

18 Cowen, A., Hames, G., Monk, D. et al. (2011) SOIMUMPs design handbook. Revision 8.0. http://www.memscapinc.com (accessed 6 September 2011).

19 Partridge, A., Reynolds, J.K., Chui, B.W. et al. (2000). A high-performance planar piezoresistive accelerometer. *J. Microelectromech. Syst.* 9: 58–66.

20 Dao, D.V., Nakamura, K., Bui, T.T., and Sugiyama, S. (2010). Micro/nano-mechanical sensors and actuators based on SOI-MEMS technology. *Adv. Nat. Sci.: Nanosci. Nanotechnol.* 1: 013001.

21 Petersen, K., Barth, P., Poydock, J. et al. (1988). Silicon fusion bonding for pressure sensors. IEEE Technical Digest on Solid-State Sensor and Actuator Workshop, Hilton Head Island, SC, USA, 144–147.

22 Petersen, K., Gee, D., Pourahmade, F. et al. (1991). Surface micromachined structures fabricated with silicon fusion bonding. TRANSDUCERS '91: 1991 International Conference on Solid-State Sensors and Actuators. Digest of Technical Papers, San Francisco, CA, USA, 397–399.

23 Wang, Y., Zheng, X., Liu, L., and Li, Z. (1991). A novel structure of pressure sensors. *IEEE Trans. Electron Devices* 38 (8): 1797–1802.

24 Parameswaran, L., Hsu, C., and Schmidt, M.A. (1995). A merged MEMS-CMOS process using silicon wafer bonding. Proceedings of International Electron Devices Meeting, Washington, DC, USA, 613–616.

25 Luoto, H., Henttinen, K., Suni, T. et al. (2007). MEMS on cavity-SOI wafers. *Solid-State Electron.* 51 (2): 328–332.

References

26 Yeh, C. and Najafi, K. (1997). A low-voltage tunneling-based silicon micro accelerometer. *IEEE Trans. Electron Devices* 44: 1875–1882.

27 Najafi, K. and Suzuki, K. (1989). A novel technique and structure for the measurement of intrinsic stress and Young's modulus of thin films. Proceedings IEEE Workshop on Microelectromechanical Systems (MEMS 89), 96–97.

28 Gianchandani, Y.B. and Najafi, K. (1996). Bent-beam strain sensors. *J. Microelectromech. Syst.* 5 (1): 52–58.

29 Gianchandani, Y.B. and Najafi, K. (1997). A silicon micromachined scanning thermal profiler with integrated elements for sensing and actuation. *IEEE Trans. Electron Devices* 44 (11): 1857–1868.

30 Kourepenis, A., Borenstein, J., Connelly, J. et al. (1998). Performance of MEMS inertial sensors. IEEE 1998 Position Location and Navigation Symposium (Cat. No.98CH36153), Palm Springs, CA, USA, 1–8.

31 Weinberg, M., Connelly, J., Kourepenis, A., and Sargent, D. (1997). Micro electro-mechanical instrument, and systems development at the Charles Stark Draper Laboratory, Inc. 16th DASC. AIAA/IEEE Digital Avionics Systems Conference. Reflections to the Future. Proceedings, Irvine, CA, USA, 8.5–33.

32 Mason, A., Yazdi, N., Chavan, A.V. et al. (1998). A generic multielement microsystem for portable wireless applications. *Proc. IEEE* 86 (8): 1733–1746.

33 Gianchandani, Y.B., Ma, K.J., and Najafi, K. (1995). A CMOS dissolved wafer process for integrated P^{++} microelectromechanical systems. Proceedings of the International Solid-State Sensors and Actuators Conference - TRANSDUCERS '95, Stockholm, Sweden, 79–82.

34 Juan, W. and Pang, S.W. (1996). Released Si microstructures fabricated by deep etching and shallow diffusion. *J. Microelectromech. Syst.* 5: 18–23.

35 Juan, W. and Pang, S.W. (1998). High-aspect-ratio Si vertical micromirror arrays for optical switching. *J. Microelectromech. Syst.* 7 (2): 207–213.

36 Shaw, K.A., Zhang, Z.L., and MacDonald, N.C. (1994). SCREAM I: a single mask, single-crystal silicon, reactive ion etching process for microelectromechanical structures. *Sens. Actuators A* 40: 63–70.

37 MacDonald, N.C. (1996). SCREAM micro electro mechanical systems. *Microelectron. Eng.* 32: 49–73.

38 de Boer, M.J., Tjerkstra, R.W., Berenschot, J.W. et al. (2000). Micromachining of buried micro channels in silicon. *J. Microelectromech. Syst.* 9: 94–103.

39 Lee, S., Park, S., and Cho, D. (1999). The surface/bulk micromachining (SBM) process: a new method for fabricating released MEMS in single crystal silicon. *J. Microelectromech. Syst.* 8: 409–416.

40 Mizushima, I., Sato, T., Taniguchi, S., and Tsunashima, Y. (2000). Empty-space-in-silicon technique for fabricating a silicon-on-nothing structure. *Appl. Phys. Lett.* 77: 3290–3292.

41 Sato, T., Mizushima, I., Taniguchi, S. et al. (2004). Fabrication of silicon-on-nothing structure by substrate engineering using the empty-space-in-silicon formation technique. *Jpn. J. Appl. Phys.* 43: 12–18.

42 Armbruster, S., Schafer, F., Lammel, G. et al. (2003). A novel micromachining process for the fabrication of monocrystalline Si-membranes using porous silicon.

TRANSDUCERS '03. 12th International Conference on Solid-State Sensors, Actuators and Microsystems. Digest of Technical Papers (Cat. No.03TH8664), Boston, MA, USA, 246–249.

43 Melzer, F. *Consumer MEMS – A Technology Play*. Bosch Sensor Tec.

44 Wang, J. and Li, X. (2011). Single-side fabricated pressure sensors for IC-foundry compatible high-yield and low-cost volume production. *IEEE Electron Device Lett.* 32 (7): 979–981.

45 Wang, J. and Li, X. (2011). A single-wafer-based single-sided bulk-micromachining technique for high-yield and low-cost volume production of pressure sensors. Transducers'2011, Beijing, China, 410–413.

46 Seidel, H., Csepregi, L., Heuberger, A., and Baumgärtel, H. (1990). Anisotropic etching of crystalline silicon in alkaline solutions. *J. Electrochem. Soc.* 137 (11): 3612–3626.

47 Wang, J., Xia, X., and Li, X. (2012). Monolithic integration of pressure plus acceleration composite TPMS sensors with a single-sided micromachining technology. *J. Microelectromech. Syst.* 21 (2): 284–293.

3

Enhanced Bulk Micromachining Based on MIS Process

Xinxin Li and Heng Yang

Shanghai Institute of Microsystem and Information Technology, State Key Lab of Transducer Technology, Chinese Academy of Sciences, 865 Changning Road, Shanghai 200050, China

3.1 Repeating MIS Cycle for Multilayer 3D structures or Multi-sensor Integration

3.1.1 Pressure Sensors with PS^3 Structure

Borrowing ideas from medical minimally invasive surgery (MIS), Xinxin Li et al. developed a micro-holes interetch and sealing (MIS) process for MEMS pressure sensors. As is addressed at the last part of Chapter 2, MIS process is single-wafer based and single-sided performed; thereby it is compatible with standard analog IC foundry manufacturing concept. Besides, unlike its silicon-on-nothing (SON) process counterparts of empty-space-in-silicon (ESS) and advanced porous silicon in membrane (APSM) where only one-layer diaphragm-cavity structure can be built, the MIS steps can be repeatedly employed to form more complex multilayer 3D structures. In this way the single-sided fabrication capability for 3D MEMS is much enhanced.

Figure 3.1 shows the 3D structure diagram and process steps for an advanced piezoresistive pressure sensor, where the vacuum cavity and diaphragm structure is suspended from silicon substrate by using a cantilever-like structure to accommodate the pressure sensing part. Such a formed structure is named PS^3, i.e. packaging stress suppressing suspension [1]. It can be seen from the process flow that the MIS process cycle is repeated twice to form the absolute pressure sensor and the suspension structure, respectively. During the second MIS cycle, the micro-hole trenches are replaced by the narrow-ditch trench to define the suspended structure, and the final vacuum sealing step is cancelled.

Such a PS^3 structure is for the purpose of depressing thermal-mismatch stress from the packaging substrate, which is a common problem when the pressure sensor chip is attached on a non-silicon substrate, and time- and labor-consuming compensation for the stress-induced thermal drift is generally needed to decrease the inaccuracy of the sensor. With the PS^3 structure however, the packaging stress can be eliminated.

3D and Circuit Integration of MEMS, First Edition. Edited by Masayoshi Esashi.
© 2021 WILEY-VCH GmbH. Published 2021 by WILEY-VCH GmbH.

3 Enhanced Bulk Micromachining Based on MIS Process

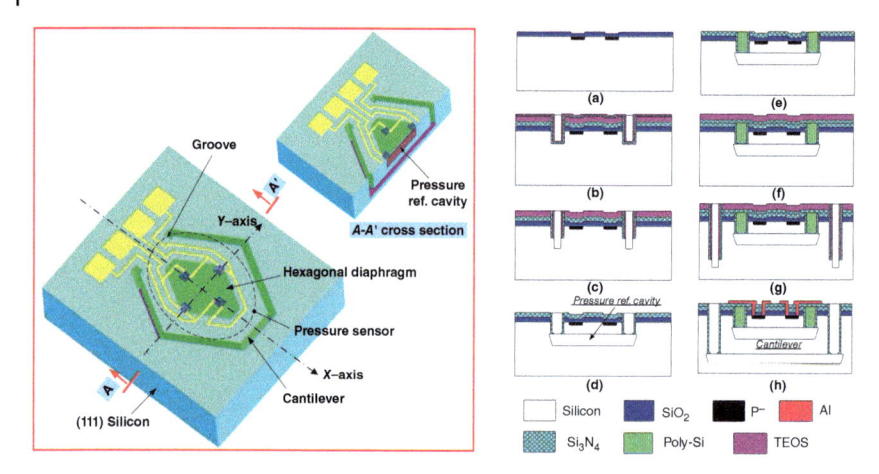

Figure 3.1 (left) 3D schematic of the pressure sensor integrated monolithically with the PS^3 structure. (right) Process steps for the PS^3 integrated pressure sensor, with the shown cross section cut along the dotted line of A-A' in (left).

The finite-element simulation results shown in Figure 3.2 well verify the function of the PS^3 structure.

As is shown in Figure 3.3, the multilayer structure of PS^3 is built by repeating the MIS process cycle. The packaged PS^3 sensor is tested and compared with a fabricated counterpart sensor without PS^3. The PS^3 structure successfully decreases the temperature coefficient of offset (TCO) of the pressure sensor. Without any compensation for thermal-drift used, the temperature drift of the sensor with the PS^3 structure is merely 0.016%/°C FSO (Full Scale Output), which is about 15 times better than that of the sensor without the PS^3 structure, where the tested temperature range is from −40 to +120 °C.

In order to further decrease TCO of the sensor, which may be induced from other factors like the thermal stress of the surface insulation layer (e.g. SiN$_x$ and/or SiO$_2$), a dual-unit PS^3 sensor structure is designed and fabricated with the MIS process [2]. Figure 3.4 shows the concept of the dual-unit sensor compared with two single-unit sensors with or without the PS^3 configuration, respectively. In the dual unit sensor, one unit is pressure sensitive but another dummy unit is pressure insensitive by connecting the air in the cavity with that of outside the cavity. It is worth to mention that the surface insulating-layer-induced thermal stress distribution on the diaphragms of the two units is identical. It is just this point that can be used to realize self-compensation for TCO of the sensor.

The fabricated dual-unit PS^3 sensor is shown in the figure with every piezoresistor denoted as R_1-R_4 on the sensing diaphragm and R'_1-R'_4 on the dummy diaphragm. Based on that the stress at the top and bottom corners of the diaphragm is with opposite sign to the stress at the left and right corners, and a piezoresistor on the sensitive diaphragm is connected in serial with its opposite-signed counterpart on the dummy diaphragm. Then a cross-unit Wheatstone bridge is formed as is shown in the figure. In this way the resistance change in any arm of the bridge, which is

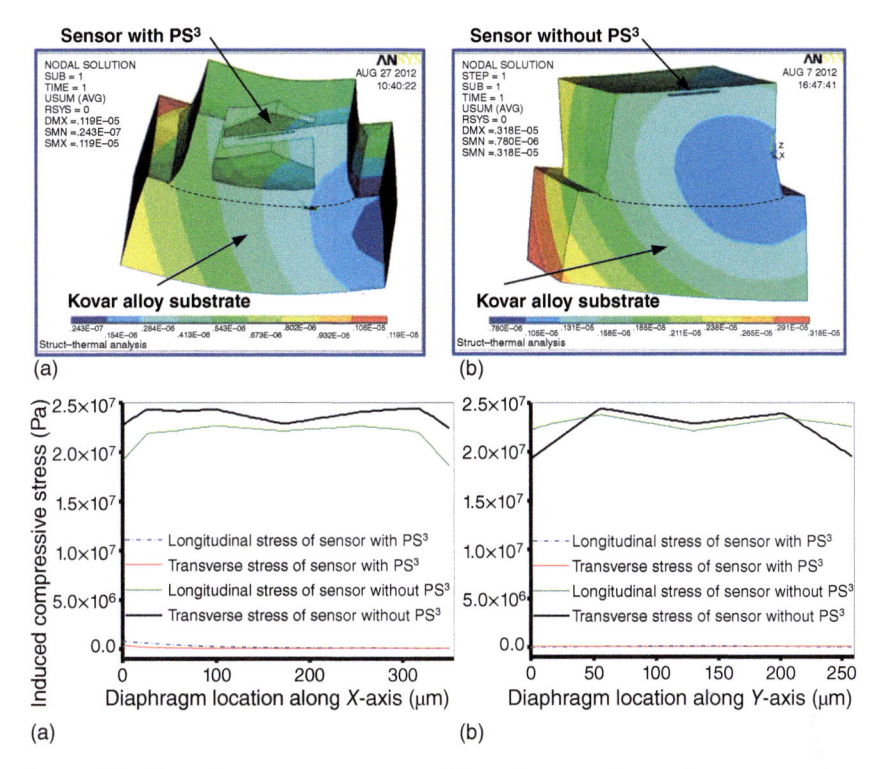

Figure 3.2 (Top) After the pressure sensor chip packaged on Kovar-alloy substrate, the packaging thermal-mismatch-induced structure-bending effect (representing the packaging-stress) on the sensing-diaphragm deflection is simulated, resulting in the cross-sectional deflection shape for the sensors embedded in the PS^3 (a) and that for the sensor without the PS^3 (b). (Bottom) Comparison of the simulated packaging thermal-mismatch-induced diaphragm bending stress (distributed on the top surface of the diaphragm) between the sensor with PS^3 and another without PS^3. (a) Packaging-induced diaphragm stress along X-axis (denoted in Figure 3.1a). (b) Packaging-induced diaphragm stress along Y-axis (denoted in Figure 3.1a).

due to the surface insulator stress, has a self-compensation property, and thereby the relevant temperature drift is expected to be largely depressed.

TCO of the three types of sensors (single-unit non-PS^3 sensor, single-unit PS^3 sensor, and dual-unit PS^3 unit) are tested within a wide temperature range from −40 to +125 °C for comparison. The three types of pressure sensors are all packaged in market-available Kovar-alloy sensor shells. The measured TCO for the three sensors are plotted together in Figure 3.4, where there is no any extra compensation method used for the thermal drift. The dual-unit PS^3 sensor with correct cross-unit Wheatstone-bridge configuration is with the tested TCO of merely 0.002%/°C FSO, which is about 11 times better than that of the single-unit PS^3 sensor and about 72 times better than that of the single-unit non-PS^3 sensor. Obviously, the suspended dual-unit pressure sensor can effectively suppress on-chip unbalancing factors induced residual stress, thereby improving temperature stability. Without any extra

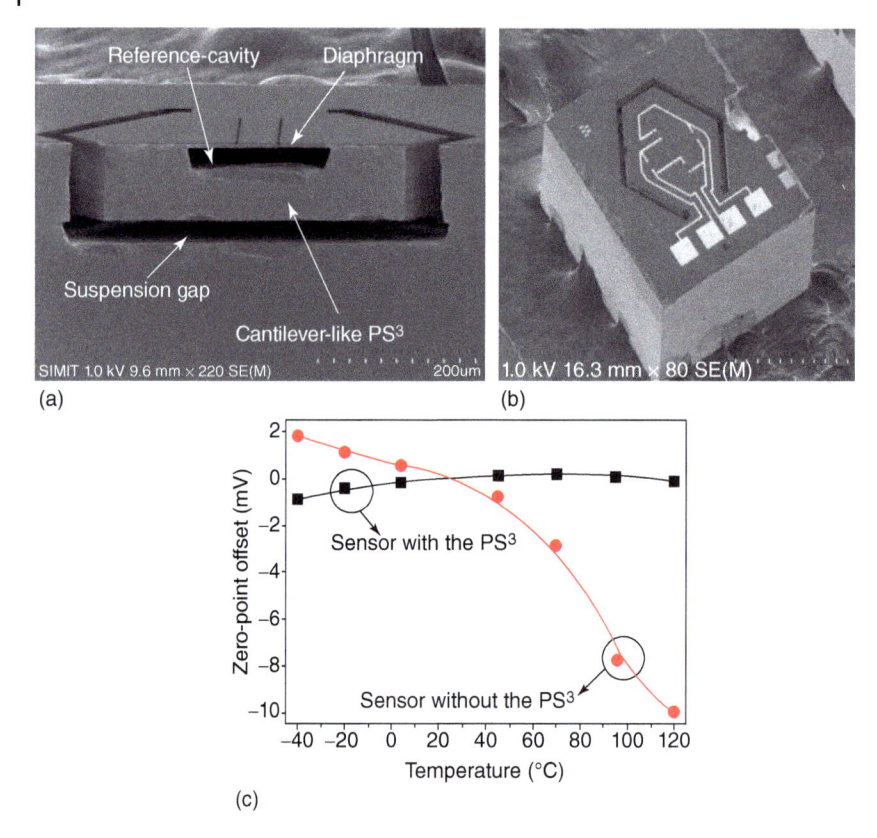

Figure 3.3 (a) SEM image showing the cross-sectional cutting view of the fabricated sensor with the PS^3. (b) SEM image of the PS^3 pressure sensor chip. (c) Comparison in temperature drift of zero-point offset between the sensor with PS^3 and the sensor without PS^3.

compensation method used, so low the TCO datum of 0.002/°C that it can already well meet the requirement of most low-cost pressure sensor applications.

3.1.2 P+G Integrated Sensors

Multiple operation of the MIS process cycle can be used to monolithically fabricate composite sensors in one chip. A typical example is integrated P+G sensor for automotive tire pressure monitoring system (TPMS) application, where a pressure sensor and an accelerometer are normally needed for monitoring tire pressure and wheel rotation. If the MIS process is used to integrate both the pressure sensor and the accelerometer, the difficulty lies in that the thickness of the seismic mass for the accelerometer has to be the same as that of the pressure sensing diaphragm; thereby the mass will be too small to reach the required sensitivity to acceleration. In this way a thick layer of electroplated metal mass has to be formed on top of the thin silicon plate to meet the sensitivity requirement, which largely increases the fabrication complexity and cost [3]. Recently the updated version of TPMS sensors are required to integrate both x- and z-axis accelerometers (i.e. detecting both in-plane

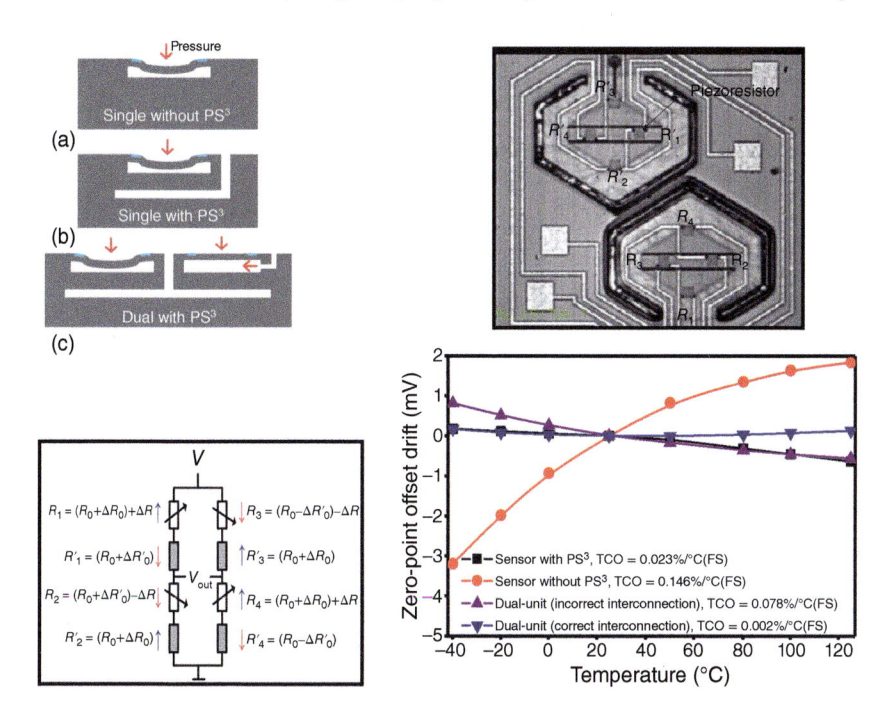

Figure 3.4 (Top left) Cross-sectional schematics of the three types of pressure sensor structures for performance comparison. (a) Conventional single-unit pressure sensor. (b) Single-unit PS³ sensor. (c) Dual-unit PS³ sensor structure for further on-chip elimination of thermal instability induced by other factors like the stress in the surface insulating layer. (Top right) top-view infrared micro-images of the dual-unit PS³ sensor with the piezoresistors denoted. (Bottom left) Cross-unit Wheatstone-bridge interconnection for on-chip TCO self-compensation. (Bottom right) Tested TCO results of the three types of sensors are plotted together for comparison. The measured TCO results are listed in the inset labels.

and vertical accelerations) to add the function of automatically identifying a certain wheel to obtain more tire information from individual wheels. Formation of such a complex monolithic structure is really challenging. By repeating the MIS process cycle, the P+x&zG integrated sensor is fabricated in one (111) wafer, and the fabrication is only performed from front side of the wafer [4]. Figure 3.5 shows the structure of the P+2G sensor chip (2 × 2 mm in size) and the fabrication flow with repeatedly processed MIS cycle. Besides the PS³ pressure sensor, an in-plane X-axis cantilever-mass accelerometer and a vertical Z-axis cantilever-mass accelerometer are monolithically integrated in the same chip.

For further decreasing the chip size of the P+G TPMS sensors, a PinG configuration is realized by repeating the MIS process cycle, where the PS³ pressure sensor is laid inside the seismic mass of the cantilever-mass accelerometer [5]. Figure 3.6a shows the designed PinG sensor chip, where the PS³ pressure sensor is put inside the mass of the Z-axis accelerometer. Now look at the cross-sectional schematic of the PinG structure. The cantilever of the accelerometer, the mass bottom and over-range stop gap, as well as the pressure-sensing diaphragm and its over-range stop distance (i.e. the vacuum cavity height), all these micromechanical structures have different

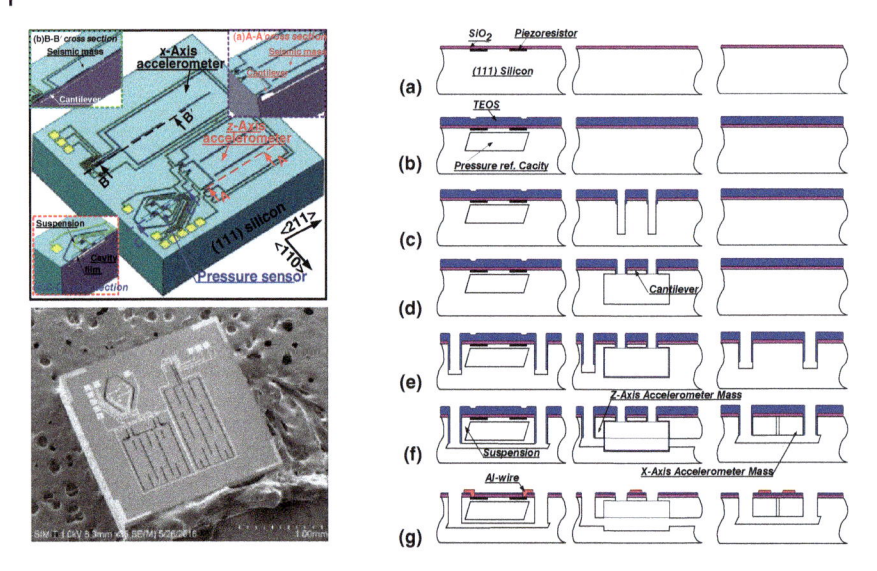

Figure 3.5 (left) 3D scheme and SEM image of the P + 2G TPMS sensor chip. (right) Fabrication steps by repeatedly using the MIS process cycle, where the cross-sections are cut along the lines marked in the 3D scheme of (left).

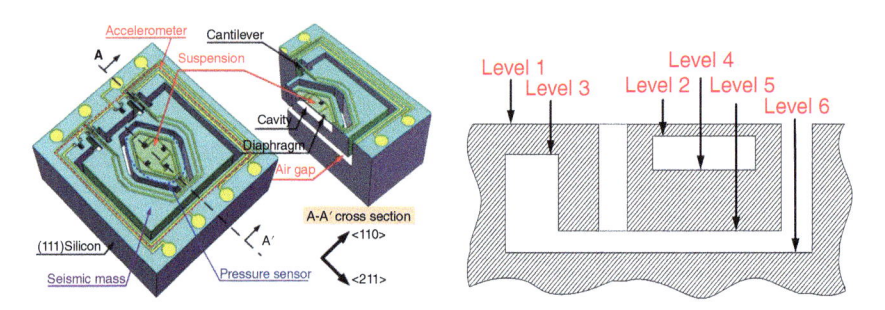

Figure 3.6 3D structure and cross-sectional schematic of the single-side processed PinG sensor.

silicon thickness or etching depth, thereby requiring to build six 3D structure levels, which is denoted in Figure 3.6b.

Figure 3.7 shows the repeatedly cycled MIS process to sequentially form the pressure sensor, the accelerometer's mass bottom, and its cantilever. And the fabricated PinG sensor is demonstrated with different SEM images. Processed only from the front-side of a single (111) wafer, such a sophisticated 3D micromachining structure, which contains six levels, it is really fabricated.

3.2 Pressure Sensor Fabrication – From MIS Updated to TUB

With the MIS process, absolute pressure sensors can be well fabricated. For gauge or differential pressure sensors, however, another pressure rather than that introduced

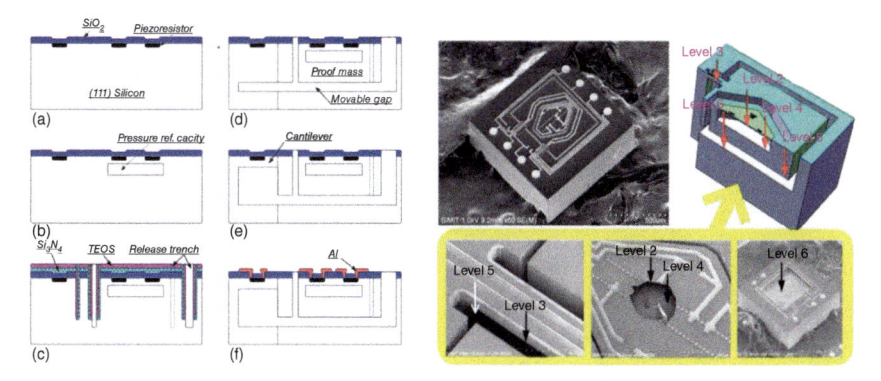

Figure 3.7 Process flow and the views of the fabricated PinG sensor chip.

from the front side of the diaphragm is indispensable. If only simply opening an inlet from the backside of the wafer by using deep reactive ion etching (DRIE), the etching-rate non-uniformity throughout the whole wafer will easily cause damage of the very thin silicon diaphragm (several microns in thickness), thereby significantly lowering the fabrication yield. In order to secure the fabrication yield when opening the backside pressure inlet, Xinxin Li's group recently modifies the MIS process to develop its updated version, which is named TUB (abbreviation of thin-film under bulk). Figure 3.8a shows 3D structure of a differential pressure sensor by using the TUB process. A twin-island reinforced beam-diaphragm structure is employed to concentrate pressure-induced stress into the beam and thereby to realize both high sensitivity and high linearity [6]. Piezoresistive Wheatstone bridge is laid out on the central and side positions of the single-crystalline silicon beam for sensing signal readout, while stress-released poly-silicon thin film (with the smallest thickness as 1μm) serves as diaphragm. Two backside pressure inlet holes are opened to the cavity under the diaphragm, and each hole is just located beneath each island.

In order to ensure the uniformity of the very small thickness for realizing high sensitivity, the poly-silicon deposition is expected to be automatically stopped when the desired thickness is reached. Deposition poly-silicon through the micro-holes in MIS process is a smart idea to automatically stop the deposition, where the diameter (or half side length) of the holes are designed as the same as the desired poly-silicon thickness [7]. In this way the deposition is stopped when the micro-holes are fulfilled and blocked when the deposited thickness of the poly-silicon thin-film reaches the desired thickness (e.g. 1–3 μm). However, a new problem needs to be addressed: how to remain the poly diaphragm when the single-crystalline silicon is etched from the front side to form the beam-island diaphragm? In the work of [7], a very thin layer of thermal oxidization is formed before the poly-silicon deposition. The oxidization is performed by feeding oxygen gas into the cavity through the micro-holes. Since the formed SiO_2 thin layer is sandwiched between the single-crystalline silicon and the following deposited poly-silicon, it can act as an etching-stop layer for the following front-side silicon etching to form the single-crystalline silicon beam-island structure, while protecting the poly-silicon diaphragm from etching. Besides, the SiO_2 layer at the bottom of the cavity serves as the etching-stop layer to prevent

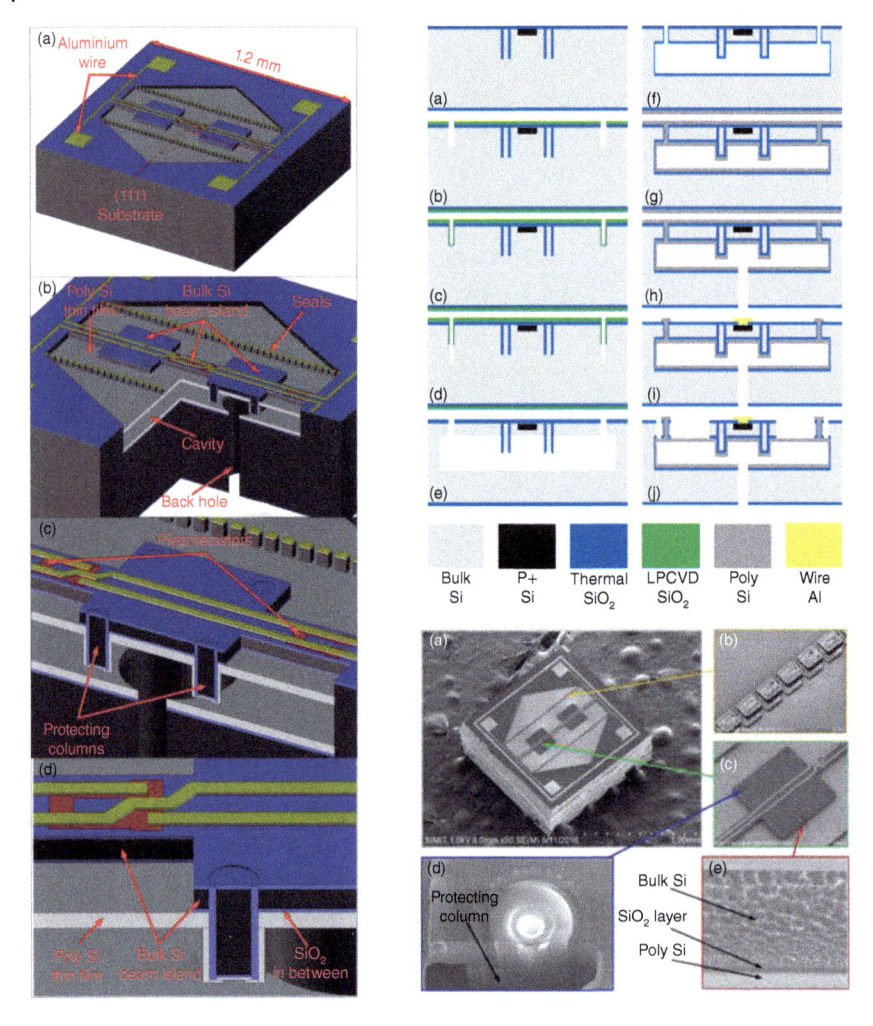

56 | *3 Enhanced Bulk Micromachining Based on MIS Process*

Figure 3.8 (left) Structure schematic of the different pressure sensor by using the TUB process, with sequentially magnified local views at beneath. (right) Process steps for the sensor and the SEM imaged of the fabricated sensor structures.

over deep-etching from the backside to form the pressure-inlet openings. Due to that the backside through-wafer DRIE is quite deep and the etching-rate nonuniformity throughout the whole wafer is not negligible, any over etch would damage the very thin beam-island-diaphragm silicon structure from the backside. The pre-buried SiO_2 layer can successfully stop the silicon deep etching when the backside etching reaches the cavity. After the front-side shallow etching for the structure is completed, the SiO_2 on the diaphragm regions will be removed, with an exception at the beam-island area. The mismatch stress due to the remained SiO_2 beneath the silicon beam-island can be largely compensated by the SiO_2 insulating layer on top of the beam-island structure. Therefore, with the TUB process steps, the poly-silicon thin-film diaphragm is formed under the single-crystalline bulk silicon layer for the beam-island structure. In Figure 3.8b not only the process details are shown but also the fabricated sensor structure is imaged with SEM. It is worthy pointing out that four micro-feet at backside are formed at the four corners of each rectangular island for sustaining the structure under over-range pressure. Attributed to the TUB process, as low as 1 kPa pressure measure-range is realized, and the sensor is tested to feature high sensitivity, high linearity, and over 100 times over-range protecting capability.

With the updated TUB process, the pressure sensor chip can be fabricated into a very small size since the poly-silicon diaphragm can be so thin as 1 μm by setting the diameter of the opened hole trenches as 2 μm. Then the TMAH wet etching to form the very small cavity (under the diaphragm) can be completed very fast. Figure 3.9a shows the TUB fabricated barometric pressure sensor, which has the chip size as small as 0.4 × 0.4 mm. Due to the very rapid etching to form the small-sized cavity, the micro-trench openings no longer need to be laid in the diaphragm area. In contrast the micro-holes can be laid only at the single-crystalline silicon islands and the regions surrounding the diaphragm. Such micro-hole layout helps to form a "scar-free" poly-silicon diaphragm [8]. Such small sensors can be wafer-level high-yield fabricated by using the TUB process, and the fabricating cost for one die can be so low as one US cent. So cheap sensors still feature high sensitivity of about 1 mV/1 kPa/3.3 V and fine linearity of about 0.3% FS. The infrared microscopic images in Figure 3.9b are taken at sequential moments of the TMAH lateral etching to form the hollowed cavity under the diaphragm.

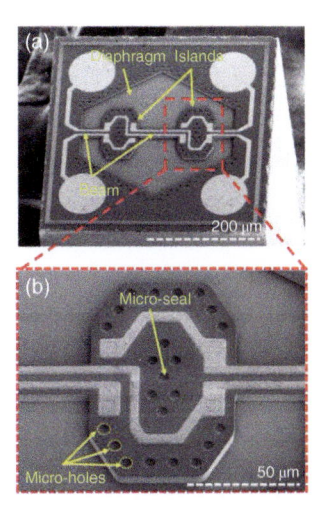

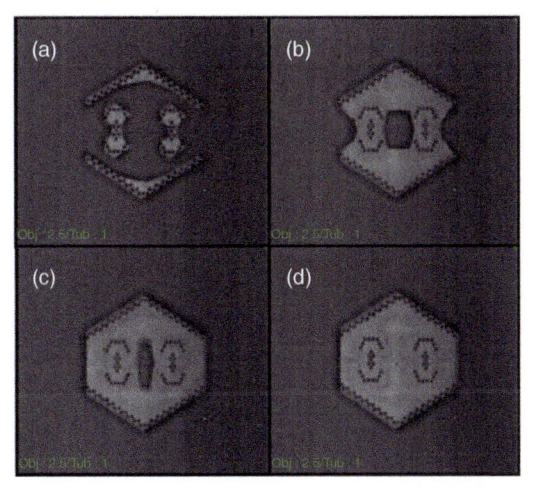

Figure 3.9 (left) Top view SEM of the "scar-free" pressure sensor chip, where all the micro-seals for the micro-holes are outside the diaphragm. (right) Infrared micrographs recording the moments during the cavity etching process with TMAH. (a)–(d) sequentially show the etching for 0.5, 1, 1.5, and 2 hours, respectively. The micro-holes in the island help to divide the hexagonal etching area into several small ones, thereby significantly shortening the etching time.

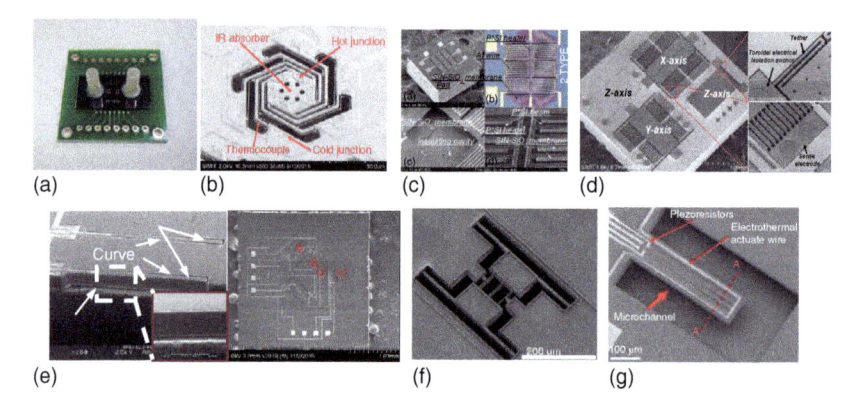

Figure 3.10 Various MEMS sensors fabricated with flexibly modified MIS process. (a)–(g) differential-pressure flow sensor, infrared sensor, thermopile gas flow sensor, tri-axis accelerometer, shock sensors, resonant chemical sensor, and resonant-cantilever particle detector, respectively. Source: (a) Liu et al. [9], (b) Li et al. [10], (c) Xue et al. [11], (d) Chen et al. [12], (e1) Wang and Li [13], (e2) Cai et al. [14], (f) Yu et al. [15], (g) Bao et al. [16].

3.3 Extension of MIS Process for Various Advanced MEMS Devices

The basic concept of the MIS process, i.e. excavating a cavity under a silicon diaphragm by previously opening vertical micro-trenches of desired depth in (111) wafer, can be flexibly employed with suitable modification to fabricate various types of MEMS device structures.

Some developed MEMS structures are sequentially demonstrated in Figure 3.10, which are (a) differential-pressure micro-flow-sensor chip for direct surface mounting technique (SMT) packaging on printed circuit board (PCB) [9], (b) thermopile infrared sensor with high-sensitivity thermocouple between single-crystalline-silicon and metal [10], (c) single-side fabricated p^+Si/Al thermopile-based gas flow sensor [11], (d) tri-axis capacitive accelerometer [12], (e) single-axis and monolithically integrated tri-axis high-shock accelerometers [13, 14], (f) dog-bone micro-resonator for detection of trace-level chemical vapor [15, 17], and (g) resonant-cantilever particle (PM2.5) sensor for environmental air-pollution monitoring [16]. These devices always feature either better performance or lower fabrication cost, which is attributed to the unique advantages of the MIS process.

References

1 Wang, J. and Li, X. (2013). Package-friendly piezoresistive pressure sensors with on-chip integrated packaging-stress-suppressed suspension (PS^3) technology. *J. Micromech. Microeng.* 23: 045027.

2 Wang, J. and Li, X. (2014). A dual-unit pressure sensor for on-chip self-compensation of zero-point temperature drift. *J. Micromech. Microeng.* 24 (8): 085010.

3 Wang, J., Xia, X., and Li, X. (2012). Monolithic integration of pressure plus acceleration composite TPMS sensors with a single-sided micromachining technology. *IEEE J. Microelectromech. Syst.* 21 (2): 284–293.

4 Wang, J., Ni, Z., Zhou, J. et al. (2017). *Pressure + X/Z Two-Axis Acceleration Composite Sensors Monolithically Integrated in Non-SOI Wafer for Upgraded Production of TPMS (Tire Pressure Monitoring System)*, 1359–1362. Las Vegas: IEEE MEMS.

5 Wang, J. and Li, X. (2015). Single-side fabrication of multilevel 3-D microstructures for monolithic dual sensors. *J. Microelectromech. Syst.* 24 (3): 531–533.

6 Bao, M. (2000). *Micro Mechanical Transducers-Pressure Sensors, Accelerometers and Gyroscopes*. Amsterdam, The Netherlands: Elsevier.

7 Zou, H., Wang, J., and Li, X. (2017). High-performance low-range differential pressure sensors formed with a thin-film under bulk micromachining technology. *J. Microelectromech. Syst.* 26 (4): 879–885.

8 Ni, Z., Jiao, D., Zou, H. et al. (2017). 0.4mm×0.4mm Barometer sensor-chip fabricated by a scar-free 'MIS' (minimally invasive surgery) process for 0.01US$/die product. *Transducers* 2017: 774–777.

9 Liu, J., Wang, J., and Li, X. (2012). Fully front-side bulk-micromachined single-chip micro flow-sensors for bare-chip SMT (surface mounting technology) packaging. *J. Micromech. Microeng.* 22: 035020.

10 Li, W., Ni, Z., Wang, J., and Li, X. (2019). A front-side microfabricated tiny-size thermopile infrared detector with high sensitivity and fast response. *IEEE Trans. Electron Devices* 66 (5): 2230–2237.

11 Xue, D., Song, F., Wang, J., and Li, X. (2019). Single-side fabricated p$^+$ Si/Al thermopile-based gas flow sensor for IC-foundry-compatible, high-yield, and low-cost volume manufacturing. *IEEE Trans. Electron Devices* 66 (1): 821–824.

12 Chen, F., Zhao, Y., Wang, J. et al. (2018). A single-side fabricated triaxis (111)-silicon microaccelerometer with electromechanical sigma–delta modulation. *IEEE Sens. J.* 18: 1859–1869.

13 Wang, J. and Li, X. (2010). A high-performance dual-cantilever high-shock accelerometer single-sided micromachined in (111) silicon wafers. *J. Microelectromech. Syst.* 19 (6): 1515–1520.

14 Cai, S., Li, W., Zou, H. et al. (2019). Design, fabrication, and testing of a monolithically integrated tri-axis high-shock accelerometer in single (111)-silicon wafer. *Micromachines* 10 (4): 227.

15 Yu, F., Wang, J., Xu, P., and Li, X. (2016). A tri-beam dog-bone resonant sensor with high-Q in liquid for disposable 'test-strip' detection of analyte droplet. *IEEE J. Microetecteomech. Syst.* 25 (2): 244–251.

16 Bao, Y., Cai, S., Yu, H. et al. (2018). A resonant cantilever based particle sensor with particle-size selection function. *J. Micromech. Microeng.* 28: 085019.

17 Yu, F., Xu, P., Wang, J., and Li, X. (2015). Length-extensional resonating gas sensors with IC-foundry compatible low-cost fabrication in non-SOI single-wafer. *Microelectron. Eng.* 136: 1–7.

4

Epitaxial Poly Si Surface Micromachining

Masayoshi Esashi

Tohoku University, Micro System Integration Center (μ SIC), 519-1176 Aramaki-Aza-Aoba, Aoba-ku, Sendai, 980-0845, Japan

4.1 Process Condition of Epi-poly Si

Epi-poly Si (epitaxial poly Si) was developed in Fraunhofer Institute for Silicon Technology (ISIT) in Germany and Uppsala University in Sweden [1]. The epi-poly Si in thickness of 20 μm or more, which is around 10 times thicker than that of conventional poly Si, can be deposited with controlled film stress [2]. Doping for n^+ layer can be carried out by adding PH_3 (phosphine) to the gas. The thick-doped epi-poly Si is suitable especially for making capacitive sensors with comb structures. This is because the influence of a stray capacitance is reduced, owing to its relatively large capacitance, and hence the chip for the capacitance detection circuit need not to be monolithically integrated with the MEMS sensor. This enables design flexibility and low cost.

The fabrication process is as follows. Thin (125 nm thick) poly Si is deposited by low-pressure chemical vapor deposition (LPCVD) on SiO_2 at 650 °C. This process makes seeds for the epi-poly Si. Low-stress (3 MPa) poly Si with vertical column can be grown at 1000 °C in Si epitaxial reactor. High deposition rate (0.55 μm min^{-1}), small surface roughness, and small stress (3 MPa) can be achieved at 1000 °C in reduced pressure.

The SEM photograph of cross section of the epi-poly Si layer is shown in Figure 4.1 [3]. The vertical column in the photograph is effective to reduce the stress.

4.2 MEMS Devices Using Epi-poly Si

Monolithic integrated capacitive accelerometer was developed by the company Robert Bosch using the epi-poly Si [4]. The process flow of the integrated accelerometer is shown in Figure 4.2. Buried diffusion layer is formed for electrical interconnection between the MEMS accelerometer and the circuit. Sacrificial oxide layer is made ((a) 1). 10 μm thick epi-poly Si is deposited on the sacrificial oxide

3D and Circuit Integration of MEMS, First Edition. Edited by Masayoshi Esashi.
© 2021 WILEY-VCH GmbH. Published 2021 by WILEY-VCH GmbH.

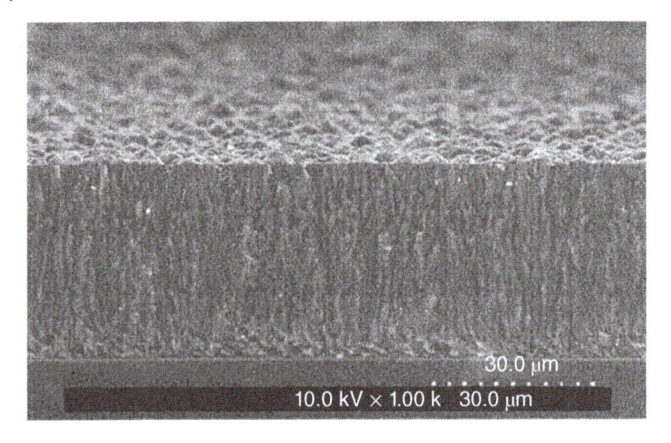

Figure 4.1 Cross-sectional SEM image of epi-poly Si [3].

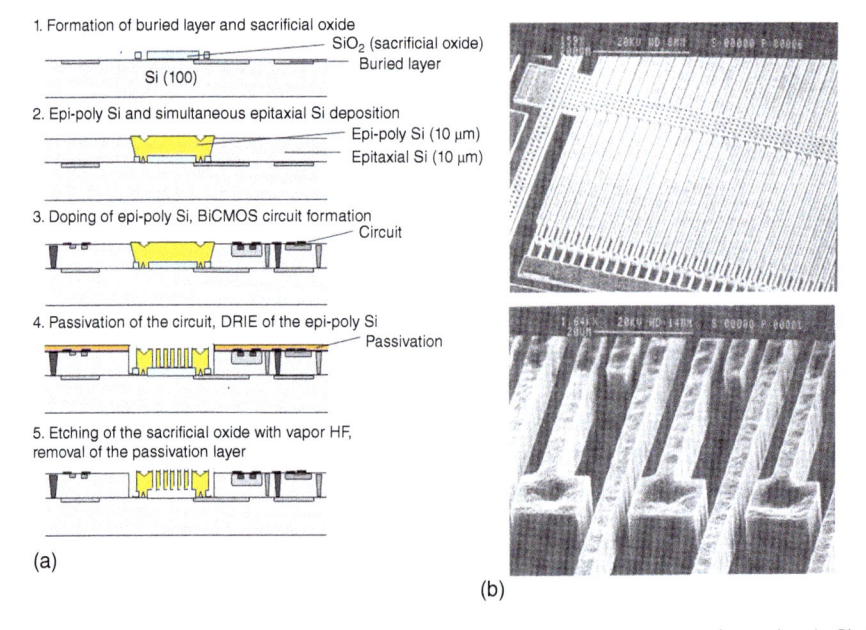

Figure 4.2 Process flow and photographs of integrated accelerometer using epi-poly Si. (a) Process flow. (b) Photographs. Source: Offenberg et al. [4].

layer, and simultaneously monocrystalline Si is epitaxially grown on the Si surface ((a) 2). Doping of the epi-poly Si is carried out by ion implantation. Bipolar and CMOS (BiCMOS) circuit is fabricated on the epitaxially grown Si layer ((a) 3). The circuit is protected by a passivation layer and the epi-poly Si is patterned using deep reactive ion etching (DRIE) ((a) 4). The photographs of a comb-like structure fabricated by the DRIE is shown in Figure 4.2b. The sacrificial oxide is etched out with vapor HF. This vapor etching is effective to prevent the sticking of the epi-poly Si microstructure to the bottom Si surface. If wet etching is used, the sticking can

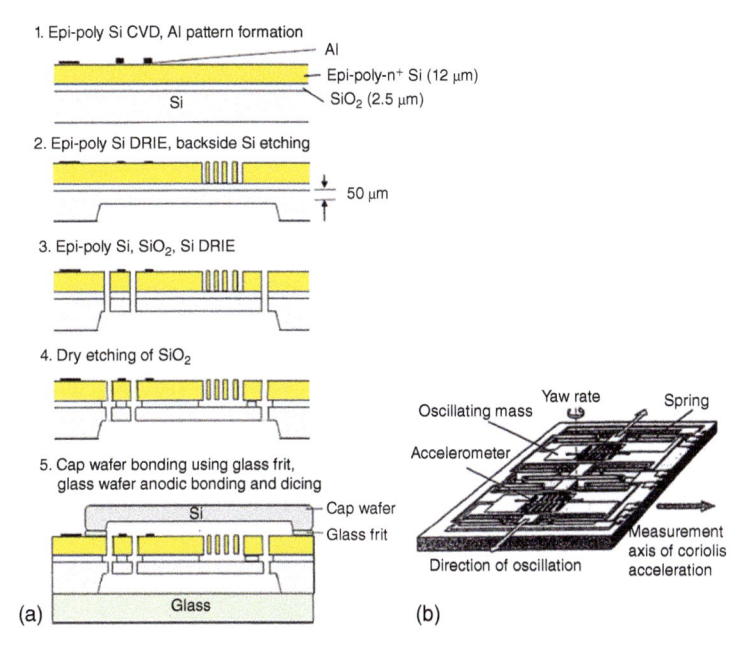

Figure 4.3 Resonant gyroscope using epi-poly Si. (a) Process flow. (b) Structure of resonant gyroscope. Source: Lutz et al. [5].

happen by surface tension of rinsing water when dried. The passivation layer is etched out finally ((a) 5).

Figure 4.3 shows a resonant gyroscope fabricated using the epi-poly Si in Robert Bosch. The process flow is explained in Figure 4.3a [5]. Phosphorous doped epi-poly n^+ Si (12 μm thick) is formed on 2.5 μm thick SiO_2 which is deposited for the purpose of sacrificial layer on a Si wafer. Al is deposited and patterned ((a) 1). Trenches are made in the epi-poly Si by DRIE. Si is etched from the backside of the wafer ((a) 2). Epi-poly Si, SiO_2, and Si are etched through the thickness by the DRIE ((a) 3). The sacrificial SiO_2 exposed trough the trench in the epi-poly Si is etched by vapor HF ((a) 4). Cap Si wafer is bonded in vacuum by glass frit (low-melting-point glass) for vacuum packaging and finally diced ((a) 5). The structure of the resonant gyroscope is shown in Figure 4.3b. Two masses are suspended with springs and oscillate electromagnetically in opposite directions. The oscillation is driven by AC current on the mass and outer permanent magnets. When the sensor chip rotates around perpendicular axis, the yaw rate generates Coriolis force and mass motion by the Coriolis force is detected by the capacitive accelerometers on the oscillating masses.

The epi-poly Si has been applied to epi-seal in Figure 4.4 to fabricate monocrystalline Si oscillator sealed in high vacuum [6]. The process flow is explained in Figure 4.4a. Silicon on insulator (SOI) wafer is used ((a) 1). Trenches are made in the surface active layer of the SOI wafer by DRIE ((a) 2). The trenches are filled with CVD SiO_2 and the SiO_2 is patterned ((a) 3). The epi-poly Si is deposited and trenches are made in it by the DRIE ((a) 4). SiO_2 is etched through the trenches ((a) 5). The vent from the trench is sealed with SiO_2 and then Al pads are formed

64 | *4 Epitaxial Poly Si Surface Micromachining*

1. SOI wafer

   Si
   SiO$_2$
   Si

2. Trench etching by DRIE

3. Trench refill with SiO$_2$ and patterning

   SiO$_2$

4. Epi-poly Si CVD and trench etching by DRIE

   Epi-poly Si

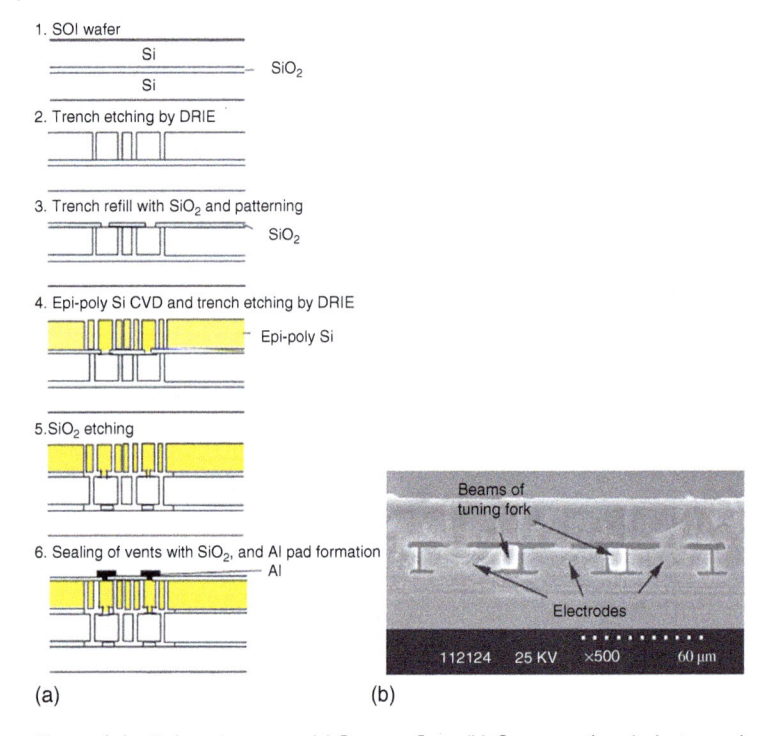

5. SiO$_2$ etching

6. Sealing of vents with SiO$_2$, and Al pad formation

   Al

(a)   (b)

Beams of tuning fork
Electrodes
112124 25 KV ×500 60 μm

Figure 4.4 Epi-seal process. (a) Process flow. (b) Cross-sectional photograph. Source: Candler et al. [6].

by Al deposition and patterning ((a) 6). Cross-sectional photograph is shown in Figure 4.4b.

MEMS sensors used in smartphones and other microsystems have been fabricated with the epi-poly Si. The process flow in Figure 4.5 is developed in STMicroelectronics [7] and applied to fabricate three axes gyroscope [8] and other MEMS devices. The process sequences are as follows. SiO$_2$ (2 μm) is formed on the surface of a Si wafer by thermal oxidation (1 in the figure). Thin poly Si (650 nm) is deposited and patterned for the purpose of electrical interconnection (2 in the figure). Sacrificial oxide (2 μm) is deposited by plasma enhanced (PE) CVD and patterned (3 in the figure). Thick epi-poly Si (15 μm) is deposited (4 in the figure). Trenches are made by DRIE (5 in the figure). The sacrificial oxide and the underlined thermal oxide are etched in dry process (6 in the figure).

The MEMS by epi-poly Si are used for SiP MEMS by connecting with circuit chips. There are stacked configuration and side-by-side configuration in the SiP as shown in Figure 1.5 (Chapter 1). The stacked hybrid configuration has an application-specific integrated circuit (ASIC) chip on a packaged MEMS as shown in Figure 4.6. The fabrication process is as follows. A MEMS wafer and a cap wafer having glass frit are prepared (1 in the figure). Cap wafer is bonded to the MEMS wafer using the glass frit (2 in the figure). Hermetic sealing can be performed because the melted glass frit can cover the nonplanar surface made by the lateral

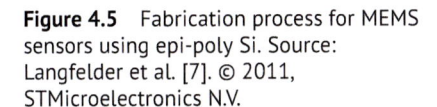

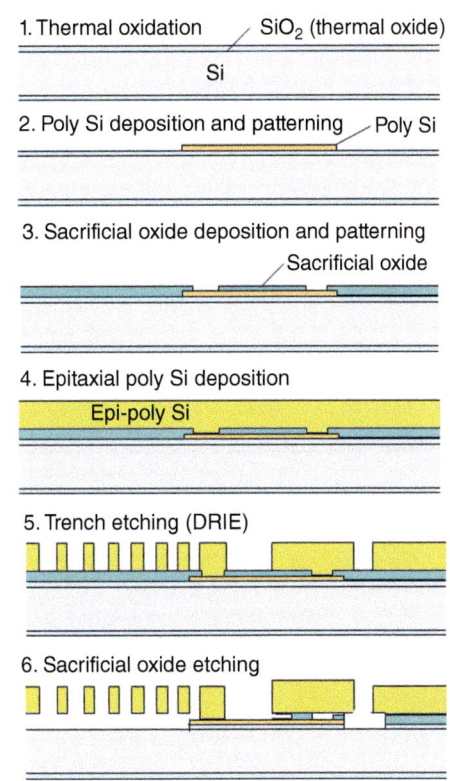

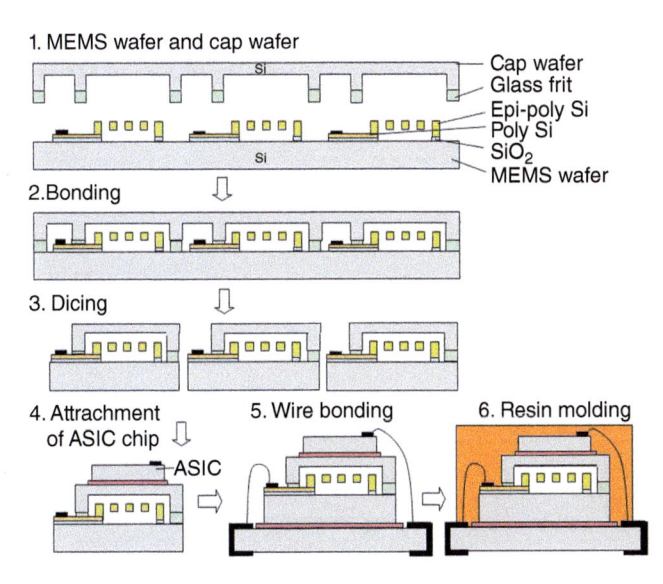

Figure 4.5 Fabrication process for MEMS sensors using epi-poly Si. Source: Langfelder et al. [7]. © 2011, STMicroelectronics N.V.

1. Thermal oxidation — SiO$_2$ (thermal oxide) / Si
2. Poly Si deposition and patterning — Poly Si
3. Sacrificial oxide deposition and patterning — Sacrificial oxide
4. Epitaxial poly Si deposition — Epi-poly Si
5. Trench etching (DRIE)
6. Sacrificial oxide etching

1. MEMS wafer and cap wafer — Cap wafer, Glass frit, Epi-poly Si, Poly Si, SiO$_2$, MEMS wafer
2. Bonding
3. Dicing
4. Attachment of ASIC chip — ASIC
5. Wire bonding
6. Resin molding

Figure 4.6 Fabrication process of stacked hybrid configuration (ASIC on packaged MEMS).

66 | *4 Epitaxial Poly Si Surface Micromachining*

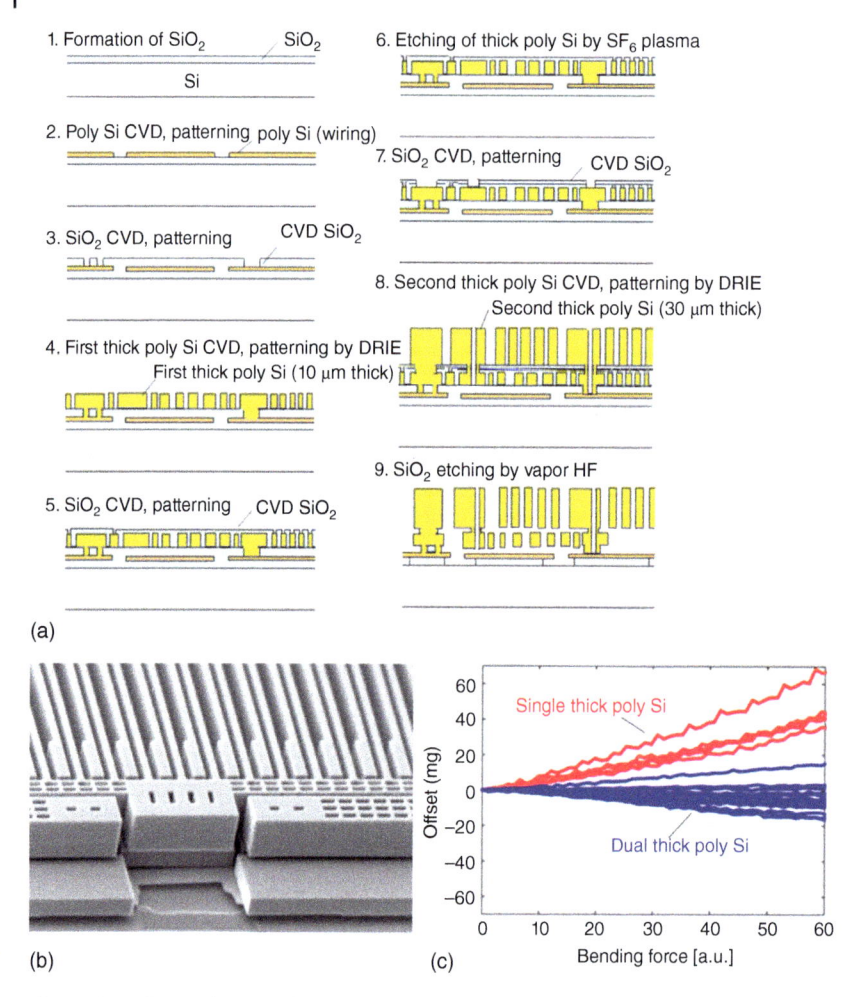

Figure 4.7 Dual thick poly Si accelerometer. (a) Process flow. (b) SEM photograph. (c) Offset shift as a function of bending force. Source: Classen et al. [9]. © 2017, IEEE.

feedthrough (3 in the figure). We can get packaged MEMS by dicing the cap wafer to open the bonding pads. After the dicing the MEMS wafer for chip separation, the ASIC chip is attached on the packaged MEMS chip (4 in the figure). The ASIC chip and the MEMS chip are connected by wire bonding on a substrate (5 in the figure) and molded with resin (6 in the figure).

MEMS sensor which is not influenced by packaging stress was developed using dual thick poly Si in Robert Bosch [9]. Figure 4.7 shows the fabrication process in (a) SEM photograph in (b) and offset shift as a function of bending force in (c). The fabrication process is explained in the following. After formation of SiO_2 by thermal oxidation ((a) 1), thin poly Si for wiring ((a) 2) and SiO_2 ((a) 3) are deposited by CVD and patterned. First thick poly Si (10 μm thick) is deposited by CVD and etched by DRIE ((a) 4). SiO_2 is deposited by CVD and patterned ((a) 5). The poly Si is etched by isotropic SF_6 plasma ((a) 6). CVD SiO_2 is formed and patterned ((a) 7). Second thick poly Si (30 µm thick) is deposited by CVD and etched by DRIE ((a) 8). Sacrificial SiO_2 is etched out by vapor HF ((a) 9). As shown in (c) the offset shift by the bending force is reduced roughly by a factor of 4 using the dual thick-layer poly Si with a suitable support structure.

References

1 Kirsten, M., Wenk, B., Ericson, F., and Schweitz, J.A. (1995). Deposition of thick doped polysilicon films with low stress in an epitaxial reactor for surface micromachining applications. *Thin Solid Films* 259 (2): 181–187.
2 Lange, P., Kirsten, M., Riethmüller, W. et al. (1996). Thick polycrystalline silicon for surface-micromechanical applications: deposition, structuring and mechanical characterization. *Sens. Actuat. A* 54: 674–678.
3 Suzuki, Y., Totsu, K., Watanabe, H. et al. (2012). Low-stress epitaxial polysilicon process for micromirror devices. The 29th Symposium on Sensors, Micromachines and Applied Systems, SP2-7, Kitakyushu City, Japan (22–24 October 2012), 548–553.
4 Offenberg, M., Lärmer, F., Elsner, B. et al. (1995). Novel process for a monolithic integrated accelerometer. The Eighth International Conference on Solid State Sensors and Actuators, and Eurosensors IX (Transducers'95 Eurosensors IX), Stockholm, Sweden (25–29 June 1995), 589–592.
5 Lutz, M., Golderer, W., Gerstenmeier, J. et al. (1997). A precision yaw rate sensor in silicon micromachining. 1997 International Conference on Solid-State Sensors and Actuators (Transducers'97), Chicago, USA (16–19 June 1997), 847–850.
6 Candler, N., Park, W.-T., Li, H. et al. (2003). Single wafer encapsulation of MEMS Devices. *IEEE Trans. Adv. Packaging* 26 (3): 227–232.
7 Langfelder, G., Longoni, A.F., Tocchio, A., and Lasalandra, E. (2011). MEMS motion sensors based on the variations of the fringe capacitances. *IEEE Sens. J.* 11 (4): 1069–1077.
8 Prandl, L., Caminada, C., Coronato, L. et al. (2011). A low-power 3-axis digital-output MEMS gyroscope with single drive and multiplexed angular rate readout. 2011 IEEE International Solid-State Circuit Conference (ISSCC 2011), San Francisco, USA (20–24 February 2011), 104–105.
9 Classen, J., Reinmuth, J., Kälberer, A. et al. (2017). Advanced surface micromachining process – a first step towards 3D MEMS. The 30th IEEE International Conference on Micro Electro Mechanical Systems (MEMS '17), Las Vegas, USA (22–26 January 2017), 314–316.

5

Poly-SiGe Surface Micromachining

Carrie W. Low[1], Sergio F. Almeida[2], Emmanuel P. Quévy[3] and Roger T. Howe[4]

[1] TDK InvenSense, 1745 Technology Dr., San Jose, CA 95110, USA
[2] DiDi Labs, 450 National Ave., Mountain View, CA 94043, USA
[3] ProbiusDx Inc., 39355 California St., Suite 207, Fremont, CA 94538, USA
[4] Stanford University, Department of Electrical Engineering, 330 Jane Stanford Way, Stanford, CA 94305, USA

5.1 Introduction

In recent years, micro-electro mechanical system (MEMS) technologies have enabled chip-scale sensors, actuators, and resonators, which are fabricated using the tools of semiconductor manufacturing. However, most MEMS products are made with two-chip solutions: through the assembly of the MEMS chip with a separate integrated circuit (IC) into a multichip module. Although this approach is convenient, since the nonstandard MEMS processes, such as deep reactive ion etching, can be done in boutique MEMS foundries, it fails to leverage the highly controlled processes in mainstream complementary metal oxide semiconductor (CMOS) foundries. In addition, the two-chip solution can degrade the performance of some MEMS sensors and resonators due to higher parasitic capacitances. The two-chip solution is necessary for using standard low-pressure chemical vapor deposition (LPCVD) poly-Si as a structural material, since it is deposited at higher temperatures than CMOS wafers can tolerate. LPCVD poly-SiGe was investigated for MEMS applications at the University of California (UC), Berkeley, [1–3] and Interuniversity Microelectronics Centre (IMEC) [4–6] since the late 1990s, with IMEC starting a long-term project on plasma-enhanced CVD poly-SiGe in the early 2000s. A new application for poly-SiGe technology is nano-mechanical relays for ultralow-power computing, which have been the subject of intensive research at UC Berkeley since 2008 [7].

In 2004, Silicon Clocks was founded to commercialize UC Berkeley's LPCVD poly-SiGe technology for fabricating resonators on top of foundry CMOS for timing applications. In 2010 Silicon Labs acquired Silicon Clocks and introduced CMEMS® technology for timing. Semiconductor Manufacturing International Corporation (SMIC) was the foundry partner for CMEMS® process. Using low-temperature LPCVD poly-SiGe as the structural material, MEMS resonators can be built directly on top of an advanced mixed-signal CMOS, achieving MEMS + CMOS integration

3D and Circuit Integration of MEMS, First Edition. Edited by Masayoshi Esashi.
© 2021 WILEY-VCH GmbH. Published 2021 by WILEY-VCH GmbH.

in a single chip. Smaller size, better performance, lower cost and greater scalability are the advantages of MEMS + CMOS integration.

5.1.1 SiGe Applications in IC and MEMS

Silicon-germanium has been studied extensively as the base material for heterojunction bipolar junction transistors [8, 9], as the gate, source/drain, or channel material for CMOS devices [10–12] and as the absorption material for optical or thermal electronics [13, 14].

Poly-SiGe has also been investigated as alternative structural material for surface micromachining. Poly-SiGe has material properties, which are similar to those of poly-Si. In contrast to poly-Si, poly-SiGe can be deposited and crystallized at very low temperatures with good stability, which makes it promising for post-CMOS integration of MEMS. Applications of SiGe MEMS include gyroscopes [4], optical devices [15], resonators [16], bolometers [5], and pressure sensors [6]. Poly-Ge is an attractive sacrificial layer since it can be etched with high selectivity with respect to poly-$Si_{1-x}Ge_x$ alloys with Ge concentration less than 60% [2].

5.1.2 Desired SiGe Properties for MEMS

The desired SiGe properties for MEMS applications are very different from those of electronic device applications. In general, a film thicker than 2 μm is needed for lateral capacitive sensing. For post-CMOS processing, the deposition temperature of poly-SiGe is limited to below 450 °C. Deposition rate and crystallinity of the film can be improved with higher germanium content. However, the etch selectivity of a pure germanium sacrificial layer to a poly-SiGe structural layer for H_2O_2 etching decreases with increasing germanium content in the poly-SiGe film. A germanium content of 60% is desired for reasonable deposition rate and crystallinity with adequate resistance to H_2O_2 etching. In order to have good electrical connection to the electronics, the desired resistivity is below 10 mΩ-cm for RF MEMS applications. For inertial sensor applications with long suspension lengths, low residual stress and stress gradient are required. To avoid buckling of a clamped-clamped beam, a small tensile residual stress is desired. However, films with compressive stress can be used for some applications, if it is compensated in the suspension design. Low strain gradient is the most critical requirement for inertial sensor applications. The typical strain gradient specification for inertial sensors is less than 1×10^{-5} μm^{-1}, which results in less than 5 μm tip deflection for a 1 mm long beam. In addition to the aforementioned materials requirements, developing a high throughput, high yield, and repeatable process is critical for large volume production.

5.2 SiGe Deposition

5.2.1 Deposition Methods

Several approaches to depositing poly-SiGe for MEMS applications have been investigated by various research groups: atmospheric- or reduced-pressure chemical vapor deposition (APCVD or RPCVD) [17], LPCVD [1, 2, 18], plasma-enhanced chemical vapor deposition (PECVD) [19–21], and pulsed laser deposition (PLD) [22]. The deposition rate for the APCVD or RPCVD processes is about 4 nm min^{-1} at 520 °C, which is too low to be economical at CMOS compatible temperatures. Films deposited by PLD have high particle density and require addition annealing for crystallization.

The deposition rate for the PECVD process is about 100 nm min^{-1} at 450 °C [19], which is about six times higher than that of the LPCVD process at the same temperature. While the LPCVD process has lower deposition rate, it can have a large batch size for higher throughput and lower cost. Another major advantage of LPCVD process is its conformal coverage of all surfaces, which can also be used for planarization and gap filling. Both PECVD and LPCVD poly-SiGe processes are promising for post-CMOS integration; research and development are focused on pushing down the thermal budget, fine tuning the material properties, and developing a robust process for high volume production.

5.2.2 Material Properties Comparison

Poly-SiGe is an attractive MEMS structural material for post-CMOS applications as it provides excellent mechanical properties at CMOS compatible temperatures. Table 5.1 lists ballpark numbers for material properties of poly-SiGe film deposited by PECVD and LPCVD methods.

PECVD SiGe process has been studied at IMEC [19–21]. In IMEC's SiGe technology platform, silane (SiH$_4$) and germanium (GeH$_4$) in hydrogen are the precursor gases; diborane (B$_2$H$_6$) in hydrogen is used as the in situ doping gas. The baseline poly-SiGe structural layer is 4 µm thick, deposited at 450 °C as a stack of a 400 nm CVD layer, a 1.6 µm PECVD layer, and another 2 µm PECVD layer. The CVD layer

Table 5.1 Material properties of PECVD [20] and LPCVD poly-SiGe films [23].

Property	Unit	CVD/PECVD SiGe at 450 (°C)	LPCVD SiGe at 410 (°C)
Film thickness	µm	4	2
Ge content	–	~70%	~60%
Density (ρ)	kg m^{-3}	4400	4130
Young's modulus (E_y)	GPa	130	135.5
Poisson's ratio (ν)	–	0.22	0.283
Residual stress (σ_0)	MPa	40	−153
Strain gradient	µ/m	<2 × 10^{-5} (4 µm thick film)	~3 × 10^{-4} (2 µm thick film)
Coefficient of thermal expansion (CTE)	K^{-1}	5 × 10^{-6}	4 × 10^{-6}
Thermal conductivity (κ)	W m^{-1} K	11	9.6
Resistivity (ρ)	mΩ-cm	1–100	0.6

Source: Based on Gonzalez and Rottenberg [20].

serves as the polycrystalline seeding to induce crystallization in the PECVD layer. High deposition rate is achieved due to energy supplied by the plasma. Hydrogen dilution in the process gases provide an alternate source of energy for crystallization at low temperature. As a result, PECVD SiGe films typically have very high hydrogen content. The hydrogen evolves with excimer laser annealing and leaves small pores in the film [24]. In comparison, excimer laser annealing does not result in pores in LPCVD films [25].

LPCVD SiGe processes have been studied by UC Berkeley, Silicon Clocks, Inc., and Silicon Labs, Inc. In the LPCVD process, pure silane (SiH_4) and pure germanium (GeH_4) are the precursor gases; boron trichloride (BCl_3) diluted in helium is used as the in situ doping gas. The baseline poly-SiGe structural layer is 2 μm thick, deposited at 410 °C with a thin layer of amorphous silicon seeding. The material properties of PECVD and LPCVD SiGe films are in the same range. The residual stress of the LPCVD SiGe film is not as favorable as that of the PECVD film, but it is sufficiently low for some applications. The strain gradient of the 2 μm LPCVD SiGe film is also higher than that of 4 μm PECVD SiGe.

5.2.3 Cost Analysis

One major drawback of poly-SiGe compared to poly-Si is its cost. Germane (GeH_4) is an expensive gaseous precursor needed for poly-SiGe deposition. The cost of a bottle containing 2600 g of pure GeH_4 is $20 604 [26]. Calculated with Ideal Gas Law, the cost per standard cubic centimeter (scc) of GeH_4 is about $0.027, which is about 80× the cost of SiH_4 per scc. In addition, more complex equipment is required for poly-SiGe depositions. Poly-SiGe deposition needs to be optimized to increase throughput and reduce material cost. Table 5.2 as follows compares the cost of the poly-SiGe deposited by the PECVD and LPCVD methods. Equipment throughput and GeH_4 material cost are both considered for a 4 μm-thick poly-SiGe structural layer deposition.

The batch size of PECVD process is typically a single wafer per chamber. A 4 μm-thick SiGe film is deposited as a layer stack of 400 nm CVD film, 1.6 μm PECVD film, and another 2 μm PECVD film. Chamber cleaning is required for every 2 μm of SiGe deposition to avoid excessive particles [20]. The wafer is removed from the process chamber during the in situ chamber clean, and an oxide is formed on the surface of the SiGe film. A quick CF_4 clean is performed to remove the unwanted oxide at the layer interface. The total process time for the 4 μm baseline PECVD SiGe wafer is approximately 37 minutes [20].

For LPCVD process, the batch size for a production tool is 141 wafers. Excluding dummy wafers at the ends of the boat, about 100 product wafers can be loaded per batch. A 4 μm-thick film can be deposited in single pass without running the chamber cleaning cycle. Particle monitor and recovery can be done after the deposition, if needed. The total process time for a 4 μm baseline LPCVD SiGe deposition is approximately 1285 minutes, including wafer loading, deposition time, and particle recovery after unloading. The numbers in Table 5.2 show that LPCVD process has higher throughput and lower material cost. However, in the case of an unexpected

Table 5.2 Cost comparison of PECVD [20, 21] and LPCVD poly-SiGe films.

	IMEC PECVD poly-SiGe	SMIC LPCVD poly-SiGe
Equipment	AMAT Centura CxZ	Kokusai Electric DJ835V
Substrate size (mm)	200	200
Number of product wafer per batch	1	100
Deposition temperature (°C)	450	410
Deposition rate (nm/min)	36 (CVD); 164.4 (PECVD)	6.4
Film thickness (μm)	4	4
Ge content	77% (CVD); 71% (PECVD)	60%
Deposition time (min)	11 (CVD); 22 (PECVD)	625
Total process time (min)	37	1285
Throughput (unit per hour)	1.62	4.67
GeH_4 flow rate (sccm)	332	118
GeH_4 gas concentration	10% in H_2	100%
GeH_4 cost per wafer	$29.84	$20.08

Sources: Gonzalez and Rottenberg [20], Guo et al. [21].

process problem, 100 product wafers could all be scrapped for the LPCVD process. Also, production PECVD tools can have multiple chambers to improve throughput.

5.3 LPCVD Polycrystalline SiGe

5.3.1 Vertical Furnace

Vertical LPCVD furnaces are commonly used in the industry. Compared with horizontal furnaces, vertical furnaces have both better process uniformity and larger throughput [27]. The SiGe production furnace used for the CMEMS® process at SMIC Shanghai was made by Hitachi Kokusai Electric Inc., Model DJ835V. This vertical furnace has slots for 141 wafers (200 mm diameter). The operating pressure range is from 100 mTorr to a few Torr. The operating temperature range is from 300 °C to 800 °C. This furnace was initially designed for LPCVD poly-Si deposition and later retrofitted for the poly-SiGe process, as shown in Figure 5.1. This tool represents the third generation of SiGe deposition tools, which have been retrofitted by the CMEMS® process engineering team and benefited from the lessons learned in the earlier furnaces.

The retrofit focused mainly on the gas panel and gas distribution inside the chamber. Precursor gases of the SiGe furnace include pure disilane (Si_2H_6), silane (SiH_4), and germane (GeH_4). Disilane is mainly used for a thin amorphous silicon deposition as an incubation layer. Silane and germane are the precursors for SiGe deposition. Boron trichloride (BCl_3) diluted in helium (He) is used as the dopant

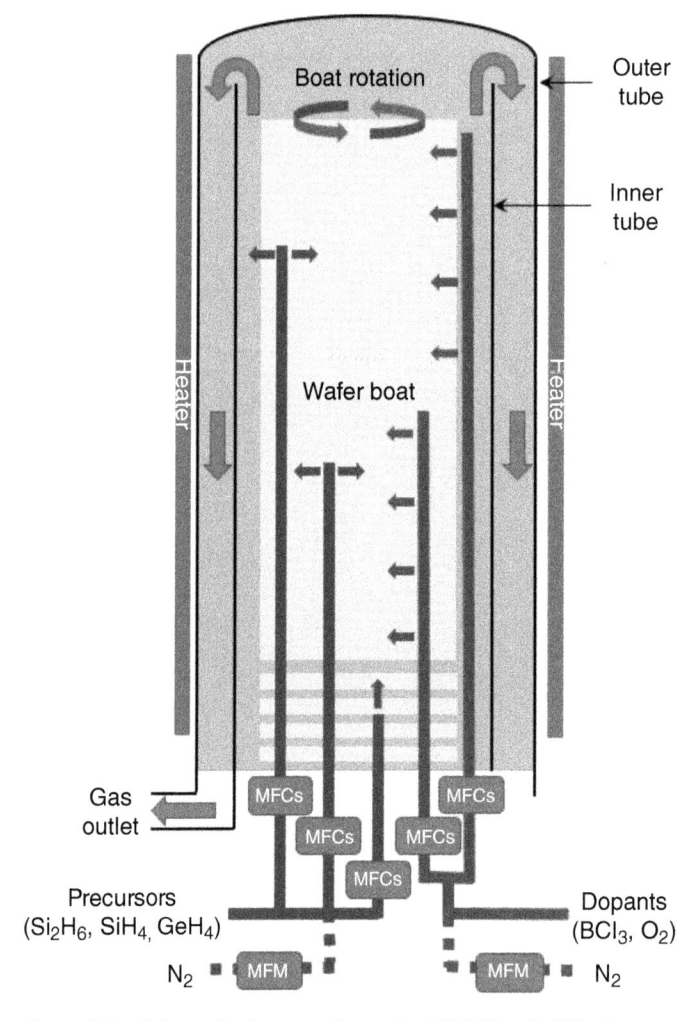

Figure 5.1 Schematic diagram of a vertical LPCVD poly-SiGe furnace.

gas. Oxygen (O_2) diluted in helium (He) is also connected to the dopant line for experimental purposes.

All process gases are introduced inside the inner tube from the bottom, travel to the top of the chamber, and are pumped out at the bottom through the outer tube to the gas outlet. There are three precursor injectors located at various height in the chamber, each with a gas outlet only at its end. Since the precursors travel through the hot zone of the furnace, deposition will occur inside the precursor injectors. If the precursor gases were introduced via a multi-pore injector, deposition inside the injector and on the pores will result in nonuniform gas distribution over the lifetime of the injector. Multiple precursor injectors are used in this design to keep gas distribution consistent over the duration of a maintenance cycle. Also, each precursor injector has reasonably large diameter and large outlets at the top to prevent in-use clogging. A previous study showed that the dopant gas boron trichloride does not decompose to clog up the injector [28]. Two multi-pore dopant injectors are used to uniformly distribute the dopant gas. The specially designed gas distribution system helps to improve cross-load uniformity, and the wafer boat rotation helps to improve cross-wafer uniformity. Also, every gas introduced via each of the injectors is individually controlled by a mass flow controller (MFC), allowing the process engineer to tune the gas flow to further improve uniformity.

5.3.2 Particle Control

As deposition builds up inside the process chamber and on the boat, films can peel off due to thermal cycles and become a source of particles. Compared to quartz, silicon carbide (SiC) has closer thermal expansion compared to poly-SiGe. The use of a SiC boat and liner minimizes thermal strain, film peeling, and particle generation, which lengthens the maintenance cycle. Having a nitrogen-controlled mini-environment at the load lock is also critical for particle control. The nitrogen environment significantly reduces the oxidation of the boat and the chamber during loading and unloading, which improves particle performance.

5.3.3 Process Monitoring and Maintenance

The process monitor of the SiGe LPCVD furnace includes periodic MFC qualification, particle monitoring, and deposition rate and uniformity monitoring. The deposition rate and the thin film's mechanical and electrical properties strongly depend on the germanium and the dopant content, which in turn depend on the outputs of each of the MFCs. It is necessary to monitor closely the performance of the MFCs for process control. There are three MFCs for each of the precursor gases and two MFCs for each of the dopant gases. The most important attribute of an MFC for achieving run-to-run repeatability is not the accuracy but the consistency. Mass flow meters (MFMs) are built into the gas panel for in situ MFC monitoring (Figure 5.1). The MFM line is bypassed during regular deposition, and it is activated for MFC monitoring. In the MFC test recipe, nitrogen is directed through each of the MFCs, one by one. The nitrogen flow is regulated by each MFC and quantified by the MFM. The MFM is calibrated to a primary standard with nitrogen, and a correction factor is used to convert the actual nitrogen output from the MFC being tested.

Cleanliness is essential for particle control in a production environment. Due to the nature of thick SiGe and Ge depositions for MEMS applications, particle performance is monitored frequently via mechanical runs in the SiGe furnace. The substrate for the mechanical runs is a bare silicon wafer, which is pre- and post-scanned in the particle counter. Nitrogen flow is used to circulate loose particles, which are detected by counting particles added on the bare silicon wafers. If the particle count is out of specification, a nitrogen purge cycle is used to clean the furnace.

The total deposition thickness on the furnace walls is also monitored. After a critical thickness (~80 µm), a tube pull and wet clean of the hardware will be needed to prevent the furnace from suffering a particle failure. An unexpected tube cooldown due to a power outage could also trigger a tube pull before the critical thickness. Films could peel off and generate particles due to temperature cycles. The tube pull maintenance procedure includes disassembly, hardware (boat, liner, and other quartz) cleaning, installation, temperature profile, and MFC qualification. To recover the process, a 500 nm a-Si film is first deposited to coat the bare hardware wall, and then a SiGe deposition and a Ge deposition follow to ensure the process is recovered. To monitor the process stability of the SiGe furnace, baseline SiGe depositions are done regularly. Film thickness, germanium content, stress, and resistivity are collected at the top, center, and bottom of the device slots.

5.3.4 In-line Metrology for Film Thickness and Ge Content

Poly-SiGe film's thickness can be measured with cross-sectional Scanning Electron Microscopy (SEM), and germanium content can be measured by Secondary Ion Mass Spectroscopy (SIMS). In production environments, in-line and nondestructive characterizations of the poly-SiGe film are required for process control.

Poly-SiGe film thicknesses for MEMS applications are usually in the μm range. It is difficult to measure the opaque and rough poly-SiGe film using optical measurement tools, such as optical probes or ellipsometers. X-ray fluorescence (XRF) spectrometry provides a fast and nondestructive solution for film thickness and germanium content measurements. The XRF measurement time is about 100 s per spot, and the spot size is about 40 mm in diameter. For a $Si_{1-x}Ge_x$ film, the Ge count ($count_{Ge}$) of the XRF measurement depends on the film thickness (t_{SiGe}) and the germanium content ($at\%_{Ge}$). The Si count ($count_{Si}$) of the XRF measurement depends on the film thickness (t_{SiGe}), the germanium content ($at\%_{Ge}$), and also the silicon detected from the underlying SiO_2 film. The relationships can be described with the following equations:

$$t_{SiGe} \times at\%_{Ge} = A \times count_{Ge}$$

$$t_{SiGe} \times (1 - at\%_{Ge}) = B \times count_{Si} - C \times count_{O_2}$$

where A, B, and C are constants and $count_{O_2}$ is the XRF oxygen count. To find constants A, B, and C, a set of $Si_{1-x}Ge_x$ reference wafers are used for calibration. The $Si_{1-x}Ge_x$ reference wafers have various thicknesses (0.5–5 μm) and germanium contents (50–70% and 100%). The film thickness and the germanium content are measured by SEM and SIMS, respectively.

For SiGe film in the range of 0–5 μm, the Ge count and the film thickness are relatively linear. As the film gets thicker, the Ge X-ray generated at greater depths is absorbed by the overlying SiGe film. It would take a SiGe film thickness of 20 μm to absorb about half of the X-rays generated. Also, for SiGe films thicker than 2 μm, the constant C is negligible due to Si X-ray absorption by the overlying SiGe film. For SiGe films with thickness in the range of 2–5 μm and germanium content in the range of 50–70%, film thickness and germanium content can be expressed as

$$t_{SiGe} = A \times count_{Ge} + B \times count_{Si}$$

$$\text{at\%}_{Ge} = \frac{A \times \text{count}_{Ge}}{A \times \text{count}_{Ge} + B \times \text{count}_{Si}}$$

For pure germanium films in the thickness range of 0–5 µm, film thickness can be expressed as

$$t_{Ge} = A \times \text{count}_{Ge}$$

5.3.5 Process Space Mapping

The desired SiGe film for the CMEMS® process needs to be polycrystalline, conductive, conformal and have ~60% germanium content. The desired Ge film needs to be both uniform and conformal. In production environment, good cross-load and cross-wafer uniformity are critical for high throughput.

SiGe and Ge process development for CMEMS® production was based on previous academic research [18] and pre-production experience. SiGe deposition rate increases with temperature, pressure, total precursor gas flow, and germanium content. The resistivity of the film mainly depends on dopant gas flow rate. Average residual stress of the SiGe film is compressive in the temperature range of 400–450 °C and more compressive at the lower part of this range. Strain gradient of the SiGe film depends strongly on the film's texture, which is very sensitive to all the process conditions. Etch rate in hydrogen peroxide solutions mainly depends on the germanium content in the film and thus increases inversely with SiH_4:GeH_4 ratio [2].

To characterize a new furnace, multiple experiments were run. Baseline conditions from the previous furnace generation were first tested. Higher pressure was used on the second run to boost deposition rates and challenge uniformity. Next, the first pass of the design of experiment (DOE) focused on getting the desired 60% germanium content, reasonable deposition rate, and resistivity. Seeding conditions, total SiH_4 and GeH_4 flow, temperature, and pressure were fixed based on previous knowledge, with the SiH_4:GeH_4 ratio and dopant gas flow rate as the variables. The SiH_4:GeH_4 gas flow ratio should be around 2.4 : 1 to achieve 60% germanium content, with a deposition rate of around 6 nm min^{-1}. The resistivity of the SiGe film is around 1 mΩ-cm. Within wafer germanium content, deposition rate, and resistivity were reasonable, with boat rotation.

The second pass of the experiment focused on improving the cross-load uniformity. This SiGe furnace was designed to have flexibility to tune cross-load uniformity. First, the precursor gases are distributed through top, center, and bottom injectors. SiH_4 and GeH_4 outputs at each location can be individually adjusted with MFCs. Secondly, there are five heating zones in the furnace, allowing small temperature variations to be applied to different zones. Lastly, the loading pattern can also be used to adjust uniformity. There are 141 wafer slots in this SiGe furnace. In areas with lower deposition rate, wafers can be loaded at every-other-slot to reduce surface area.

There are three layers of SiGe deposition in the CMEMS process; two of the layers are electrical interconnects and one of them is the mechanical layer for the resonator.

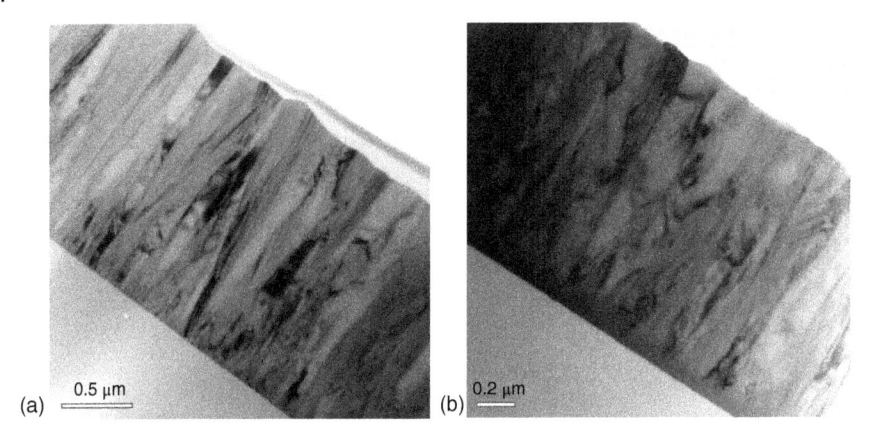

Figure 5.2 TEM images: (a) Poly-SiGe electrical interconnect layer and (b) Poly-SiGe mechanical film. Source: Low [18].

The electrical interconnect layers need to be fully crystallized for good conductivity. A quick (5–10 min) SiGe seeding layer with low pressure, low SiH_4 and GeH_4 flow rates, and high BCl_3 flow are used to enhance crystallization. At low deposition rates, gas molecules have more time to settle down at low energy crystal lattice sites on the wafer surface before the next gas molecules are absorbed. Once the wafer is covered with fully crystallized seeding layer, pressure and gas flows are raised up to enhance deposition rate. The later deposited film follows the polycrystallinity of the seeding layer, and the entire film thickness is fully crystallized, as showed in Figure 5.2a.

The mechanical layer needs to have low strain gradient (less than $1 \times 10^{-4}\ \mu m^{-1}$) for resonator applications. A previous study [18] showed that fine grain seeding results in conical microstructures with high strain gradient. In order to achieve film with low strain gradient and good uniformity, there should be a thin amorphous region at the bottom of the SiGe film while keeping the columnar grain texture. For the mechanical layer, a thin layer of amorphous silicon was used as the seeding layer. The later deposited SiGe film has about 0.1 μm amorphous region at the bottom, and the rest of the thickness is crystallized, as showed in Figure 5.2b.

The germanium film in the CMEMS® process is used as sacrificial layer, a spacer layer for defining the submicron gap, a filling layer for planarization, and the bonding material on the cap wafer. The main requirement for its deposition is conformal step coverage and good uniformity. Conformal step coverage is achieved with low gas flow rate and low process temperature of 350 °C. Good uniformity is achieved with a few degrees of temperature tilting across the load.

5.4 CMEMS® Process

The term CMEMS® comes from the contraction of two acronyms: CMOS + MEMS. CMEMS® technology enables the modular post-processing of MEMS devices directly on top of advanced RF/mixed-signal CMOS circuitry (0.18 μm and below).

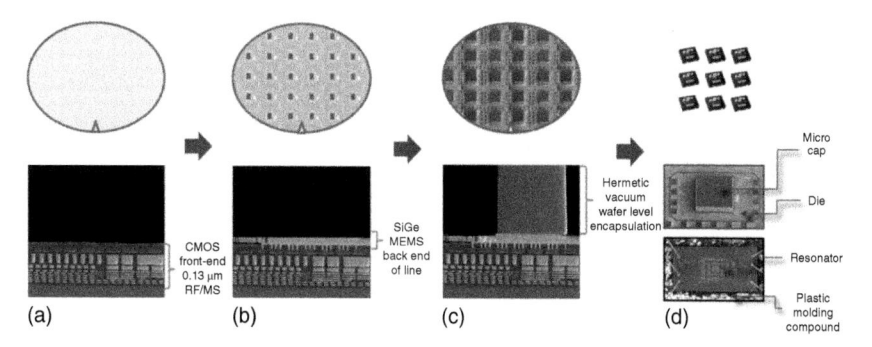

Figure 5.3 Basic sequence of the CMEMS manufacturing flow. (a) CMOS starting material, (b) polycrystalline SiGe MEMS devices, (c) wafer-level encapsulation, and (d) die singulation and standard small-size packaging assembly. Source: Quevy [29].

As shown in Figure 5.3, (a) a passivated and planar CMOS wafer is the starting material, on top of which (b) polycrystalline SiGe is surface-micromachined into integrated MEMS devices, which are (c) encapsulated in a vacuum using wafer-level bonding. The completed wafer continues to probing, (d) die singulation, and standard small-size packaging assembly, just like a standard CMOS product. This approach leverages the scalability of state-of-the-art CMOS manufacturing as a modular back-end-of-line option, potentially attached to the same manufacturing line used to fabricated advanced CMOS wafers. The latest generation of CMEMS® manufacturing allows integration of 0.2 μm space and 0.5 μm line feature-size MEMS device on top of 0.13 μm CMOS with eight metal layers.

Figure 5.4 shows an SEM of a plate resonator. This resonator features ohmic contacts between the CMOS top metal and the SiGe MEMS layers, oxide-filled slits on the resonator plate for temperature compensation, submicron (0.2 μm) electrostatic transducer gap, and eutectic wafer-to-wafer bonding for vacuum hermetic encapsulation. There are six masks used to build this MEMS device and two more masks for the encapsulation.

5.4.1 CMOS Interface Challenges

Building the MEMS structure on top of existing CMOS wafers involves several challenges. To be CMOS compatible, the highest temperature of the MEMS process is at most 410 °C. Thermal cycles of the manufacturing flow is a challenge for materials with different stress and coefficient of thermal expansion (CTE) mismatch. Poly-SiGe is compressive with a CTE of single-digit ppm/°C as opposed to the metal stack underneath, which is usually tensile, having a CTE in the 10–20 ppm/°C range. The CMOS metallization usually has dummy fill patterns to maintain uniform pattern density. However, depositing the poly-SiGe film on large blocks of dummy metal can result in film delamination due to thermal mismatch. Breaking the dummy metal blocks into small pieces allows the materials to coexist peacefully.

Within 1–2 mm of edge of a CMOS wafer, there is usually no pattern. However, the material stack is not well defined, due to multiple steps of lithography with

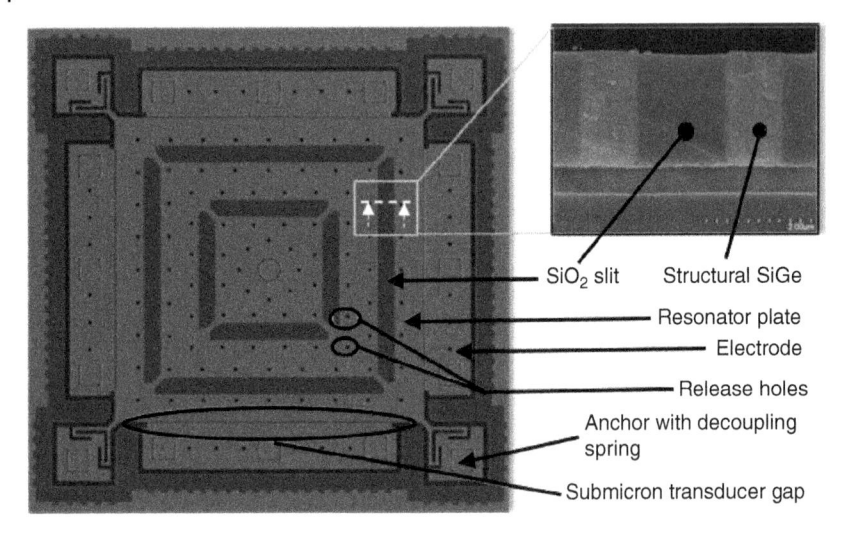

Figure 5.4 Top and cross-sectional views of a plate resonator, mechanically compensated for temperature stability. Source: Quevy [29].

edge bead removal, as well as deposition in tools, which use clamping rings. The material on the wafer edge is usually thick passivation oxide and nitride. Thermal mismatch between the passivation and the poly-SiGe can result in edge-film delamination. Doing a wafer edge etch after the SiGe deposition significantly reduces edge delamination and particle issues.

Substrate outgassing is also a concern. MEMS resonators require a stable vacuum environment to operate. However, volatile molecules can diffuse out of the metal stack and change the vacuum level of the encapsulated cavity. Careful study of the physical chemistry of the backend stack materials [30] is needed to achieve stable device performance.

5.4.2 CMEMS Process Flow

Figures 5.5–5.19 illustrate a perspective, cross-sectional view of the CMEMS® manufacturing technology [31]. A disk resonator is used to illustrate the process flow simulation. The process flow is classified into multiple modules: top metal, plug, structural SiGe, slit, structure, spacer, electrode, and pad.

5.4.2.1 Top Metal Module

The top metal module is the last metallization of the CMOS process flow, and it is also the electrical interface to the MEMS devices. Figure 5.5 shows the simplified CMOS circuitry substrate with oxide dielectric layer, topmost copper layer, and oxide/nitride passivation layers. The oxide/nitride passivation layers are etched to reach the copper layer for interconnect. In Figure 5.6, an aluminum metal layer is deposited and patterned into the conductive portion of the MEMS electrodes. Isolation layer silicon oxide is then deposited and planarized with chemical mechanical polishing (CMP).

5.4 CMEMS® Process | 81

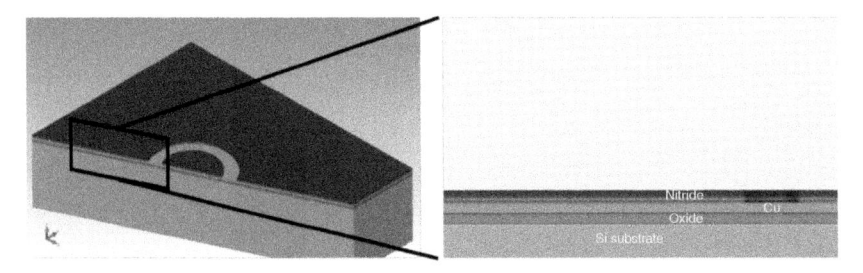

Figure 5.5 CMOS top metal. Source: Quevy [31, 32].

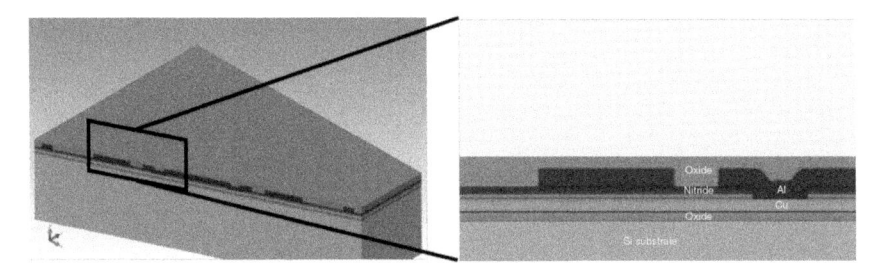

Figure 5.6 CMOS top passivation. Source: Quevy [31, 32].

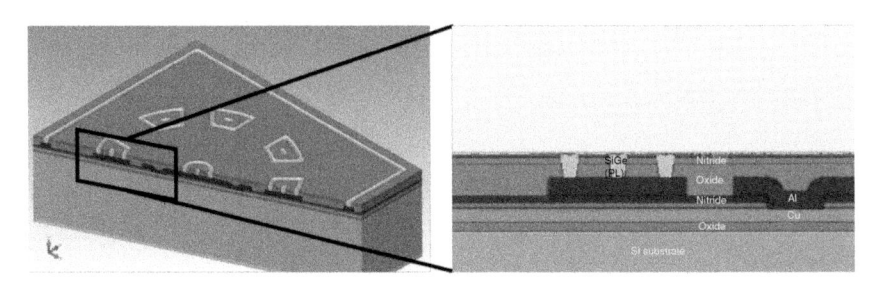

Figure 5.7 Via formation to CMOS. Source: Quevy [31, 32].

Figure 5.8 Sacrificial layer, etch stop layer, and structural layer depositions. Source: Quevy [31, 32].

Figure 5.9 Slit etch. Source: Quevy [31, 32].

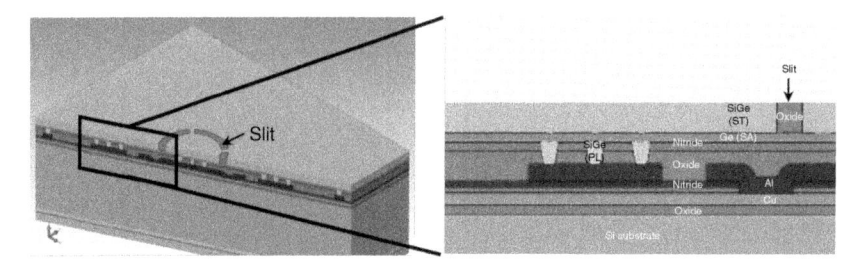

Figure 5.10 Slit formation. Sources: Quevy et al. [31, 32].

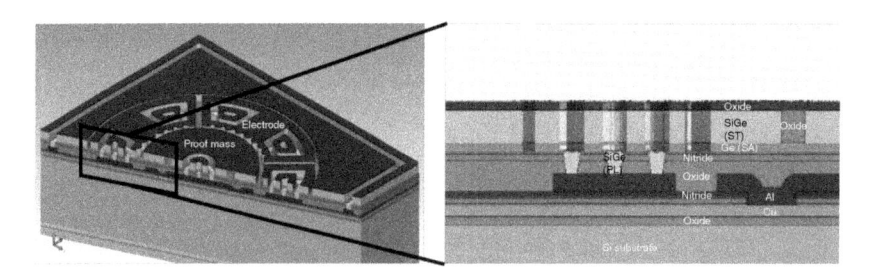

Figure 5.11 Structural layer etch. Sources: Quevy et al. [31, 32].

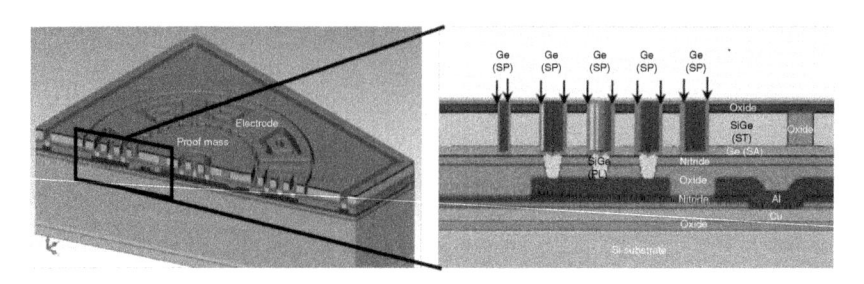

Figure 5.12 Spacer layer deposition. Sources: Quevy et al. [31, 32].

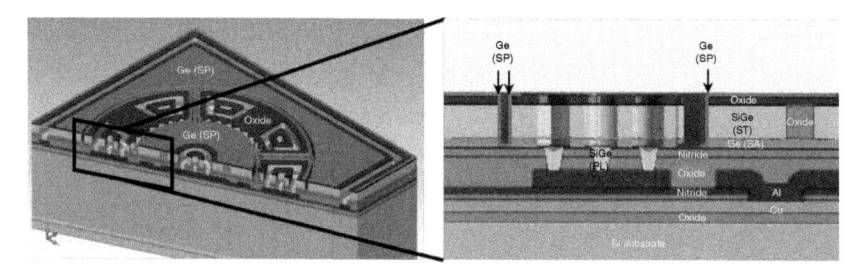

Figure 5.13 Spacer layer etch. Sources: Quevy et al. [31, 32].

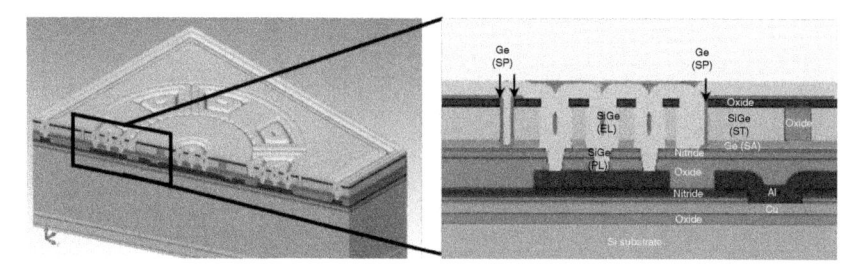

Figure 5.14 SiGe(EL) layer deposition. Sources: Quevy et al. [31, 32]..

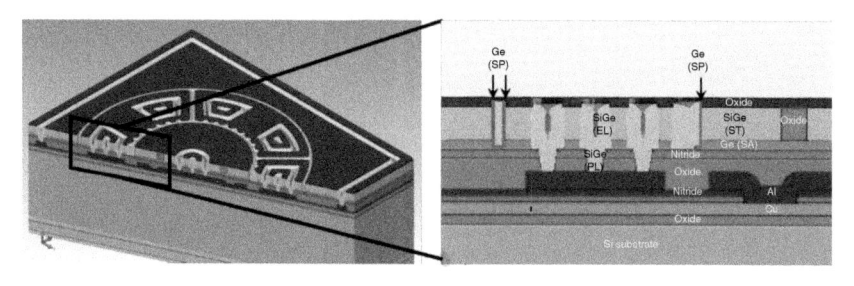

Figure 5.15 SiGe(EL) layer etch back. Sources: Quevy et al. [31, 32]..

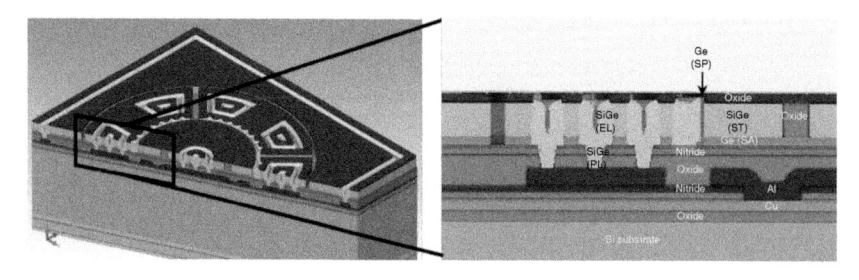

Figure 5.16 Electrode formation etch. Sources: Quevy et al. [31, 32]..

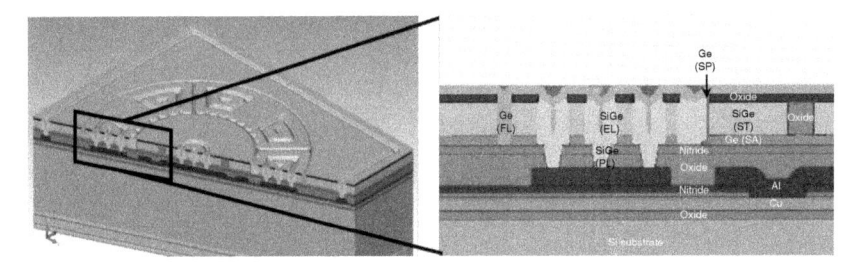

Figure 5.17 Ge(FL) deposition. Sources: Quevy et al. [31, 32]..

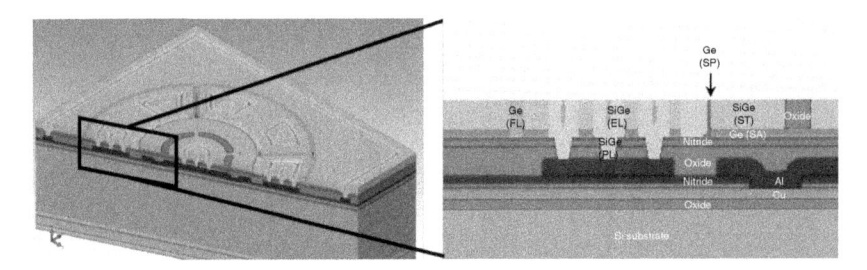

Figure 5.18 Ge(FL) etch back and hard mask removal by CMP. Sources: Quevy et al. [31, 32].

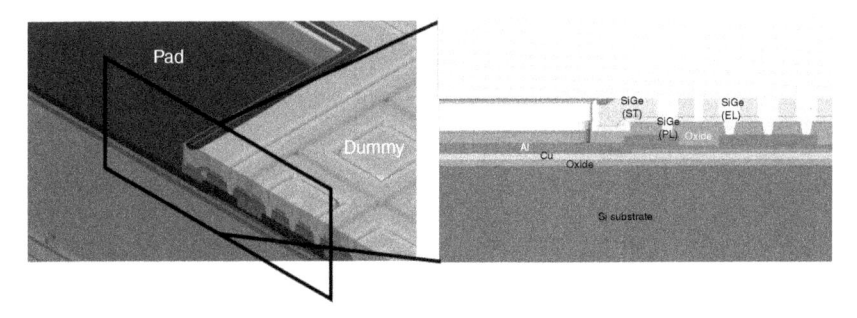

Figure 5.19 Pad and seal ring etch. Sources: Quevy et al. [31, 32].

5.4.2.2 Plug Module

The plug module is the interface of the CMOS circuitry and the MEMS devices, creating electrical contacts and mechanical anchors. After the oxide CMP, additional nitride and oxide layers are deposited as passivation layers. As shown in Figure 5.7, the passivation layers are patterned and etched to reach the aluminum layer. Titanium and titanium nitride layers are deposited as the glue layer (Ti) and the barrier layer (TiN). Then 1 μm thick, boron-doped poly-SiGe plug layer SiGe(PL) is deposited to conformally fill the vias. Having sufficient thickness of the TiN barrier is important to prevent eutectic reaction between the aluminum and the SiGe. SiGe(PL), TiN, and Ti layers are then removed by etch back processes which endpoint on the oxide passivation layer, leaving the material only inside the vias, thereby forming electrical contacts.

5.4.2.3 Structural SiGe Module

The structural SiGe module deposits the sacrificial and structural materials for the MEMS resonators. In Figure 5.8, 0.5 μm of germanium sacrificial layer Ge(SA), 0.1 μm of oxide etch stop layer, and 2.5 μm of poly-SiGe structural layer SiGe(ST) are deposited. CMP is then used to polish the SiGe(ST) layer to remove surface roughness.

5.4.2.4 Slit Module

SiGe has a temperature coefficient of Young's modulus which is on the same order as single-crystal silicon (~−60 to −80 ppm/°C), which means it gets softer as temperature increases. Fortunately, SiO_2 becomes stiffer as temperature increases. This increase in Young's modulus can be used to counteract the softening of the SiGe [33–35]; the slit module embeds inlays of SiO_2 into the structural SiGe layer to achieve this goal.

In Figure 5.9, an oxide hard mask layer is deposited. The oxide hard mask layer and the SiGe(ST) layer are patterned and etched. The etch stops at the oxide etch stop layer above the Ge(SA) layer. In Figure 5.10, high-density plasma (HDP) oxide is deposited to conformally fill the slit area. The HDP oxide is then polished with CMP until SiGe(ST) structural layer is exposed, leaving the oxide only in the slit area.

5.4.2.5 Structure Module

The structure module defines the resonator's proof mass, electrodes, and release holes. In Figure 5.11, a hard mask oxide layer is first deposited. In multiple dry etch steps, the oxide hard mask, SiGe(ST) layer, oxide etch stop layer, and Ge(SA) sacrificial layer are removed, exposing the interconnects SiGe(PL) formed in the plug module, as well as oxide passivation in other areas.

5.4.2.6 Spacer Module

The spacer module defines the submicron transducer gap. In Figure 5.12, a 0.2 μm thick germanium spacer layer Ge(SP) is deposited to conformally cover the patterned structural device, on the top and bottom surfaces, as well as on all sidewalls. The thickness of the germanium spacer is critical as it defines the gap size of the electrostatic transducer. In the lithography step, photoresist covers the spacer on the proof mass sidewall while leaving the electrode sidewall opened for etch. With more than 2 μm topology, over exposure, double puddle, and double rinse are needed to generate the desired photoresist profile. Also, a crenellation pattern is used at the electrode edge to give more clearance for the photoresist to clear out at the bottom of the trench. In Figure 5.13, an isotropic etch is used to clear the germanium on the electrode sidewall with some over etch into the SiGe(ST) layer. After photoresist removal, the germanium spacer is left at the proof mass sidewall and the resist-covered top surface.

5.4.2.7 Electrode Module

The electrode module defines the mechanical anchor and electrical connections of the MEMS device. In Figure 5.14, a layer of 1.5 μm thick poly-SiGe electrode

SiGe(EL) is deposited to fill all the trenches opened by the structural layer etch. This SiGe(EL) layer is in contact with the structural SiGe(ST) layer as the mechanical anchor and electrical contact. To ensure good electrical contact, the wafer needs to go through sputter clean and dilute hydrofluoric acid dip to remove native oxide on the sidewall before the SiGe(EL) deposition. In Figure 5.15, the SiGe(EL) layer is etched back from the top surface, end-pointing on the oxide hard mask layer. Then, the electrodes and anchor area are covered by photoresist. As shown in Figure 5.16, excess portions of SiGe(EL) layer are etched to separate the MEMS devices from the dummy structures and to separate electrodes from each other. Also, the SiGe(EL) layer inside the release holes is removed.

Referring to Figure 5.17, a 0.5 μm germanium filling layer Ge(FL) is conformally deposited on the wafer to fill in all the gaps opened by the electrode etch. The germanium on the top surface is then etched back to expose the oxide hard mask layer. In Figure 5.18, the hard mask layer is removed by CMP. Filling the trenches with the Ge(FL) layer prevents CMP particles from getting trapped in the trenches.

5.4.2.8 Pad Module

The pad module (Figure 5.19) opens the aluminum bond pads and seal ring areas. With multiple dry etch steps, SiGe(ST), SiGe(EL), oxide etch stop, Ge(SA) layer, SiGe(PL) plug layer, and passivation layers on top of the aluminum bond pads and seal ring areas are removed.

5.4.3 Release

Figure 5.20 shows the MEMS device after release. All of the germanium layers Ge(SA), Ge(SP), and Ge(FL) are removed in heated (90 °C) hydrogen peroxide solution. The hydrogen peroxide release will slightly oxidize the SiGe and the aluminum surfaces. A quick dip in 50 : 1 diluted hydrofluoric acid is used to refresh the wafer surface and make it hydrophobic. To minimize release stiction, the wafers then go through a Marangoni Effect dryer (Dai Nippon Screen Model WS-820L). Wafers are slowly pulled out of a bath of deionized (DI) water with an isopropyl alcohol (IPA) layer on the top surface. With the surface tension gradient, the water is pushed out from the wafer, leaving the released devices stiction free.

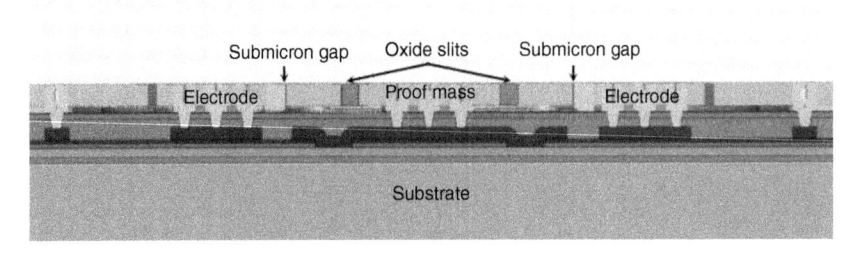

Figure 5.20 Released device. Sources: Quevy et al. [31, 32].

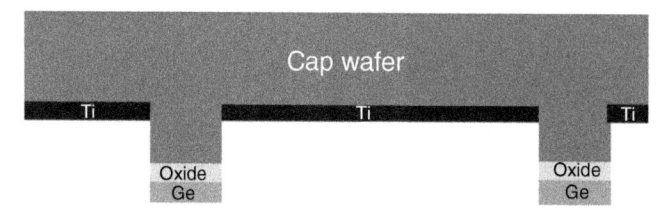

Figure 5.21 Cap cross-sectional view.

5.4.4 Al–Ge Bonding for Microcaps

The MEMS device is encapsulated with wafer-to-wafer boning. Figure 5.21 illustrates the cross-sectional view of the cap wafer. A layer of oxide isolation and 0.5 µm of germanium are first deposited. Then the seal ring is defined and the cavity is created with deep reactive ion etch. A layer of metallic titanium (non-poison mode) [36] is sputtered into the cavity, which serves as the getter material to stabilize the vacuum level in the sealed cavity. The cap wafer and the device wafer are bonded together with the eutectic reaction of the germanium and the aluminum at 435 °C at the seal ring, as shown in Figure 5.22. An annealing step is needed to activate the Ti getter after bonding. In Figure 5.23, the bonded wafer pair is ground down on both the cap side and the device side. The cap wafer is then diced to expose the bond pads.

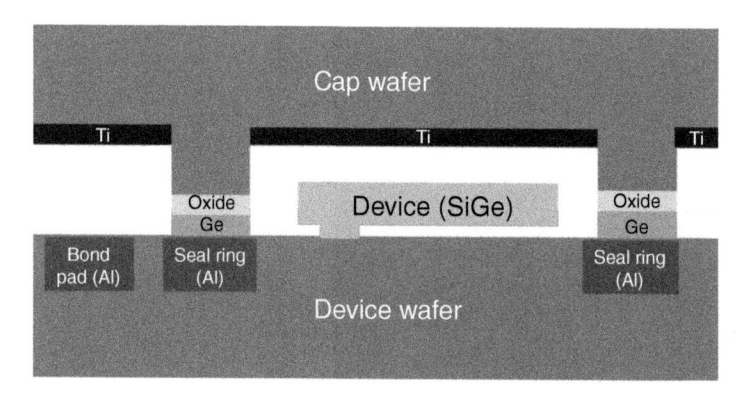

Figure 5.22 Wafer to wafer bonding with Al–Ge eutectic reaction.

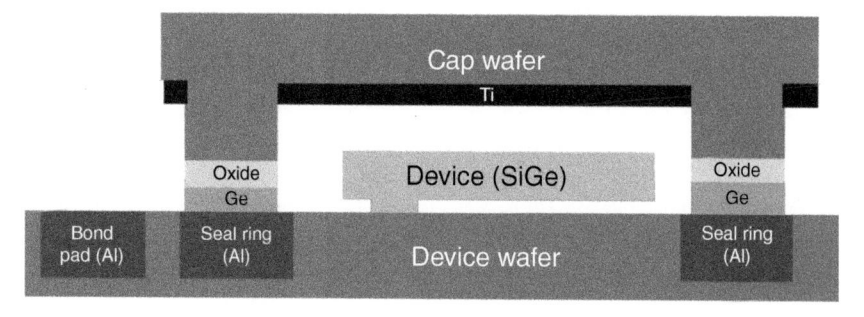

Figure 5.23 Final grinding and dicing.

5.5 Poly-SiGe Applications

5.5.1 Resonator for Electronic Timing

CMEMS® technology was first aimed at frequency control products to compete with the 100 year-old quartz crystal resonator technology. Almost every type of electronic product relies on a tiny piece of quartz crystal to generate the reference frequency. Quartz-based oscillators have been the standard for frequency control with their stabilities, which can reach parts per billion (ppb). Over the years, quartz manufacturers have been continuously refining their process to make quartz-based oscillators more stable, accurate, and available in smaller packages. Over the past decade, two-chip MEMS-based oscillators [37] entered the frequency control market with advantages in lead time, supply stability, device size, and price. On the other hand, MEMS resonators entered the frequency control market with some shortcomings. First, MEMS devices cannot be physically trimmed as can their quartz counterparts to achieve parts per billion accuracy. Out of the fab resonator typically has ±0.2% 6-sigma range of target frequency. Second, standard MEMS materials (single-crystal silicon, poly-silicon, and poly-SiGe) have mechanical properties that drift significantly with temperature. They become softer as temperature rises, which translates into a frequency temperature coefficient as large as −30 ppm/°C. In contrast, quartz crystals have orientations that yield near-zero temperature coefficients of frequency.

To address trimming and initial temperature accuracy, the CMEMS® oscillator architecture has a MEMS-stabilized voltage-controlled oscillator (VCO), resulting in a device with a separate oscillator locked to the MEMS reference through a phase-locked loop (Figure 5.24). The loop forces a predetermined ratio between the MEMS oscillator and the VCO so that the output frequency of the VCO is corrected if there is any inaccuracy present in the MEMS reference. This approach also allows the programmability of the output frequency. By adding a temperature sensor to the system, the trim values affecting the MEMS-to-VCO frequency ratio can be varied over temperature to compensate for the MEMS oscillator temperature drift. With circuit compensation, the output clock signal is accurate and stable without trimming the MEMS oscillator.

To address the temperature drift of the MEMS resonators, CMEMS® technology embeds the poly-SiGe structural material with regions of silicon dioxide (SiO_2), which has a compensating behavior over temperature (Figure 5.4). As temperature increases, poly-SiGe becomes softer while SiO_2 becomes harder. The compensation SiO_2 is placed at the maximum stress point in the form of small slits of oxide, as described in Section 5.4.2.4. Figure 5.25 shows the native frequency temperature characteristic of the CMEMS® plate resonator, which lies between ±2 ppm/°C due to manufacturing tolerances, which is very close to the behavior of quartz crystals.

With thermal drift compensated by oxide slits and careful design of the temperature sensor, CMEMS® oscillators exhibit short-term stability in the range of a few ppb. At the system level, temperature calibration further compensates the device's temperature stability (Figure 5.26).

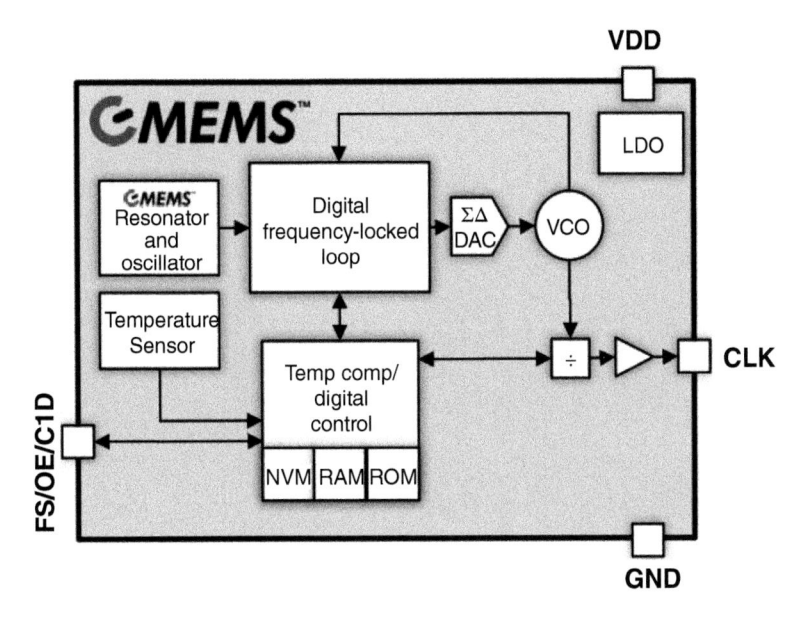

Figure 5.24 CMEMS oscillator architecture and block diagram. Source: Quevy [29]. © 2013, Silicon Laboratories.

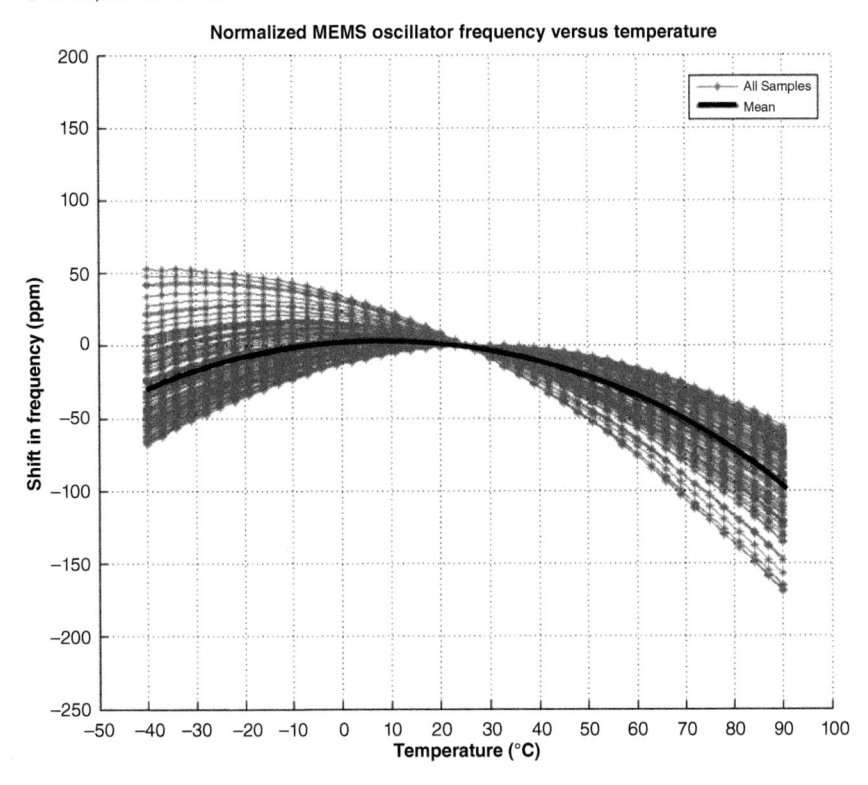

Figure 5.25 Native frequency temperature characteristic of the plate resonator shown in Figure 5.4. Source: Quevy [38].

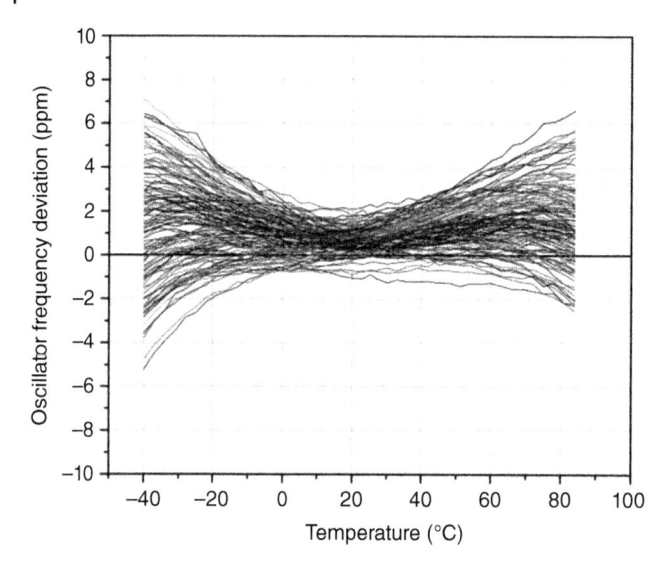

Figure 5.26 Temperature characteristic of fully calibrated and compensated CMEMS® oscillator with solder shift included (sample size ~100). Source: Quevy [29].

CMEMS® oscillators have better performance in thermal slew compared to both two-chip MEMS oscillators and quartz oscillators. The three architectures (two-chip MEMS, quartz, and CMEMS®) were tested in a freeze spray/heat gun experiment. As shown in Figure 5.27, the CMEMS® oscillator remains stable with large thermal slew, while the other solutions clearly show temperature-slew sensitivity. The thermal path for two-chip MEMS oscillators includes CMOS chip, wire bonds, die attach epoxy, MEMS chip, and the whole package, which makes it difficult for the system to compensate for the changes in the temperature. Quartz has a smaller temperature sensitivity but a larger thermal lag from its ceramic package. CMEMS® oscillators have small temperature sensitivity with mechanical compensation, also short thermal path, and small thermal constants due to monolithic integration. System integration and thermal transport are both important for stability in uncontrolled environments.

CMEMS® oscillators also have superior long-term aging performance when compared with two-chip MEMS oscillators and quartz oscillators. Figure 5.28 provides a comparison of several quartz, two-chip MEMS and CMEMS® oscillators in an aging test. In this plot, the quartz oscillators were aged at 70 °C, while all two-chip MEMS and CMEMS® devices were aged at 125 °C and then extrapolated to the same duration. The CMEMS® device shows greater stability over time over existing MEMS technology approaches.

In conclusion, CMEMS® oscillators combines the advantages of MEMS-based timing solutions while retaining and improving many of the best characteristics of traditional quartz crystals. CMEMS® oscillators have superior manufacturability, fast lead time, and competitive performance compared to traditional quartz and two-chip MEMS oscillators.

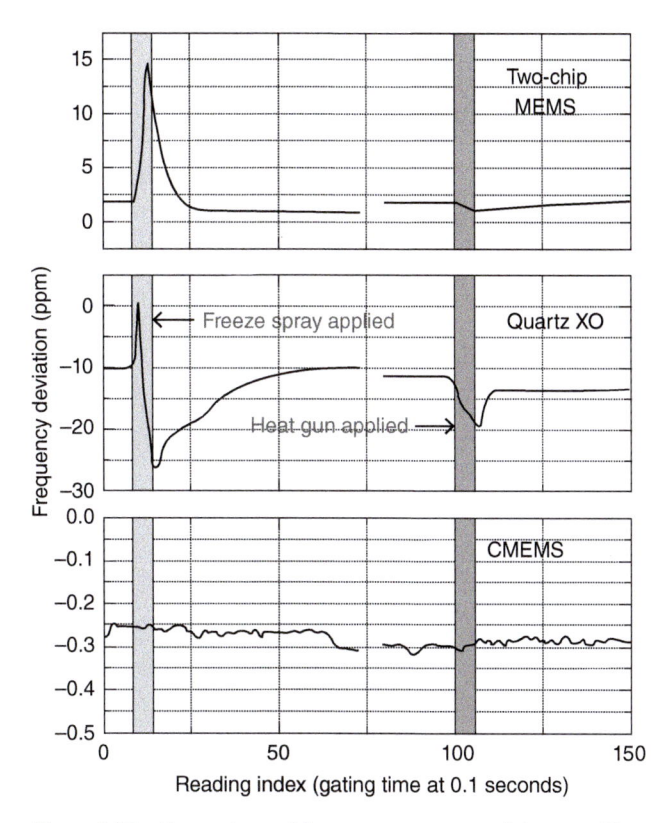

Figure 5.27 Comparison of frequency response of three oscillator architectures to fast thermal transients. Source: Quevy [29].

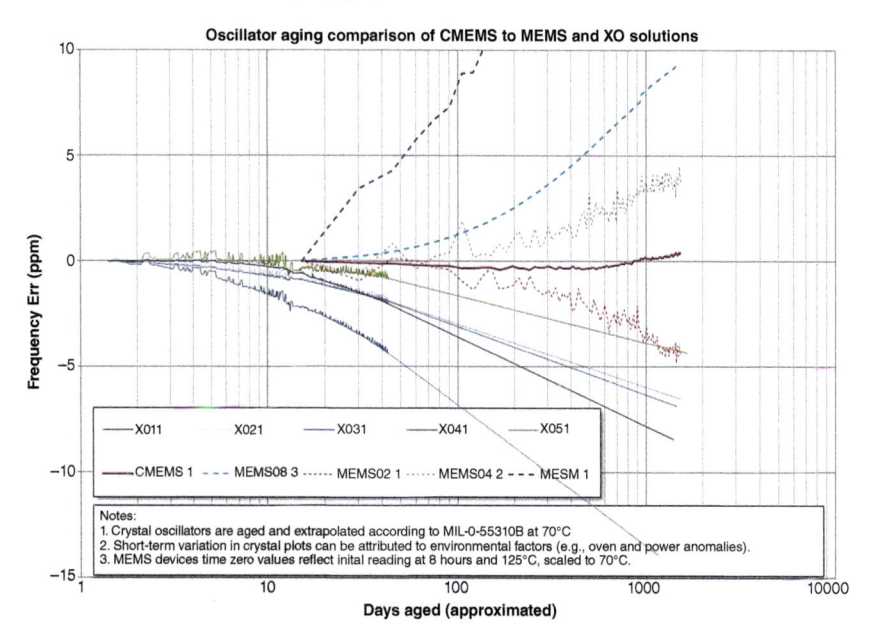

Figure 5.28 Comparison of aging results of crystal, two-chip MEMS, and CMEMS® oscillators. Source: Quevy [38].

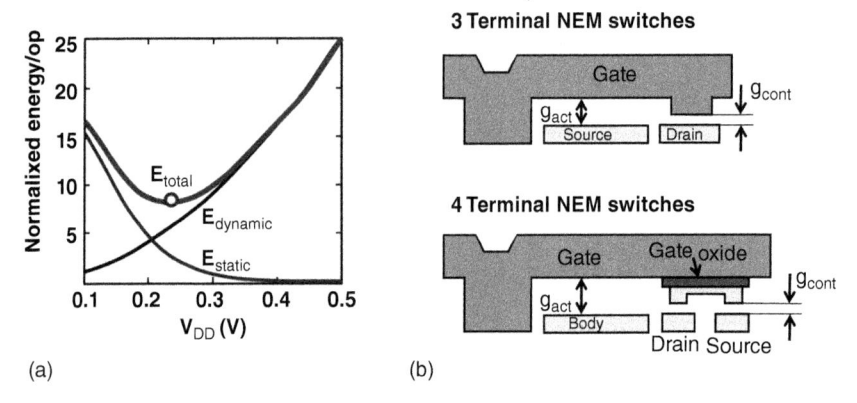

Figure 5.29 (a) Normalized energy versus operating voltage of a CMOS transistor. Source: Nathanael [39]. (b) schematic of 3 and 4 terminals NEM switches.

5.5.2 Nano-electro-mechanical Switches

Poly-SiGe Nano-Electro-Mechanical (NEM) switches emerged as an alternative to overcome the drawbacks of MOS transistors for ultralow-power applications. Even though the transistor has been scaling down in size, the operating voltage (V_{DD}) has not been following the same trend. Due to the fact that V_{DD} has not been scaled down commensurately with the transistor's dimensions, the power density of chips has become the dominant challenge for continued improvement on CMOS technology. The transistor's total energy consumption is comprised by the sum of the dynamic and the static energies. The minimum energy consumed by a transistor during switching decreases quadratically with V_{DD}; however, the off-state leakage current increases exponentially as V_{DD} decreases. Thus, the total energy versus V_{DD} has a minimum, as illustrated in Figure 5.29a. On the other hand, NEM switches offer zero static energy (no leakage current), and so energy consumption is only due to switching. Therefore, a 10× reduction on V_{DD} represents a 100× reduction in total energy consumption.

Typical NEM switches consist of two parallel metallic plates separated by an air gap (or actuation gap); one is fixed on a substrate and the other is suspended in the air by springs. When a voltage is applied across the plates, an electrostatic force is created and brings both into proximity. When the device is in off state, there is zero leakage current due to the air gap separating the contacts (contact gap), whereas there is a sharp transition to the on state due to metal-to-metal contact. There are two main configurations of NEM switches: three terminal switches where the electrostatic force is between the source and gate and four terminal switches where the electrostatic force is between the gate and the body (4th terminal), as shown in Figure 5.29b. For the latter, the output is independent from the input and allows using the body terminal to reduce the gate voltage.

UC Berkeley's NEM poly-SiGe switches are four terminal devices, which have been used to make complementary logic circuits operating at 50 mV [40] and have shown a mechanical endurance of 10^{10} cycles [41]. Its fabrication process flow [40]

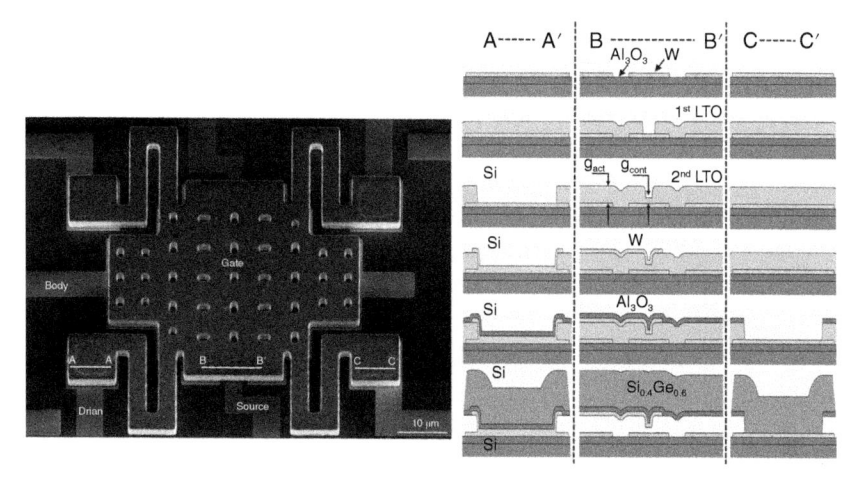

Figure 5.30 Fabrication process for NEM switches.

consists of six lithography steps, as shown in Figure 5.30. The process starts by thermal atomic layer deposition (ALD) of Al_2O_3 at 300 °C in order to isolate the devices from the Si substrate. Then tungsten is sputtered at room temperature and dry etched to form the body, source, and anchor electrodes. Low-temperature SiO_2 (LTO) is deposited by LPCVD at 400 °C to form the contact (g_{cont}) and actuation (g_{act}) gaps. Each deposition is followed by a dry etching step to form a contact dimple and a via for the source electrode, which is consecutively deposited and patterned similarly as the first metal. The gate oxide is made of Al_2O_3 and deposited by plasma-enhanced ALD at 270 °C and patterned by dry etching, creating a via for the anchors. The structural layer consists of a poly-$Si_{0.4}Ge_{0.6}$ film, which is deposited by LPCVD at 410 °C. The Poly-SiGe is in situ doped in order to obtain a highly conductive film while maintaining a low stress gradient. Thereafter, a low-temperature CVD oxide (LTO) layer is deposited as a hard mask for SiGe etching. As a final step, the device is released by selectively removing the LTO using vapor HF. The process flow is kept at temperatures below 410 °C, which is compatible with post-processing most CMOS technologies.

Figure 5.31a,b shows the waveform of an AND gate and a 2 : 1 MUX operating at 50 mV. NEM switches are able to operate as p-type or n-type depending on the body bias, as shown in the schematics. The actuation gap to contact gap ratio and the adhesive force between contacts are the limiting factors for V_{DD} reduction in NEM switches. V_{DD} is proportional to the difference between the actuation and the contact gap. However, as the gaps get smaller, the fabrication yield decreases drastically due to stiction during release. To overcome this limitation, body bias is used to artificially reduce the actuation gap without sacrificing yield. UC Berkeley has shown a yield higher than 98% using this technique. Another important consideration is the ratio g_{cont}/g_{act}. Due to the nonlinearity of the electrostatic force, parallel-plate switches collapse when the gap between the plates reduces to 1/3 of the total gap. As result, if the ratio is >1/3 the devices operate in the pull-in mode regime, and the voltage

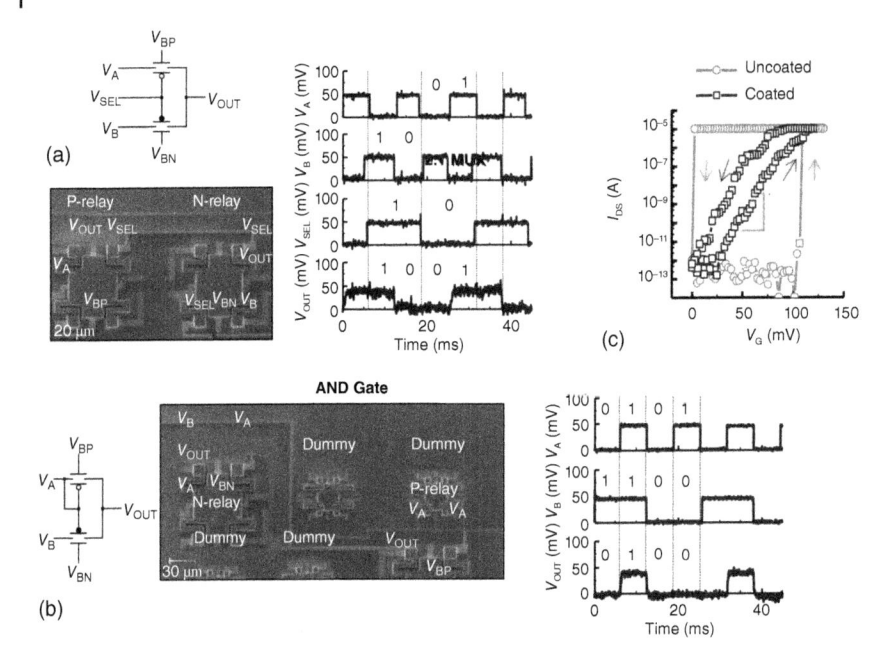

Figure 5.31 Schematic circuit diagrams, plan-view SEM images, and waveforms of (a) 2 : 1 MUX and (b) AND gate operating at 50 mV. (c) I_{DS}–V_G curve of NEM switch with and without molecular coating. Source: Ye et al. [40]. © 2018, IEEE.

needed to turn off the switch (the release voltage) decreases drastically, leading to a large hysteresis voltage which constrains V_{DD}. In the case where the ratio <1/3, the relay operates in a non-pull-in mode where the hysteresis voltage is low and only depends on the contact adhesive force. For this reason, the control of the stress gradient of the structural film is very important. Poly-SiGe with a composition of 60% Ge and 40% Si with a thickness of 1.8 µm has shown to be optimal. Additionally, the addition of a fluorinated self-assembled monolayer on the contacting surfaces has shown a 71% reduction in the hysteresis voltage [42]. In conclusion, the proper spring constant, body bias, gap ratio, and molecular coating contribute to achieving a V_{DD} smaller than 50 mV [40], as shown in Figure 5.31c.

References

1 Franke, A.E. (2000). Polycrystalline silicon-germanium films for integrated microsystems. Ph.D. Thesis, University of California at Berkeley.
2 Heck, J.M. (2001). Polycrystalline silicon germanium for fabrication, release, and packaging of microelectromechanical systems. Ph.D. Thesis, University of California at Berkeley.
3 Low, C.W., King Liu, T.-J., and Howe, R.T. (2007). Characterization of polycrystalline silicon-germanium film deposition for modularly integrated MEMS applications. *IEEE/ASME J. Microelectromech. Syst.* 16 (1): 68–77.

4 Witvrouw, A., Mehta, A., Verbist, A. et al. (2005). Processing of MEMS gyroscopes on top of CMOS ICs. Proceedng of 52nd IEEE International Solid-State Circuits Conference, San Francisco, CA, (6–10 February 2005), 88–89.

5 Sedky, S., Fiorini, P., Baert, K. et al. (1999). Characterization and optimization of infrared poly SiGe bolometers. *IEEE Trans. Electron Devices* 46 (4): 675–682.

6 Gonzalez, P., Rakowski, M., San Segundo, D. et al. (2012). CMOS-Integrated poly-SiGe piezoresistive pressure sensor. *IEEE Electron Device Lett.* 33 (8): 1204–1206.

7 Qian, C., Peschot, A., Osoba, B. et al. (2017). Sub-100 mV computing with electro-mechanical relays. *IEEE Trans. Electron Devices* 64 (3): 1323–1329.

8 Iyer, S.S., Patton, G.L., Delage, S.S. et al. (1987). Silicon-germanium base heterojunction bipolar transistors by molecular beam epitaxy. Proceedings of International Electron Devices Meeting, New York, 874–876.

9 Cressler, J.D. and Niu, G. (2003). *Silicon-Germanium Heterojunction Bipolar Transistors*. Artech House.

10 Kistler, N. and Woo, J. (1993). Symmetric CMOS in fully-depleted silicon-on-insulator using P^+-polycrystalline SiGe gate electrodes. Proceedings of International Electron Devices Meeting, 727–730.

11 Takeuchi, H., Lee, W.-C., Ranade, P., and King, T.-J., (1999). Improved PMOSFET short-channel performance using ultra-shallow $Si_{0.8}Ge_{0.2}$ source/drain extensions. Proceedings of International Electron Devices Meeting, 501–504.

12 King, T.-J. and Saraswat, K.C. (1994). Polycrystalline silicon-germanium thin-film transistors. *IEEE Trans. Electron Devices* 41 (9): 1581–1591.

13 Strasser, M., Aigner, R., Franosch, M., and Wachutka, G. (2002). Miniaturized thermoelectric generators based on poly-Si and poly-SiGe surface micromachining. *Sens. Actuators A* 97–98: 535–542.

14 Van Gerwen, P., Slater, T., Chévrier, J.B. et al. (1996). Thin film boron-doped poly-crystalline $silicon_{70\%}$-$germanium_{30\%}$ for thermopiles. *Sens. Actuators A* 53: 325–329.

15 Lin, B.C.-Y, King, T.-J., and Muller, R.S. (2006). Poly-SiGe MEMS actuators for adaptive optics. Photonics WEST, sponsored by SPIE, The International Society for Optical Engineering, Conference, San Jose, CA, (25 January 2006) 6113–6128.

16 Quévy, E.P., San Paulo, A., Basol, E. et al. (2006). Back-end-of-line Poly-SiGe disk resonators. 19th IEEE Micro Electro Mechanical Systems Conference (MEMS-06), Istanbul, Turkey, January 2006, 234–237.

17 Sedky, S., Fiorini, P., Caymax, M. et al. (1998). Structure and mechanical properties of polycrystalline silicon germanium for micromachining applications. *IEEE/ASME J. Microelectromech. Syst.* 7 (4): 365–372.

18 Low, C.W. (2007). Novel processes for modular integration of silicon-germanium MEMS with CMOS electronics. Ph.D. Thesis, University of California at Berkeley.

19 Mehta, A., Gromova, M., Czarnecki, P. et al. (2005). Optimization of PECVD poly-SiGe layers for MEMS post-processing on top of CMOS. Proceedings of 13th International Conference on Solid-State Sensors, Actuators and Microsystems (Transducers 05), Seoul, Korea (5–9 June 2005) 1326–1329.

20 Gonzalez, P. and Rottenberg, X. (2015). Thin films on silicon: poly-SiGe for MEMS-above-CMOS applications. In: *Handbook of Silicon Based MEMS Materials and Technologies*, 2nde (eds. M. Tilli, M. Paulasto-Krockel, T. Motooka and V. Lindroos), 141–154. Elsevier.

21 Guo, B., Severi, S., Bryce, G. et al. (2010). Improvement of PECVD silicon-germanium crystallization for CMOS compatible MEMS applications. *J. Electrochem. Soc.* 157 (2): 103–110.

22 Sedky, S., El Defrar, I., and Mortagy, O. (2006). Pulsed laser deposition of boron doped $Si_{70}Ge_{30}$. Proceedings of Materials Research Society Meeting, San Francisco, CA.

23 Silicon Clocks, Inc. (2008). Documentation. http://www.Siliconclocks.com.

24 Sedky, S., Gromova, M., Van der Donck, T. et al. (2006). Characterization of KrF excimer laser annealed PECVD Si_xGe_{1-x} for MEMS post-processing. *Sens. Actuators A* 127: 316–323.

25 Sedky, S., Howe, R.T., and King, T.-J. (2004). Pulsed laser annealing, a low thermal budget technique for eliminating stress gradient in poly-SiGe MEMS structures. *IEEE/ASME J. Microelectromech. Syst.* 13 (4): 669–675.

26 Air Liquide (2019). Private communication. https://www.airliquide.com.

27 Ogawa, K., Mino, Y., and Ishihara, T. (1989). Performance of a new vertical LPCVD apparatus. *J. Electrochem. Soc.* 136 (4): 1103–1108.

28 Low, C.W., Wasilik, M.L., Takeuchi, H. et al. (2000). In-situ doped poly-SiGe LPCVD process using BCl_3 for post-CMOS integration of MEMS devices. Proceedings of Electrochemical Society SiGe Materials, Processing, and Devices Symposium, Honolulu, HI (3–8 October 2000) 1021–1032.

29 Quevy, E.P. (2013). *CMEMS® Technology: Leveraging High-Volume CMEMS Manufacturing for MEMS-Based Frequency Control*. A White Paper Published by Silicon Laboratories.

30 Howe, R.T., Quevy, E.P., and Gu, Z. (2015). Gas diffusion barriers for MEMS encapsulation. US Patent 9,018,715 B2, 28 April 2015.

31 Quevy, E.P. (2009). IC-compatible MEMS structure. US Patent 7,514,760, 7 April 2009.

32 Quevy, E.P., Low, C.W., Hui, J.R., and Gu, Z. (2014). Technique for forming a MEMS device. US Patent 8,852,984 B1, 7 October 2014.

33 Quevy, E.P. and Bernstein, D.H. (2007). Method for temperature compensation in MEMS resonators with isolated regions of distinct material. US Patent 7,639,104, 9 March 2007.

34 Bernstein, D.H., Howe, R.T., and Quevy, E.P. (2007). MEMS structure having a compensated resonating member. US Patent 7,591,201, 9 March 2007.

35 Howe, R.T., Quevy, E.P., and Bernstein, D.H. (2007). MEMS structure having a stress inverter temperature-compensated resonating member. US Patent 7,514,853, 10 May 2007.

36 Fouad, O.A., Rumaiz, A.K., and Shah, S.I. (2009). Reactive sputtering of titanium in Ar/CH_4 gas mixture: Target poisoning and film characteristics. *Thin Solid Films* 51 (19): 5689–5694.

References

37 Lam, C.S. (2008). A review of the recent development of MEMS and crystal oscillators and their impacts on the frequency control products industry. 2008 IEEE Ultrasonics Symposium, Beijing, China, November 2008, 694–704.

38 Quevy, E.P. (2013). *CMEMS® Oscillator Architecture*. A White Paper Published by Silicon Laboratories.

39 Nathanael, R. (2012). "Nano-electro-mechanical (NEM) relay devices and technology for ultra-low energy digital integrated circuits. Ph.D. Thesis, University of California at Berkeley.

40 Ye, Z.A., Almeida, S., Rusch, M. et al. (2018). Demonstration of 50-mV digital integrated circuits with microelectromechanical relays. Proceedings of 2018 IEEE International Electron Devices Meeting (IEDM), San Francisco, CA, USA (December 2018), 4-1.

41 Chen, Y., Nathanael, R., Yaung, J. et al. (2013). Reliability of MEM relays for zero leakage logic. Proceedings of Reliability, Packaging, Testing, and Characterization of MOEMS/MEMS and Nanodevices XII. International Society for Optics and Photonics, San Francisco, CA, USA, March 2013. Vol. 8614, 861404.

42 Fathipour, S., Almeida, S.F., Ye, Z.A. et al. (2019). Reducing adhesion energy of nano-electro-mechanical relay contacts by self-assembled Perfluoro (2, 3-Dimethylbutan-2-ol) coating. *AIP Adv.* 9 (5): 055329.

6

Metal Surface Micromachining

Minoru Sasaki

Toyota Technological Institute, Department of Advanced Science and Technology, 2-12-1 Hisakata, Tenpaku-ku, Nagoya 468-8511, Japan

6.1 Background of Surface Micromachining

The typical surface micromachining uses polysilicon (poly Si) as the structure material. There is the residual stress in the deposited poly Si film. However, the high-temperature annealing can remove this stress, since the device substrate is usually the crystal Si. The effect for getting the stress–free film is obtained because the deposited Si has almost the same property, such as the coefficient of the thermal expansion, with that of the substrate. This becomes important especially when the device size is large. The thin structure bends when the film stress exists especially with the stress gradient against the thickness. Against this restriction, the advantages of the surface micromachining using thin layers are the design flexibility for realizing actuators with soft springs. This can be seen from the fact that the spring constant is proportional to the cubic of the film thickness.

When the metal is selected as the structure material, the high-temperature annealing cannot remove the film stress since the metal has much larger thermal expansion against the Si substrate. From this viewpoint based on the conservative material selection, the surface micromachining is recommended to use the poly Si films on crystal Si substrate. One memorial product is the accelerometer ADXL50 (released in 1991 by Analog Devices, Inc.) realized as the monolithic chip [1]. The metal structural layers on Si substrate is considered to be rather unreasonable. In addition, the device which uses the contact mechanism is considered to be avoided, since the same material will diffuse easily with each other, causing the adhesion problem, which fixes the micro electro-mechanical systems (MEMS) actuator. ADXL50 doesn't have the contact mechanism.

Against the aforementioned design guide line, there are some devices fabricated by the metal surface micromachining including the company products at present. The motivation of the metal usage is the metal line structures, which are already included in complementary metal oxide semiconductor (CMOS) with the established process. For connecting the transistors and the gathered units, the metal lines are constructed

3D and Circuit Integration of MEMS, First Edition. Edited by Masayoshi Esashi.
© 2021 WILEY-VCH GmbH. Published 2021 by WILEY-VCH GmbH.

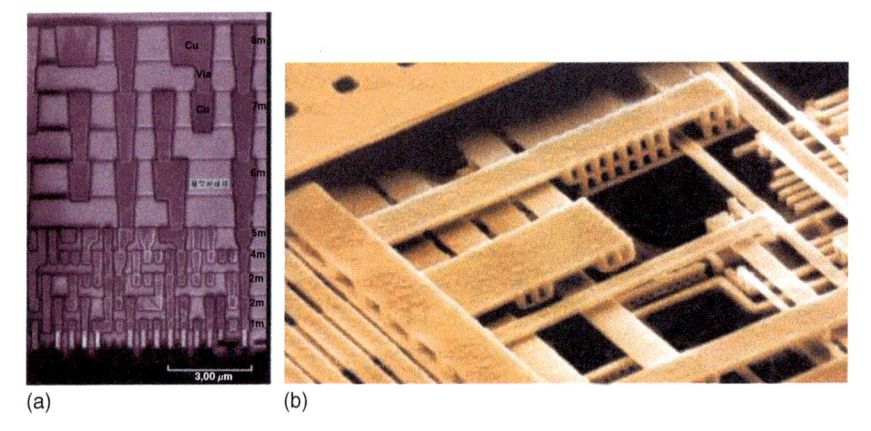

Figure 6.1 (a) Cross section of 8-level Cu interconnection system for 130 nm technology node: fluorinated silicate glass is used as interlayer dielectric. (b) Multilevel interconnect metal lines in LSI. Source: (a) Ohsaki [2]. Copyright 2003 and 1999, the Japan Society of Applied Physics, (b) Kawasaki [3]. Copyright 2003 and 1999, the Japan Society of Applied Physics.

with many layers. Figure 6.1 shows the example. The lower layer is for connecting the transistors near each other. The upper metal layer is for connecting the units with each other. So, the lower layers are thinner, and the upper layers are thicker [2]. In the integrated circuit, the metal lines are with the interlayer dielectric. If these layers are removed making the air gap as shown in Figure 6.1b [3], the structure becomes very similar to the movable MEMS device. The mask design for the metal layers can be tuned for realizing the variety of structures.

6.2 Static Device

The relatively thick (>10 μm) metal lines are fabricated starting from the preparation of the sacrificial mold (polymer and metal) followed by the electroplating of the metal. There has been an approach to realize 3D photoresist molds in order to obtain 3D metal microstructures. This method has been applied to fabricate solenoid inductors. radio frequency (RF) or microwave applications require the reduction of the size of the electrical elements as well as the substrate coupling and Ohmic loss. X-ray Lithographie, Galvanoformung, Abformung (LIGA) [4] and LIGA-like ultraviolet (UV) technologies had been introduced to build microstructures on the substrate. The need of the inductor is high in RF region. Figure 6.2 shows the inductors fabricated using the metal surface micromachining based on the multi-exposure and single development of the thick photoresist of the mold [5]. The suspended flat and relatively thick structure is obtained over the substrate by 100 μm.

One version of the metal surface micromachining is making the layer thin ultimately. This will minimize the thermal conduction through the film and give the strong advantage for the bolometer using the temperature change of the suspended structure caused by the absorbed infrared being incident on. Figure 6.3 shows

Figure 6.2 Spiral inductor suspended by 100 μm over the substrate supported. Source: Yoon et al. [5]. Figure 6. © 2002, IEEE.

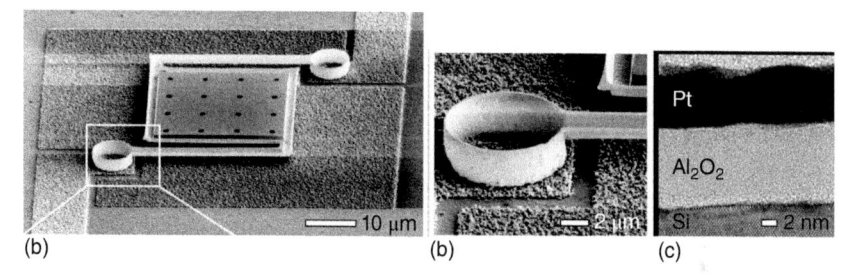

Figure 6.3 Bolometer having nm-thin metal structure layer. (a) Whole view. (b) Magnified view of the anchor and 3 μm wide suspension beam. (c) Detailed TEM image of the Pt/Al$_2$O$_3$ membrane. Source: Purkl et al. [6] Copyright 2013, IEEE.

the uncooled bolometer having the nm-thin platinum layers with the underlying Al$_2$O$_3$ layer (about 10 nm thick) with rim [6]. The sacrificial layer is 3.5 μm thick poly Si removed by XeF$_2$ gas etching. The thickness of Pt is 5.5 nm, realized by the atomic layer deposition using the precursors of (methylcyclopentadienyl) trimethyl-platinum and oxygen activated by the plasma. Usual thick metal has high reflectivity against the infrared. This is the disadvantage for using the bolometer. However, when Pt film becomes thinner than 7 nm, this property changes to show the higher absorption. Such thin platinum can work for both the absorber and the temperature-dependent resistor. Bolometers with a platinum thickness of 5.5 nm and an absorber area of 30×30 μm^2 exhibit a thermal conductance of 1.1×10^{-7} W K^{-1} and a responsivity of 2×10^7 V WA^{-1} with the bias current of 200 μA. Noise Equivalent Temperature Difference is reported to be 163 mK.

6.3 Static Structure Fixed after the Single Movement

Figure 6.4 shows the micromirror developed in the MicroStar project of Lucent Technologies, Inc. for realizing the large-scale optical switch. The first demonstration was at the conference Telecom 99, in October 1999. Figure 6.4a is the first version of the three micromirror [9]. The diameter of the tilting mirror is 500 μm arrayed with

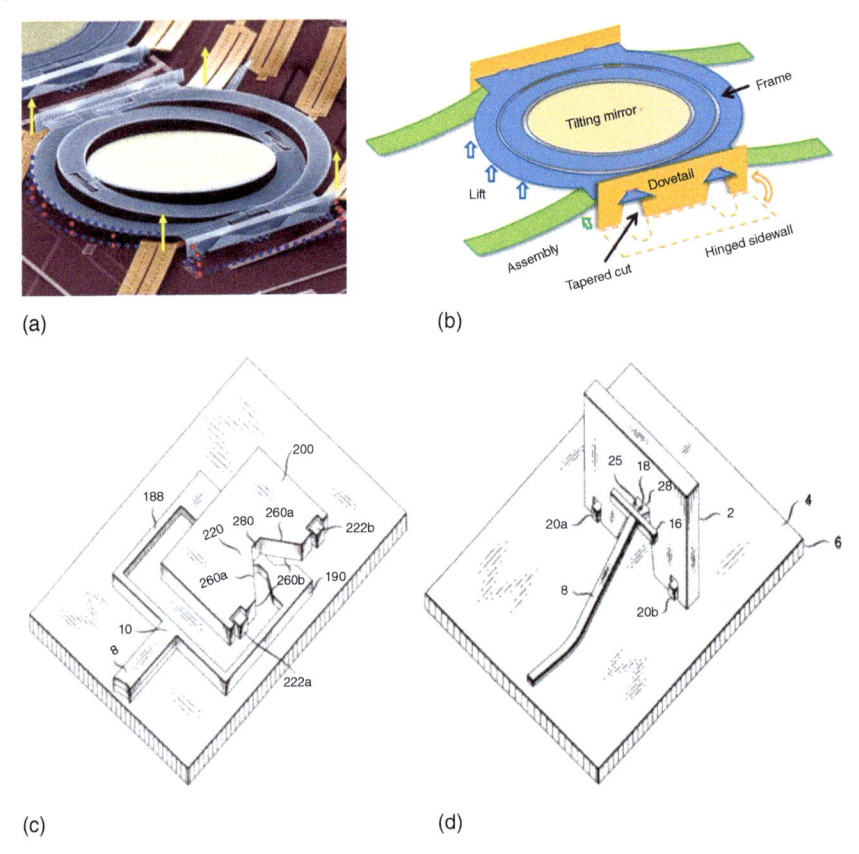

Figure 6.4 (a) Self-assembled beam-steering micromirror. (b) Schematic drawings illustrating the 3D layered structure. Self-assembly mechanism using the hinge illustrating (c) before and (d) after the bend-up motion of the assembly arm. Source: (a, b) Based on Aksyuk et al. [7], (c, d) Aksyuk and Bishop [8] Reused with permission of Nokia Corp. and AT & T Archives.

1 mm pitch [7]. The top surface is coated by gold film for obtaining reflectivity. The thickness of poly Si is 3.5 µm. As the noticeable feature, this tilting mirror is lifted by the surrounding four springs (400 µm long and 100 µm wide) from the substrate by 50 µm. This large lift is essential for the large tilt angle of the mirror. The parallel plate-type actuator with the underlying four electrodes is adopted. The tilting angle is about ±9°. The residual stress in the metal layer on the poly Si layer is used as the energy source to perform the lifting assembly. The self-assembly is accomplished during the release step in the process sequence without the external power supply. The mechanical energy is stored during the metal deposition on four assembly arms. The metal name and its deposition condition are not open. After the release of the assembly arms, the tensile stress in this layer causes them to bend up and push the mirror frame, lifting it in place above the substrate automatically. Figure 6.4b shows the simplified cartoon. Figures 6.4c and 6.4d show the magnified hinged elements before and after the bending-up motion of the assembly arm, respectively [8]. The

tapered cuts in the hinged sidewalls are engaged with the dovetail structures at the frame edge, and as the frame is raised, the sidewalls are rotated 90° out of their initial positions within the substrate plane. The final vertical position of the sidewall is provided by the lithographically defined stop, locking the frame precisely.

The plastic deformation of the metal layer is also reported to tune MEMS layer after the fabrication. This deformation is the same with the failure mechanism, which includes a phenomenon wherein the devices don't return to the original position after repeated actuation. The 10 μm-thick electroplated Ni lever beam is Joule heated to tune the electrostatic actuation performance [10].

6.4 Dynamic Device

6.4.1 MEMS Switch

The reed relays and the solid state switches are used at frequencies from 0 Hz to 100 GHz. The working frequencies are for wireless communications (<3000 GHz), radar systems for automotive (24, 60, and 77 GHz), and satellite communication (12-35 GHz). The conventional relay has the disadvantage of the narrower band of the operation frequency; the shorter the lifetime, the limited the channel and the larger the package size. Over 10 GHz, the performance of the semiconductor switch degrades. MEMS switch has the advantages of near-zero power consumption, high isolation, and low insertion loss [11].

MEMS switch structure has the source, gate, and drain electrodes. Figure 6.5a shows a schematic drawing for explaining the operation [12]. Case A shows OFF position of the switch beam without the contact to the drain electrode. When the DC voltage is applied to the gate, the electrostatic pull down force bends the switch beam. The positive and negative charged plates attract each other. When the gate voltage is high enough, the attraction force (red arrow) overcomes the spring force of the cantilever beam, and the beam bends down to touch the drain as shown in case B. This connects the circuit between the source and the drain, and the switch becomes ON position. When the gate voltage is returned to 0 V as shown in case C, the electrostatic attraction force disappears, and the switch beam acts as a spring with the sufficient restoring force (blue arrow) to open the connection between the source and the drain, returning to OFF position. Figure 6.5b shows four main steps of the fabrication sequence [12]. (i) The switch is constructed on a high resistivity Si wafer, which has a thick dielectric layer deposited on top to provide the electrical isolation between the substrate below. A standard back-end CMOS interconnect process is used to realize the interconnections to the MEMS switch. (ii) Low resistivity metal and poly Si are used to make an electrical connection and are embedded into the dielectric layer [13]. Metal vias marked in red are used to provide a connection to the switch input and output and the gate electrode to wire bond pads as shown in Figure 6.5c. The cantilever beam is surface micromachined using the sacrificial layer to create the air gap underneath. Figure 6.5c shows the structure with important dimensions of the moving parts. The switch consists of a gold beam 7.2 μm thick with a contact gap of 0.3 μm and an actuation gap of 0.7 μm. (iii) The cantilever

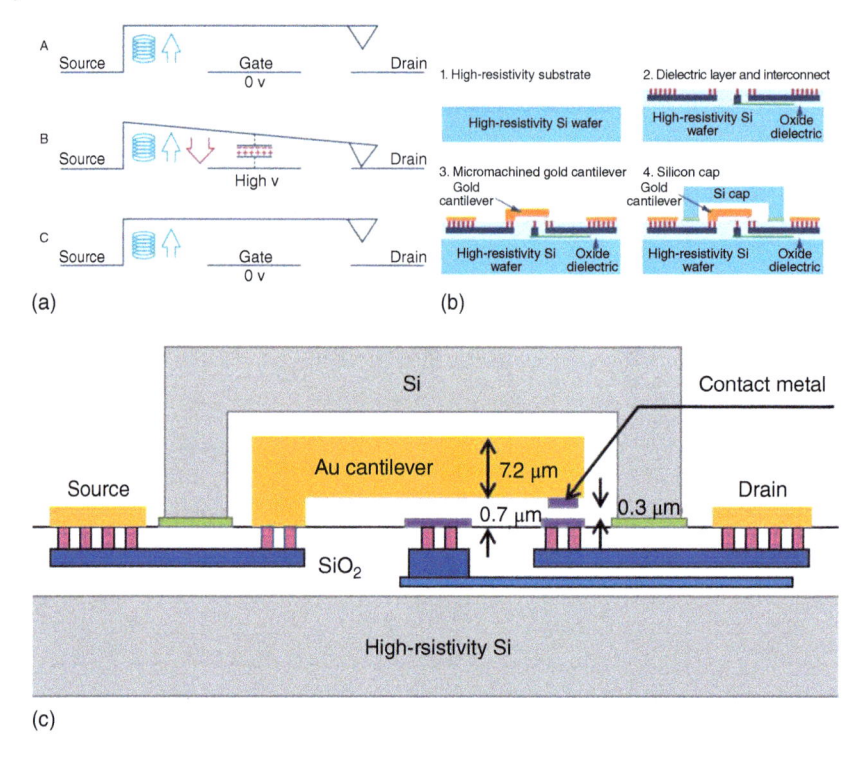

Figure 6.5 (a) Schematic drawing showing the MEMS switch operation. (b) Overview of MEMS switch fabrication. (c) Cross section of the MEMS switch die. Source: (a, b) Carty et al. [12]. © 2016, Analog Devices, Inc., (c) Modified from Ceccarelli et al. [13].

switch beam and the bond pads are made of gold. Switch contact and gate electrodes are formed using a low-resistance thin metal deposited on the dielectric surface. This metal material is not open. The metal should be hardwearing conductive material. According to the patent [14], examples are ruthenium or platinum for the contact.

Figure 6.6a shows the magnified cartoon of four MEMS switches in a single-pole four-throw multiplexer configuration. Each switch beam has five Ohmic contacts in parallel to reduce the resistance and increase the power handling capability when the switch is ON. Figure 6.6b shows SEM image of the MEMS switch capped by a high-resistivity Si chip to form a hermetic protective housing. This Si cap is shown in Figure 6.5c. The environmental robustness and the cycle lifetime of the switch increase. The switch die is on the left and the driving ASIC is on the right. Gold wire bonding is used to connect MEMS die to the metal lead frame, being encapsulated in the plastic package.

6.4.2 Digital Micromirror Device

The digital micromirror device (DMD) developed by Texas Instruments, Inc. is a unique MEMS device mainly used in the projection display. The digital operation means that the micromirror tilts to the mechanical limit in +/− directions having

(a) (b)

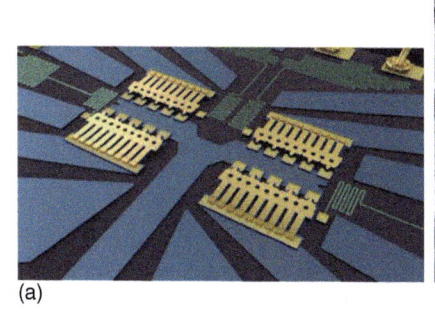

Figure 6.6 (a) Magnified cartoon of four MEMS switch cantilevers. (b) Capped MEMS switch and driving ASIC in the same plastic package. Source: (a) Carty et al. [12] Figure 4. ©, Analog Devices, Inc., (b) Ceccarelli et al. [13] Figure 2(b). © 2016, IEEE.

the maximum angle of about 10-12°. The other angle is not used. The angle control becomes open loop, making the operation simple. Figure 6.7a shows the schematic drawing of the device following 0 or 1 operation command [15]. The mirror at ON state reflects light to the project lens and the corresponding pixel becomes bright in the projected image. The mirror at OFF state reflects light to the light dump and the pixel becomes dark in the image. The reflected light from the micromirror gives white or black information with the high contrast. The gray scale is realized by switching white and black pixel information faster than the response time of the human eye (about 60 Hz). When the white and black ratios are both 50%, the human recognizes its pixel as 50% gray brightness. This time-sequenced control gives the advantage of the accurate brightness control since the timing signal and the duty-ratio control give high accuracy of the relative brightness. The response time for tilting the micromirror is <5 µs, giving enough margin for switching three colored (red, green, and blue) lights within 1/60 s for making one flame image. So, although the actual pixel information is digital 0 or 1 of single color, the time-averaged information of the mirror array is integrated in the human brain to give the full-color image.

The aforementioned device operation means that the micromirror has to move landings to the base electrode all the time even when the corresponding pixel expresses the same color continuously. This will require severe lifetime guarantee against the moving parts. Figure 6.7b shows the see-through drawing of the micromirror metal structure layers on CMOS static random access memory circuit. Since the single micromirror element is small, having an area of about $10 \times 10 \, \mu m^2$, the bending of the mirror plane does not become the problem. The high fill factor of the arrayed micromirror plates is realized, since the underlying microactuator can be hidden.

Figure 6.8 shows the micromirrors observed from the side after the long-time operation removing the bias driving voltage. The micromirror surfaces have different inclinations since the operation tilting command has clustered to one side. This built-in angle is called as the hinge memory. This angle requires the additional driving voltage for tilting the mirror to the opposite side. If this angle becomes too large,

106 | *6 Metal Surface Micromachining*

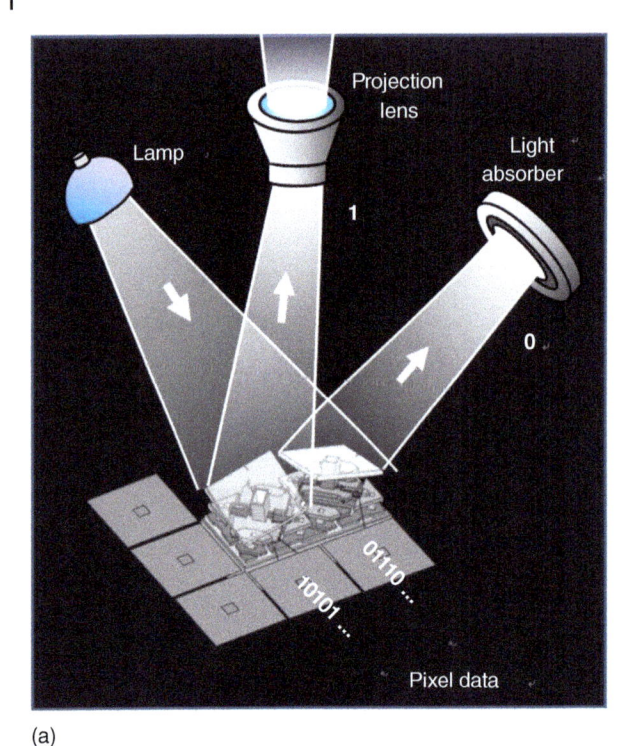

Figure 6.7 (a) OFF and ON positions of the micromirrors during their operation. (b) See-through DMD superstructure. Source: (a) Based on Gong and Hogan [15], (b) Sontheimer [16]. © 2002, IEEE.

the mirror will no longer be able to tilt to the opposite side, resulting in a failure. Surprisingly, the hinge memory in Figure 6.8 is not the failure level but the allowable one. Although this wide allowable margin realized by the optical setup is the wisdom supporting the DMD, the hinge memory is the largest hurdle limiting the device lifetime. Remembering that the ductile Al is used as the structure material,

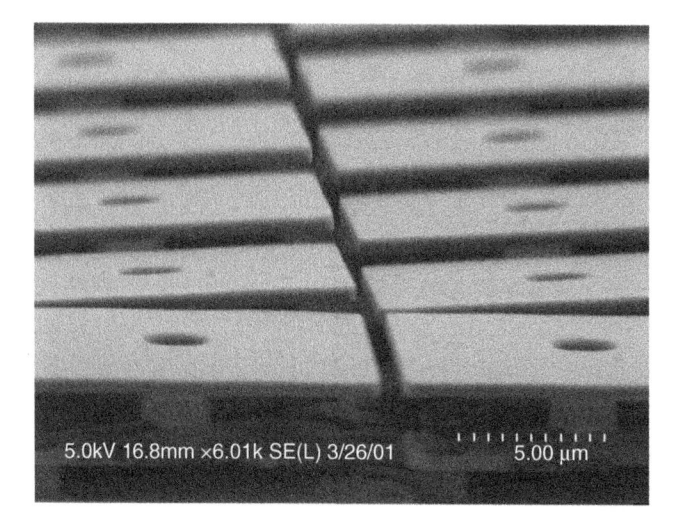

Figure 6.8 Micromirrors exhibiting the hinge memory. The first row at the bottom of the image is in the normal flat unbiased state. The second and subsequent rows are tilted to the minus side. Source: Sontheimer [16]. Figure 2. © 2002, IEEE.

Table 6.1 Mechanical properties of bulk Al alloys and $TiAl_3$-O film reported.

	Tensile strength	Notes
Pure Al	40–50 MPa (With work hardening: ~100 MPa)	FCC crystal structure
Al with 3% Si	120–130 MPa	40–50 MPa of 0.2% proof stress
Extra super duralumin including Zn	580 MPa	Used in airplane in the 1940s
Ti-6Al-4V alloy	1000–1200 MPa	$\alpha+\beta$ type
	Strain 6–11%	Used in airplane at present
$TiAl_3$-O used in hinge	3200 MPa	Derived from Figure 6.9 [17]
	Strain 2%	

occurrence of such deformation problem is quite natural. Al has face centered cubic (FCC) crystal structure. Its deformation is generally easy having the slipping feature of atoms. At the same time, the mechanical characteristics of Al is known to change when the alloy elements are added. The work hardening is also known to change the performance as listed in Table 6.1.

In 1992, Texas Instruments, Inc. started the Digital Imaging Venture Project for evaluating the DMD in the market [17]. At that state, DMD lifetime was about 100 h operating at 65 °C. The device structure was originally made of low-alloy Al (Al-1%Si-0.2%Ti) for both the mirror and the hinges. This material is for the electrical connection and not for the mechanical element. The performance was poor in low-yield stress, has small linear elastic region on the stress-strain curve, no fatigue limit, and a relatively large grain size (100-500 nm) compared to hinge width, creep, and so on. The new hinge material was searched for solving the problems. This search started from the metals used in the semiconductor manufacturing, since the hinge material has to be integrated into the existing DMD fabrication process with the high yields. The materials tried include pure Ti, W alloy with about 7.5wt%Ti, AlN, Al-2%Cu, and Al/Al_2O_3 composites.

TiAl alloy is the intermetallic compound, whose regular atomic structure gives the higher creep resistance. The heat resistance is expected to improve from the nature of Ti. From the advantage of being light weight (specific gravities are 4.5 for Ti and 2.7 for Al), TiAl alloy has been applied to a turbine wheel in an automotive turbocharger and a turbine blade in an aero jet engine [18], improving mechanical loss and also combustion efficiency. On the other hand, TiAl has several inherent weak points such as brittleness, low oxidation resistance, and manufacturing difficulties. As for the bulk material, these problems are overcome by adding alloying elements with adequate balance, microstructure control, and some innovative processes.

The hinge material candidate moved to amorphous $TiAl_3$ deposited by the sputtering. The quality of the amorphous phase was reported to influence the performance. When the film was deposited in the fixed wafer system in the sputtering machine, the inside cell structure with about 50 nm size was produced. This cell is generated because the island or 3D growth is produced where a second layer of atoms begin to build before the first layer is completely populated. The cellular film showed the worse mechanical properties compared to the pure amorphous one. Although the cell is not the clear grain structure, the cell boundaries in nanostructured materials are considered to be similar to grain boundaries in the conventional materials and may be responsible for creep resistance and possibly hinge memory.

To improve the hinge memory, different combinations of nitrogen and oxygen were added to the $TiAl_3$ during its growth. Incorporation of about 4% oxygen in the amorphous $TiAl_3$-O was adopted for the hinge. Figure 6.9 shows the stress-strain curve of 50 nm thick film. The curve is much different from that of the ductile material. The curves show a large elastic range until the microfracturing initiation at the strain of about 2%. The lateral axis includes the strain levels occurred in the DMD hinge and the spring tip. They are well below 2%. The offset of the vertical axis may relate the stress relaxation. The stress relaxation at 0.5% strain was measured for films produced with various gas combinations. The amorphous $TiAl_3$-O was found to produce the lowest stress relaxation. As an additional contribution of amorphous $TiAl_3$-O, this alloy forms spring tips on the sides of the yoke as shown in Figure 6.10. These springs, being newly introduced, allow the gentle bounce of the yoke, helping the release from the touched surface. This reduces the problem caused by the contact adhesion, which disables the pixel permanently. As listed in Table 6.1, the inclination of the stress-strain curve in Figure 6.9 gives the elastic modulus of about 160 GPa, which is surprisingly the same level with that of the crystal Si. The aforementioned hinge alloy improvement progressed from 1994 to 1996. $TiAl_3$-O hinge,

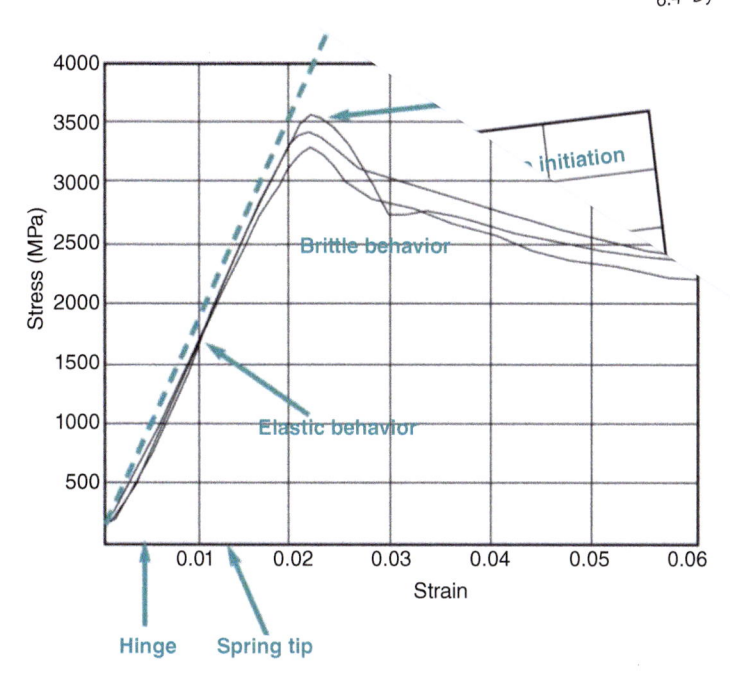

Figure 6.9 Stress-strain curve of 50 nm thick amorphous TiAl$_3$–O film deposited on a polyimide substrate. Source: Tregilgas [17]. Reprinted with permission of ASM International. All rights reserved. www.asminternational.org.

Figure 6.10 Hidden structure below the reflection mirror plate. H-shaped yoke tilts and lands to the base of its edge. TiAl$_3$–N is used for H-shaped yoke. TiAl$_3$–O is used for hinges and spring tips. Source: Van Kessel et al. [19]. Figure 11. © 1998, IEEE.

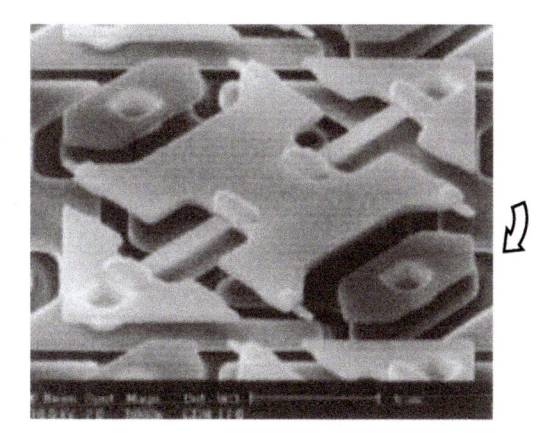

found as the higher-strength material with fewer slip systems, was used in production continuously at least until 2005 [17]. The lifetime of DMD extends to more than 100 000 h which is much longer than the target of 5000 h.

Figure 6.11 shows the fabrication sequence of the DMD superstructure using the aforementioned hinge material. The process starts from the completed CMOS static random access memory circuit having two metal layers. A thick SiO$_2$ is deposited

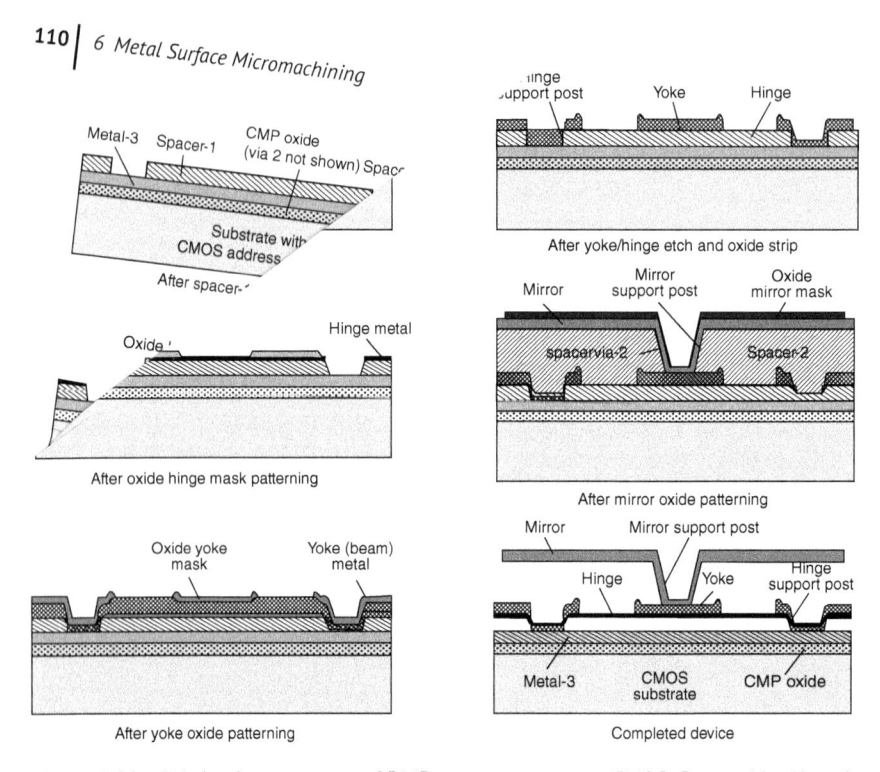

Figure 6.11 Fabrication sequence of DMD superstructure on CMOS. Source: Van Kessel et al. [19]. © 1998, IEEE.

over Metal-2 and then planarized using a chemical mechanical polish. The superstructure process begins with deposition and patterning of Metal-3 layer. An organic sacrificial layer (Spacer-1) of the photoresist is spin coated and hardened. This spacer defines the gap for tilting the yoke as shown in Figure 6.10. The hinge metal (TiAl$_3$–O) and the thicker yoke metal (TiAl$_3$–N) are sputter deposited covering the sidewalls of the spacer. Hinges are typically about 1000 nm wide, about 4000 nm long, and 50–70 nm thick. These TiAl$_3$ layers are etched using the plasma-deposited SiO$_2$ masks, which are patterned in advance. The etching process is the single step having the benefit that the hinge metal becomes continuous everywhere under the yoke metal layer. A second organic sacrificial layer (Spacer-2) is spin coated, patterned, and hardened. The aluminum layer is sputter-deposited over Spacer-2 to form the mirror planes. This patterning also uses the oxide etch mask in the same manner as that for the yoke layer. An air gap under the mirror is formed by plasma etching of the photoresist sacrificial layers. After this, the movable DMD superstructure becomes delicate against the inside adhesion and the particles from outside. A thin, self-limiting, anti-stick layer is deposited to make the surface energy of the contacting parts lower. Similar passivation is repeated after the bond wire attachment in their individual packages. Then, a lid with the optical window is welded to the package to ensure a clean environment for the DMD superstructure.

6.5 Summary

Remembering that the basic surface micromachining uses the standard CMOS materials of poly Si layers as the structure material, further development for introducing the metal material is as reasonable as the technical direction. In 1991, the accelerometer ADXL50 was released by Analog Devices, Inc. The following year, Texas Instruments, Inc. started the Digital Imaging Venture Project. The standard CMOS circuit has metal lines for connecting the transistors and the gathered units. The variety of structures can be built by changing the mask design. As seen from the fact that there are commercially available devices prepared by the metal surface micromachining, some metal structures have been found to allow the mechanical movement and the contact of the MEMS device, solving the problems relating with the creep and the adhesion. This extracts the MEMS-based advantages in the higher level, such as the major changes of conductivity or reflected light intensity, realizing the performances which cannot be obtained by methods based on electrical or material modulations.

References

1 Riedel, B. (1993). Surface-machined monolithic accelerometer. *Analog Dialogue* 27 (2): 3–7.
2 Ohsaki, A. (2003). Trends in multilevel interconnections for ULSI devices. *OYO BUTURI* 72 (9): 1162–1166, and "OYO BUTURI" cover page.
3 Kawasaki, H. (1999). Electromigration phenomena in ULSI multilevel metallization. *OYO BUTURI* 68 (11): 1226–1236, and "OYO BUTURI" cover page.
4 Hirata, Y., Inagaki, S., Chiba, Y. et al. (2012). Commercialization of ultra micro fabrication using electroplating. *J. Jpn. Soc. Precis. Eng.* 78 (12): 1025–1029.
5 Yoon, J.-B., Kim, B.-I., Choi, Y.-S., and Yoon, E. (2002). 3-D lithography and metal surface micromachining for RF and microwave MEMS. Proceedings of the IEEE International Conference on Micro Electro Mechanical Systems, 673-676.
6 Purkl, F., English, T., Yama, G. et al. (2013). Sub-10 nanometer uncooled platinum bolometers via plasma enhanced atomic layer deposition. Proceedings of the IEEE International Conference on Micro Electro Mechanical Systems, 185-188.
7 Aksyuk, V.A., Pardo, F., Carr, D. et al. (2003). Beam-steering micromirrors for large optical cross-connects. *J. Lightwave Technol.* 21 (3): 634–642.
8 Aksyuk, V.A. and Bishop, D.J. (1999). Self-assembling micro mechanical device. US Patent 5,994,159, Lucent Technologies, Inc.
9 Aksyuk, V.A., Simon, M.E., Pardo, F., and Arney, S. (2002). Optical MEMS design for telecommunications applications. Solid-State Sensor, Actuator and Microsystems Workshop Hilton Head Island (2-6 June 2002), 1–6.
10 Yoon, Y.-H., Han, C.-H., Lee, J.-S., and Yoon, J.-B. (2018). A proactive plastic deformation method for fine-tuning of metal-based MEMS devices after fabrication. *J. Microelectromech. Syst.* 27 (6): 1124–1134.

11 Rebeiz, G.M. (2003). RF MEMS switches: status of the technology. Proceedings of the 12th International Conference on Solid State Sensors, Actuators and Microsystems, Boston (8-12 June 2003), 1726-1729.

12 Carty, E., Fitzgerald, P., and McDaid, P. The Fundamentals of Analog Devices' Revolutionary MEMS Switch Technology. Analog Devices Technical Article. https://www.analog.com/en/technical-articles/fundamentals-adi-revolutionary-mems-switch-technology.html#.

13 Ceccarelli, E.M., Heffernan, C., Browne, J., and Fitzgerald, P. (2016). Intrinsic reliability characterization for stand-alone MEMS switch technology. Proceedings of IEEE International Integrated Reliability Workshop, 80-82.

14 Macnamara, J.G., Fitzgerald, P.L., Goggin, R.C., and Stenson, B.P. (2018). MEMS switch device and method of fabrication. US Patent 9,911,563 B2.

15 Gong, C. and Hogan, T. (2014). CMOS compatible fabrication processes for the digital micromirror device. *IEEE J. Elect. Devices Soc.* 2 (3): 27–32.

16 Sontheimer, A.B. (2002). Digital micromirror device (DMD) hinge memory lifetime reliability modeling. Proceedings of IEEE International Reliability Physics Symposium, 118-121.

17 Tregilgas, J. (2005). Amorphous hinge material. *Adv. Mater. Process.* 163 (1): 46–49.

18 Koyanagi, Y. (2017). Technology evolution for commercial expansion of TiAl alloys as a light weight heat resistant material. *Senkiseiko* 88 (2): 77–84. (in Japanese).

19 Van Kessel, P.F., Hornbeck, L.J., Meier, R.E., and Douglass, M.R. (1998). A MEMS-based projection display. *Proc. IEEE* 86 (8): 1687–1704.

7

Heterogeneously Integrated Aluminum Nitride MEMS Resonators and Filters

Enes Calayir[1], Srinivas Merugu[2], Jaewung Lee[2], Navab Singh[2], and Gianluca Piazza[1]

[1]*Carnegie Mellon University (CMU), Department of Electrical and Computer Engineering, 5000 Forbes Avenue, Pittsburgh, PA 15213, USA*
[2]*A*STAR Institute of Microelectronics (IME), Singapore, 11, Science Park Road, Singapore Science Park II, Singapore 117685, Singapore*

7.1 Overview of Integrated Aluminum Nitride MEMS

With the introduction of thin-film aluminum nitride (AlN) and especially the commercial success of the thin-film bulk acoustic resonator (TFBAR) or film bulk acoustic resonator (FBAR), interest in piezoelectrics for micro electro-mechanical systems (MEMS) blossomed. The investments in the development of repeatable physical vapor deposition techniques for the growth of AlN films on silicon have spurred a great deal of activities in the field of piezoelectrics and especially AlN MEMS. Practically, most conventional MEMS devices that used to be made out of silicon have recently been reproduced (in most cases with enhanced performance) by using AlN thin-film piezoelectric technology. For example, resonators [1–4], filters [5–7], switches [8–10], energy harvesters [11–13], ultrasonic transducers [14, 15], microphones [16, 17], strain sensors [18], chemical sensors [19], and accelerometers [20] have been demonstrated using AlN thin films.

AlN MEMS also stands out as a preferred candidate for the synthesis of integrated MEMS–CMOS silicon chips, which will enable the More-than-Moore vision. In this vision in which MEMS and circuits coexist on the same chip or are stacked to form a single component, AlN MEMS would be used pervasively to interface with the analog world and perform sensing and actuation, signal processing, and computing.

Various approaches have been used to integrate AlN MEMS with complementary metal oxide semiconductor (CMOS) electronics [21–27]. Each approach has its own unique advantages and drawbacks. We can classify CMOS–MEMS integration methods mainly as monolithic integration and hybrid (heterogeneous) integration in terms of the number of substrates used in the final integrated technology [28]. In the case of monolithic integration, a single substrate is involved and both technologies are fabricated on a single wafer. In the case of hybrid integration, two or more substrates form the final chip stack. Each technology is processed individually

3D and Circuit Integration of MEMS, First Edition. Edited by Masayoshi Esashi.
© 2021 WILEY-VCH GmbH. Published 2021 by WILEY-VCH GmbH.

up to a certain step, and the multiple substrates are then integrated via bonding. Based on the targeted application, any of these approaches could be adopted [29, 30]. Hybrid integration is advantageous when the two technologies need to be developed, modified, and upgraded independently from one another. This feature does not only decrease the fabrication complexity, development time, and cost of integration, but also provides more flexibility in choosing or advancing the technology of each chip in the stack [31]. Furthermore, when a size mismatch exists between the CMOS and the MEMS technology, as in the case of the specific applications presented herein, hybrid integration is more economical as it uses more efficiently the area of both substrates.

In this chapter we present a method of integrating a specific class of AlN MEMS resonators and filters [32] with CMOS circuits using flip-chip bonding. We describe the wafer level processing steps required to build a robust AlN MEMS platform, which includes device packaging and redistribution layers (RDLs). These wafer level steps are followed by singulation of the MEMS devices into individual chips, which are then flip-chip bonded to an advanced CMOS node. Although the example reported herein is specific to a particular class of AlN MEMS resonators and a demonstration vehicle based on self-healing filters, the 3D heterogeneous integration process is broadly applicable to any other AlN MEMS technology.

7.2 Heterogeneous Integration of Aluminum Nitride MEMS Resonators with CMOS Circuits

Heterogeneous integration of AlN MEMS resonators with CMOS electronics comes with various challenges, which are specific to the target application. In general, when a large array of MEMS devices are integrated with CMOS electronics, a size mismatch between the two chips exist, which means that routing through the CMOS layers is impractical and uneconomical [33]. Therefore a RDL should be developed on the MEMS chip. In order to enable such RDL layer, also the MEMS device encapsulation should occur at the wafer level. Thin-film encapsulation (TFE) is important to protect the AlN MEMS device against environmentals and during the flip-chip bonding step [34].

In this chapter we report an overview of the process flow developed on an 8" Si wafer in collaboration with A*STAR Institute of Microelectronics (IME), Singapore, for the making of an AlN MEMS resonators and filters, a TFE process, an RDL, and a bumping (performed by Tag and Label Manufacturers Institute) and flip-chip bonding to a 28 nm CMOS chip from Samsung. The cross-sectional overview of the developed AlN MEMS platform with 3D hybrid integration to CMOS is shown in Figure 7.1. The development of this platform can be divided into four main stages: (i) fabrication of MEMS devices (i.e. AlN resonators and filters), (ii) TFE of the same, (iii) fabrication of RDL on MEMS for efficient and low-loss signal routing, and (iv) solder bumping of CMOS bonding pads and flip-chip bonding of the same to the AlN MEMS chip. In the following subsections, we provide a description of the unique challenges that were overcome to develop all these different processes to enable

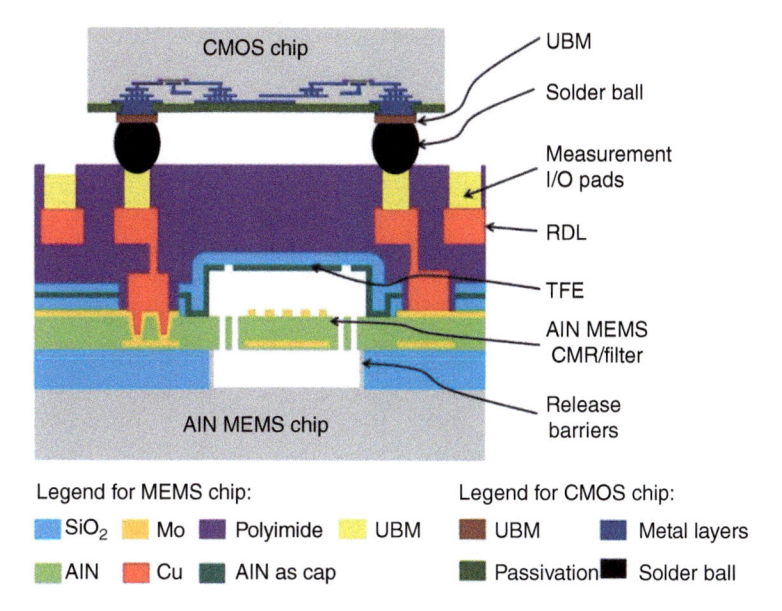

Figure 7.1 Cross-sectional cartoon view of the developed AlN MEMS platform with its 3D hybrid integration to a CMOS chip.

high performance 3D heterogeneously integrated resonators and filters. The wafer level fabrication steps executed at A*STAR IME are shown in Figure 7.2 and will be described in details in the following subsections. The CMOS bumping and flip-chip bonding steps will be shown and described separately in the following subsections.

7.2.1 Aluminum Nitride MEMS Process Flow

The MEMS fabrication starts with the deposition of a 3.5 μm thick plasma enhanced chemical vapor deposition (PECVD) SiO_2 on an 8" standard high-resistivity (HR) Si wafer. This oxide layer is later used as sacrificial material to release the device from the substrate. In order to isolate the released area, Si barriers of 2 μm width are defined inside the SiO_2 layer. This is done by deep trench etching of the oxide layer followed by poly Si filling using a low-pressure chemical vapor deposition process. Then, excess Si is removed with a chemical mechanical planarization (CMP) step, which also minimizes the surface roughness of the oxide layer (i.e. Step 1 in Figure 7.2). Next, the first layer of molybdenum (Mo) in the amount of 150 nm is deposited on a 20 nm AlN seed layer and then patterned to be used as bottom metal plate for the AlN MEMS resonators (i.e. Step 2 in Figure 7.2). After the formation of the bottom metal electrode, a 1 μm thick piezoelectric AlN is deposited to serve as the device layer (i.e. Step 3 in Figure 7.2). Via holes to provide electrical connection between top and bottom Mo layers are defined in AlN with a Cl_2-based etch process. The second Mo layer of 150 nm is then deposited and patterned to define a set of top interdigitated electrodes for driving the resonators. The vias in the AlN layer are also filled with Mo (i.e. Step 4 in Figure 7.2). Note that this Mo layer is also used

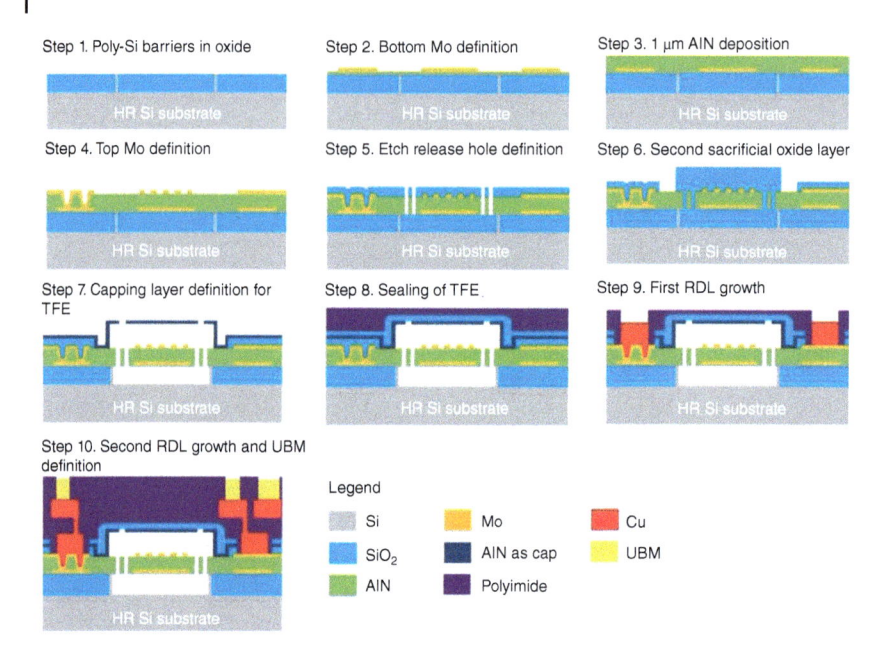

Figure 7.2 The fabrication process flow of AlN MEMS platform developed at A*STAR IME, Singapore (TFE, thin-film encapsulation, UBM, under bump metallization, and RDL, redistribution layer).

to interconnect resonators to form the filters, which consist of three series cascaded AlN MEMS resonators in this work. Then, the release holes are defined in the AlN device layer to be able to release the bottom sacrificial material placed underneath the resonator body (i.e. Step 5 in Figure 7.2). The MEMS fabrication completes at this step as the sacrificial material is released after the encapsulation of the AlN MEMS devices.

7.2.2 Encapsulation of Aluminum Nitride MEMS Resonators and Filters

In order to protect the AlN-suspended structures during the flip-chip bonding process and provide hermetic sealing to environmental conditions, it is of paramount importance to encapsulate the AlN MEMS devices. The main goal is to develop a cost-effective wafer-level encapsulation process for large-scale integration (LSI) of AlN MEMS resonators.

The TFE of AlN MEMS contains two major layers, which serve two distinct purposes. The first one is a thin layer that caps the entire moving resonator body. A second layer is used to hermetically seal the holes in the cap after the release. This layer also includes thick polyimide to provide for additional structural rigidity and an overall smoothened topography across the wafer.

In order to make the capping layer, first a layer of SiO_2 is deposited on the MEMS device to form the top sacrificial layer and ultimately a gap between the cap and the suspended resonator body. After that, the SiO_2 layer is etched to define the anchor for the cap (i.e. Step 6 in Figure 7.2). These anchors lie outside the release barriers defined in the AlN MEMS process in order to ensure sturdiness in the encapsulation after sacrificial material release. At the same time this layer defines the release area on top of the resonator. In other words, the anchor surrounds the sacrificial release layer on top of the resonator body, and the trenches forming the anchor act as etch stop barriers for the top oxide. After the formation of the capping layer (mostly made out of AlN with some other A*STAR IME-proprietary interface layers) is completed with the definition of the capping release holes, all the devices on the wafer are simultaneously released by dry vapor hydrofluoric acid (HF) (i.e. Step 7 in Figure 7.2). With the etch stop barriers surrounding top and bottom sacrificial material, devices with different dimensions can be fully released without the concern of undercutting any area under the pads and interconnects. From a design standpoint, this is a very important aspect since the effective anchor of the resonator can be accurately defined and devices can be densely packed. After the release, the wafer is coated with SiO_2 to seal the openings in the capping layer formed for sacrificial material release. The release holes in the capping layer are placed in such a way that any material that might deposit inside the capping layer during the sealing process would go in regions, which do not impact the device performance. In order to restrict the amount of dielectric over such areas, these holes are designed to be quite small so that they are rapidly sealed during the deposition of SiO_2. Since the top Mo is now below a stack of dielectric materials, vias are defined through the seal and the cap layers at the locations where interconnects need to be built. Then, a layer of photo-definable packaging polyimide is spin coated and cured on the sealed MEMS wafer with a final thickness of 5–7 µm. The commercial grade wafer-level coating of polyimide is used to enhance the mechanical and chemical robustness of the overall MEMS wafer. With the deposition of the polyimide, the TFE process completes (i.e. Step 8 in Figure 7.2).

The main challenge in the development of TFE for AlN Contour Mode Resonators (CMRs) was to ensure the control of the stress levels in the thin films to minimize bending of the capping layer. The level to which the stress can be controlled poses limitations on the device size and aspect ratio and effectively constrains the device optimization. The materials for the cap and seal also have to be compatible with the device operation at radio frequencies and should minimize signal feedthrough [35].

Several sources of failures were encountered during the development of TFE, such as insufficient release of sacrificial material, breaking of thin films, downward buckling of the capping layer, and/or upward buckling of device layer AlN, causing the cap to touch the resonator body. To address these challenges, the etch release holes in both the AlN device layer and capping layer were made as small as possible and distributed as uniformly as possible without interfering with the resonator active area. Also, these two sets of etch release holes should not be overlapped in order to ensure minimal topography in the capping layer, and thus higher structural rigidity. Additionally, the thickness of the sacrificial material should be optimized to minimize residual stresses in the TFE layer and also ensure it can be fully etched during the release step. In order to set the thicknesses of the sacrificial layer and material stack forming the cap, a design of experiment (DoE) was performed.

The results of this DoE highlighted that in order to increase the resilience of the capping layer in TFE and enhance overall device yield, a material stack instead of only AlN is needed to be deposited (information on these layers and their thicknesses are proprietary of A*STAR IME). Another outcome of the study was that the sacrificial oxide layer should be made at least 3 μm thick to ensure that the device can be properly released and any residual bending does not cause contact of the capping layer with the resonator.

The same DoE also suggested some additional constraints on the maximum length the capping layer could span (<110 μm), and the minimum width the anchor of the capping layer should be 10 μm. Clearly, these two constraints put a limitation on the maximum device size, which can be encapsulated as well as the maximum device density, which can be achieved. Some SEM images of a successfully developed TFE process for AlN MEMS resonators and filters are shown in Figure 7.3.

7.2.3 Redistribution Layers on Top of Encapsulated Aluminum Nitride MEMS

The RDL development is guided by the goal of synthesizing low-loss and low parasitics metallization layers for signal interconnects and routing on the AlN MEMS platform with limited additional fabrication complexity. This approach is helpful in utilizing the full potential of AlN MEMS resonators as building blocks in RF front-end applications, as well as efficient use of the CMOS chips. In addition to the universal bump metallization (UBM) layer, RDL includes two copper (Cu) metal layers to ensure adequate flexibility in routing for the 3D integrated systems.

The RDL fabrication starts with via openings in the first polyimide layer, which is deposited as the last step of the TFE fabrication. These via openings provide the electrical connection between the first redistribution metal layer and top Mo. Then, a 3 μm thick Cu is deposited and patterned as first signal rerouting layer for the AlN MEMS platform (i.e. Step 9 in Figure 7.2). Next, another set of polyimide and Cu is deposited and patterned to form the second signal routing layer. After also covering this second Cu layer with polyimide, pad openings are defined and a standard 3 μm thick UBM formed by a stack Cu/Ni/Au is deposited and patterned with lift-off process to facilitate flip chip solder-bump bonding from these MEMS chips to the CMOS chips. The polyimide layers in each step provide natural planarization for the platform and smooth the overall wafer topography. After dicing the wafer into the individual MEMS dies, the full AlN MEMS fabrication process completes (i.e. Step 10 in Figure 7.2).

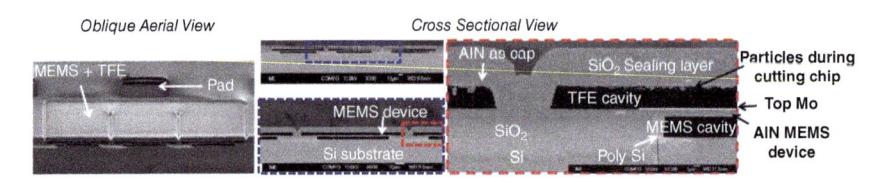

Figure 7.3 Oblique aerial and cross-sectional SEM images of TFE for AlN MEMS resonators.

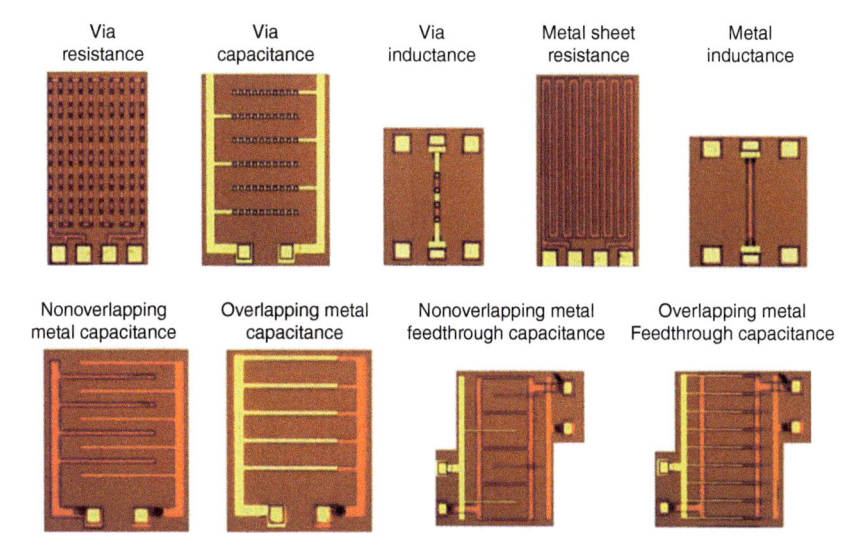

Figure 7.4 Optical images of a subset of test structures used in the characterization and modeling of RDL fabrication steps.

The development of RDL mostly consisted in setting the appropriate thicknesses for the metal layers and the inter-metal dielectrics with the goal of minimizing the resistive losses and capacitive and inductive parasitics associated with routing of arrayed devices and circuits. Minimization of the fabrication complexity is also important in setting of the layer thicknesses. In this respect, a combination of finite element analysis and DoE were used to drive the layer thickness selection. Different test structures were designed to model and extract overlapping, nonoverlapping and feed-thru capacitances as well as sheet resistances and inductances of all the metal layers and inter-metal vias (see Figure 7.4). In order to de-embed any contact resistance in the test structures and improve modeling accuracy, the I/O pads of resistive structures were designed to be compatible with four-point probe measurement techniques as suggested in Refs [36, 37]. For test structures that model capacitive and inductive behaviors of RDLs, three different lengths of the same shape were designed in order to make recursive least square fitting [38] to increase accuracy in the parameter extraction. Experimental results were in close agreement with finite element analysis (FEA) designs, which lead to the selection of 3 μm thick Cu routing metal layers and 5–7 μm thick polyimide layers. Since the polyimide layer provides natural planarization for the platform, its thickness varies around the metal strip lines where they overlap (see Figure 7.5).

7.2.4 Selected Individual Resonator and Filter Frequency Responses

A specific resonator geometry (Figure 7.6) was used to verify the impact of encapsulation and RDL on the performance of individual devices and filters. The admittance response of the single resonator and filter frequency S-parameter response

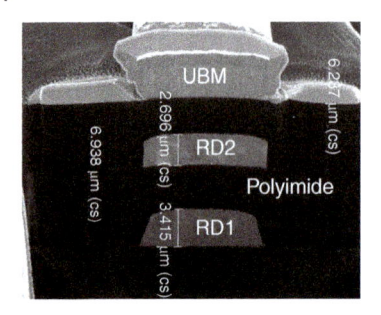

120 | *7 Heterogeneously Integrated Aluminum Nitride MEMS Resonators and Filters*

Figure 7.5 A cross-sectional SEM image of RDL fabrication after a laser FIB cut.

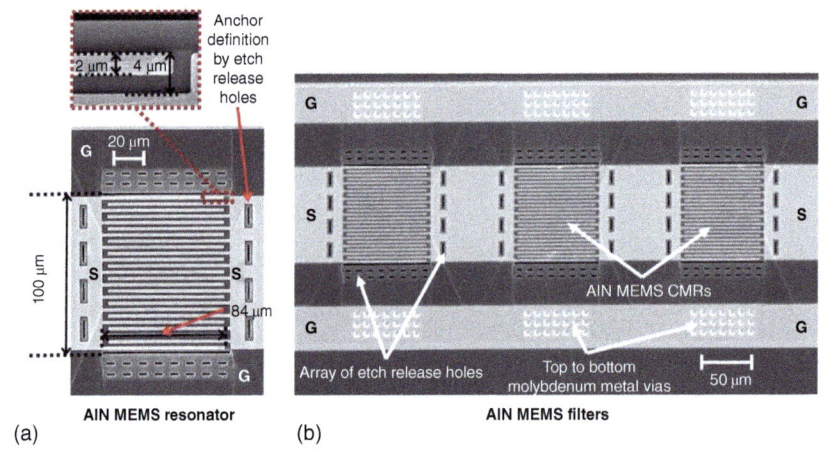

Figure 7.6 Aerial SEM image of (a) two port AlN MEMS resonator and (b) filter formed by three cascaded resonators. The images are taken from a die that went through the AlN MEMS only fabrication (Steps 1–5 in Figure 7.2).

comprised of three cascaded resonators are plotted in Figure 7.7 to compare the impact of TFE and RDL processes on the AlN MEMS device performance.

The tabulated performance comparison for AlN MEMS only devices versus encapsulated resonators and filters are provided in Table 7.1 and Table 7.2, respectively.

Table 7.1 Extracted performance comparison of AlN MEMS resonators after MEMS only and full fabrication processes.

Fabrication type	Extracted performance comparison of AlN MEMS CMRs			
	f_o (GHz)	k_t^2 [%]	Q	C_o (fF)
MEMS only process	1.164	1.68	1423	303
Full process (MEMS+TFE+RDL)	1.152	1.61	1850	311

Note that in the MEMS only there is no UBM formation. Four critical resonator parameters are needed to characterize the behavior of the AlN MEMS CMRs [39], which are f_o (resonant frequency), k_t^2 (electromechanical coupling coefficient), Q (quality factor), and C_o (electrostatic device capacitance).
Source: Based on Calayir et al. [39].

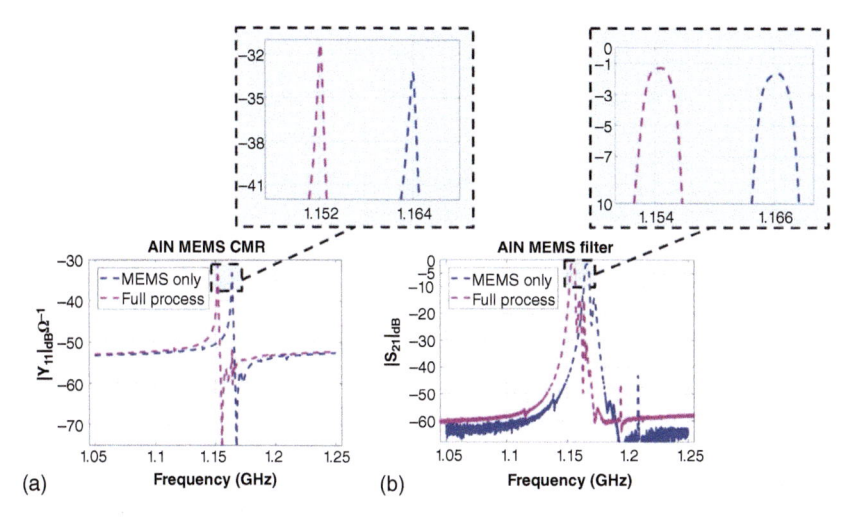

Figure 7.7 Selected (a) two-port resonator frequency response and (b) filter frequency response of three two-port resonators cascaded in series for both MEMS only fab and encapsulated fab runs. Tabulated performance comparison for MEMS only fab versus encapsulated fab is provided in Table 7.1 and Table 7.2, respectively, for the resonators and filters. Note that for the filter results, the s-parameter termination is set to 200 Ω in software to match the response.

Table 7.2 Extracted performance comparison of AlN MEMS filters after MEMS only and full fabrication processes.

Extracted performance comparison of AlN MEMS filters				
Fabrication type	f_o (GHz)	IL (dB)	BW (MHz)	OBR(dB)
MEMS only process	1.166	1.62	3.47	22.26
Full process (MEMS+TFE+RDL)	1.164	1.30	3.38	23.58

Note that in the MEMS only there is no UBM formation. Four parameters are defined here to characterize the filter behavior as in Ref. [40], which are f_o (pass-band center frequency), IL (insertion loss), BW (3-dB bandwidth of passband), and OBR (out of band-rejection at frequencies 1.5 BW away from the f_o).
Source: Based on Calayir et al. [40].

This comparison clearly reveals that the resonators and filter built with TFE and RDL steps (Figure 7.8) do not suffer any performance degradation due to the additional layers and processing. Actually, the filter losses reduce with the minimization of resistive losses in the interconnects.

7.2.5 Flip-chip Bonding of Aluminum Nitride MEMS with CMOS

This section reports on the steps associated with the 3D hybrid integration of a 2 × 2 mm AlN MEMS filter array chip fabricated using the process presented herein with a 1.35 × 1.35 mm CMOS chip fabricated in a 28 nm process line at Samsung,

7 Heterogeneously Integrated Aluminum Nitride MEMS Resonators and Filters

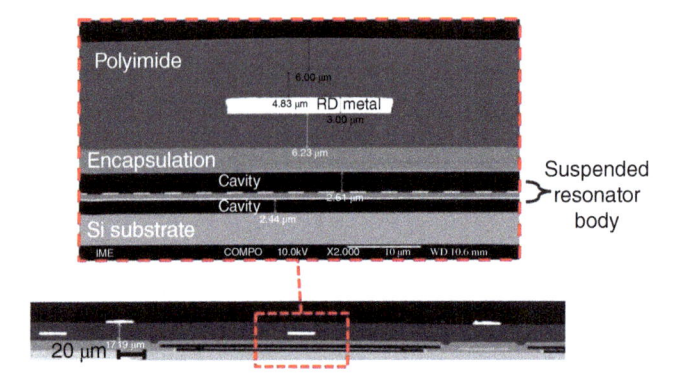

Figure 7.8 Cross-sectional SEM image of a device after full fabrication process. The image is taken from the side after focused-ion beam cut and mechanical polishing.

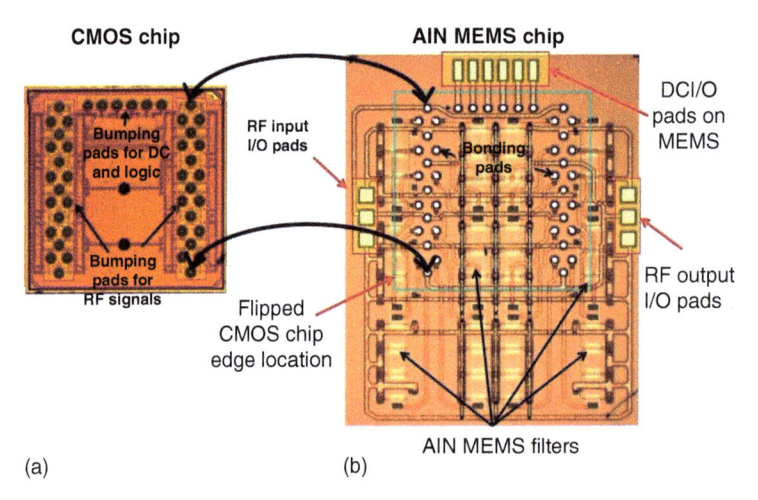

Figure 7.9 Optical microscope images of (a) the fabricated CMOS chip (28 nm Samsung technology with solder bumps) and (b) the designed AlN MEMS chip, which is meant to be flip-chip bonded to the CMOS chip.

South Korea. 50 μm diameter solder balls were placed on the CMOS chips by TLMI, Texas, United States. The optical microscope images of the fabricated standalone AlN MEMS and CMOS chips before the integration are shown in Figure 7.9. The final chip integration was done via flip-chip solder bump-bonding process at A*STAR IME, Singapore. The flip-chip bonding process included three steps:

1. Sipping solder balls in flux to eliminate any oxide layer on solder bumps and also to improve bonding between pads.
2. Pick and place MEMS chip on top of integrated circuit (IC) chip. This process is done using FC300 high-precision die/flip chip bonder, and it includes accurate alignment of the chips.

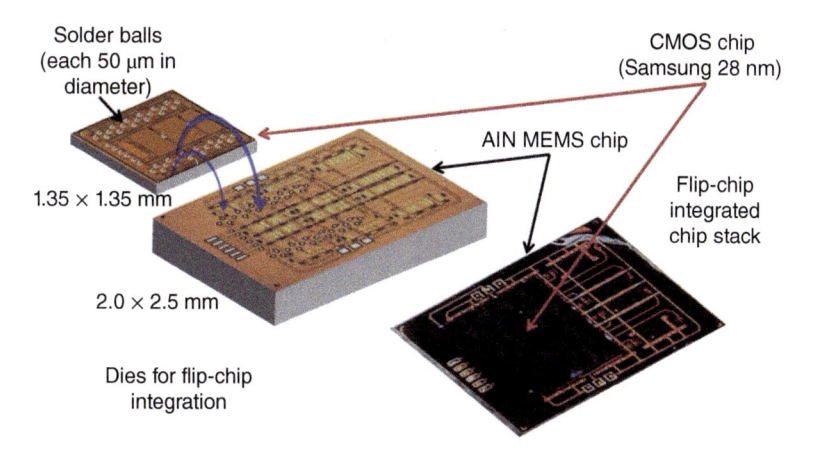

Figure 7.10 Rendering of the flip-chip integration process of an AlN MEMS chip with the CMOS die.

3. The dies are then kept in BTU's Pyramax@'s reflow oven so that the solder balls reflow on the pads and make strong bond holding the MEMS and IC chips together.

Figure 7.10 mimics the flip-chip bump-bonding process of an AlN MEMS chip and a CMOS die. Since RDL offers low-loss signal routing and flexibility in the size of AlN MEMS chip for I/O pad placement, the MEMS chip was made as the substrate chip where all the I/O pads are placed for probe landing and electrical testing. The necessary DC power and digital logic signals, as well as the interconnects between the CMOS components and AlN MEMS filters, are going from the MEMS to the CMOS through solder-bump bonding pads.

7.3 Heterogeneously Integrated Self-Healing Filters

In this section, we describe a proof-of-concept demonstration of an innovative self-healing filtering approach, which is uniquely enabled by the 3D hybrid integration of AlN MEMS with CMOS (Figure 7.11).

7.3.1 Application of Statistical Element Selection (SES) to AlN MEMS Filters with CMOS Circuits

The practical implementation of narrowband AlN MEMS filters is hindered by fabrication-induced process and mismatch variations [41].

Table 7.3 summarizes the die-level performance statistics of standalone AlN MEMS filters comprising three cascaded two-port AlN MEMS resonators, which were described in the previous sections. The preeminent filter variation is in f_o, although insertion loss (IL), bandwidth (BW), and out of band-rejection OBR) also vary due to resonator process-induced variations. Variations in f_o is more critical in narrowband filters because ±0.02% change in f_o corresponds to 13.9% of the

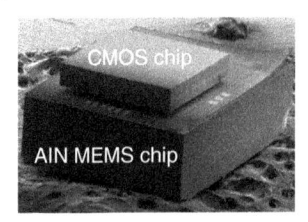

Figure 7.11 Flip-chip bonded AlN MEMS and CMOS chip stack.

Table 7.3 Measured statistics of encapsulated standalone AlN MEMS filters.

	f_o	IL	BW	OBR
Mean	1.152 GHz	1.27 dB	3.31 MHz	23.48 dB
STD as of % of mean	0.02%	3.27%	0.51%	0.60%

To get the statistics of AlN MEMS filters, 12 identically designed filters were placed within 2 × 2 mm chip area as 3 × 4 matrix. Each filter here comprises of three cascaded two-port AlN resonators with the geometry described in the previous section.

BW of these filters. These variations prevent the filters from achieving optimal performance and hinder the implementation of more complex circuits where arrays of these filters are used. Therefore, it is crucial to develop highly reliable and robust systems, which can tolerate these variations.

In order to address the challenge posed by intra-die variations, we borrow the concept of statistical element selection (SES) technique presented in Refs [41–43]. In order to apply the SES algorithm, we divide the ultimate desired filter into smaller versions (sub-filters formed by higher impedance/smaller resonators) and create a bank of them by adding identical redundant elements. Via series CMOS switches placed at the RF input and output of the AlN MEMS sub-filters, a subset k from the bank of N nominally identical sub-filter elements, is combinatorially selected and connected in parallel in order to construct a high-yield, self-healing filter. Figure 7.12 shows the conceptual circuit diagram of the self-healing filter array. For even a modest array size (N) and selection size (k), a large number of combinations is available, for example, $^{12}C_4 = 495$.

In order to illustrate the beneficial effect of SES on improving the statistical variations of f_o, we generate the probability density function (PDF) of a standalone filter versus application of SES with N of 12 and k of 4, hence providing 495 unique, selectable filtering components. In this comparison, a typical filter is designed to have f_o at 1.15 GHz and BW of 3.8 MHz. When we require the f_o of these filters to be within 100 kHz of the targeted value, it can be easily observed that the SES technique provides a dramatic increase in the yield with respect to that of a standalone filter (less than 36%), as illustrated by Figure 7.13.

7.3.2 Measurement of 3D Hybrid Integrated Chip Stack

We practically demonstrated the concept of SES by integrating an array of 12 identical AlN MEMS sub-filters with CMOS switches. The two chips were heterogeneously integrated according to the process steps described in the previous sections.

Figure 7.12 Conceptual circuit diagram for self-healing AlN MEMS filters using CMOS switches at the RF input and output. In this diagram k is 4 and N is 12.

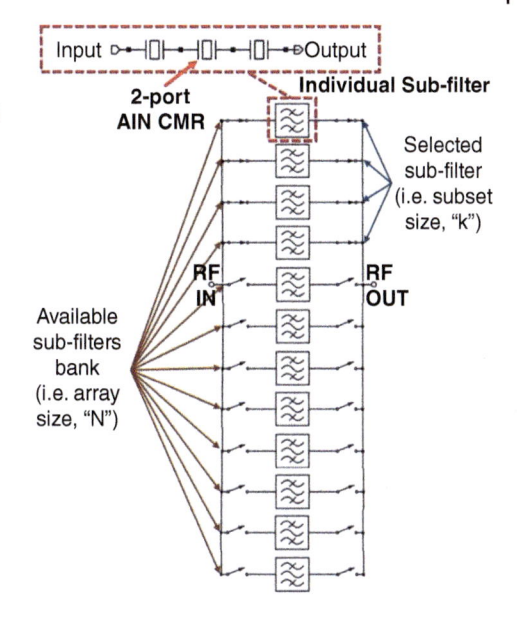

Figure 7.13 PDF of measured center frequency offset (Δf_o) of standalone filters versus simulated distribution for a self-healing filter. The normal fit in the plot is drawn for standalone filters in order to verify the frequency distribution of the sampled filter responses.

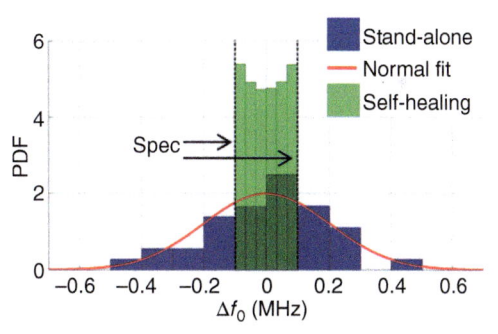

The responses of several possible combinations based on the measurements taken from one of the 3D integrated chip stacks are shown in Figure 7.14. An IL as low as 3.15 dB and an OBR of as high as 25.1 dB were achieved. In the same plot, we compare the response of the self-healing filters to that of stand-alone filters prior to integration with CMOS.

The additional 2 dB of IL in the self-healing filters with respect to that of a standalone filter comes from the CMOS switch in-series resistance (~10.7 Ω per switch and dominant term), the partial signal routing on CMOS, and the signal interconnect parasitics between the chips via the solder balls. The out-of-band performance of the integrated chips (i.e. the response at frequencies further away from the passband of the filters) could also be further improved and made to be closer to the individual standalone filter response by improving the quality of ground connections in the two chips.

Despite the reduced performance of the integrated filters, it was still possible to prove the concept of SES. The application of the SES technique on the experimental data taken from a 3D hybrid integrated chip stack is shown in Figure 7.15.

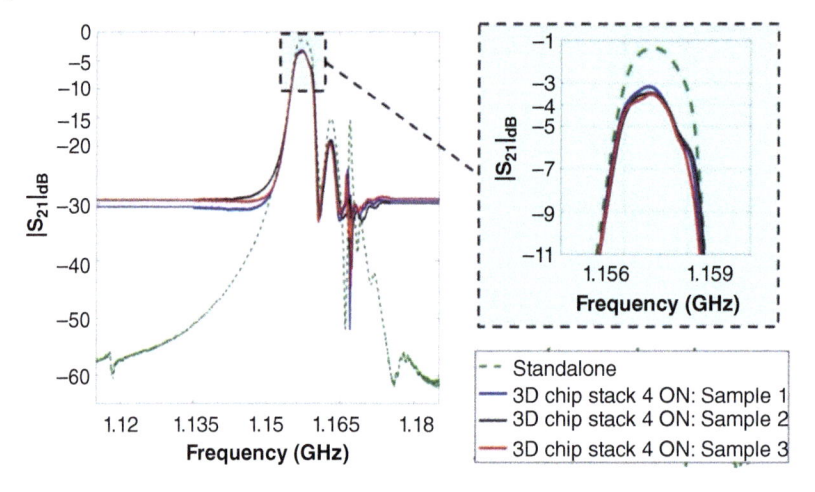

Figure 7.14 Matched frequency response of three possible self-healing filters (red, blue, and black) versus a standalone filter (dashed green).

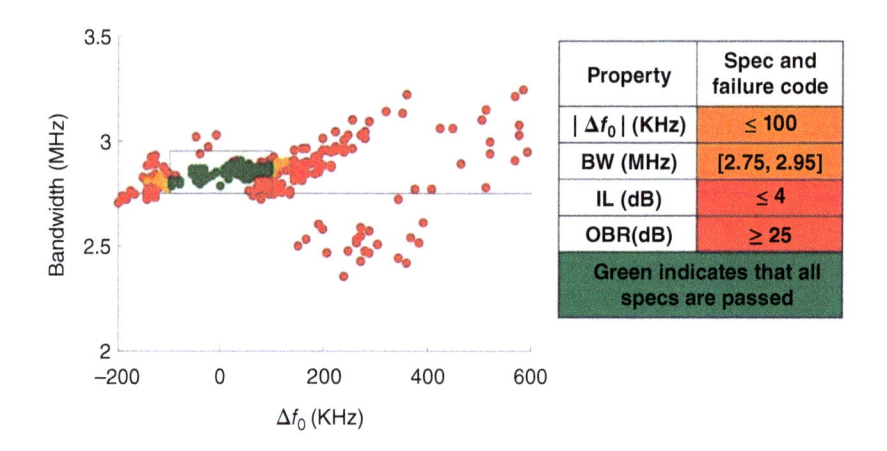

Figure 7.15 Application of SES algorithm on self-healing AlN MEMS filters with $N = 12$ and $k = 4$. Sub-filters on the AlN MEMS chip are combinatorially selected through CMOS switching matrix controlled by a chain of D flip-flops connected in series.

Considering the frequency responses of the self-healing filters, we set the parameter specifications on the filter performance as $\Delta f_o < 100$ kHz, $IL < 4$ dB, BW between 2.75 and 2.95 MHz, and $OBR > 25$ dB. In the figure, the data points marked as green represent the combination of sub-filters that yield a resulting filter, which meets all the desired specifications. The orange points indicate the filters that fail as regards to f_o and BW specifications but pass the IL and OBR metrics. The red points represent filters that fail as regards to either IL or OBR specifications.

In addition to the clear yield improvements, SES also offers other benefits as regards to tuning of some of the filter parameters. For example, in this specific case, we can see that a tuning range of 300 kHz for f_o and 250 kHz for BW can be achieved.

References

1. Piazza, G., Stephanou, P.J., and Pisano, A.P. (2006). Piezoelectric aluminum nitride vibrating contour-mode MEMS resonators. *J. Microelectromech. Syst.* 15: 1406–1418.
2. Ruby, R. C., Bradley, P., Oshmyansky, Y. et al. (2001). Thin film bulk wave acoustic resonators (FBAR) for wireless applications. 2001 IEEE Ultrasonics Symposium. Proceedings. An International Symposium (7–10 October 2001), 813–21.
3. Harrington, B.P. and Abdolvand, R. (2011). In-plane acoustic reflectors for reducing effective anchor loss in lateral-extensional MEMS resonators. *J. Micromech. Microeng.* 21: 085021 (11 pp.).
4. Bjurstrom, J., Katardjiev, I., and Yantchev, V. (2005). Lateral-field-excited thin-film Lamb wave resonator. *Appl. Phys. Lett.* 86: 154103 (3 pp.).
5. Chengjie, Z., Sinha, N., and Piazza, G. (2010). Very high frequency channel-select MEMS filters based on self-coupled piezoelectric AlN contour-mode resonators. *Sens. Actuators, A* 160: 132–140.
6. Rinaldi, M., Zuniga, C., Chengjie, Z., and Piazza, G. (2010). Super-high-frequency two-port AlN contour-mode resonators for RF applications. *IEEE Trans. Ultrason. Ferroelectr. Freq. Control* 57: 38–45.
7. Ruby, R., Bradley, P., Larson, J., III, et al. (2001). Ultra-miniature high-Q filters and duplexers using FBAR technology. 2001 IEEE International Solid-State Circuits Conference. Digest of Technical Papers (5–7 February 2001), 120–1.
8. Mahameed, R., Sinha, N., Pisani, M.B., and Piazza, G. (2008). Dual-beam actuation of piezoelectric AlN RF MEMS switches monolithically integrated with AlN contour-mode resonators. *J. Micromech. Microeng.* 18: 105011 (11 pp.).
9. Sinha, N., Wabiszewski, G.E., Mahameed, R. et al. (2009). Piezoelectric aluminum nitride nanoelectromechanical actuators. *Appl. Phys. Lett.* 95: 053106 (3 pp.).
10. Sinha, N., Jones, T.S., Zhijun, G., and Piazza, G. (2012). Body-biased complementary logic implemented using AlN piezoelectric MEMS switches. *J. Microelectromech. Syst.* 21: 484–496.
11. Elfrink, R., Kamel, T.M., Goedbloed, M. et al. (2009). Vibration energy harvesting with aluminum nitride-based piezoelectric devices. *J. Micromech. Microeng.* 19: 094005 (8 pp.).
12. Elfrink, R., Renaud, M., Kamel, T.M. et al. (2010). Vacuum-packaged piezoelectric vibration energy harvesters: damping contributions and autonomy for a wireless sensor system. *J. Micromech. Microeng.* 20: 104001 (7 pp.).
13. Ting-Ta, Y., Hirasawa, T., Wright, P.K. et al. (2011). Corrugated aluminum nitride energy harvesters for high energy conversion effectiveness. *J. Micromech. Microeng.* 21: 085037 (9 pp.).
14. Guedes, A., Shelton, S., Przybyla, R. et al. (2011). Aluminum nitride pMUT based on a flexurally-suspended membrane. TRANSDUCERS 2011 - 2011 16th International Solid-State Sensors, Actuators and Microsystems Conference (5–9 June 2011), 2062–5.

15 Shelton, S., Mei-Lin, C., Hyunkyu, P. et al. (2009). CMOS-compatible AlN piezoelectric micromachined ultrasonic transducers. 2009 IEEE International Ultrasonics Symposium (20–23 September 2009), 402–5.

16 Littrell, R. and Grosh, K. (2012). Modeling and characterization of cantilever-based MEMS piezoelectric sensors and actuators. *J. Microelectromech. Syst.* 21: 406–413.

17 Williams, M.D., Griffin, B.A., Reagan, T.N. et al. (2012). An AlN MEMS piezoelectric microphone for aeroacoustic applications. *J. Microelectromech. Syst.* 21: 270–283.

18 Goericke, F.T., Chan, M.W., Vigevani, G. et al. (2011). High temperature compatible aluminum nitride resonating strain sensor. TRANSDUCERS 2011 - 2011 16th International Solid-State Sensors, Actuators and Microsystems Conference (5–9 June 2011), 1994–7.

19 Zuniga, C., Rinaldi, M., Khamis, S.M. et al. (2009). Nanoenabled microelectromechanical sensor for volatile organic chemical detection. *Appl. Phys. Lett.* 94: 223122 (3 pp.).

20 Olsson, R.H. III, Wojciechowski, K.E., Baker, M.S. et al. (2009). Post-CMOS-compatible aluminum nitride resonant MEMS accelerometers. *J. Microelectromech. Syst.* 18: 671–678.

21 Gokhale, V.J., Figueroa, C., Tsai, J.M.L., and Rais-Zadeh, M. (2015). Low-noise AlN-on-Si resonant infrared detectors using a commercial foundry MEMS fabrication process. 2015 28th IEEE International Conference on Micro Electro Mechanical Systems (MEMS), Estoril, 73–76.

22 Podoskin, D., K. Brückner, M. Fischer et al. (2015). Multi-technology design of an integrated MEMS-based RF oscillator using a novel silicon-ceramic compound substrate. 2015 German Microwave Conference, Nuremberg, 406–409.

23 Patterson, A., Calayir, E., Fedder, G.K. et al. (2015). Application of statistical element selection to 3D integrated AlN MEMS filters for performance correction and yield enhancement. 2015 28th IEEE International Conference on Micro Electro Mechanical Systems (MEMS), Estoril, 996–999.

24 Kochhar, A., T. Matsumura, G. Zhang et al. (2012). Monolithic fabrication of film bulk acoustic resonators above integrated circuit by adhesive-bonding-based film transfer. 2012 IEEE International Ultrasonics Symposium, Dresden, 1047–1050.

25 Horsley, D.A., Y. Lu, H.Y. Tang et al. (2016). Ultrasonic fingerprint sensor based on a PMUT array bonded to CMOS circuitry. 2016 IEEE International Ultrasonics Symposium (IUS), Tours, 1–4.

26 Wojciechowski, K.E., Olsson, R.H., Tuck, M.R. et al. (2009). Single-chip precision oscillators based on multi-frequency, high-Q aluminum nitride MEMS resonators. TRANSDUCERS 2009 - 2009 International Solid-State Sensors, Actuators and Microsystems Conference, Denver, CO, 2126–2130.

27 Dubois, M.-A., Carpentier, J.F., Vincent, P. et al. (2006). Monolithic above-IC resonator technology for integrated architectures in mobile and wireless communication. *IEEE J. Solid-State Circuits* 41 (1): 7–16.

28 Mansour, R.R. (2013). RF MEMS-CMOS device integration: an overview of the potential for RF researchers. *IEEE Microwave Mag.* 14 (1): 39–56.

29 Qu, H. (2016). CMOS MEMS fabrication technologies and devices. *Micromachines* 7 (1): 14.

30 Witvrouw, A. (2008). CMOS-MEMS integration today and tomorrow. *Scr. Mater.* 59 (9): 945–949.

31 Ramm, P., A. Klumpp, J. Weber et al. (2010). 3D integration technology: status and application development. 2010 Proceedings of ESSCIRC, Seville, 9–16.

32 Piazza, G., Stephanou, P.J., and Pisano, A.P. (2007). Single-chip multiple-frequency ALN MEMS filters based on contour-mode piezoelectric resonators. *J. Microelectromech. Syst.* 16 (2): 319–328.

33 Cardoso, A., L. Dias, E. Fernandes et al. (2017). Development of novel high density system integration solutions in FOWLP-complex and thin wafer-level SiP and wafer-level 3D packages. 2017 IEEE 67th Electronic Components and Technology Conference (ECTC), Orlando, FL, 14-21.

34 Soon, J.B.W., Singh, N., Calayir, E. et al. (2016). Hermetic wafer level thin film packaging for MEMS. 2016 IEEE 66th Electronic Components and Technology Conference (ECTC), Las Vegas, NV, 857-862.

35 Najafi, K. (2003). Micropackaging technologies for integrated microsystems: applications to MEMS and MOEMS. Proceedings SPIE Micromachining and Microfabrication Process Technology III, 1-19.

36 Newman, M.W., S. Muthukumar, M. Schuelein et al. (2006). Fabrication and electrical characterization of 3D vertical interconnects. 56th Electronic Components and Technology Conference 2006, San Diego, CA, 394-398.

37 Baodong, L., Pengfei, W., and Xinfu, L. (2016). Micro-area sheet resistance measurement system of four-point probe technique based on LabVIEW. 2016 International Symposium on Computer, Consumer and Control (IS3C), Xi'an, 998-1001.

38 Ismail, M.Y. and Principe, J.C. (1996). Equivalence between RLS algorithms and the ridge regression technique. Conference Record of The Thirtieth Asilomar Conference on Signals, Systems and Computers, Pacific Grove, CA, USA, Vol. 2, 1083-1087.

39 Calayir, E., Piazza, G., Soon, J.B.W., and Singh, N. (2016). Analysis of spurious modes, Q, and electromechanical coupling for 1.22 GHz AlN MEMS contour-mode resonators fabricated in an 8″ silicon fab. 2016 IEEE International Ultrasonics Symposium (IUS), Tours, 1–4.

40 Calayir, E., Xu, J., Pileggi, L. et al. (2017). Self-healing narrowband filters via 3D heterogeneous integration of AlN MEMS and CMOS chips. 2017 IEEE International Ultrasonics Symposium (IUS), Washington, D.C. (6–9 September 2017).

41 Wang, F., G. Keskin, A. Phelps et al. (2012). Statistical design and optimization for adaptive post-silicon tuning of MEMS filters. DAC Design Automation Conference 2012, San Francisco, CA, 176-181.

42 Liu, R. and Pileggi, L. (2015). Low-overhead self-healing methodology for current matching in current-steering DAC. *IEEE Trans. Circuits Syst. II: Express Briefs* 62 (7): 651–655.

43 Keskin, G., Proesel, J., and Pileggi, L. (2010). Statistical modeling and post manufacturing configuration for scaled analog CMOS. IEEE Custom Integrated Circuits Conference 2010, San Jose, CA, 1-4.

8

MEMS Using CMOS Wafer

Weileun Fang[1], Sheng-Shian Li[1], Yi Chiu[2] and Ming-Huang Li[1]

[1] National Tsing Hua University, Department of Power Mechanical Engineering, Kuang-Fu Road, Hsinchu 300044, Taiwan
[2] National Chiao Tung University, Department of Electrical Engineering, Ta-Hsueh Road, Hsinchu 30010, Taiwan

8.1 Introduction: CMOS MEMS Architectures and Advantages

The semiconductor industries celebrated the 60th anniversary of integrated circuits (IC) in 2018. By using the semiconductor fabrication processes including thin film deposition, patterning, and etching, millions to billions of electronic components can be fabricated and integrated on a single chip in several millimeter square. Moreover, hundreds to tens of thousands of chips can be batch fabricated on an 8–12-in wafer. According to the business model established in the semiconductor industry, many fabless design houses (for example, Qualcomm, MediaTek, etc.) could implement their IC devices using the fabrication processes provided by foundries (for example, TSMC, UMC, etc.). In general, foundries could offer standard fabrication processes and related design rules to fabless customers to save their development time. Thanks to the aforementioned business model, it enables the birth and growth of many small and medium enterprises in semiconductor industries. It would be beneficial if the micro-electro mechanical systems (MEMS) industries could also leverage such business model to accelerate the development and commercialization of MEMS products and also increase the number of MEMS fabless design houses.

Presently, the planar fabrication technology (such as the semiconductor fabrication processes) has been extensively exploited to fabricate and integrate electronic, mechanical, optical, bio, etc. devices on various substrates [1–3]. Since there are many types of mechanical components with diversified design considerations, for example, the flexible spring and the stiff proof mass, MEMS structures have different process requirements. In some cases, fabrication processes are developed to offer functional materials as well as thin films with better mechanical properties [4]. Thus, it is well known that MEMS industries have the challenge of "one

3D and Circuit Integration of MEMS, First Edition. Edited by Masayoshi Esashi.
© 2021 WILEY-VCH GmbH. Published 2021 by WILEY-VCH GmbH.

product, one process" (Bosch Sensortec. 3-Axis accelerometer. https://www.bosch-sensortec.com/bst/home/home_overview). To date, tons of fabrication processes have been developed to realize MEMS devices of different applications (MEMSCAP. MUMPs® process. http://www.memscap.com/products/mumps; TDK InvenSense, https://www.invensense.com/technology/). Many fabrication platforms and multi-project wafer (MPW) processes, including bulk, surface, silicon on insulator (SOI), etc. micromachining technologies, have also been established in order to follow the successful model of the semiconductor industry (Silex Microsystems, https://silexmicrosystems.com/mems-foundry/; Teledyne DALSA, https://www.teledynedalsa.com/en/home/; Asia Pacific Microsystems, http://www.apmsinc.com/eng/home; Sensornor, http://www.sensornor.com/). For instance, the bulk micromachining technology shown in Figure 8.1 has been exploited by SensorNor to offer the MPW process with Si MEMS structures (Sensornor, http://www.sensornor.com/). Moreover, as displayed in Figure 8.2, the well-known MUMPs surface micromachining process with two to three poly Si layers demonstrates its capability to realize passive and active micro-mechanical components [5–8] and to further integrate these components to form complicated devices and microsystems [9–11].

The surface micromachining processes with the thick epi-poly silicon layer for mechanical structures have been adopted by MEMS industries as inertial sensor platforms (STMicroelectronics, http://www.st.com/)[12], as depicted in Figure 8.3. The SOI micromachining technology shown in Figure 8.4 is employed by Tronics to realize the MPW process with suspended MEMS structures [13]. The process platforms could fabricate and monolithically integrate various MEMS structures, sensors, and actuators on a single chip so that microsystems can be achieved using the system on chip (SoC) approach. Moreover, the concept of "combo sensors" or "sensor hubs" has attracted attention in the recent years [14]. For example, the inertial hub consists of accelerometer, gyroscope, and magnetometer (TDK InvenSense, https://www.invensense.com/technology/), and the environmental

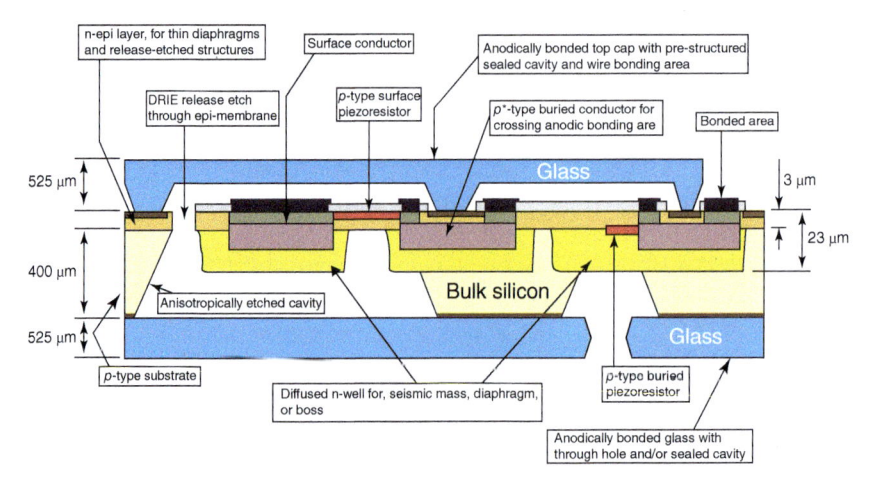

Figure 8.1 MPW process with Si MEMS structures by SensorNor Source: Sensornor, http://www.sensornor.com/.

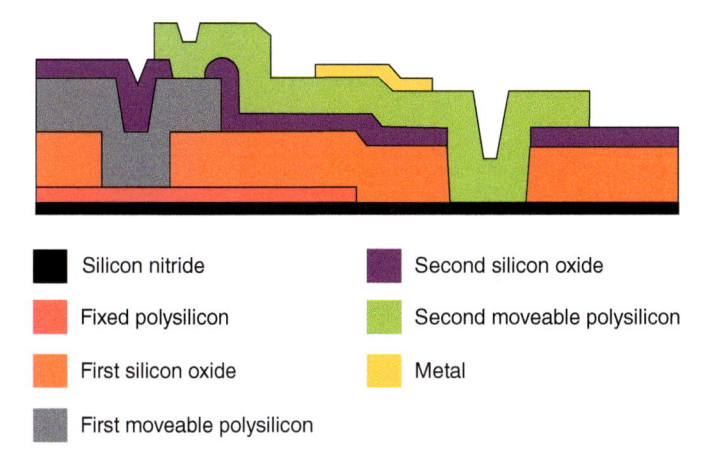

8.1 Introduction: CMOS MEMS Architectures and Advantages | **133**

- ■ Silicon nitride
- ■ Fixed polysilicon
- ■ First silicon oxide
- ■ First moveable polysilicon
- ■ Second silicon oxide
- ■ Second moveable polysilicon
- ■ Metal

Figure 8.2 MUMPs surface micromachining MPW process. Source: PolyMUMPs Design Handbook, a MUMPs® process, Allen Cowen, Busbee Hardy, Ramaswamy Mahadevan, and Steve Wilcenski MEMSCAP Inc. Revision 13.0 MEMSCAP, MUMPs® process, http://www.memscap.com/products/mumps. © 2011, MEMSCAP.

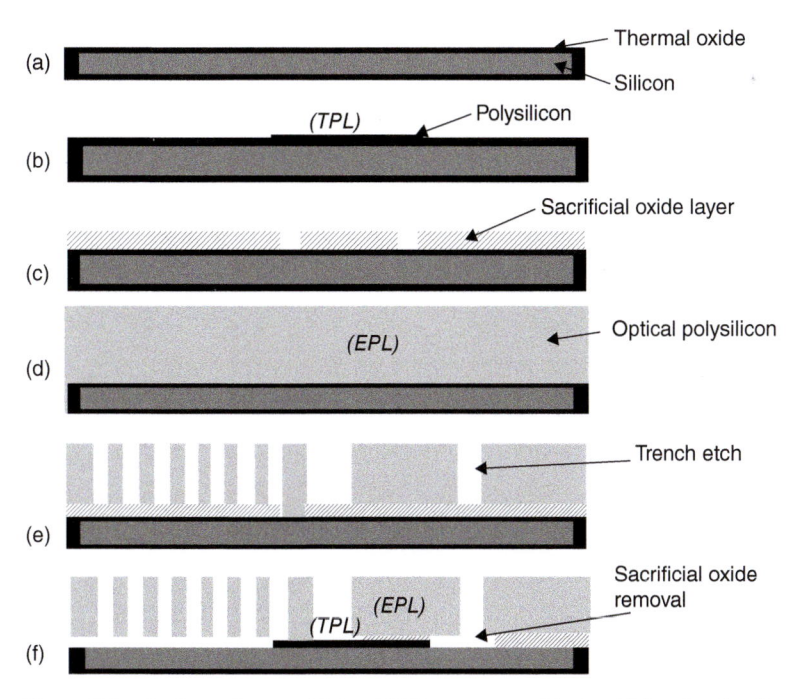

(a) Thermal oxide / Silicon
(b) (TPL) Polysilicon
(c) Sacrificial oxide layer
(d) (EPL) Optical polysilicon
(e) Trench etch
(f) (TPL) (EPL) Sacrificial oxide removal

Figure 8.3 Surface micromachining MPW processes with the thick epi-poly silicon layer. Source: Langfelder et al. [12]. © 2012, IEEE.

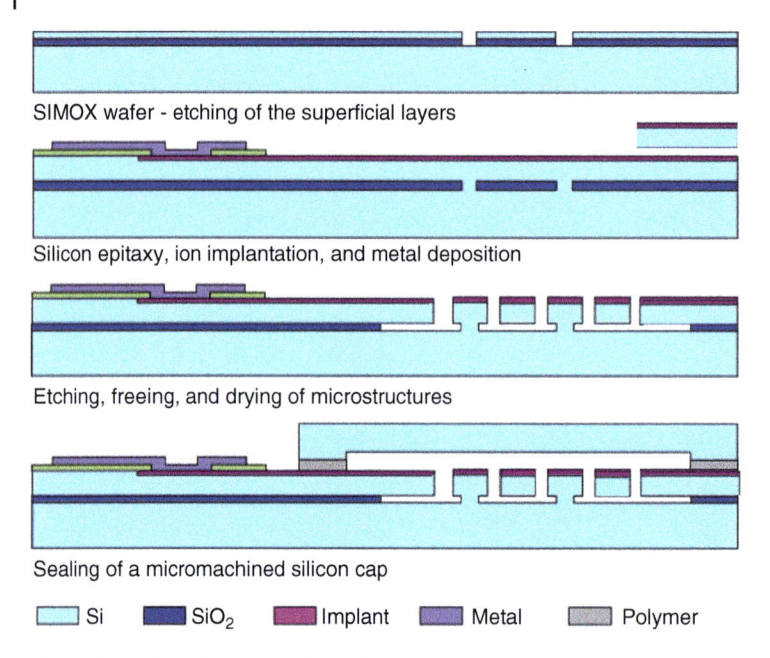

Figure 8.4 SOI micromachining MPW processes by Tronics. Source: Renard [13]. © 2000, IOP Publishing.

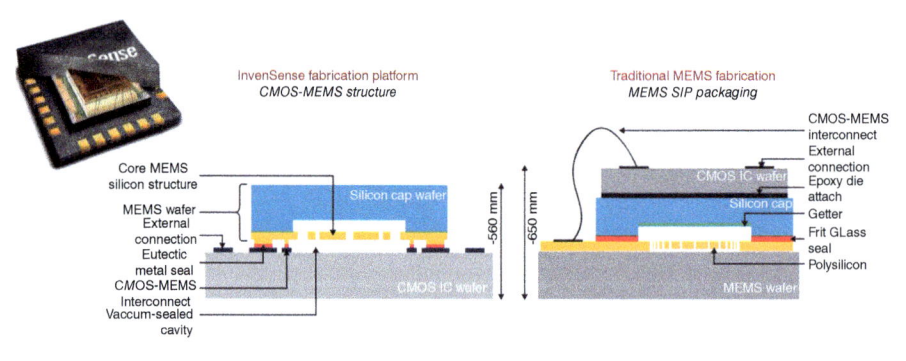

Figure 8.5 Process platform for combo sensors of TDK InvenSense. Source: TDK InvenSense, https://www.invensense.com/technology/.

hub consists of temperature, pressure, and humidity sensors (Bosch Sensortec, BME680 integrated environmental units. https://www.bosch-sensortec.com/bst/home/home_overview). Thus, as shown in Figure 8.5, the process platforms have the potential to enable the formation of sensor hubs using the SoC approach (TDK InvenSense, https://www.invensense.com/technology/). In short, the development time for MEMS devices could significantly be reduced if (i) the foundry could establish several standard and stable process platforms and related design rule and (ii) the designers (in fabless, design houses, etc.) could select a proper and existing process platform to design the MEMS device. In addition, the foundry could provide full wafer as well as MPW services through the standard process platform.

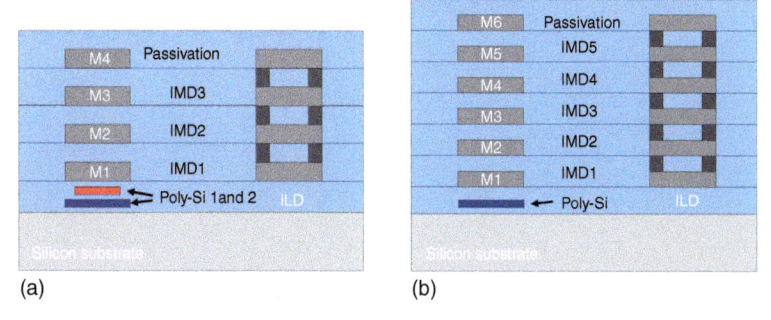

Figure 8.6 Two existing standard CMOS processes. (a) 0.35 µm 2P4M process and (b) 0.18 µm 1P6M process.

Finally, the process platform enables design houses to implement combo sensors or microsystems on a single chip.

The complementary metal oxide semiconductor (CMOS) transistor is the key building block for microelectronic devices and systems. Thus, the fabrication processes to fabricate CMOS devices (named CMOS processes) have been developed and improved. Presently various standard CMOS processes, for example, the 0.35 µm 2P4M (two poly Si layers and four metal layers) process shown in Figure 8.6a, the 0.18 µm 1P6M (one poly Si layer and six metal layers) process shown in Figure 8.6b, the bipolar and complementary metal oxide semiconductor (BiCMOS) process, and so on, have been established in foundries. By following the well-known Moore's Law [15], the size and density of CMOS components will be continuously improved. The foundries also significantly enhance the yield and reliability of the devices and meanwhile reduce the process cost. In short, CMOS processes are mature and available in many existing foundries. Thus, it would be a cost-effective approach to leverage the existing CMOS fabrication technologies to implement MEMS devices. Moreover, the 8 inch or even 6 inch CMOS process technologies could meet most of the size requirements for MEMS devices. In this regard, the situation of no depreciation cost for most of the 8-inch foundries is another advantage for the CMOS MEMS technology. As a result, the CMOS-based micromachining process technology offers a promising approach to implement MEMS devices.

Figure 8.7 displays a simple concept to extend the CMOS technology to implement MEMS devices, which indicates that suspended MEMS structures can be realized by adding a few etching processes, such as etching of metal films, dielectric films, and silicon substrate, after the completion of CMOS chips in Figure 8.6. Since the CMOS process is mainly used to implement IC, it is easy to monolithically fabricate and integrate the microelectronics and micro-mechanical components using the CMOS MEMS technology. According to the concept in Figure 8.7, many post-CMOS MEMS processes have been developed to fabricate CMOS-based MEMS devices, for instance, inertial sensors [17, 18], microphones [19], pressure sensors [20, 21], chemical sensors [20, 22], humidity sensors [23], actuators [24], resonators [25], etc. Several successful CMOS MEMS devices have also been commercialized, such

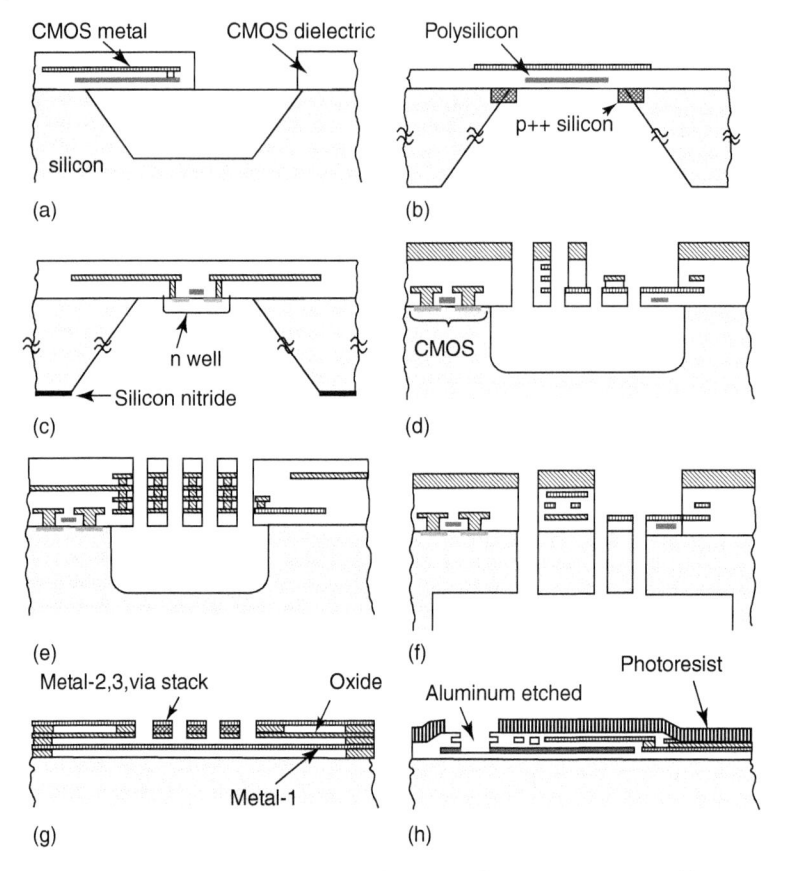

Figure 8.7 Summary of various post-CMOS thin film and substrate etching processes to fabricate CMOS-based MEMS devices. Source: Fedder [16]. © 2005, IEEE.

as the inertia sensor (MEMSIC, http://www.memsic.com/memsic/), flow sensor (Sensirion, https://www.sensirion.com/cn/), etc. Thus, some of the post-CMOS processes for the implementation of MEMS devices are summarized in [16, 26]. However, since many of these CMOS MEMS devices are fabricated using different post-CMOS processes, it is not straightforward to realize the MPW. Moreover, it is also challenging to monolithically integrate CMOS MEMS devices with different post-CMOS processes to achieve sensor hubs on a single chip.

As displayed in Figure 8.8, generic post-CMOS processes have been established to show the possibility of fabricating and integrating diverse CMOS MEMS devices on a wafer [27]. Through this post-CMOS process technology, CMOS MEMS-based MPW and microsystems can be realized, and CMOS MEMS sensor hubs are also achieved using the SoC approach. The generic fabrication platform in Figure 8.8 consists of the standard 0.35 μm 2P4M (two poly Si and four metal layers named M1–M4) process prepared by TSMC and the in-house post-CMOS processes developed in [27–30]. Based on the process platform, different sensors are demonstrated and also monolithically integrated, for instance, as shown in the scanning electron

Figure 8.8 Generic post-CMOS processes to fabricate different MEMS devices for the applications of combo sensors or MPW. (a) CMOS layer stacking, (b) backside silicon etching, (c) metal set etching, (d) etching of dielectric layer, (e) etching of silicon substrate, and (f) additional process (e.g., polymer dispensing). Source: Fang et al. [27]. © 2013, IEEE.

microscope (SEM) micrographs of Figure 8.9. Figure 8.9a displays the integration of a pressure sensor and a single-axis accelerometer on a single chip [28]. The second example shown in Figure 8.9b is the monolithic integration of three single-axis acceleration sensing units to form a tri-axis accelerometers [29]. Moreover, as exhibited in Figure 8.9c, integration of the pressure sensor, the tri-axis accelerometer, and the temperature sensor is achieved to realize a tire pressure monitoring system (TPMS) [30]. In summary, the existing standard CMOS processes are mature

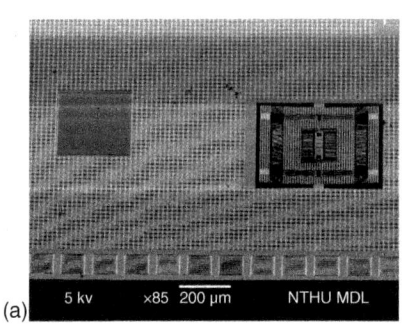

Figure 8.9 Examples to show the implementation and monolithic integration of CMOS MEMS sensors, including (a) integration of a pressure sensor and a single-axis accelerometer. Source: Sun et al. [28] © 2009, IOP Publishing, (b) integration of three single-axis acceleration sensing units. Source: Tsai et al. [29] © 2009, IOP Publishing, and (c) integration of pressure sensor, temperature sensor, and accelerometer to form the TPMS. Source: Sun et al. [30]. © 2009, IEEE.

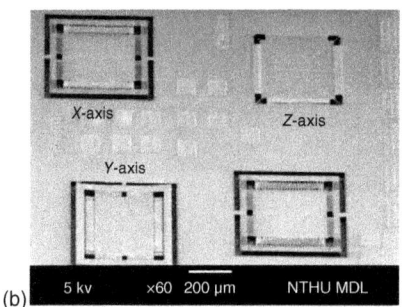

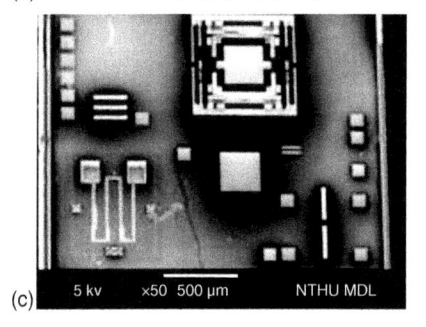

tools to implement the stacking and patterning of thin film layers. By adopting the post-CMOS processes, MEMS devices can be fabricated and integrated on the wafer, which is prepared by foundries through available CMOS processes.

In this chapter, several fabrication technologies, including different CMOS and post-CMOS processes, to implement CMOS MEMS devices, will be introduced. Firstly, the post-CMOS process modules to define multilayer thin films of back end of line (BEOL) and to remove the silicon substrate will be introduced in Section 8.2. After that, Sections 8.3 and 8.4, respectively, elaborate two foundry available standard CMOS processes as examples, including the 0.35 μm 2P4M (two poly Si and four metal layers) and the 0.18 μm 1P6M (one poly Si and six metal layers) CMOS platforms, to show the implementation of various MEMS devices by integrating different post-CMOS process modules. Moreover, Section 8.5 will introduce several process technologies to deposit functional materials, such as the polymer, or to assemble functional structures, such as the magnetic ball on the CMOS chip, to enhance the performances of MEMS devices. To show the capability of CMOS MEMS, Section 8.6 further demonstrates the monolithic integration of multiple sensors and sensing circuits to exhibit the capability of CMOS MEMS for the realization of sensing hubs. Finally, Section 8.7 will point out several concerns when extending the existing CMOS technology to achieve MEMS devices.

8.2 Process Modules for CMOS MEMS

This section provides several key process modules (or process building blocks) to realize the aforementioned CMOS MEMS devices. The post-CMOS technology has been implemented for many years to attain monolithic integration with the IC technology and avoid thermal budget limitation simultaneously. The standardization of the CMOS process allows design flexibility and accessibility in the selection of the CMOS foundry as well as the technology node depending on system requirement and cost [31–33]. As the name reveals, in post-CMOS MEMS, all the MEMS process steps are carried out only after the CMOS fabrication steps are finished. The post-CMOS MEMS fabrication process can be broadly categorized into two methods – one is the oxide etching step and the other is metal removal. Both of these can be implemented in the 2-poly-4-metal (2P4M) 0.35 μm and the 1-poly-6-metal (1P6M) 0.18 μm CMOS technologies. The geometry of the MEMS devices is defined by the layout of BEOL layers with critical dimensions <1 μm. The final release step can be done by various etching techniques, such as XeF_2-based isotropic silicon etching or KOH-based anisotropic etching.

The conceptual plot of a cross-sectional schematic of fabricated MEMS SoC device without any lithography steps realized for 0.35 μm (2P4M) and 0.18 μm (1P6M) CMOS technology is shown in Figure 8.10. Most of the material layers that are embedded in the front end of line (FEOL) and BEOL in the conventional IC architecture have specific roles. The on-chip passive and active elements are implemented through the polysilicon and silicon substrate in the FEOL portion of

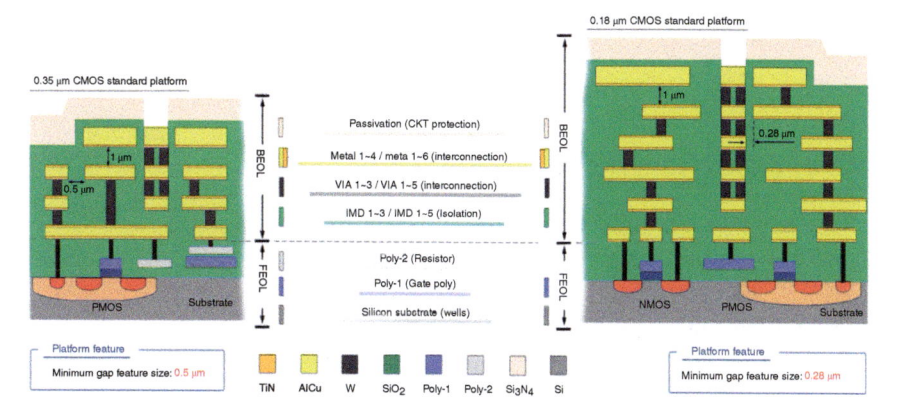

Figure 8.10 Cross-sectional view of unreleased CMOS MEMS chip realized in 0.35 μm (2P4M) and 0.18 μm (1P6M) CMOS technology, showing a specific stacking configuration and an inherent monolithic capability. Source: Chen et al. [33]. © 2018, MYU K.K.

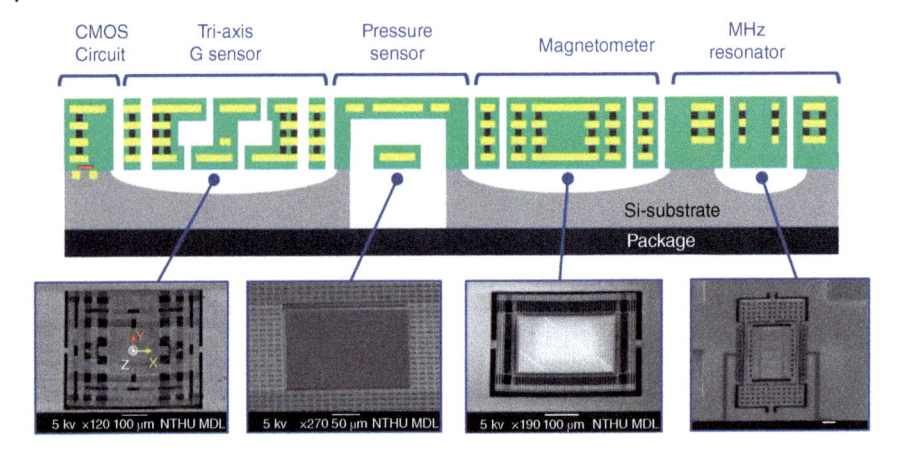

Figure 8.11 Configuration showing the arrangement of different sensors and resonator devices designed on one CMOS MEMS platform. Source: Courtesy of Prof. Weileun Fang, NTHU, TW.

the CMOS IC. The multilevel metal layers are used to establish a wiring connection via the BEOL processing, aiming to connect those individual elements in all spatial dimensions.

Multilevel stacking configuration of CMOS MEMS allows designers to define the mechanical devices lithographically with flexible signal routing, thus enabling a range of mechanical structure design with various transduction mechanisms (like capacitive and piezoresistive transduction) in commercial foundry service. Figure 8.11 shows the various ways of realizing different sensors and resonators in the CMOS MEMS platform. Using the existing layers in CMOS allows cost-effective fabrication, accessible codesigning MEMS platform, and fast lead time by removing the additional masks and lithography steps for MEMS fabrication [34].

8.2.1 Process Modules for Thin Films

Sensors and resonators have to be released to enable the mechanical motion of transducers, such as cantilever beam, bridge structure, and membrane. As a result, this section presents three micromachining approaches on CMOS BEOL to achieve the required thin film structures while the substrate is intact. Based on different BEOL sacrificial material removal, they are categorized as metal sacrificial, oxide sacrificial, and TiN-composite process modules.

8.2.1.1 Metal Sacrificial

As the name denotes, this post-CMOS process uses metal or stack of metal from the BEOL as the sacrificial metal to realize the CMOS sensors/resonators [29]. The sacrificial metal is etched using a piranha etchant containing a 1:3 mixture of hydrogen peroxide (H_2O_2) and heated sulfuric acid (H_2SO_4) to form an oxide-rich structure or a composite structure, depicted in Figure 8.12a. This step is followed by

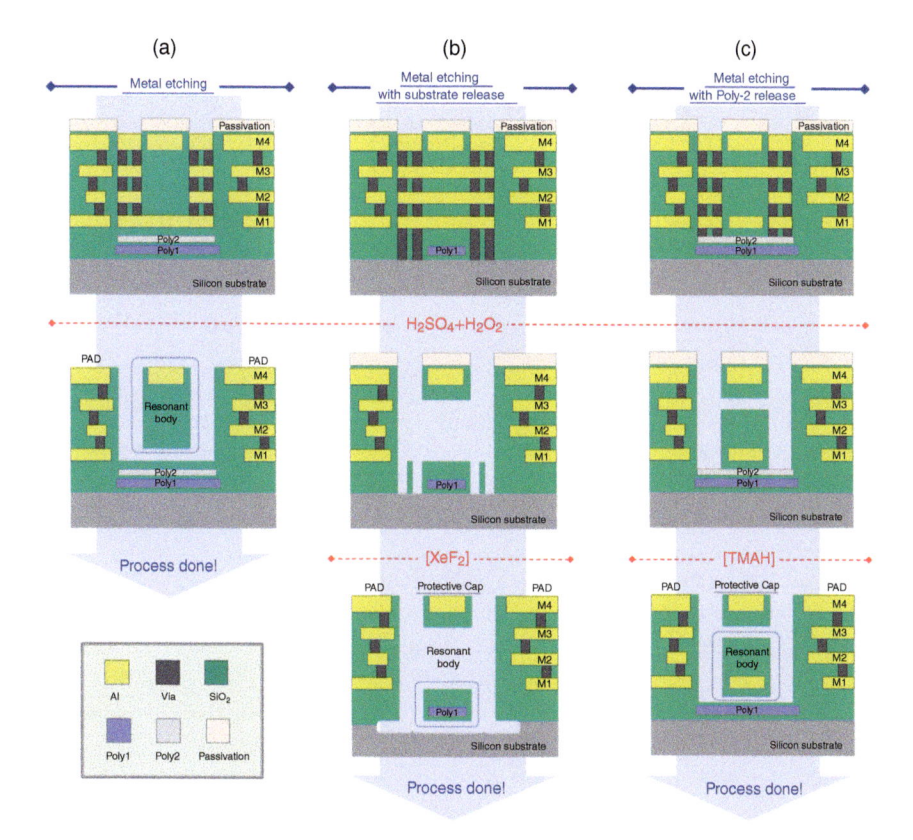

Figure 8.12 Comprehensive post-process flows of oxide-rich devices, including (a) general metal etching, modified metal etching with (b) Si substrate release, and (c) poly-2 layer release. Source: Chen et al. [33]. © 2018, MYU K.K.

a single-step reactive ion etching (RIE) (usually dry oxide etch) to open the probing pads for device testing or wire bonding. This maskless etching process provides good selectivity between the oxide layers and the metal and offers nearly a 100% device yield in both 2P4M and 1P6M CMOS platforms [35, 36]. The oxide-rich structures obtained by this post-process features lower degradation of the quality factor since the oxide layers exhibit high intrinsic Q [37–39]. Bulk-mode vibration of an oxide-rich stacking structure can provide Q in above 15 000 [40]. The BEOL SiO_2 also provides passive temperature compensation, thus further enhancing the frequency stability over a wide thermal range. The standard polysilicon layer belonging to FEOL can also be included as a part of the mechanical structure using the usual metal etching and XeF_2 substrate release process as depicted in Figure 8.12b. This technique is advantageous as the polysilicon actuating and sensing regions can be fully decoupled to avoid feedthrough caused by low structural resistance in thermal piezoresistive resonators (TPR) [41]. This enables low thermal capacitance, enhancement of transduction efficiency, and sensitivity [42].

Although the CMOS MEMS metal-removal process shows excellent etching selectivity to form oxide-rich structures with simple process steps, the motional resistance

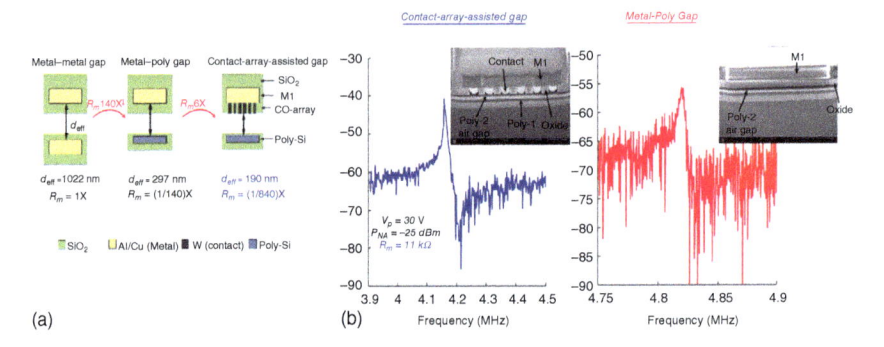

Figure 8.13 (a) Prediction of reduced capacitive motional resistance (R_m) with effective gaps of 1 μm, 297 nm, and 190 nm, respectively. (b) Frequency response measured under different gap configurations based on poly-2 release process, showing a noticeable improvement of R_m through a contact array design. Source: Chen et al. [33]. © 2018, MYU K.K.

(R_m) usually lies in the MΩ level due to a larger effective gap spacing in capacitive transducers. To overcome this, a two-step wet released process is suggested to create a resonator-to-electrode gap of 180 nm for strong electrostatic coupling by creating a submicron gap deep inside [43]. As illustrated in Figure 8.12c, the post-fabrication starts from metal wet etching through a mixture of H_2SO_4 and H_2O_2 to remove the sacrificial metal, with high etching selectivity to silicon dioxide [44, 45]. The sacrificial polysilicon layer (poly-2) in 2P4M platform is then released by tetramethylammonium hydroxide (TMAH) solution, thus providing an air-gap spacing of 180 nm. To open the aluminum pad for wire bonding and probing, the front-side RIE is employed to etch the Si_3N_4 passivation layer as the final step.

Figure 8.13a shows the cross-sectional view of different possible arrangements obtained by the metal-removal process. As the effective gap is reduced, a reduction in capacitive motional resistance is also observed. The poly-2 etching approach creates an effective gap of 297 nm between metal and polysilicon, providing a reduction in R_m by 140 times as compared to the conventional metal etch approach, enabling monolithic implementation of filtering and timing building blocks [45, 46]. A sixfold reduction in motional resistance can be attained by designing a pillar-like arrangement of tungsten under the enclosed metal region, making an equivalent gap of 190 nm [47]. Figure 8.13b shows good agreement with formula predictions, with the contact array approach providing lower motional resistance than the metal-poly gap approach. Here, the minor discrepancy between experimental and analytical result comes from the smaller transduction area implemented in contact array design, owing to the fabrication limitation of the CMOS process. The inset SEM image shows the FIB view for both contact-array- and metal-to-polysilicon gap approaches.

8.2.1.2 Oxide Sacrificial

There are two ways to implement the oxide removal release process: one is the wet etching (liquid-based process) and the other is dry etching (plasma-based). In the wet etching process, the via walls (i.e. tungsten) should be placed at both sides of

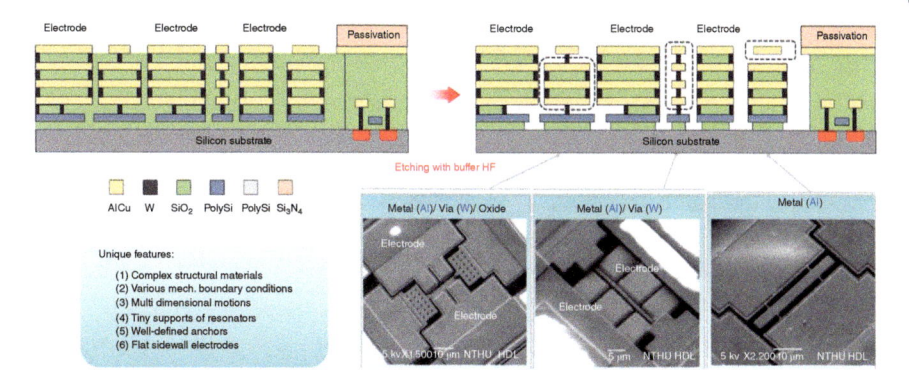

Figure 8.14 Post-process steps of various CMOS MEMS resonators realized by standard 0.35 μm 2P4M CMOS technology with the oxide-removal approach. Source: Chen et al. [33]. © 2018, MYU K.K.

the device edge, creating a metal-rich structure, to protect the embedded oxide. Figure 8.14 shows the fabrication flow for the oxide removal release process where the metal-rich structures can be retained by etching the exposed dielectric using available SiO_2 etchant like buffer oxide etch (BHF/BOE) solution, Silox Vapox III, etc. These etchants have very high selectivity towards metal layers, contacts (i.e. tungsten), and vias. Lopez et al. [48] have demonstrated the smallest air gap of 40 nm using the dedicated 0.35 μm CMOS platform. Although the dielectric wet etch process is simple and straightforward, the etching time of the process should be carefully controlled to prevent undercut to avoid a poor device yield. As compared to the dielectric wet etching, the anisotropic oxide dry etching enables more functionalities by keeping polysilicon (viz. micro-oven, thermal actuation, and piezoresistive sensing) [49, 50]. Thus, it is possible to design resonators (including both in-plane and out-of-plane structures) in this fabrication platform and use various combinations of materials, such as metal-rich or metal/oxide composite, offering better transducer efficiency and more design flexibility suited for sensor and RF applications [51, 52]. All of these can be achieved by using a passivation layer to protect the transistor circuits. The minimum electrode-to-resonator gap (0.5 μm in 2P14 and 0.28 μm in 1PM6 platform) achieved in this configuration is limited by the design rule check (DRC) of the 0.35/0.18 μm CMOS technology node, and the motional resistance of the designed resonators is in several MΩ due to their lower coefficient of electromechanical coupling. Also, the quality factor Q of the metal-rich resonators is limited because aluminum is treated as a highly lossy acoustic material [53]. This can be largely reduced by opting for the metal release process or going for composite-material resonator structures.

8.2.1.3 TiN-composite (TiN-C)

Although the oxide-rich feature can address most of the issues like Q-factor and device yield existing in metal-rich counterparts for resonant application, the dielectric charging issue still possesses a significant challenge for capacitive MEMS resonators. For pure silicon devices, by relying on the positive TC_E (temperature coefficient of elastic modulus) value of SiO_2, the TC_f (temperature coefficient of frequency) value of the dielectric layer can be effectively engineered [54].

However, in capacitive transduction, when a DC bias is applied, the induced charge gets accumulated and trapped in the dielectric layer. This will cause an undesired charging effect, affecting the electrostatic spring constant (k_e) and causing the resonant frequency to drift over time. This charging issue creates problems in terms of frequency stability, especially for oxide-rich resonators, whose geometry is limited by the fixed stacking material configuration from the standard CMOS MEMS platform [55]. Therefore, a novel post-CMOS fabrication process, namely, the TiN-C CMOS MEMS platform is established by preserving the TiN layer as the electrode and removing the AlCu metal core [56]. This helps the device to achieve a decent electromechanical coupling along with charge elimination for capacitive transduction.

The TiN-C structure mainly uses the BEOL materials employed in the standard 0.35 μm CMOS technology. Figure 8.15 illustrates the post-CMOS release process. The structural profile is decided by releasing the metal sacrificial layers as per the process described in Section 8.2.1.1. Next, the exposed oxide and TiN materials are etched away by front-side dielectric and metal RIE processes, respectively. After the dry etching process, a commercial Al etchant is used to remove the AlCu metal core of the CMOS interconnect for releasing MEMS structure. Finally, a two-step RIE process is used again to open the probing pad region for the device testing.

Figure 8.16 shows the SEM image of a TiN-C free–free beam (FFB) resonator, having a transduction gap of 400 nm, where the oxide fin structure provides the electrical isolation capability for multiport operation. The frequency spectra for the first 30 minutes were recorded to study the effect of charging on the resonant

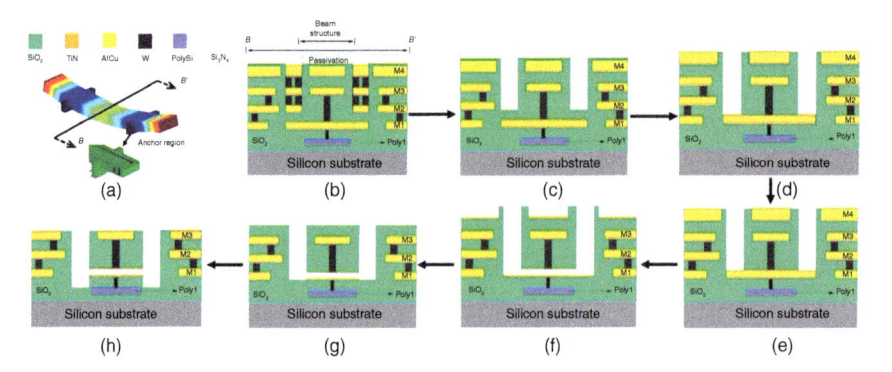

Figure 8.15 Complete post-fabrication process for TiN-C CMOS MEMS platform. (a) Device geometry, (b) unreleased CMOS MEMS chip, (c) metal wet etching process for structure definition, (d, e) dry released process for removing the exposed oxide and TiN layers, (f) structure released by Al etchant, and (g, h) probing pad opening by two-step RIE processes. Source: Chen et al. [33]. © 2018, MYU K.K.

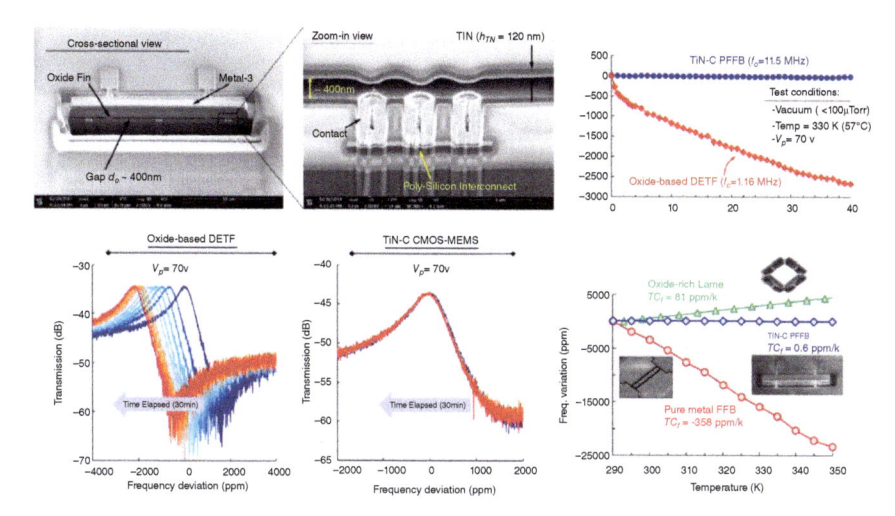

Figure 8.16 Measured results of frequency drift over time for both traditional oxide-rich and TiN-C resonators. Source: Chen et al. [33]. © 2018, MYU K.K.

behavior of the capacitive transducer. There was no significant drift for the TiN-C FFB resonator. The oxide-rich double-ended tuning fork (DETF) resonator shows more than 2000 ppm frequency deviation due to the built-in voltage induced by inter-gap charge [57]. Unlike the pure-metal FFB [25] and oxide-rich Lamé mode resonators [37], which have a very negative and positive TC_f value, the TiN-C FFB resonator achieves the lowest TC_f, considering all the passive temperature compensation techniques used in CMOS MEMS technology to date. Despite the limitation on BEOL material choice and control on their physical properties (i.e. Young's modulus, residual stress, etc.), this post-process platform integrating "acoustic," "mechanical," and "electrical" domains would lead to a single-chip MEMS-based solution for a signal processor in future wearable/IoT electronics.

8.2.2 Process Modules for the Substrate

Another way to release the devices is to etch the silicon substrate to free the structures for large-sized devices, such as accelerometers, microphones, and pressure sensors. This section also elaborates on three micromachining approaches on the substrate, which covers dry and wet isotropic and anisotropic silicon etching.

8.2.2.1 SF$_6$ and XeF$_2$ (Dry Isotropic)

When etchants are in vapor phase the etching process can be termed as "dry". The etching can either be isotropic or anisotropic. The dry isotropic etching depends on neither direction of ion bombarding nor orientation of the substrate material. The dry isotropic etching is performed with fluorine free radicals which are generated by plasma using a gas such as SF_6 or by plasmaless decomposition of XeF_2. These etching are carried out in non-passivated or non-covered regions [58].

The isotropic dry etching of silicon is performed using xenon difluoride (XeF_2). This vapor-phase etching process enables the fabrication of large structure undercut and exhibits excellent etch selectivity, enabling the use of various etch masks like aluminum, photoresist, silicon dioxide, and silicon nitride. However, the roughness of resulting etched silicon surfaces tends to be more. Since this is vapor phase etching, stiction between the released structure and the substrate is minimized. The silicon etch rates for XeF_2 largely depends on surface area of the silicon exposed to the etchant, resulting in typical values of $\sim 1\,\mu m\,min^{-1}$. Isotropic XeF_2 etch along with the metal sacrificial wet etch or oxide sacrificial etch can be used to realize a variety of MEMS transducers on the CMOS platform [59, 60]. The etching process with XeF_2 poses a safety hazard. Proper care should be ensured because SiF_4 is a byproduct of the process. Breathing too much XeF_2 or SiF_4 can result in chemical burns to the respiratory tract.

8.2.2.2 KOH and TMAH (Wet Anisotropic)

The most common silicon substrate release technique is anisotropic wet etching, which can be used to fabricate membrane and beam structures. The wet etchants etch the bulk silicon substrate with different etch rates along different crystal directions because of the inherent anisotropy of the crystal. As soon as the (111) silicon plane or a silicon dioxide (or silicon nitride) layer is reached, the etching by wet anisotropic silicon etchants ceases, i.e. the etch rate is reduced by at least one to two orders of magnitude. Reproducible etching results can be produced only if reliable etch stop techniques are used. The most common anisotropic silicon etchant potassium hydroxide (KOH) solution is typically used in the fabrication of membrane structures, e.g. pressure sensors. The silicon nitride films often act as an etching mask since the etch rate of silicon dioxide in KOH solutions is fairly high (for thermal oxide $\sim 1\,\mu m\,h^{-1}$ in 6 M KOH solution [22]). KOH solutions are relatively inexpensive and are very stable, yielding reproducible etching results. The silicon etch rate is greatly reduced if a highly doped p-type region is encountered in a CMOS substrate during KOH etching process. One of the main disadvantages of KOH solution is that it can also significantly etch the SiO_2 and Al regions, thus requiring protection for circuitry. Etching with KOH is performed from the backside of the wafer, while a mechanical protective film protects the front side of the wafer [61]. Alternative silicon etchants are ammonium hydroxide-based compounds, such as tetramethylammonium hydroxide (TMAH). The aluminum metallization etch rates can be reduced by adjusting the pH value [62], thereby making TMAH one of the candidate etchants for releasing the designed microstructures from the front portions of the CMOS substrate [45, 46, 63]. The etch rate using TMAH gets reduced in highly boron-doped regions ($N_A \geq 10^{19}\,cm^{-3}$). The so-called electrochemical etch-stop technique (ECE) can also be used to stop the etching at a p–n junction [64]. This etch stop technique is often used to fabricate the silicon membrane as well as n-well structures.

8.2.2.3 RIE and DRIE (Front-side RIE, Backside DRIE)

Deep reactive ion etching (DRIE) is one of the important steps used for dry etch process. It can be used to create very high aspect ratio microstructures. DRIE systems rely on the alternate process of etching and polymer-coated sidewall protection steps using high-density plasma sources. In the Bosch process, a mixture of argon gas and trifluoromethane is used for polymer deposition. Photoresist and silicon dioxide layers can be used as etch masks. The DRIE system is very expensive and can process only one wafer at a time in comparison to the simple wet etching step. But it can provide very high anisotropy independent of the orientation of the crystal. DRIE or a combination of both (RIE and DRIE), performed on the backside or front side, or both sides of the substrate, has made the fabrication of CMOS MEMS devices possible on a wide spectrum [59, 60]. Different types of MEMS devices have been developed and commercialized by performing front-side RIE on the BEOL CMOS thin film structural materials.

MEMS structures can be released and defined by a sequence of processes consisting of isotropic etching of SiO_2, with DRIE of silicon followed by anisotropic silicon etching undercut [18]. The top metal layer serves as a hard mask, forms the MEMS structures, and also provides protection to the associated CMOS circuitry. Multilayer CMOS stack made of dielectric and metals can form the proof masses and mechanical springs of the inertial sensors fabricated using thin film technology. The comb fingered structure is used to sense the change in sidewall capacitance. The electrical connection inside the mechanical structures can be established using efficient routing of layers embedded in the FEOL and BEOL, enabling different sensing schemes. The residual stress in the thin film CMOS layer stack can cause large vertical curling and lateral buckling of the suspended MEMS structures, posing a major challenge for optimum device performance as well as fabrication. Although for such devices with a smaller form factor like RF MEMS and thermal sensors, the structural curling can be tolerated; for inertial sensors and similar devices, which relatively need a large size, curling can have severe impact and may require compensation technique [17]. Moreover, the size and mass of the devices are limited because of the need for etch holes in the fabrication. To increase the mass and robustness and to overcome the structural curling of MEMS structures, the single-crystal silicon (SCS) can be used underneath the CMOS BEOL material stacks. Xie et al. [18] illustrate a process flow for SCS silicon structures formed directly from the silicon substrate using DRIE in the 0.35 µm CMOS platform. The use of DRIE in CMOS MEMS technology has benefited the fabrication of relatively large MEMS devices, e.g. micromirrors [65]. This technology has also been used to demonstrate a CMOS MEMS gyroscope with a low-noise floor [50]. To achieve a higher signal-to-noise ratio (SNR), SCS is attached underneath the CMOS stack comb fingers, which helps to increase the sensing capacitance of capacitive sensors. A backside silicon DRIE step is added to define the thickness of the silicon structure to be included in the design. Thus, to define the MEMS microstructure regions, an additional lithography step on the backside of the wafer is defined.

8.3 The 2P4M CMOS Platform (0.35 µm)

The 0.35 µm CMOS platform can be used to realize different MEMS structures by the use of different post-CMOS process modules described in Section 8.2. The structure of the CMOS platform was outlined at the beginning of Section 8.2. The technology node consists of different CMOS BEOL and FEOL materials, which comprises of two polysilicon layers and four metal layers for realizing different MEMS devices. This section explains a few sensors and devices such as resonators, accelerometers, and pressure sensors to show the integration of different post-CMOS process modules in Section 8.2 to implement MEMS devices on the wafer prepared by the standard 2P4M CMOS platform.

8.3.1 Accelerometer

The most important parameter for the design of an accelerometer is its sensitivity, which can be optimized by increasing the proof mass of the accelerometer [29, 66]. The sensitivity of accelerometers can also be enhanced by increasing the number of sensing metal electrodes and the area of overlapping comb-type electrodes. Figure 8.17 shows the method by which the in-plane and out-of-plane gap structures can be realized in TSMC 0.35 µm 2P4M process. The in-plane and the out-of-plane features are necessary to realize the tri-axis sensitivity of the accelerometer.

The sacrificial metal post-CMOS etch and silicon substrate release is used to release this structure as shown in the fabrication steps in Figure 8.18. The steps involve the metal wet etching followed by the RIE to remove the passivation layer. The Si substrate is etched using the XeF_2 isotropic etching to release the structure. The important fabrication advantage that is received by the design of the CMOS MEMS accelerometer is that the mass can be increased using the total thickness close to 7 µm through this process. The metal undercut demonstrated to realize the out-of-plane structure cannot be achieved using anisotropic dry etching. The sensing electrodes with a sub-micron gap can be achieved by this process, and also the absence of metal hard mask for dry etching reduces the parasitic capacitance.

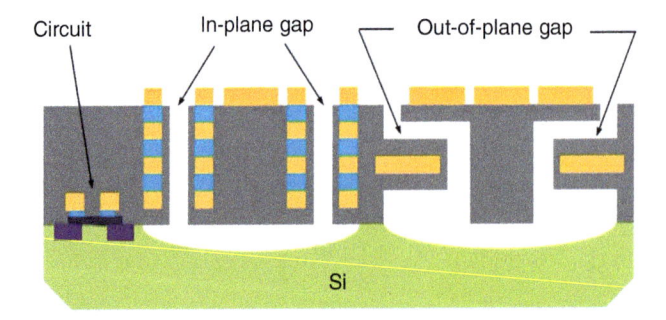

Figure 8.17 In-plane and out-of-plane gap-closing capacitive accelerometer cross-sectional view realized using TSMC 0.35 µm 2P4M foundry process with the use of metal wet etch and substrate release. Source: Tsai et al. [29]. © 2009, IOP Publishing.

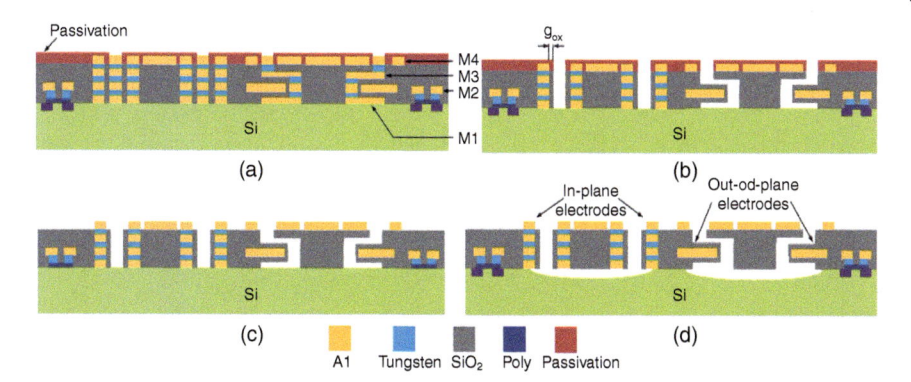

Figure 8.18 Accelerometer fabrication flow. (a) CMOS chip obtained from the TSMC foundry, (b) sacrificial metal etch, (c) passivation layer removal using RIE, and (d) XeF$_2$ isotropic etching for Si substrate release. Source: Tsai et al. [29]. © 2009, IOP Publishing.

Figure 8.9 shows the SEM images of the designed tri-axis accelerometer. The device offers a sensitivity of 11.5 mV g^{-1} (in-plane, x-axis) and 7.8 mV g^{-1} (out-of-plane, z-axis).

8.3.2 Pressure Sensor

The pressure sensor is realized in CMOS 0.35 μm process using a double-sided fabrication process. As depicted in Figure 8.19, it consists of a deformable diaphragm where the pressure is applied. The suspended structures are achieved by the backside post-process, while the sensing electrodes are formed by the metals

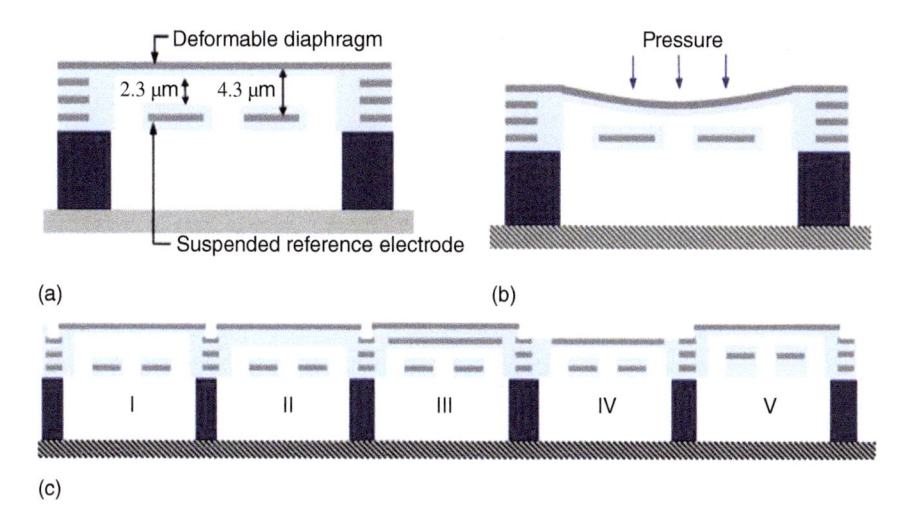

Figure 8.19 Cross-section of CMOS MEMS 0.35 μm pressure sensor SEM. (a) Pressure sensor without applied pressure, (b) pressure applied on the diaphragm, and (c) different pressure ranges realized by changing the thickness. Source: Sun et al. [28]. © 2009, IOP Publishing.

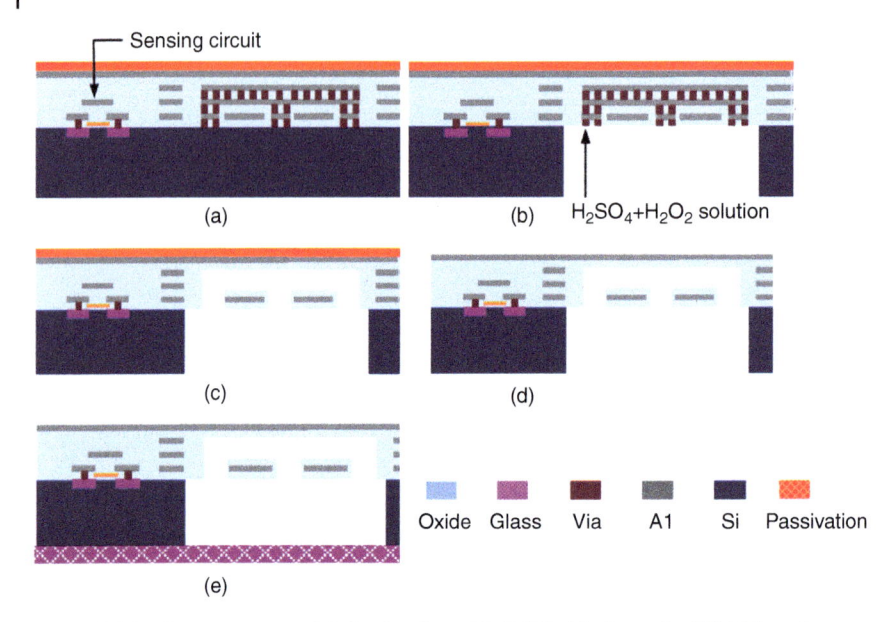

Figure 8.20 Pressure sensor fabrication flow. (a) CMOS chip from the TSMC foundry, (b) backside DRIE, (c) metal wet etching using $H_2SO_4 + H_2O_2$, (d) passivation layer removal by RIE, and (e) backside sealing using Pyrex7740. Source: Sun et al. [28]. © 2009, IOP Publishing.

which are embedded with the dielectric films. The suspended structure also reacts to applied pressure and is referred to as the reference electrode. The resultant pressure on both sides of the diaphragm is sensed using the change of gap and corresponding to the change in capacitance, which occurs between the two sensing electrodes. The presence of oxide dielectric over the sensing electrodes prevents short circuits of the sensing electrodes. Different pressure sensitivities can be achieved using different metals to attain varying diaphragm thickness and sensing gaps as shown in Figure 8.19c.

The pressure sensor process flow is shown in Figure 8.20. The process starts with backside DRIE to release the structures at the later stage. The lift-off patterned aluminum acts as the mask for DRIE getting the hole as shown in Figure 8.12b. The metal release process to etch the sacrificial layers of aluminum and tungsten-vias is carried using the solution of $H_2SO_4 + H_2O_2$ from the backside and form the metal embedded electrode structures as shown in Figure 8.20c. The front-side passivation layer is removed using RIE followed by sealing the backside with Pyrex7740 glass. Figure 8.21 shows the SEM image of the fabricated pressure sensor. Measurement results show the device sensitivity ranging from 0.14 to 7.87 mV kPa^{-1} [28].

8.3.3 Resonators

Resonators realized using CMOS MEMS provide key features, which include reduced form factor, increased performance, circuit integration that reduces the parasitic stray capacitance, and increased design matrix to accommodate different

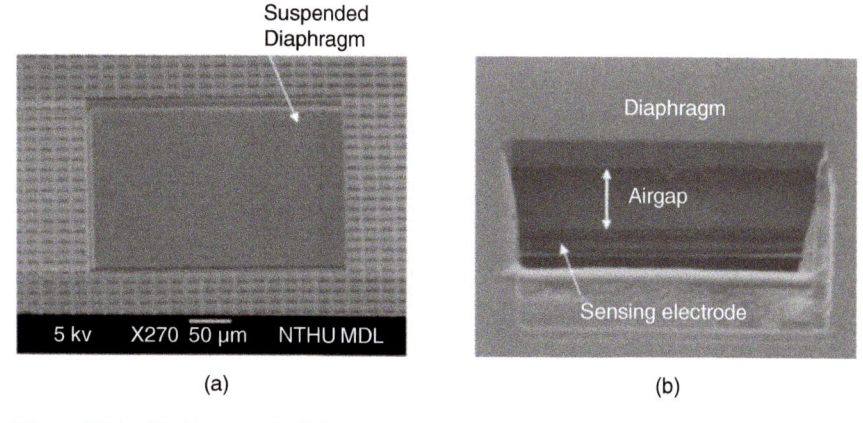

Figure 8.21 SEM images. (a) Fabricated pressure sensor and (b) FIB cut images of the diaphragm, sensing electrode, and air gap. Source: Sun et al. [28]. © 2009, IOP Publishing.

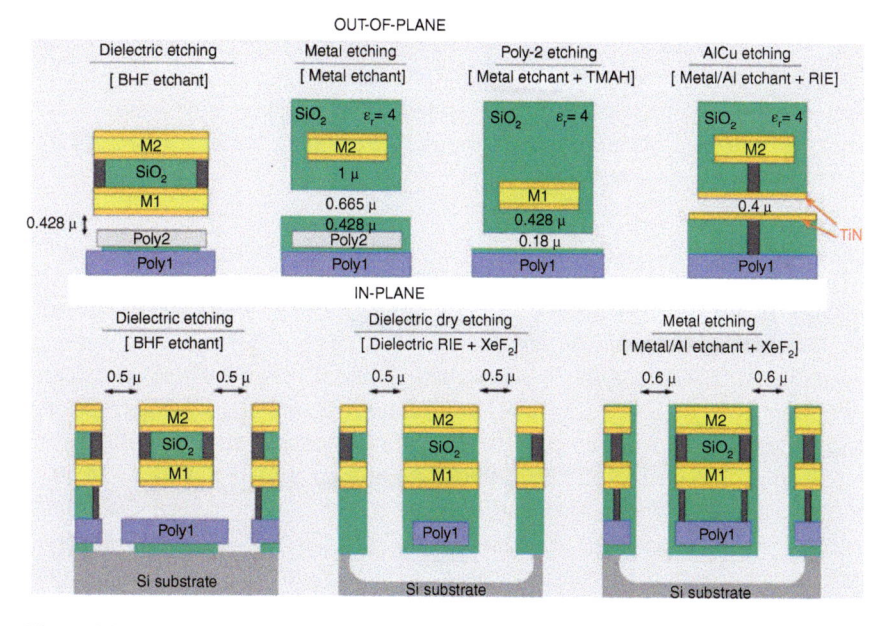

Figure 8.22 Different cross-sectional view of CMOS MEMS 0.35 μm resonator showing varied post-CMOS processes. Source: Chen et al. [34]. © 2019, IEEE.

in-plane and out-of-plane designs for applications such as oscillators and filters. Although the resonator in the CMOS process can be realized by different CMOS post-processes, they provide certain disadvantages, which lead the designer towards the TiN-C process [34, 67]. Figure 8.22 shows the different post processes, which can enable the designer to realize different resonators [34].

The resonator structures can be achieved with the combination of metal etching and XeF_2 substrate release, which resolves the problem of low-quality factors of metal-rich structures, but the major challenge that it suffers is the presence of dielectric material in between two gaps, and hence the resonator suffers from charging effect, which leads to adverse frequency drift. This process also needs improvement in the motional resistance, which can be achieved by reducing the gap further using a smaller technology node or the TiN-C process, which contains a smaller transduction gap. The resonator structures realized by dielectric etching suffers from low yield as controlling the etching time can be difficult. Even though using the dielectric dry etching can solve the problem, the use of dry etching introduces undesired re-depositions at the side walls, which is a concern. As explained before in the subsequent sections, the use of TiN composite can solve the problem generated from the charging effect, and the small gap can be realized by the process. Another TiN process described in Figure 8.23 can be used to realize structures for which the designer needs to preserve the bottom two polysilicon as a portion of mechanical structures. The main change from the TiN composite process explained in the previous section is that the structure is released by XeF_2 vapor silicon substrate etching after the first two steps of RIE as shown in Figure 8.23. The important design that needs to be considered in this process is that the bottom part of the transducer should be larger than its top to expose the sacrificial Al–Cu in metal-1 in the duration of the first two-step RIE process. Figure 8.24 shows the SEM images of the TiN-C resonator with the underlying polysilicon layer inside. With the help of controlled release time of isotropic XeF_2 etching process, the resonator circuit integration can be made very compact.

8.3.4 Others

By using the same CMOS 0.35 μm through different design mythologies but the same process sequences work, tactical sensors [68] and infrared (IR) sensors [69] can be realized. Using this process many different sensors can be integrated on the same chip to design a system, for example, integration of an accelerometer, pressure sensor, and temperature sensor can be used in the design of a smart TPMS. The device monitors the tire pressure and temperature of the wheel and along with the accelerometer can detect the centrifugal force acting on the wheel triggering the TPMS system [28]. The CMOS MEMS process comprising of the two polysilicon layer can be used to design TPR oscillator as a mass sensor, which is difficult to realize using capacitive transduction due to high loss in an ambient environment. The design of TPR in CMOS MEMS 0.35 μm [42, 70, 71] resonators can provide comparable performance in terms of high-quality factor, transduction efficiency, and mass sensitivities with the SOI–MEMS counterparts [72, 73]. It provides a distinct advantage of isolating the resistive feedthrough by using poly-1 as thermal actuation and poly-2 as piezoresistive sensors. The motional transconductance (g_m) of such resonators is 16.96 μS, which is comparable to the SOI counterparts, which are in hundreds of μS. Some of the CMOS–TPR designs can achieve quality factors of more than 10 000 [71].

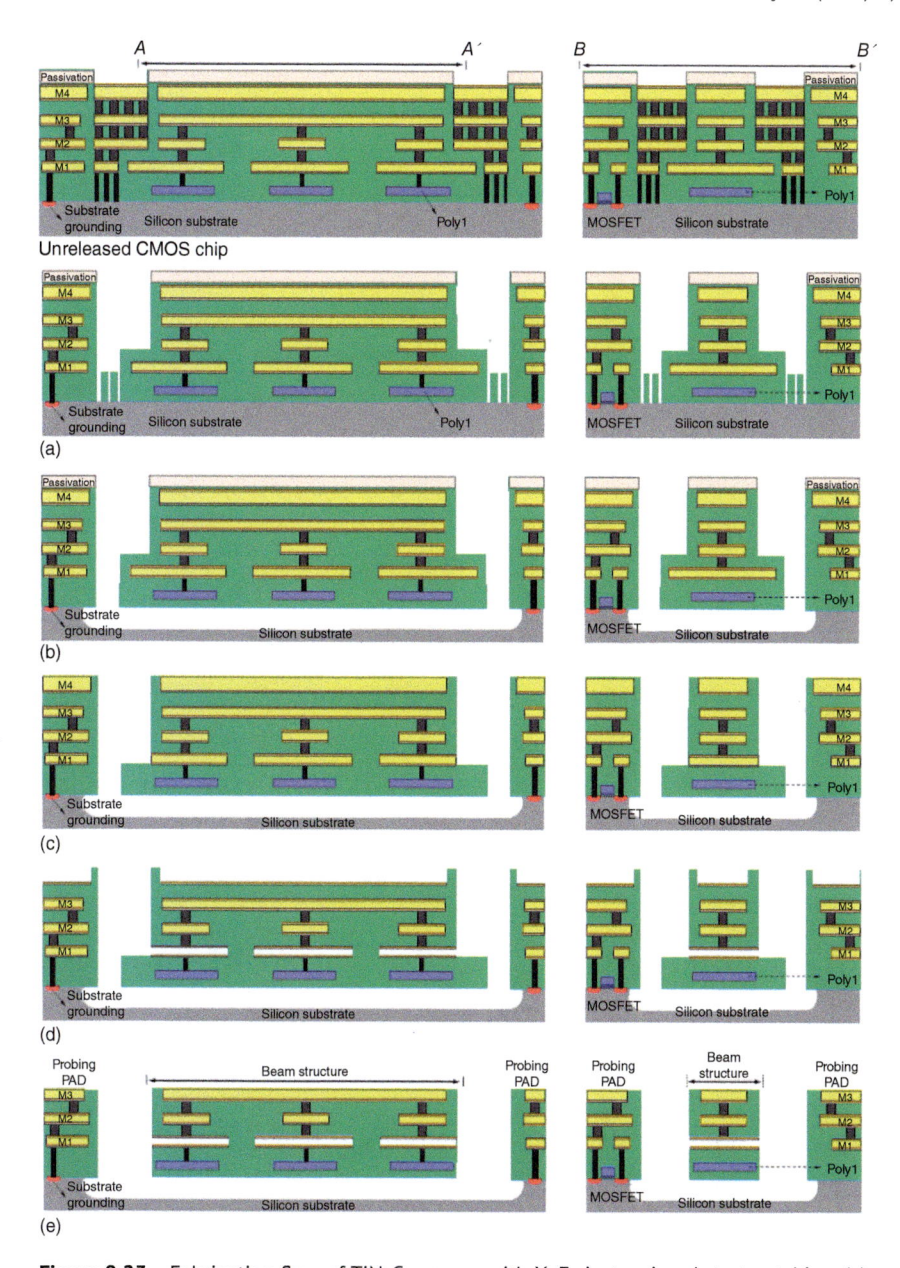

Figure 8.23 Fabrication flow of TiN-C process with XeF$_2$ isotropic substrate etching. (a) Wet-etching for metal removal with H$_2$SO$_4$ + H$_2$O$_2$, (b) dry Si etching via XeFe$_2$ for structure release, (c) dry-etching for SiO$_2$ and TiN removal (M4 as hardmask), (d) wet-etching with Al-etchant, and (e) dry-etching for oxide and TiN removal (M3 as hardmask/PAD). Source: Chen et al. [34]. © 2019, IEEE.

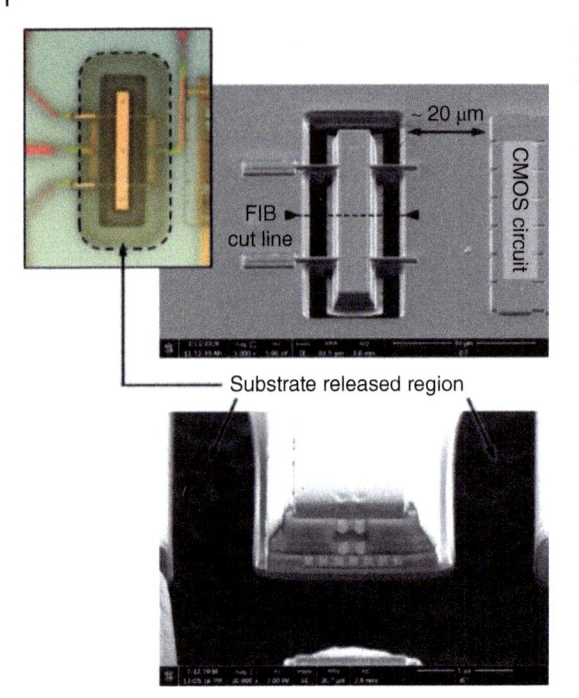

Figure 8.24 SEM images of the fabricated resonator with TiN-C process and XeF$_2$ isotropic substrate etching. Source: Chen et al. [34]. © 2019, IEEE.

8.4 The 1P6M CMOS Platform (0.18 μm)

This section explains the operating concepts and fabrication principles of tactile sensors, IR sensors, and resonators based on 1P6M CMOS technology where the process modules mentioned in Section 8.2 are utilized. Moreover, non-CMOS materials such as steel beads can also be integrated in the sensor by post-CMOS assembly to further enhance the functionality of the sensor.

8.4.1 Tactile Sensors

Tactile sensors are used to sense the touch, force, or pressure [74]. They are used in a variety of applications, such as the robotics field (fingers sensing), medical (human body weight/blood pressure measurement) [75], and consumer electronics (mobile phones). Using current MEMS technology and the standard CMOS process, their performance boosting and range of applications can be further extended by making the device small. Various sensing mechanisms are available for tactile sensing like capacitive [76, 77], piezoresistive, and piezoelectric [78, 79]. However, piezoelectric suffers from response problems and poling field-effect problems. Thus, piezoresistive and capacitive sensing mechanisms are the most common. Major parameters to optimize in tactile sensors are sensitivity and sensing range. Capacitive sensing is known to have high sensitivity but less sensing range, whereas piezoresistive offers a wide sensing range but does not offer high sensitivity. Generally, the tactile sensors are used to detect only small loads as the sensing range is mainly influenced by the stiffness of the mechanical structure, which is usually dependent on the thickness of thin films available in CMOS process. Various methods have been reported to improve the sensing range of the tactile sensors like polymer filling [76, 77] and electrorheological (ER) fluid fill-in [80]. But this again poses a challenge in increasing sensitivity along with a wide sensing range. Thus, to achieve both sensitivity and wide sensing range simultaneously, integrated capacitive and piezoresistive sensing can be used [81]. Parasitic capacitance in capacitive and environment variation sensitivity, as well as complicated fabrication in piezoresistive, is still a major challenge. Also, the capacitive and piezoresistive process requires deformation of mechanically suspended structure for sensing, which can be damaged due to the residual stress generated by the coefficient of thermal expansion (CTE) mismatch or residual stress of the thin film. Therefore, inductive sensing is in trend recently [82–85].

Most recent designs reported using inductive coil sensing consist of a chrome steel ball as the detection interface called as a tactile bump. An additional polymer layer (deformable) is deposited on the sensing chip so that by adjusting its thickness, the stiffness can be tuned, thereby tuning the sensing range [83].

As shown in Figure 8.25, a planar spiral CMOS sensing coil has a cavity which is filled by a polymer. Using the polymer encapsulation, a chrome steel ball (acting like spring and the sensing interface) is integrated over the sensing chip. Magnetic flux is introduced by applying an AC signal on the sensing coil. As the tactile force is provided over the tactile bump, the polymer gets deformed, which changes the distance between the coil and the steel chrome ball. This induced magnetic flux, thus changing the inductance of the sensing coil, which is then measured to detect the applied tactile force.

CMOS process provides a promising platform to fabricate tactile sensors by offering small-size, low-power consumption, ability for circuit integration, and flexibility in performing the post process to achieve various benefits. As shown in Figure 8.26I(a), stacking as well as patterning is done by TSMC using 0.18 μm 1P6M standard process. To generate the cavity on the coil, the metal sacrificial layer is then removed by wet etching using H_2SO_4 and H_2O_2 depicted in Figure 8.26I(b). During

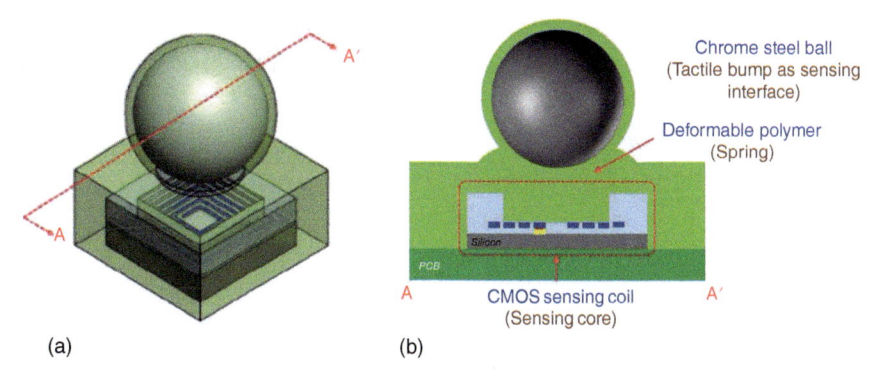

Figure 8.25 (a) Three-dimensional view and (b) cross-sectional view of inductive sensing-based tactile sensors. Source: Yeh et al. [83]. (Courtesy of Prof. Weileun Fang, NTHU, TW).

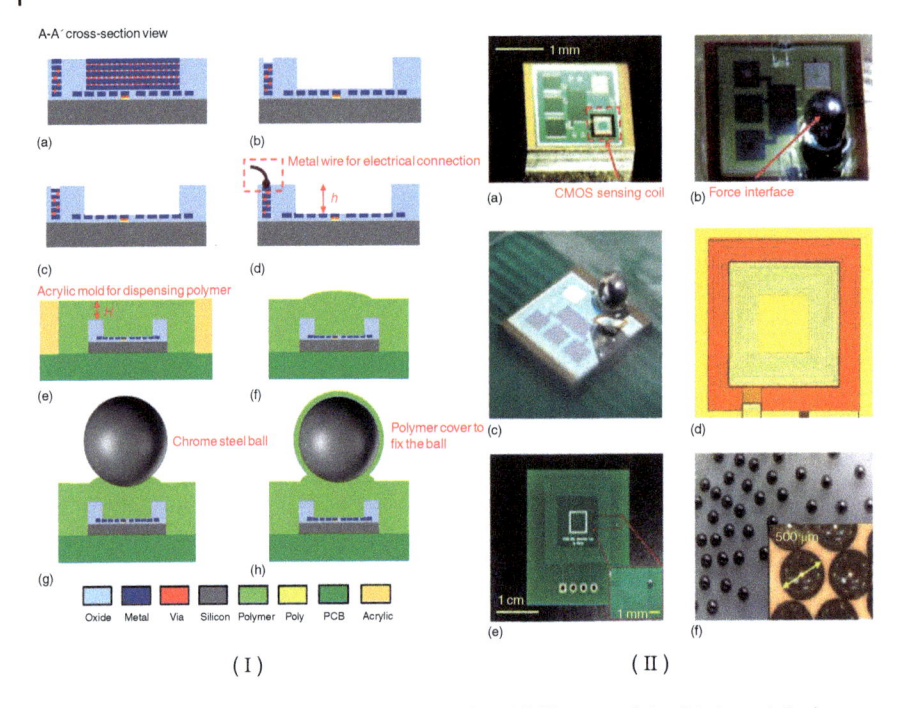

Figure 8.26 (I) Fabrication process of the device. (II) Pictures of the fabricated device. Source: Yeh and Fang [68]. (Courtesy of Prof. Weileun Fang, NTHU, TW).

the metal wet etch process, dielectric protects the sensing coil as well as the electrical routing. Figure 8.26I(c) shows the opening of the bond pads using RIE followed by the electrical connection as illustrated in Figure 8.26I(d). Figure 8.26I(e) indicates the encapsulation of the sensing chip after the polymer mold is implemented using another acrylic mold. Figure 8.26I(g) depicts the placement of the chrome steel ball over encapsulated polymer using the position table and vacuum head. Finally, the chrome steel ball is sealed by a thin layer of polymer. Figure 8.26II(a) depicts the CMOS sensing chip after metal sacrificial etch. Figure 8.26II(b) shows the integration of sensing chip with the chrome ball. Figure 8.26II(c) illustrates the polymer-coated sensing chip with the chrome steel ball present on top of the PCB along with a reading coil for wireless sensing purposes. Figure 8.26II(d) indicates the top view of the sensing coil. The inset showing the size difference between the ball and the coils depicted in Figure 8.26II(e) and II(f) finally shows the SEM micrograph of the chrome steel ball.

8.4.2 IR Sensor

IR sensors, due to their ability to detect the IR/non-visible waves, are known to be utilized in various commercial, industrial, and military applications to detect human behavior, night vision (e.g. for driving assistance), home security monitoring, noncontact temperature measurement [86, 87], etc. In general, using photon and thermal detectors is a way in which IR sensing can be done [88]. Unlike photon (quantum) detectors, thermal detectors can be used at room temperature, thus avoiding the usage of an extra cooling system [89]. Also, to analyze the reliability and performance of microsensors or microelectronic devices, it is important to characterize their thermal properties. Thus, IR sensing is usually done using thermal detectors. Further, thermal detection for IR sensing is mainly done using three sensing mechanisms: (i) bolometers, (ii) pyroelectric, and (iii) thermoelectric [90–92]. To gain advantages of no flicker noise and low power consumption, the thermoelectric effect is chosen over the others. They work on the principle of Seebeck effect, which converts the temperature difference of the sensor into an electrical output voltage.

To obtain less size and high responsivity, major considerations in thermoelectric sensor design are to achieve a good heat flow path and attain a larger temperature difference between hot and cold regions. To boost the performance of the thermoelectric IR sensors, various structures and ways are reported, including spiral thermocouple [93], stand-alone structure [94], metal-black films [95], multilayer stacking [96], serpentine thermopile [69], etc. An umbrella-like structure along with a serpentine transducer (with embedded thermocouple) is recently proposed to attain high responsivity, keeping the same device footprint [97]. To realize the mentioned configuration, TSMC 0.18 μm 1P6M standard CMOS process is used. Flexibility in electrical routing, choosing the mechanical as well as the sacrificial material, and the use of multilayer stacking in the CMOS process provides additional benefits to improve thermocouple structure design, thus improving the performance of thermoelectric IR sensors. For example, choosing polysilicon as the thermocouple material gives better performance due to high Seebeck coefficient. Figure 8.27 shows the overall sensor design. As shown, the serpentine thermopile with the embedded thermocouple is shielded with the umbrella-shaped heat

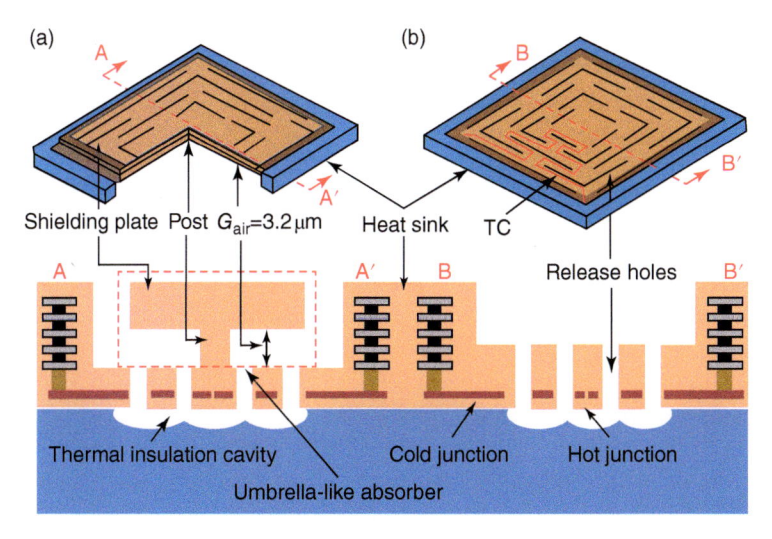

Figure 8.27 (a) Proposed IR sensor with umbrella design and (b) reference design without shielding plate absorber. Source: Shen et al. [97]. © 2019, IOP Publishing.

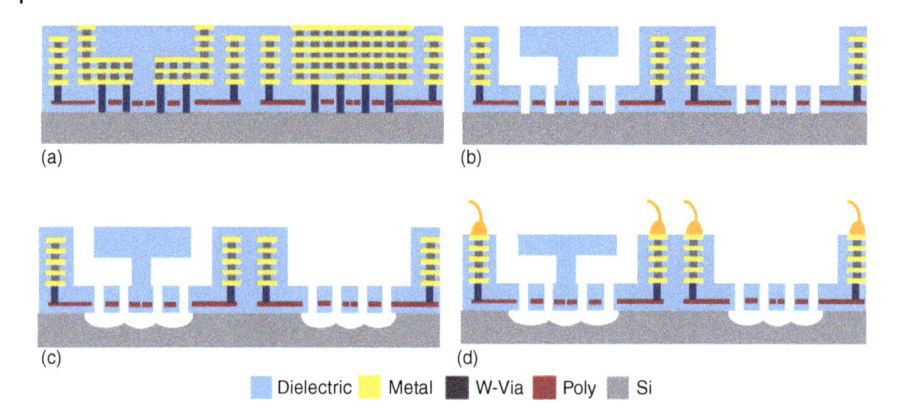

Figure 8.28 Post-CMOS fabrication process steps for the IR sensor. (a) Chip out, (b) metal wet etching, (c) XeF$_2$ released structure, and (d) RIE open PAD and wire bonded. . Source: Shen et al. [97]. © 2019, IOP Publishing.

absorber using a post. The heat generated on the umbrella absorber through the incident IR radiation is transferred to the thermocouple via post, thus increasing its temperature. This creates more temperature difference (ΔT) between cold and hot junction. Holes are made in the serpentine thermocouple to create thermal isolation with an air gap of 3.2 µm. Responsivity is further increased as no hole in the umbrella absorber increases the absorption area by 12%. Even higher responsivity can be achieved with a reduction in the sensor size.

The whole process is shown in Figure 8.28. As mentioned, the layers stacking and patterning is done by TSMC followed by some post-fabrication process to pattern and release the structure. To define the serpentine structure and umbrella-like absorber over it, metal layers are first removed using piranha treatment ($H_2SO_4 + H_2O_2$ etching solution). Next, to define thermal isolation, serpentine structure embedded with the thermocouple is suspended through dry Si bulk etching using XeF$_2$. Finally, opening the bond pads for wire bonding RIE is done.

The use of 0.18 µm 1P6M process is beneficial for the design of IR sensors in a way that its figure of merit is defined by

$$Z = \frac{\alpha^2}{\rho \kappa} T \tag{8.1}$$

where α is the Seebeck coefficient of the sensor material, ρ is the electrical resistivity, T is the temperature, and κ is the thermal conductivity, which can be improved by choosing the material with higher α from the available CMOS films (metal, dielectric, polysilicon, and tungsten). By adjusting the doping of the Si layer also, the Seebeck coefficient of the semiconductor can be increased. Final SEM micrograph after fabrication is shown in Figure 8.29.

8.4.3 Resonators

Thanks to the MEMS technology, small-size and low-cost resonators can be fabricated and used for various applications in timing reference devices and oscillators

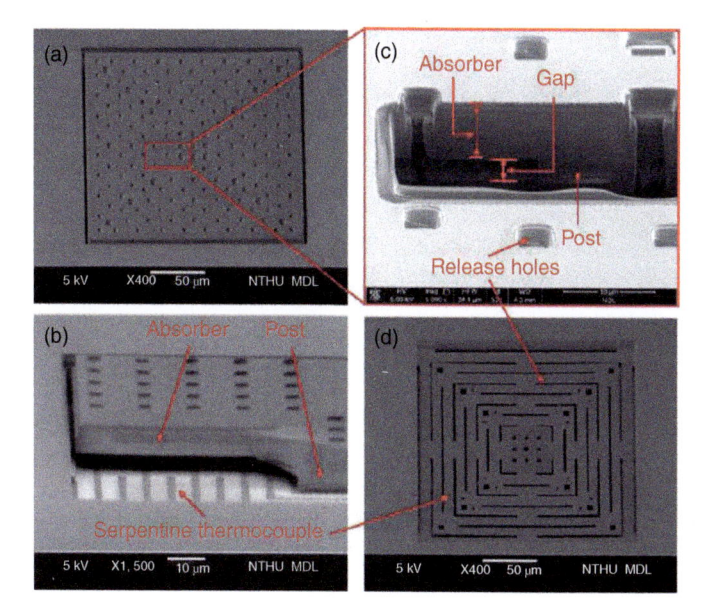

Figure 8.29 SEM image of the infrared sensors. (a, b) Proposed design with umbrella-shaped absorber, (c) FIB image showing the cross-sectional view and (d) and serpentine thermopile. Source: Shen et al. [97]. © 2019, IOP Publishing.

[98, 99]. Usage of CMOS MEMS then provides a way to integrate MEMS with the circuit, which further boosts system miniaturization to be used in a wireless transceiver system. There are mainly two key parameters to characterize the performance of a resonator: motional resistance R_m and quality factor Q. Many resonators are seen to be fabricated using CMOS 0.35 μm process [25, 100]. But the larger gap spacing and insufficient transduction area due to oxide etching [25, 101, 102] in 0.35 μm process lead to poor motional resistance R_m and lower Q. To take advantage of larger transduction area as well as smaller electrode-to-resonator gap spacing, the resonator fabrication is transferred from the CMOS 0.35 μm process to 0.18 μm process. Oxide structures embedded with the metal electrodes in the CMOS 0.18 μm 1P6M process are further utilized for the enhancement of Q-factor and temperature compensation in the resonators. Also, to suppress the feedthrough, a fully differential configuration is enabled by utilizing the flexible electrical routing provided by 0.18 μm CMOS platform.

To improve the resonator performance, other than the process benefits, the resonator structure design also plays an important role. A vertical DETF oxide resonator having higher transduction area to minimize the anchor loss is reported to have Q greater than 4800 at 10.4 MHz, with a stopband rejection of more than 20 dB [36]. As shown in Figure 8.30, an oxide DETF resonator is embedded with the metal electrodes showing electrical routing inside. Poly and M6 (Metal 6) electrodes act as differential drive ports, whereas M2 and M4 serve as differential sense ports. M1, M3, and M5 serve as the sacrificial layers during the release process. The out-of-phase mode is excited by applying DC-bias voltage to the embedded electrodes, thus

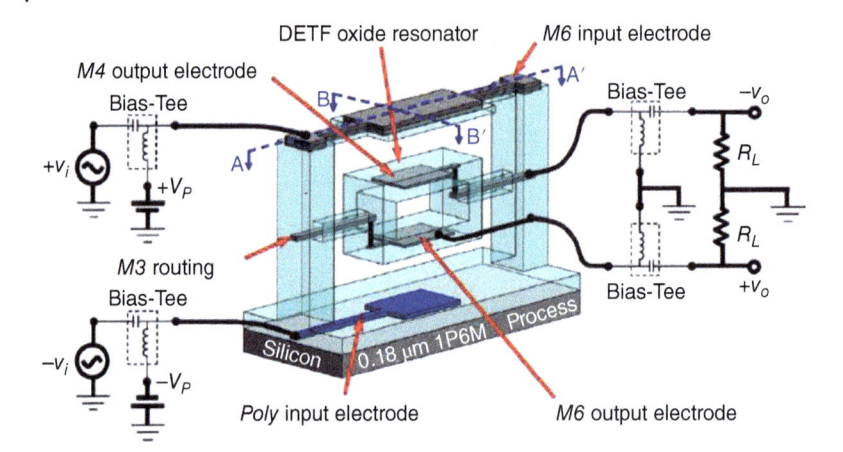

Figure 8.30 Schematic view of DETF resonator with differential drive/sense configuration. Source: Chen et al. [36]. © 2012, IEEE.

generating an electrostatic force. Due to the two time-varying capacitors generated between M4–M6 and Poly-M2, two motional currents with opposite polarity are generated at the output, which when summed together provides an enhanced motional sensing signal. This DETF resonator also suppresses the in-plane mode because of the unique mechanical design (slender beam) at the supports.

Chip obtained in Figure 8.31I(a) is manufactured using TSMC 0.18 μm CMOS platform. To obtain air gap spacing of 0.53 μm, metal sacrificial layers are then removed using wet etchant (KOH and TMAH) as shown in Figure 8.31I(b), followed by a bond pad opening using RIE as illustrated in Figure 8.31I(c). Global SEM view of the complete resonator is shown in Figure 8.31II(a). Figure 8.31II(b) depicts the cross-sectional view after FIB cut. Figure 8.31II(c) shows the zoomed-in image of the air gap with the embedded metal electrodes, and Figure 8.31II(d) illustrates the zoom-in image of the unique supporting beam.

Another example of the resonator fabricated using TSMC 0.18 μm CMOS platform but using an FFB made of metal/oxide composite structure is reported in [102] to reduce R_m (880 kΩ at 15.3 MHz) and increase Q. Figure 8.32a shows the design of the resonator along with its simulated mode shape, while Figure 8.32b illustrates the fabrication process.

After the chip is obtained from the foundry using TSMC 0.18 μm CMOS process, resonator structure is released using a highly selective oxide etchant, Silox Vapox III to define the gap spacing of 0.28 μm. Figure 8.33a illustrates the post-CMOS process indicating various types of resonators realized. Figure 8.33b depicts the FIB cross-sectional view and SEM image of the FFB and tuning fork realized using 0.18 μm 1P6M CMOS process.

8.4.4 Others

Many other types of sensors are fabricated using the 0.18 μm CMOS process. For example, both single-axis and tri-axis accelerometer using a serpentine spring

8.4 The 1P6M CMOS Platform (0.18 μm)

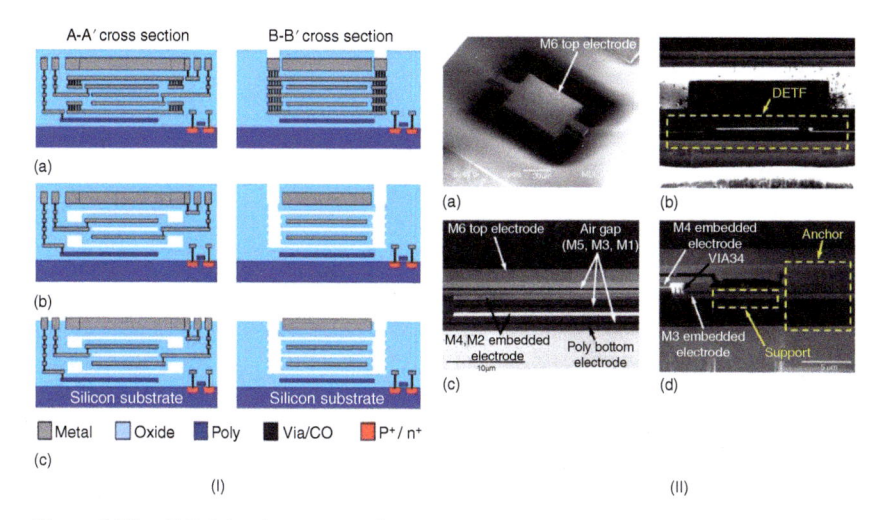

Figure 8.31 (I) Fabrication process flow and (II) global SEM and FIB cut images of the DETF. Source: Chen et al. [36]. © 2012, IEEE.

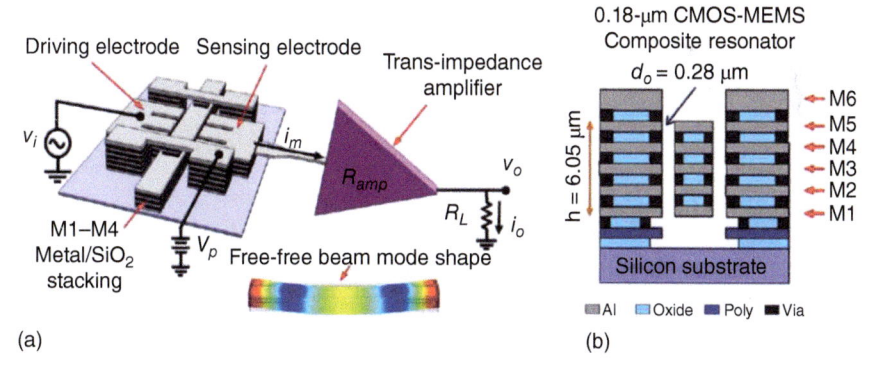

Figure 8.32 (a) Free-free beam design schematic. (b) Resonator's cross-sectional view using 0.18 μm CMOS process. Source: Li et al. [102]. © 2012, IEEE.

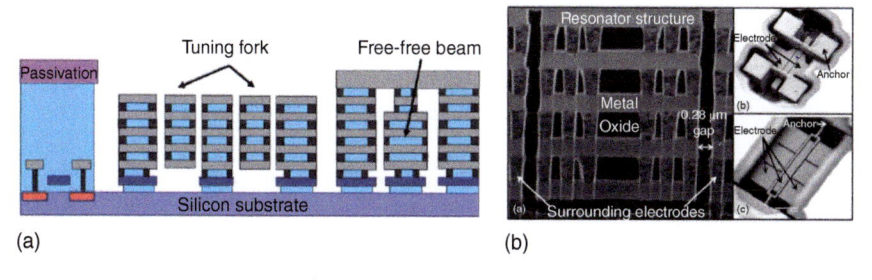

Figure 8.33 (a) Cross-sectional view after CMOS post process. (b) FIB and SEM view of resonators. Source: Li et al. [102]. © 2012, IEEE.

8 MEMS Using CMOS Wafer

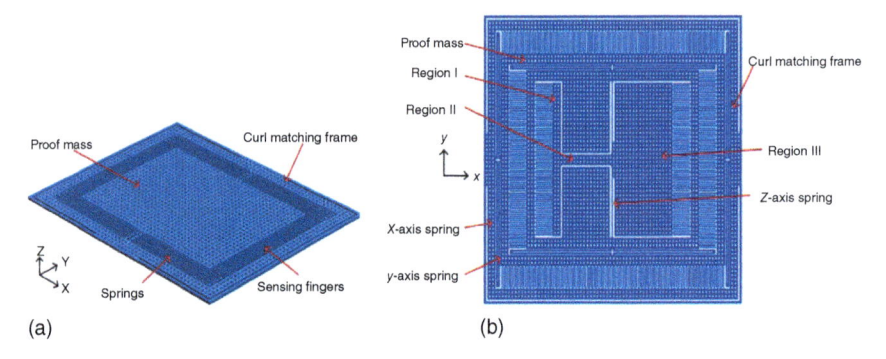

Figure 8.34 (a) Single-axis accelerometer design view. (b) Top view depicting a tri-axis accelerometer. Source: Liu and Wen [103]. CC BY 4.0.

is reported in [103] to measure the acceleration of gravity for navigation and motion detection. The schematic design of single-axis and tri-axis accelerometer is presented using Figure 8.34a and b, respectively. The fabrication process is shown in Figure 8.35. An additional metal layer (M7) is deposited after the standard 0.18 μm CMOS process is done (Figure 8.35a). A passivation layer is then patterned only over the circuit region (Figure 8.35b). This layer acts as the etch-resistant mask. Oxide dry etching (anisotropic) is then done to remove the unprotected area (Figure 8.35c), followed by isotropic silicon dry etching to finally release the microstructure (Figure 8.35d).

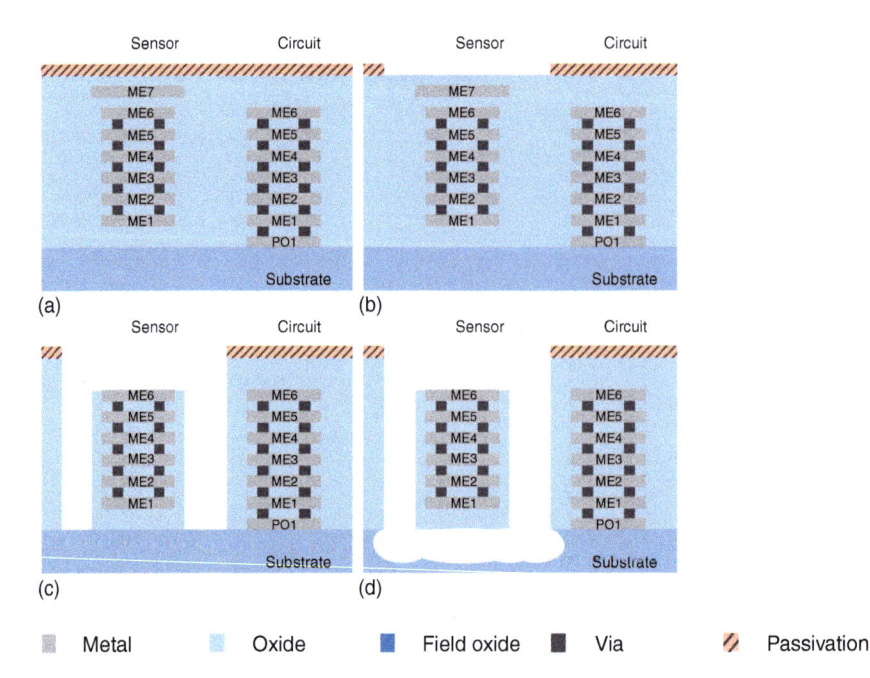

Figure 8.35 Fabrication process flow for the tri-axis accelerometer. Source: Liu and Wen [103]. CC BY 4.0.

Another example of a sensor fabricated using a 0.18 µm CMOS process is a pressure sensor [104]. Pressure sensors are used in many application areas of automobile, biomedical, and aerospace. Capacitive pressure sensors are the most common of all. With the variable separation, they measure the capacitance between the two electrodes due to the moving membrane. The cross-sectional view of a pressure sensor is shown in Figure 8.36a. The suspended membrane, as well as movable electrodes, is formed using M4 and M5 metal layers, three inter-metal dielectrics (IMD3, IMD4, and IMD5), and vias connecting M4 and M5. M2 forms the stationary electrode protected by IMD2. M3 is used as a sacrificial layer to

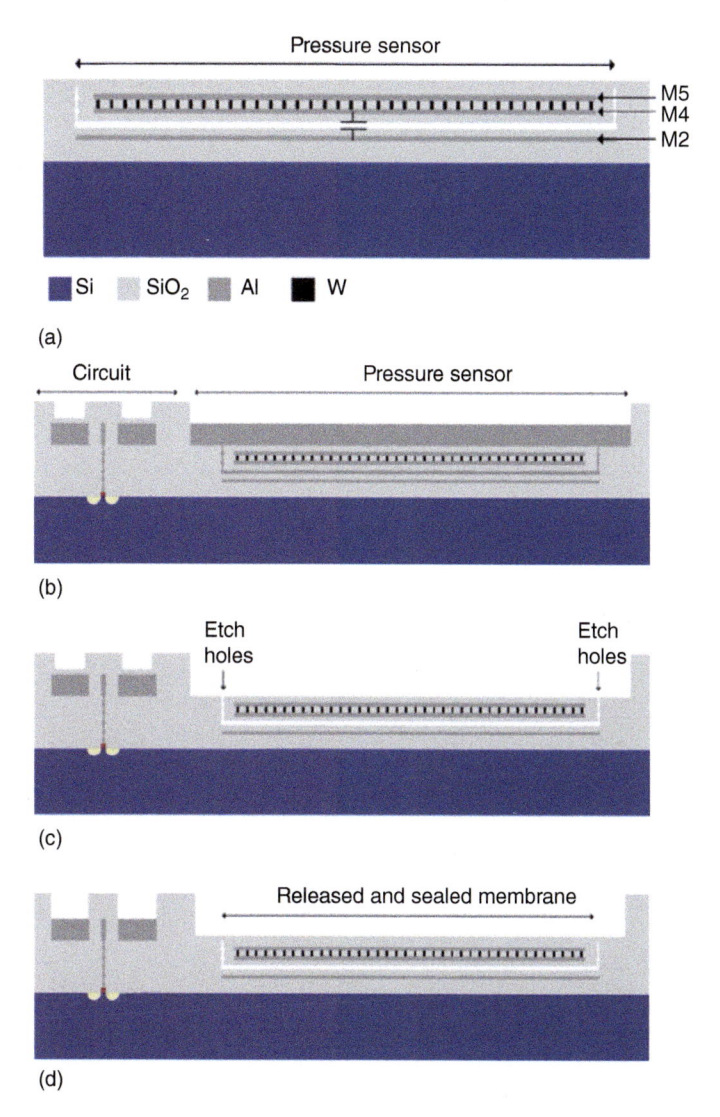

Figure 8.36 Cross-sectional view of the pressure sensor using CMOS MEMS process with its complete fabrication process flow. Source: Narducci et al. [104]. © 2013, IOP Publishing.

create a gap. After the chip is patterned using TSMC 0.18 μm CMOS process, all the area other than the sensing area is covered by the oxide layer and is patterned (Figure 8.36b). Then the sacrificial layer is etched using $H_2SO_4 + H_2O_2$ solution (Figure 8.36c). Note that electrical connections or top and bottom electrode are all protected by the oxide layer during sacrificial metal etch. The next step is the deposition of low-stress oxide to seal the etching holes followed by RIE to open the bond pads (Figure 8.36d). Table 8.1 summarizes the unique features and parameters of various sensors fabricated using the 2P4M and 1P6M CMOS MEMS platform.

8.5 CMOS MEMS with Add-on Materials

Materials used in standard CMOS processes include SCS, polysilicon, silicon-based dielectrics (SiO_2 and SiN), and metals (Al, W, and Cu). These materials are used to build transistors and interconnections for integrated circuits. However, they can also be used in MEMS structures if desired elastic or thermal properties can be designed and implemented therewith. Therefore, many physical transducers, such as accelerometers, gyroscopes, scanning mirrors, and electrothermal actuators, that employ only standard CMOS materials have been demonstrated in the CMOS MEMS technology.

To extend the scope of CMOS MEMS applications, however, more functional and active materials are needed for various sensors and actuators. These add-on materials are typically applied to the CMOS chips in the post-CMOS processes to ensure the compatibility with standard CMOS foundry processes. In the following, various CMOS MEMS devices with add-on materials are reviewed.

8.5.1 Gas and Humidity Sensors

Gas and humidity sensing is important for environmental sensing, pollution monitoring, and emission control. Standard CMOS materials usually do not interact with nor adsorb gas or moisture molecules. Therefore, additional sensing materials have to be deposited on CMOS chips to implement CMOS-based gas and humidity sensors. When the sensing films adsorb these molecules, their dielectric constants or resistivity are changed. On-chip capacitive or resistive interface readout circuits can be designed accordingly. The most commonly used sensing materials for gas and humidity sensors are metal oxides and polymers. The integration of gas sensing films with CMOS chips are discussed in the following sections.

8.5.1.1 Metal Oxide
Wide-bandgap semiconducting oxides such as tin oxide, zinc oxide, titanium oxide, and gallium oxide have been widely used for gas sensors [109]. Adsorption of gas molecules on the surface of the metal oxide induces the redox interaction and results in a change of its resistivity due to the gain or loss of electrons in the reaction. For better sensitivity, metal oxide-based gas sensors usually operate at elevated temperature around 200–400 °C. In miniaturized gas sensors, the sensing films and the micro

Table 8.1 Resonators and sensors realized using CMOS MEMS process.

Resonators and sensors realized using CMOS-MEMS process

	Reference	CMOS technology	Sacrificial layer	Structure material	Device type	Device motion	Resonant frequency	Q-Factor	Comment
Resonators	Verd et al (UAB) [100]	AMS 0.35 μm	CMOS Inter-metal dielectrics	Metal	Clamped-clamped beam	Laterally flexural	60 MHz	30	Not very high motional resistance due to limited transduction area
	Lopez et.al. (UAB) [101]	AMS 0.35 μm	CMOS Inter-metal dielectrics	Polysilicon	Clamped-clamped beam	Laterally flexural	22 MHz	4400	Very tiny transduction gap but yield problem
	Chen et.al. (NTHU) [25]	TSMC 0.35 μm	CMOS Inter-metal dielectrics	Metal	Free-free beam	Vertically flexural	3.66 MHz	1770	Multiple dimensional motion achieved due to flexible design of resonator
	Li et.al. (NTHU) [102]	TSMC 0.18 μm	CMOS Inter-metal dielectrics	Metal and CMOS inter-metal dielectrics	Free-free beam	Laterally flexural	15.3 MHz	767	Motional resistance of 880 kΩ
	Li et al. (NTHU) [105]	TSMC 0.35 μm	CMOS oxide	Metal + IMD (metal rich)	Free-free beam	Vertically flexural	10.5 MHz	2200	320 nm gap Low yield
	Li et al. (NTHU) [56]	TSMC 0.35 μm	Metal + W	Metal + IMD (oxide rich)	Overnized DETF	Laterally flexural	1.2 MHz	3029	930 nm gap, charging issue, 100% yield
	Chin et al. (NTHU) [44]	TSMC 0.35 μm	Polysilicon	Metal + IMD (oxide rich)	RGFET	Vertically flexural	4.28 MHz	1000	290 nm gap, only for two poly CMOS
	Chen et al. (NTHU) [67]	TSMC 0.35 μm	Metal (AlCu)	Metal + IMD (oxide rich)	Pseudo free-free beam	Vertically flexural	11.43 MHz	1823	400 nm gap, no charging issue, complicated post fabrication

(*Continued*)

Table 8.1 (Continued)

	Reference	CMOS technology	Sacrificial layer	Structure material	Device type	Device motion	Sensitivity (mv g^{-1})	Sensing range (g)	Comment
Accelerometers	Tsai et al. (NTHU) [29]	TSMC 0.35 μm	CMOS Inter-metal dielectrics	Metal + IMD (oxide rich)	Plate type	Laterally and vertically	z-Axis: 7.8 x-Axis/y-axis: 11.5	z-Axis: 0.01–3.0 x-Axis/y-axis: 0.01–3.0	Reduced parasitic capacitance Sub-micron sensing gap, and large sensing area
	Sun et al. (NTHU) [106]	TSMC 0.35 μm	CMOS Inter-metal dielectrics	Metal + IMD (oxide rich)	Single-proof mass with serpentine out-of-plane	Three-axis with serpentine in z-axis	z-Axis: 7.8 x-Axis: 0.53 y-Axis: 0.28	0.8–6	Reduces device footprint Suppressed cross-axis signal coupling
	Chen et al. (NCKU) [107]	TSMC 0.35 μm	CMOS Inter-metal dielectrics	Metal + IMD (oxide rich)	Single-proof mass with serpentine	x-Axis	2.4	Up to 10	Mechanical noise of 6.673 μg/√Hz, 3.1 kHz resonant frequency
	Chiang(NCYU) [108]	TSMC 0.35 μm	CMOS Inter-metal dielectrics	Metal + IMD (oxide rich)	Stress compensation frame design	Laterally and vertically	131.99	0.25–6.75	Overall noise of 579 μg/√Hz, B/W <100 Hz

	Reference	Process	Post-processing	Structure	Sensing type	Sensitivity	Range	Remarks
Tactile sensors	Lin et.al. (NTU) [75]	TSMC 0.35 μm	Oxide	Two square membranes supported by four beams at the corners	Capacitive based	2.65 Hz mmHg^{-1}	0–388 mmHg	To adjust the membrane's mechanical strength, three different via designs used
	Tu et.al. (NTHU) [81]	TSMC 0.18 μm	Metal	Capacitive membrane with four cantilever beams	Integrated capacitive and piezoresistive sensing	Capacitive: 1.8 fF mN^{-1} Piezoresistive: 9 mV N^{-1}	Capacitive 0–0.3 N Piezoresistive 0.05–0.5 N	Two-stage sensing done to enlarge the sensing range
	Yeh et.al. (NTHU) [68]	TSMC 0.35 μm	Oxide	Tri-axis tactile sensor	Inductive type	z-Axis 2.9 nH N^{-1} x-Axis 17.4^{-1} nH N^{-1} y-Axis 15.3^{-1} nH N^{-1}	—	Gap closing used for normal force detection (z-axis), area-based sensing for x- and y-axis detection, sensing coil array used
	Yeh et.al. (NTHU) [83]	TSMC 0.18 μm	Metal	Spiral sensing coil with a chrome steel ball sensing interface	Inductive type	9.22 (%/N)	0–1.4 N	Non-linearity of 2%, no mechanical suspended beam required

(Continued)

Table 8.1 (Continued)

Reference	CMOS technology	Sacrificial layer	Structure material	Device type	Sensing mechanism	Responsivity	Detectivity	Comments
Gitelman et.al.(TIIT) [88]	CMOS-SOI	CMOS inter-metal dielectrics	Silicon, polysilicon and inter-metal dielectrics	Cantilever-based mechanical structure	Thermal/uncooled sensing	40 mA W^{-1}	—	Noise equivalent temperature difference ~64mK, TCC ~4–10%
Chang et.al. (NTHU) [69]	TSMC 0.35 μm	Silicon	Oxide and polysilicon	Novel serpentine absorber membrane with strip via release hole	Thermoelectric	146.4 V W^{-1}	0.29×10^8 cm Hz$^{0.5}$W^{-1}	Average sensitivity: 2.1 mV K^{-1}
Shen et.al. (NTHU) [97]	TSMC 0.18 μm	Silicon	Oxide and polysilicon	Umbrella-like absorber over serpentine thermocouple	Thermoelectric	885.9 V W^{-1} @200 mTorr	0.058×10^8 cm Hz$^{0.5}$W^{-1}	Higher absorption area due to umbrella-like structure

IR sensors

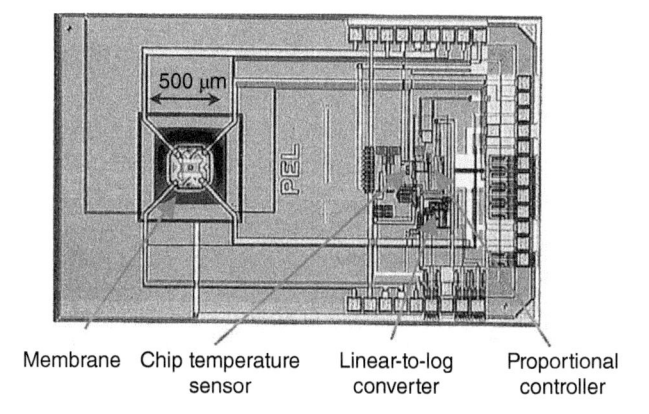

Membrane Chip temperature Linear-to-log Proportional
           sensor   converter   controller

Figure 8.37 Micrograph of integrated CMOS MEMS gas sensor. Source: Graf et al. [110]. © 2004, IEEE.

heaters can be integrated in suspended membranes. The thermally isolated design can reduce the heater power consumption and improve the sensor response time. For CMOS MEMS-based devices, the micro heaters or micro hot plates are typically implemented by using polysilicon.

Graf et al. demonstrated an integrated SnO_2-based CO sensor with on-chip control and sensing circuitry [110] (Figure 8.37). The suspended structure was released from the substrate by electrochemical anisotropic KOH etching mentioned in Section 6.2. A 5.5 μm thick *n*-well silicon island was formed at the bottom of the suspended membrane to homogenize the temperature distribution and to stiffen the structure. The sensing Pd-doped nanocrystalline SnO_2 thick film was applied to the membrane by drop coating [111]. After coating, the sensing material was annealed and sintered at 400 °C without degradation of the on-chip circuitry. Experimental results showed that the heater with a nominal resistance of 125 Ω could reach 350 °C at 5V power supply. The thermal time constant of a coated sensor was 22 ms. In the CO gas detection test, the sensor response time was less than 10 seconds and the detection limit was 5 ppm.

A similar suspended gas sensor was demonstrated by Afridi et al. [112]. The sensing SnO_2 and TiO_2 films were deposited on the CMOS chip by low-pressure chemical vapor deposition (LPCVD) at 250 and 300 °C, respectively, after the suspended membrane was released by XeF_2 silicon etching. The thermal efficiency of the micro polysilicon heater was measured to be $10\,°C\,mW^{-1}$. Various gases, including hydrogen, carbon monoxide, and methanol, were tested in the sensing experiments, and 100 ppb detection limit was demonstrated.

Dai et al. has investigated various gas and humidity sensors based on metal oxides such as CoO [113], COOOH/CNT [114], ZnO [115–117], SnO_2 [118, 119], WO_3 [120], TiO_2 [121], and Fe_2O_3 [122]. These sensors shared a common interdigital sensing electrode design. In the post-CMOS processes (Figure 8.38), the CMOS oxide in the sensing area was first etched to expose the metal or polysilicon sensing electrodes. The active sensing materials were prepared by sol–gel processes and deposited on the sensing area on the CMOS chip by using a precision micro dropper. Upon exposure

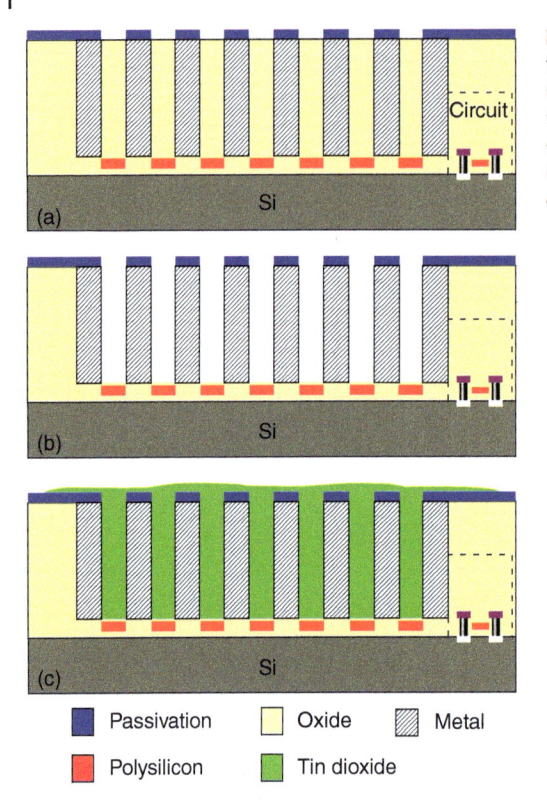

Figure 8.38 Post-CMOS fabrication processes of a typical CMOS gas sensor. (a) As-received CMOS chip, (b) sacrificial oxide etching, and (c) active sensing material deposition. Source: Yang and Dai [118]. CC BY 4.0.

to the target gas or moisture, the adsorbed gas molecules were detected by three possible mechanisms:

1. Capacitive detection of the change of dielectric constant of the sensing material between metal electrodes [113, 116, 118, 122]
2. Resistive detection of the change of resistivity of the sensing material between metal electrodes [117, 119, 121]
3. Resistive detection of the change of conductance of polysilicon electrodes [114, 115, 120]

Polysilicon micro heaters were embedded under the sensing electrodes and films if needed. However, most of these sensors were not fabricated on thermally isolated suspended membranes. Therefore higher power consumption (e.g. 1.24 W [122]) and longer response/recovery time (e.g. several tens of seconds [114–116]) were reported.

8.5.1.2 Polymer

Gas and humidity sensing can also be conducted by using polymers or polymer-based composite films. Polymers are favorable compared with metal oxides if high operation temperature is not needed.

Lazarus et al. [123, 124] demonstrated CMOS MEMS capacitive humidity sensors by filling polyimide between released comb fingers or parallel-plate electrodes.

Figure 8.39 CMOS MEMS capacitive humidity sensor. Source: Lazarus et al. [123]. © 2010, IEEE.

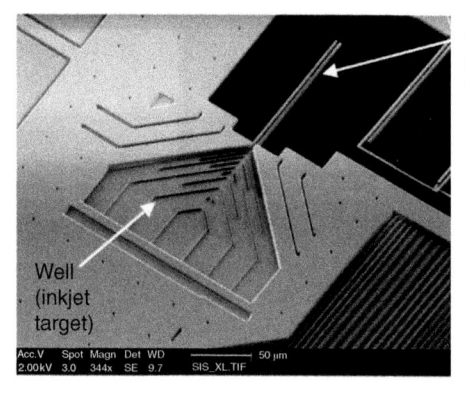

The gap between the electrodes were obtained by sacrificial oxide [123] or metal [124] etching. The released structures helped to reduce the response time because of more contact area between the sensing polyimide and the moisture (Figure 8.39). The filling of polyimide between the electrodes was carried out by ink-jetting and capillary wicking effects. In [124], the sacrificial layer of the vertical–parallel-plate (VPP) sensing capacitor was the M2 layer in the CMOS process, and the capacitor electrodes were designed in the TiW metal adhesion layers above and below M2. Such design could reduce the degradation of sensitivity due to the inter-metal oxide if M1 and M3 were used as capacitor electrodes.

Dai et al. also investigated polymer-based humidity sensors with integrated on-chip sensing circuits. In Ref. [125], a capacitive humidity sensor was fabricated by coating and patterning the polyimide sensing film on the sensing interdigital electrodes, which were exposed in the post-CMOS processes similar to that in Figure 8.38. In Ref. [126], polypyrrole was synthesized by chemical polymerization and dropped on to the CMOS sensing electrodes. The polypyrrole film was porous with a grain size of 0.3–0.5 μm. Such a nanoporous structure helped improve the sensitivity. The electrodes of the sensing capacitor were designed with a spiral pattern to increase the surface area and enhance the sensitivity (Figure 8.40).

Figure 8.40 Spiral sensing capacitor of CMOS MEMS capacitive humidity sensor. Source: Yang et al. [126]. CC BY 3.0.

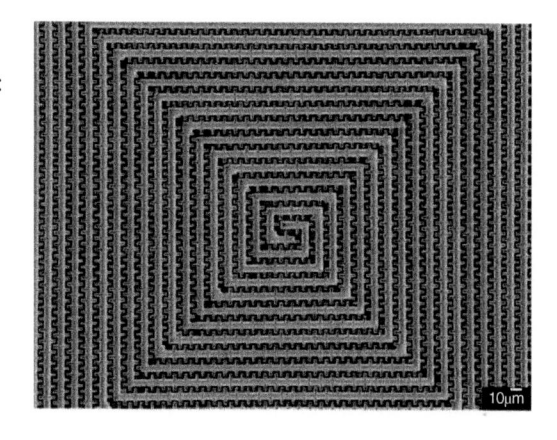

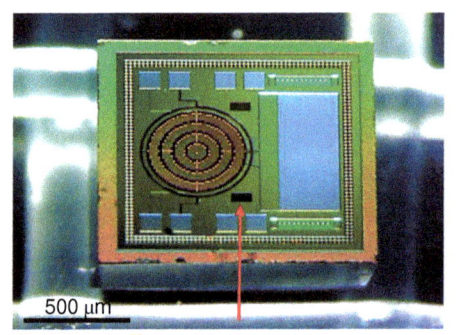

Figure 8.41 RF aerogel-based humidity sensor. Source: Chung et al. [127]. © 2015, Elsevier.

In Ref. [127], Chung et al. demonstrated a CMOS MEMS humidity sensor based on the VPP sensing capacitor similar to that in [124]. Compared with interdigital sensing capacitors, the VPP structure has less fringing capacitance due to the surrounding air, so the sensitivity can be improved. The sensing gap was filled with resorcinol-formaldehyde (RF) aerogel (Figure 8.41), which showed a twofold enhancement of humidity sensitivity compared with the commonly used polyimide.

Gas sensing was demonstrated by depositing polymer sensing materials on MEMS resonant platforms. The adsorbed gas molecules in the sensing films increase the platform mass and change its resonance frequency. Therefore the gas concentration can be measured by the shift of resonance frequency. In Ref. [128], polystyrene was deposited on the suspended platform attached to the end of a cantilever beam by ink-jet printing (Figure 8.42a). Several organic gases were tested and calibrated, including methanol, ethanol, 2-propanol, and acetone. A similar device was fabricated with polycarbosilane as the sensing film (Figure 8.42b) [129]. The device was tested for the nerve agent stimulant dimethylmethylphosphate (DMMP) in chemical weapons and a detection limit of 20 ppb, or 0.1 mg m^{-3}, was demonstrated.

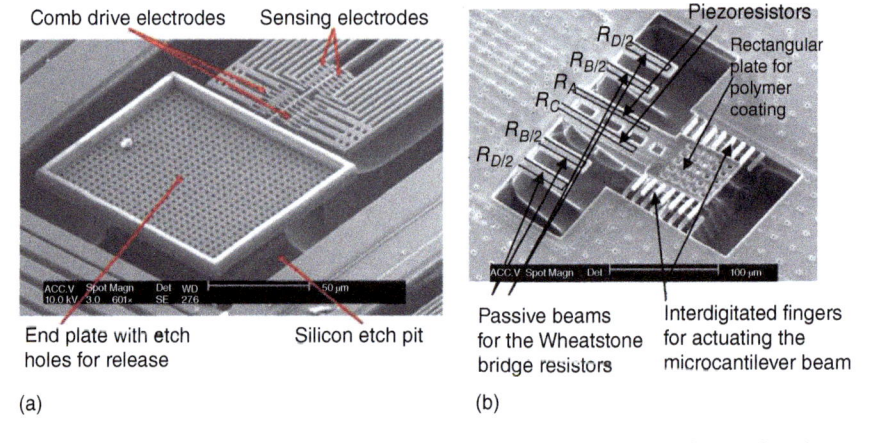

Figure 8.42 Mass-sensitive resonant gas sensors with (a) polystyrene. Source: Bedair and Fedder [128] © 2004, IEEE and (b) polycarbosilane sensing films. Source: Voiculescu et al. [129] © 2005, IEEE.

8.5.2 Biochemical Sensors

Biochemical sensing with CMOS technology typically requires coating or functionalizing the sensor surfaces with materials which can react with or absorb the target molecules. Once the target molecules are attached to the sensor surfaces, electrical output signals can be generated by various mechanisms, among which the most common ones are electrochemical detection, impedimetric measurement, mass-sensitive resonance measurement, and ion sensitive field effect transistor (ISFET)-based measurement.

In Ref. [130], a general-purpose CMOS BioMEMS platform was presented. In addition to MEMS structural release etching, an additional gold deposition was incorporated in the post-CMOS process to facilitate the binding of thiol-modified biomolecules for immobilization (Figure 8.43). Two biosensors were demonstrated. The first was an impedimetric sensor with Pt nanoparticles and silver enhancement for immunoassay. The impedance measurement with CMOS interdigital electrodes showed improved detection time compared with previous works. The second sensor was an extended-gate ISFET for creatinine detection. A creatinine-imprinted polymer film was coated on the extended gate in the top CMOS metal layer as the sensing film. In the test, the source-drain current of the ISFET was modulated by the creatinine concentration in the solution, and a ring oscillator was used to convert the current signal to a frequency output.

An open-gate ISFET was demonstrated in [131] for dopamine detection. The metal layers, inter-metal oxide, and polysilicon gate electrodes were removed by wet chemical etching in the post-CMOS processing, as shown in Figure 8.44a. The

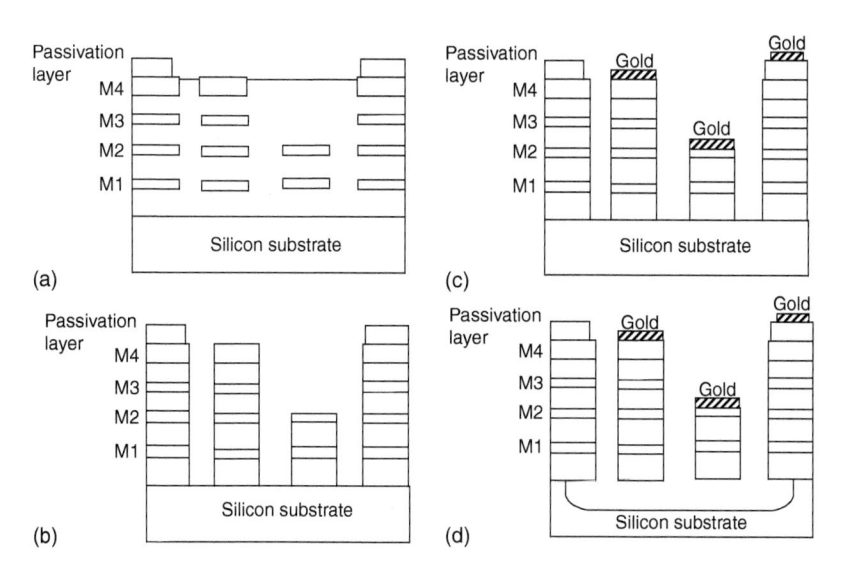

Figure 8.43 Post-CMOS processing in the CMOS BioMEMS platform. (a) As-received CMOS chip after passivation opening, (b) photolithography and etching of oxide in the sensor area, (c) patterning and deposition of Au, and (d) silicon sacrificial etching. Source: Tsai et al. [130]. © 2010, Elsevier.

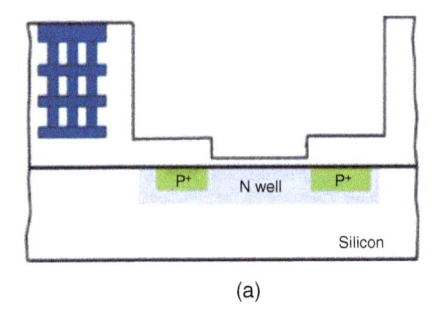

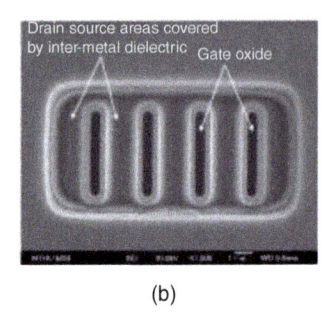

Figure 8.44 Open-gate ISFET for dopamine detection. (a) Cross section after post-CMOS processing and (b) scanning electron micrograph of a fabricated device. Source: Li et al. [131]. © 2010, IEEE.

exposed gate oxide was then functionalized for detection. Due to extremely small gate oxide thickness, the sensitivity and detection limit could be much improved. Experimental results showed a detection limit of 1–25 fM. A capacitive biosensor was demonstrated in [132] for avian influenza virus detection by using a 2P4M CMOS process. The top M4 layer was removed and the exposed oxide surface was functionalized for detection. The interdigital sensing electrodes were patterned in the M3 layer (Figure 8.45). The virus concentration was inferred from the capacitance change measured from the frequency response of the total equivalent impedance in the solution. Detection limits in the fM ranges were demonstrated.

Micro cantilevers are simple and robust devices with small dimensions and mass. Therefore sensing of loads from external force or mass can be highly effective in cantilever-based sensors. In Refs. [133, 134], capillary force acting on cantilevers was used for immunoassay. When the target molecules interacted with the functionalized cantilever surfaces, the surface energy and tension force were changed, resulting in the change of the loading force on the cantilever and its bending angle, as shown in Figure 8.46a,b [133]. The bending was detected by polysilicon piezoresistors embedded in the cantilever close to the fixed edge. The CMOS sensor

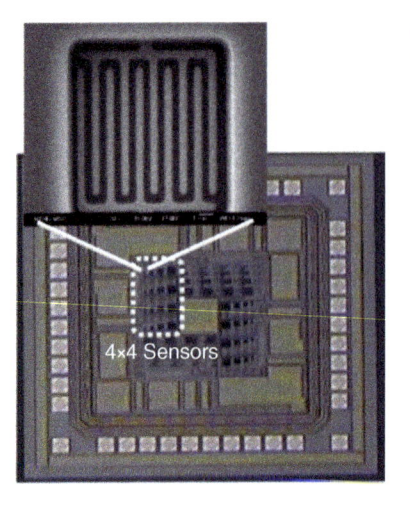

Figure 8.45 Capacitive avian influenza virus sensor. Source: Lai et al. [132]. © 2012, IEEE.

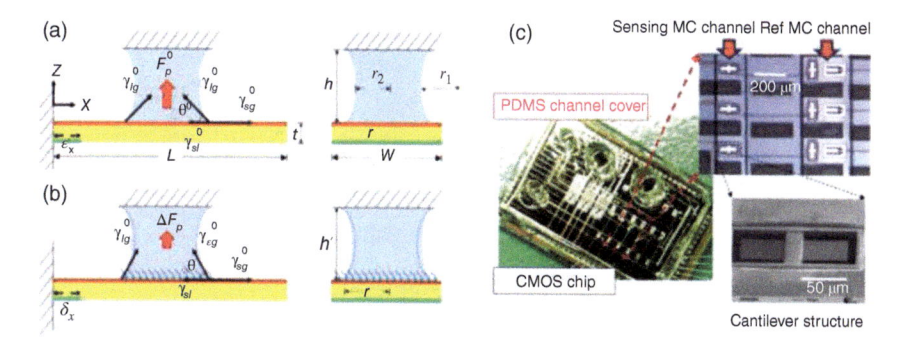

Figure 8.46 Capillary force-based biosensor. Source: Yin et al. [133, 134]. © 2011, AIP Publishing.

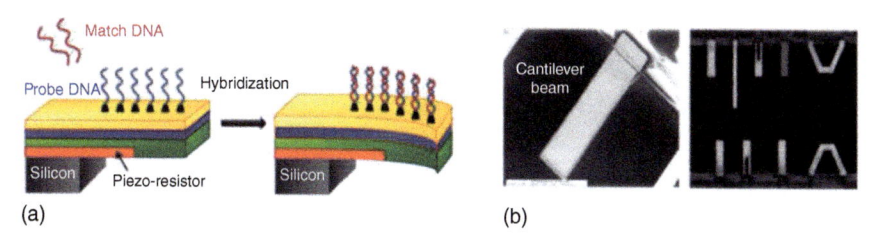

Figure 8.47 Cantilever-based HBV sensor. (a) Sensing principle and (b) fabricated cantilever. Source: Huang et al. [135]. © 2013, IEEE.

was embedded in a microfluidic chip (Figure 8.46c [134]), and a real-time and label-free immunoassay of troponin I (cTnI) was demonstrated with a detection limit of 1 pg ml^{-1}.

In Refs. [135, 136], a CMOS SoC with a cantilever-based DNA sensor was demonstrated for Hepatitis B virus (HBV) detection. As shown in Figure 8.47a, when the target DNA matches the probe DNA immobilized on the cantilever surface, the hybridization process creates a surface stress which changes the bending angle of the cantilever. In this work, the bending of the cantilever was detected by embedded piezoresistors via ring oscillators. Furthermore, a fully integrated SoC that included the sensor, controller, and wireless modulation and transmission modules was implemented and demonstrated. Experimental results showed that the sensor could detect 1 base pair DNA mismatch and achieved a detection limit of less than 1 pM.

8.5.3 Pressure and Acoustic Sensors

In CMOS pressure and acoustic sensors, the sensing membranes are usually designed and implemented using the BEOL metal/oxide stacks. If only front-side etching is used to release the sensing membranes, etching holes are opened on the membrane to facilitate the etching process. The etching holes have to be sealed after releasing to form closed membranes and isolated pressure/acoustic chambers for the sensors to function. Such sealing is usually achieved by conformal deposition of inorganic or organic thin films to cover the holes.

8 MEMS Using CMOS Wafer

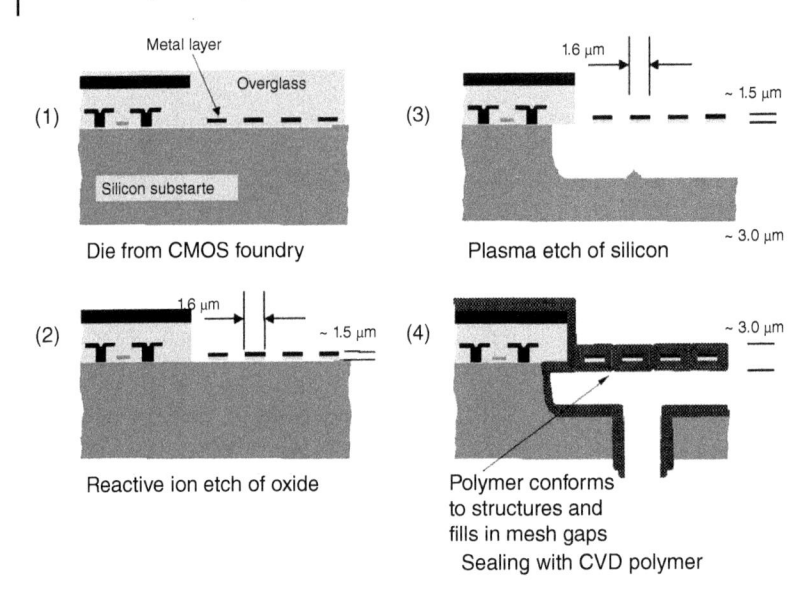

Figure 8.48 Fabrication process of the CMOS MEMS acoustic sensing membrane. Source: Neumann and Gabriel [19]. © 2002, Elsevier.

Monolithic CMOS MEMS microphones were pioneered by Akustica in the 1990s [19, 137]. The sensing membrane was fabricated by coating the metal and oxide mesh released from the CMOS substrate with Teflon-like polymer (0.5–1.0 μm) (Figure 8.48). With a membrane area of 0.61 mm^2, an A-weighted noise level of 46 dB SPL was achieved [137]. Such a single-chip solution has smaller footprint than the other two-chip solutions, which require separate microphone chips and ASIC signal conditioning chips. In [138, 139], capacitive micromachined ultrasonic transducers (CMUT) for medical imaging were demonstrated. The sensing capacitor was formed between the M4 and M2 layer in a 2P4M CMOS process. The sensing gap was obtained by removing the sacrificial M3 layer by wet chemical etching. Circular membranes with up to 100 μm diameter were successfully released. After releasing, the etching holes were sealed by silicon dioxide or parylene-D thin films (Figure 8.49). For the two types of sealed sensors, the measured sensitivities were 151 and 370 mV$_{pp}$ MPa^{-1} V^{-1}, respectively. The equivalent pressure noise floors were 3.3 and 1.35 Pa rtHz^{-1} at 1 V membrane bias, respectively. Three-dimensional imaging was demonstrated in [139].

Parylene was also used to seal etching holes in pressure sensors [140–142]. In Ref. [140], a special force–displacement transduction mechanism attached to the center of the sensing membrane was implemented in the sealed reference chamber of a pressure sensor. As shown in Figure 8.50a, the entire upper sensing capacitor plate has the same displacement, which is equal to the maximum displacement at the center of the sensing membrane. This is in contrast to the conventional design where the sensing membrane itself is the upper capacitor plate and only the central portion of the plate has large displacement (Figure 8.50b), resulting in smaller

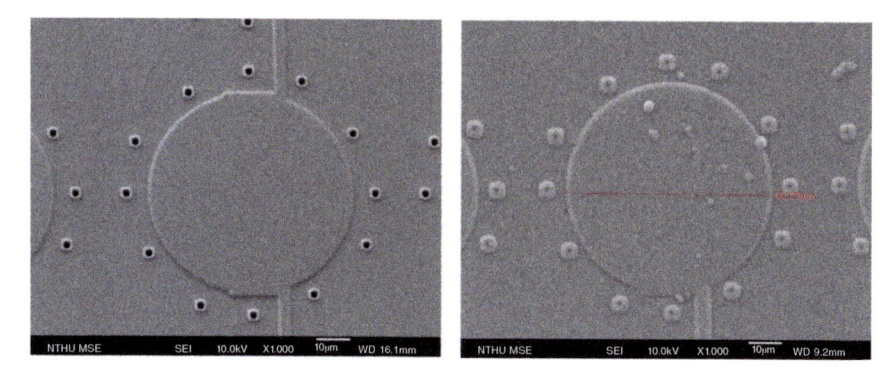

Figure 8.49 CMUT sensing membrane. (a) Before sealing and (b) after sealing of etching holes. Source: Tang et al. [138] © 2011, IOP Publishing.

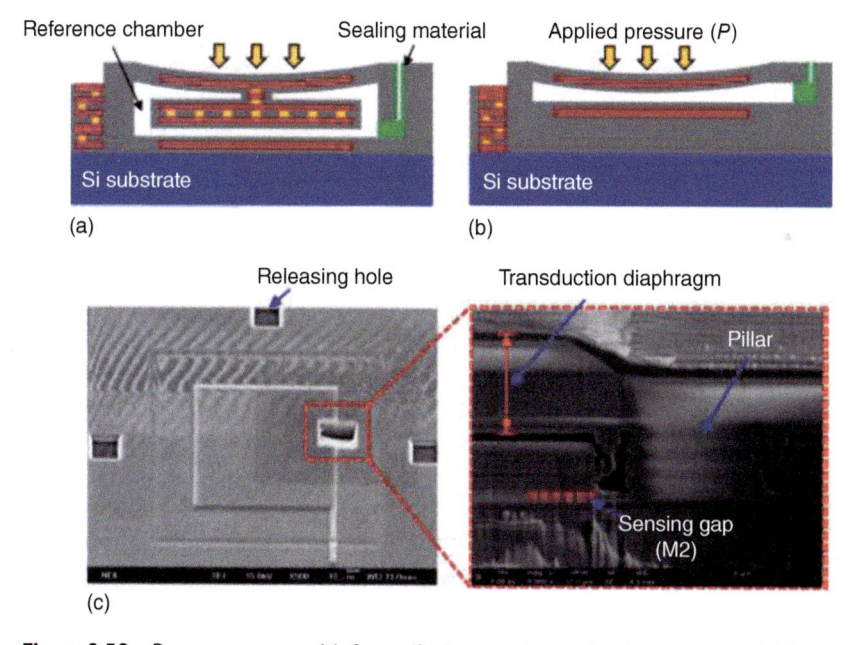

Figure 8.50 Pressure sensor with force-displacement transduction structure. (a) Proposed design, (b) conventional design, and (c) scanning electron and focused-ion-beam micrographs of fabricated device. Source: Cheng et al. [140]. © 2015, IOP Publishing.

overall capacitive signal. Experimental results showed that the proposed design enhanced the sensitivity by 126% in the 20–300 kPa pressure range with respect to the conventional design. Another design to enhance sensitivity was demonstrated in [141]. As shown in Figure 8.51, both upper and lower capacitor plates in the proposed design are deformable when subject to external pressure. This doubles the capacitance change of that in the conventional design where only the upper plate is deformable. Experimental results showed a 2.9-fold improvement of sensitivity as compared with conventional design.

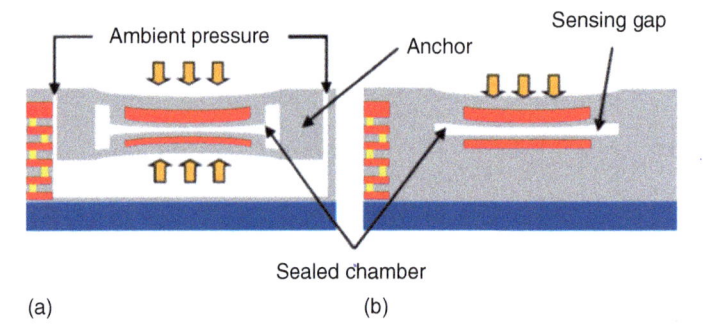

Figure 8.51 Pressure sensor with double sensing membranes. (a) Proposed design and (b) conventional design. Source: Lin et al. [141]. © 2017, IEEE.

8.5.3.1 Microfluidic Structures

Integration of CMOS chips with microfluidic structures enables direct sensing of chemical or biological species without complicated optical setup. Therefore the total volume of the system can be significantly reduced, and portability or disposability can be expected for applications such as point of care and personalized healthcare. However, due to the inherent mismatch of materials, dimensions, and planarity between the microfluidic network and CMOS chips, CMOS circuit chips are typically integrated with microfluidic networks by using embedded or molding technology on silicon [143], glass [144], poly dimethylsiloxane (PDMS) [145], epoxy mold [146], lead frames [147], or printed circuit boards (PCB) [148]. The fluidic samples in the micro channels flow horizontally over the CMOS chip surfaces [143, 144, 146–148] or vertically through the chips [145], depending on the specific device design.

In the above demonstrations, however, the CMOS sensing chips occupied only a small fraction of the total system. The fluidic handling still had to be managed externally. For a more compact and stand-along demonstration of the integration of CMOS and microfluidics, Chiu et al. [149] developed an integrated CMOS MEMS microfluidic capacitive inclinometer. Circular sensing electrodes were patterned in the metal layers in a standard 0.35 μm 2P4M CMOS process. The interdigital fringing electrodes were grouped into two differential capacitors which sensed the dielectric constant of the medium on top of the CMOS chip surface. As shown in Figure 8.52, a micro reservoir was fabricated on top of the CMOS sensing chip

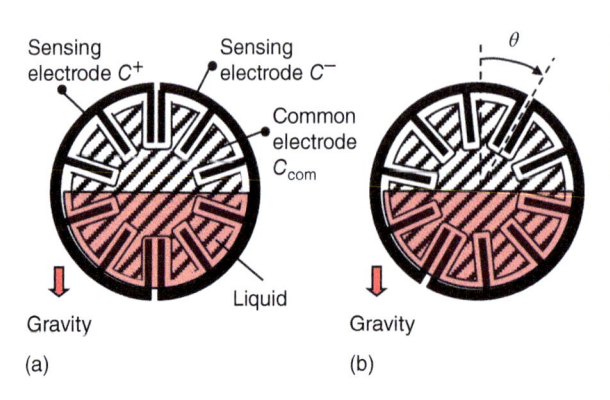

Figure 8.52 Sensing principle of the CMOS MEMS microfluidic capacitive inclinometer. Source: Chiu et al. [149]. © 2015, IEEE.

Figure 8.53 Fabrication and assembly processes of the integrated microfluidic inclinometer. Source: Chiu et al. [149]. © 2015, IEEE.

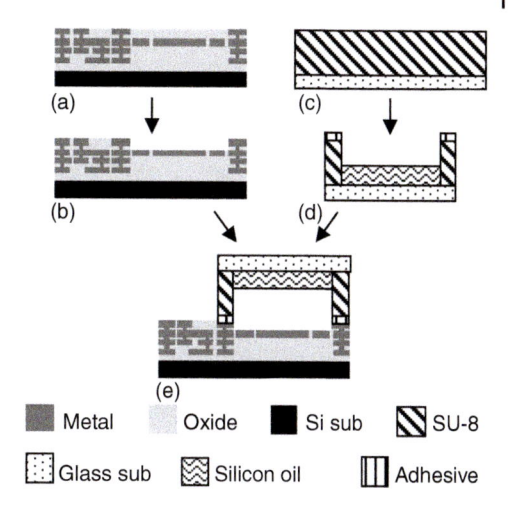

and partially filled with silicone oil. When placed vertically in the gravity field, the oil occupied the bottom half of the reservoir and covered different percentage of the two differential sensing electrodes, depending on the inclination angle of the integrated sensor in the gravity. The fabrication and packaging processes of the reservoir and the integrated sensor are shown in Figure 8.53. First, the passivation on top of the sensing electrodes in the as-received CMOS chip was removed by RIE (Figure 8.53a,b) to expose the sensing electrodes to the sensing media. The microfluidic reservoir was fabricated in 500-μm-thick dry film resist (DFR) on a glass substrate by multiple lamination and photolithography (Figure 8.53c). The reservoir chip was then diced and filled with silicone oil of half of the reservoir volume (Figure 8.53d). The CMOS chip and the reservoir chip were bonded by a flip chip bonder, as shown in Figure 8.53e. Figure 8.54 shows the fabricated reservoir chip and the air bubble in the reservoir at different inclination angles. The on-chip readout circuit consisted of a modulator and an inverting amplifier, as shown in Figure 8.55. Measurement results showed a sensitivity of 0.48 mV deg^{-1} and linear range of ±60°.

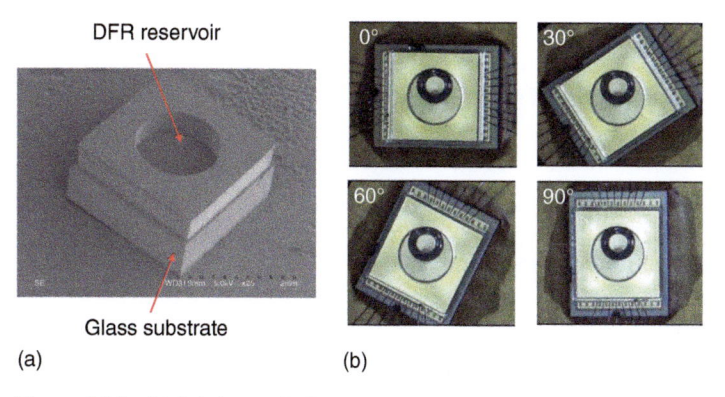

Figure 8.54 (a) Fabricated DFR reservoir and (b) air bubble inside the reservoir at different inclination angle. Source: Chiu et al. [149]. © 2015, IEEE.

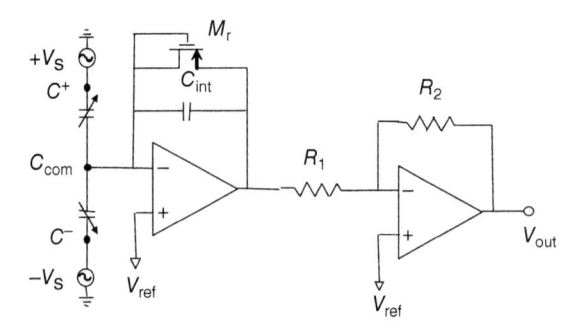

Figure 8.55 On-chip readout circuit of the microfluidic inclinometer. Source: Chiu et al. [149]. © 2015, IEEE.

8.6 Monolithic Integration of Circuits and Sensors

Most of the attraction of CMOS MEMS derives from the possibility of monolithic integration of mechanical structures with electronic circuitry by using the mature, high-volume, and high-yield CMOS foundry services. In addition, multiple metal layers in CMOS and various post-CMOS release techniques enable versatile and sophisticated device design and integration. In the following sections, multi-sensor integrations and sensor/readout circuitry integration are reviewed.

8.6.1 Multi-sensor Integration

Monolithic integration of multiple sensors on the same CMOS chip increases the functionality of the whole CMOS MEMS sensor systems. For example, multiple single-axis inertial sensors with orthogonal sensing axes can be arranged in the same accelerometer or gyroscope for multi-axes sensing. In gas sensing, arrays of gas sensors coated with different sensing films can be integrated together so that the data from different sensors can be processed by special algorithms to identify gas species in a mixture. Examples of monolithic multi-sensor integration are discussed in the following sections.

8.6.1.1 Gas Sensors

As discussed in Section 8.5.1, different sensing materials, post processing, and readout circuitry have been demonstrated in CMOS gas sensors. Since the selectivity of gas sensing by metal oxide or polymer films are not specific, multiple sensor arrays are needed in order to identify different gas species in an unknown mixture. There are two approaches to implement the sensor arrays. The first is using sensors with the same sensing mechanism but coated with different sensing films; therefore the sensors detect the same physicochemical property of different sensing films. The second approach is using sensors with different sensing mechanisms but coated with the same sensing film; therefore the sensors detect different physicochemical properties of the same sensing film when exposed to the measurand gases.

Monolithic integration of multiple gas sensors and required driving/sensing circuitry in CMOS technology was demonstrated in [150, 151]. Three different gas sen-

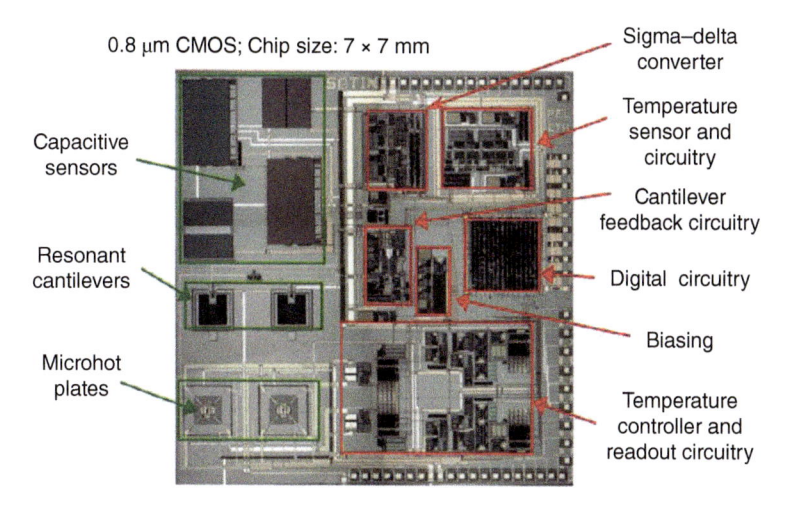

Figure 8.56 Monolithic multi-transducer CMOS MEMS gas sensor. Source: Li et al. [152]. © 2007, Elsevier.

sors (mass-sensitive resonant cantilevers, interdigital capacitive sensors, and Poly Si/Al thermoelectric calorimetric sensors) were designed and fabricated in the same CMOS chip. After post-CMOS release processes, 4 μm of poly etherurethane (PEUT) was coated on the sensing areas by air-brush spray coating. In the test, the output signals (resonant frequency shifts, capacitance values, and thermo-electric voltages) were obtained by the on-chip signal conditioning circuitry for various concentration of two volatile organic compounds (VOC), ethanol, and toluene. The detection limits of all three sensors ranged from 1 to 5 ppm.

In Ref. [152], two resonant cantilevers and two capacitive sensors for VOC sensing and two microhotplate sensors for inorganic gas sensing were monolithically integrated with sensing and control electronics (Figure 8.56). After post-CMOS micromachining, PEUT and PDMS were coated on the cantilevers and capacitive sensing electrodes by spray coating through shadow masks. Pd-doped nanocrystalline SnO_2 was deposited on the microhotplate sensors by drop coating. The cantilevers were driven by Lorentz force with piezoresistive feedback sensing. The capacitance of the interdigital capacitive sensors was converted to frequency signals by ΣΔ analog-to-digital converters. In the tests, toluene and ethanol could be easily differentiated by the array of the four VOC sensors. The co-integration of the polymer-based humidity-sensitive capacitive sensors with the SnO_2-based microhotplate CO sensors enabled the compensation of humidity effect on the CO sensing readout.

8.6.1.2 Physical Sensors

Different physical sensors such as inertial sensors, pressure sensors, and magnetic sensors can also be monolithically integrated in CMOS MEMS platforms [27, 28, 30, 81]. Fang et al. proposed a generic post-CMOS MEMS processing platform for multi-sensor integration [27]. As shown in Figure 8.8 in Section.

8.1, the process starts from a standard CMOS chip as received from the foundry (Figure 8.8a). The silicon substrate is first etched with anisotropic DRIE to expose the BEOL metal/oxide stack from the back side (Figure 8.8b). The moving structures and sensing electrodes designed and patterned in the metal/oxide stack are then partially released by using metal wet etching (Figure 8.8c) from both front and back sides. If desired, the sacrificial silicon substrate under specific devices is etched by XeF_2 isotropic etching from the front side to fully release the structures (Figure 8.8d,e). The first backside silicon etching step enables the implementation of continuous suspended membranes without etching holes and thus facilitates the design of devices such as pressure and acoustic sensors. Furthermore, post-CMOS add-on materials such as metal or polymer can be deposited prior to the etching step in Figure 8.8b or after the structure release (Figure 8.8f). Finally, another glass substrate can be bonded to the CMOS chip on the back side to form sealed cavities (Figure 8.8f). Based on this generic platform, Sun et al. demonstrated the integration of an accelerometer and a pressure sensor monolithically (Figure 8.9) [28]. The stiffness of the pressure sensing membrane and thus its sensitivity can be tailored by different membrane, electrode, and gap designs in the BEOL CMOS stacks composed of multiple metal, via, and oxide layers. In Ref. [30], a Pt-100 temperature sensor was further integrated with the accelerometer and the pressure sensor for a TPMS by depositing a 150 nm platinum layer at the beginning of the post-CMOS processing (Figure 8.9). It is noted that the front-side isotropic silicon etching in Figure 8.8e was skipped in [30], and all sensors were released by backside Si DRIE to simplify the fabrication process.

Sensors with different sensing mechanisms can also be integrated in the CMOS MEMS platform. Tu et al. [81] demonstrated vertical integration of capacitive and piezoresistive sensing in a tactile sensor to extend its sensing range. As shown in Figure 8.57a, the tactile sensor is composed of two sensing stages. The top sensor is a parallel plate capacitor with a sensing gap of 0.82 μm. Whereas the capacitive sensor has good sensitivity for small sensing gaps, its sensing range is limited by the available gap spacing in the CMOS metal/oxide stacks. Upon large force load

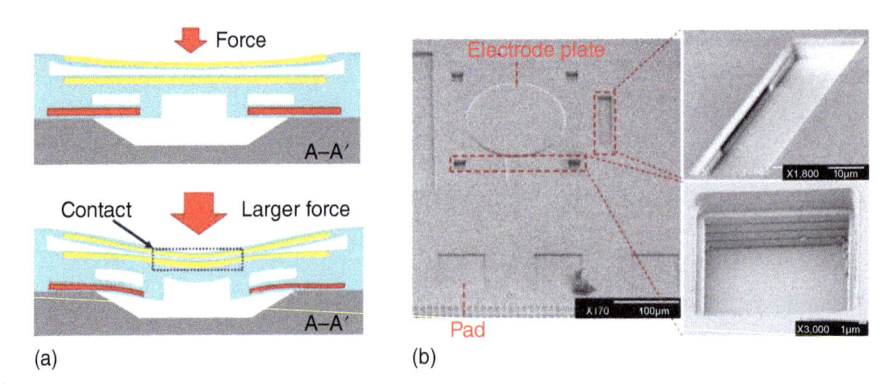

Figure 8.57 Vertical integration of capacitive and piezoresistive sensing in CMOS tactile sensors. (a) Principle and (b) SEM images showing sensing plate and release holes. Source: Tu et al. [81]. © 2017, IEEE.

that closes the capacitor gap and saturates the sensor, the second stage piezoresistive sensor is activated. The cantilever piezoresistive sensor has a much larger sensing range since the underneath silicon is removed by bulk wet etching micromachining. Thus the proposed sensor can achieve large load force range while maintaining high sensitivity at small load.

8.6.2 Readout Circuit Integration

Integration of sensors with on-chip signal-conditioning circuitry in the CMOS technology can enhance the sensor resolution by reducing parasitic effects and suppressing noise. In particular, differential sensing is frequently used to eliminate common-mode noise and drift caused by temperature variation and cross-axis sensitivity. Different types of front-end circuitry can be chosen to optimize the system performance depending on the specific sensing mechanisms.

8.6.2.1 Resistive Sensors

Piezoresistive pressure sensors, originally developed for monitoring the manifold absolute pressure (MAP) in automotive engines, are one of the first commercialized semiconductor micro sensors [153]. Therefore monolithic integration of sensors with on-chip signal conditioning circuits had long been demonstrated in both CMOS [154] and bipolar [155, 156] technologies. Piezoresistivity in semiconductor is a relatively significant effect. The change of resistance is typically detected by a fully differentially Wheatstone bridge to enhance signal amplitude and reject common-mode noise and drift [154]. In addition, the advantage of the Wheatstone bridge configuration or the single-element shear stress strain gauge design demonstrated in [155] is that only differential signals are detected. The absence of a DC offset in the sensor output alleviates the specification of analog-to-digital conversion (ADC). Since semiconductor resistivity is temperature dependent, temperature compensation is essential in the front-end signal conditioning circuitry [154, 156].

In resistive gas sensors, it is difficult to obtain differential resistive change in the sensing films. Therefore fully differential Wheatstone bridges are difficult to implement. The readout circuits in gas sensors are thus typically operational amplifier-based circuits such as inverting and non-inverting amplifiers [112, 114, 119]. Even if a Wheatstone bridge is used, only one of the resistors is sensitive to gas concentration [115, 121]. Furthermore, the resistance of the sensing films can have a range of several orders of magnitude, depending on the gas concentration. Therefore a logarithmic converter can be used to cope with such large dynamic ranges [157].

In addition to voltage-output circuitry, piezoresistive sensors, or resistive sensors in general, can also be detected by using oscillator-based circuitry of which the oscillation frequency depends on the measurand-sensitive resistance. The advantage of such frequency-output sensors is that the frequency of the output oscillators can be easily converted to digital codes by a counter without ADC. The offset issues in on-chip amplifier design can also be largely alleviated.

In Ref. [158], a single-mass tri-axis accelerometer was demonstrated by using a standard 0.35 µm 2P4M CMOS process. Both front-side and back-side post-CMOS

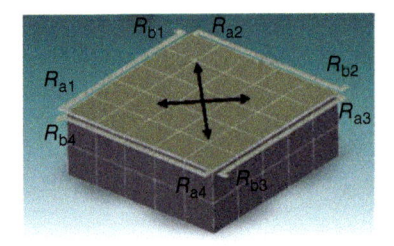

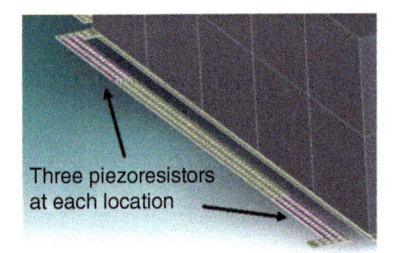

Figure 8.58 (a) 3D solid model of the accelerometer and (b) suspension beam. Source: Chiu et al. [158]. © 2014, MYU K.K.

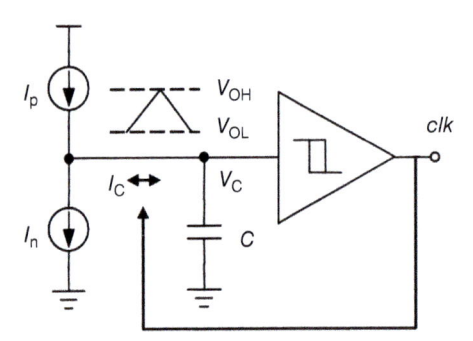

Figure 8.59 Principle of the RC relaxation oscillator for the piezoresistive accelerometer. Source: Chiu et al. [158]. © 2014, MYU K.K.

etching processes were employed to release the structure. As shown in Figure 8.58a, the silicon substrate is used as part of the proof mass to increase its mass and sensor sensitivity. The suspension beams are composed of layers from M2 to field oxide, as shown in Figure 8.58b. Differential piezoresistors are placed on both ends of the four suspension beams so that accelerations in three axes can be inferred by using a single proof mass. The readout circuit is an RC relaxation oscillator, as shown in Figure 8.59. By applying a constant voltage V_s across the differential sensing resistors $R_{p,n} = R_0 \pm \Delta R$, two sensing currents $I_{p,n} = V_s/R_{p,n}$ are generated. The difference current $I_c = I_p - I_n$ is used to charge or discharge the capacitance C. The charging/discharging cycles are controlled by the output of the Schmitt trigger. When the external acceleration causes a variation of $R_{p,n}$ and thus $I_{p,n}$, the output clk frequency, which depends on the charging current, will have a shift, which is proportional to the acceleration. Three piezoresistors are implemented at each location, as shown in Figure 8.58b. By connecting different pairs of these differential resistors, three RC oscillators can be constructed to distinguish the accelerations in three axes and eliminate the cross-axis sensitivity. The measured absolute sensitivity, relative sensitivity, and resolution along the z-axis were 198 kHz g^{-1}, 2.8 × 10^{-3} $\Delta f/f_0$ g^{-1}, and 10.9 mg rtHz^{-1}, respectively.

8.6.2.2 Capacitive Sensors

Capacitive sensing has been widely adopted in micro-motion sensors such as accelerometers [159–163]. In particular, surface-micromachined accelerometers with monolithically integrated sensor structures and readout circuits using CMOS or

BiCMOS processes were intensively investigated and demonstrated by UC Berkeley and Analog Device, Inc. in the 1990s. The advantage of capacitive sensing as compared with resistive sensing in micro sensors is that the air-gapped capacitance is much less sensitive to temperature than the semiconductor resistance. Therefore temperature-dependent drift and compensation are less essential in capacitive sensors. However, capacitive signals can be very small and susceptible to various parasitics. For example, in an accelerometer with a mechanical resonance frequency of 5 kHz, a 1 mg acceleration causes a proof mass displacement of only 0.1 Å [160]. If the sensing capacitor has a gap of 1 μm, the capacitive signal is only 10 aF. Therefore, careful design of the readout circuits is critical to pick up such small signals.

A typical surface-micromachined accelerometer is shown schematically in Figure 8.60 [161]. The external acceleration causes a displacement of the proof mass and induces a change of capacitance with respect to the stationary electrodes. For in-plane (x- or y-axis) detection, interdigital comb electrodes are commonly used, as shown in Figure 8.60. In this configuration, differential capacitance change can be easily arranged. Thus circuitry based on capacitive bridges can be used to enhance signals and reject common-mode errors [161–165]. For out-of-plane (z-axis) detection where the sensing capacitor is composed of the suspended proof mass as the upper electrode and the substrate as the bottom electrode [159, 160], differential capacitors are difficult to implement, so simpler single-ended circuits such as charge integrators are used [159].

To overcome the low-frequency amplifier errors and noises, two circuit techniques are commonly employed. Chopper stabilization is a modulation/demodulation method in which the low-frequency acceleration signal is modulated by the high-frequency carrier, as shown in Figure 8.61a. After buffering, amplification, and demodulation, a signal with enhanced SNR can be obtained. An alternative method is based on correlated double sampling as shown in Figure 8.61b. The circuit errors are sampled in the first phase of a detection cycle and subtracted from the error-containing signal in the second phase. Thus low-frequency errors and noises can be reduced to enhance the SNR.

In the earlier stage of development, force feedback was employed to keep the proof mass close to the rest position even in the presence of the acceleration-induced inertial force by applying a counter acting feedback force. In the closed-loop

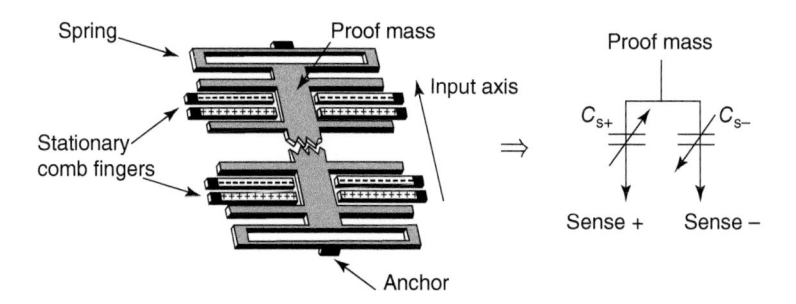

Figure 8.60 Schematic of a surface-micromachined accelerometer and its equivalent circuit model. Source: Lemkin and Boser [161]. © 1999, IEEE.

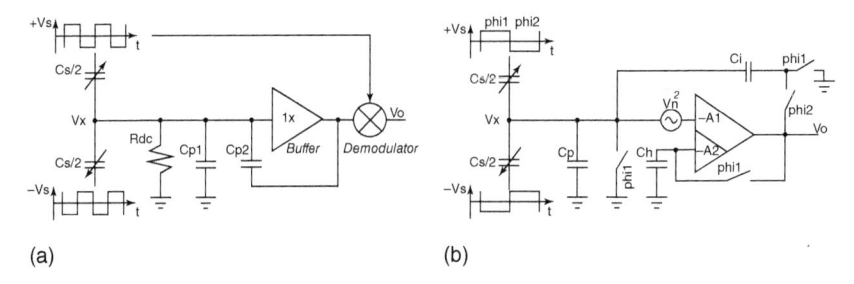

Figure 8.61 Low-frequency noise reduction techniques. (a) Chopper stabilization and (b) correlated double sampling. Source: Boser and Howe [160]. © 1996, IEEE.

force balancing scheme, the displacement of mass is minimal, thus the sensor performance is less sensitive to nonlinearity and uncertainty in both electrical and mechanical system of the sensor. Furthermore, the system bandwidth of the feedback loop can be extended beyond the natural frequency of the mass-spring system. Such a force balancing loop can be easily implemented with 1-bit electro-mechanical sigma-delta modulation, as shown in Figure 8.62 [161]. The choice of electrostatic force feedback in CMOS MEMS sensors in Figure 8.62 is natural since the feedback voltage applied to one of the electrodes in the sensing capacitor results in the feedback balancing force, and thus no extra force transducers are needed.

Digital offset trimming was employed in the capacitive sensing interface to further reduce the offset due to the capacitance mismatch at input nodes [161, 164]. As shown in Figure 8.63 [164], a binary-weighted capacitor array is added to the sensing capacitors in parallel to address their mismatch. The total capacitance of the array can be trimmed by the control bits determined in the calibration procedures.

Single sensing capacitors for vertical displacement detection, such as the z-axis sensing capacitor in [159, 160] and the flow sensing capacitor in [166], are typically detected by single-ended amplifiers such as charge integrators. They are less effective in terms of sensitivity and SNR as compared with differential capacitors. This issue has been addressed by two vertical sensing capacitor designs. The first is the vertical sensing combs demonstrated by Xie et al. [167]. As shown in Figure 8.64, multiple layers of metal electrodes are embedded in the CMOS MEMS interdigital comb fingers. By proper connection, isolated metal electrodes at different heights can be used in capacitors, which are sensitive to the vertical displacement, as

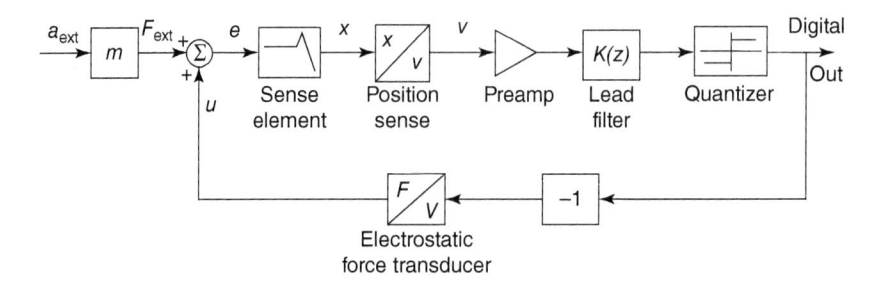

Figure 8.62 Force feedback loop. Source: Lemkin and Boser [161]. © 1999, IEEE.

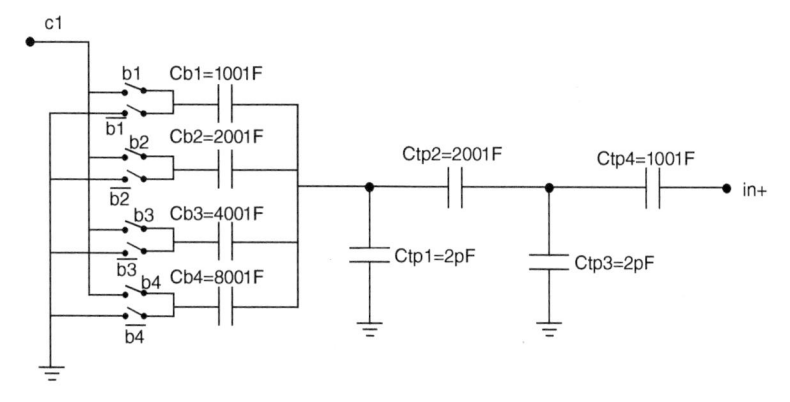

Figure 8.63 Digital trimming using binary-weighted capacitor array. Source: Tan et al. [164]. © 2011, IEEE.

shown in Figure 8.64b for a z-axis accelerometer [167]. This electrode design can be implemented by dry etching-based post processing and thus potentially has high yield. The second design is based on VPP capacitors. As shown in Figure 8.65, the M2 and M3 layers are removed by wet metal etching in the post-CMOS sacrificial release processes [106]. Thus differential VPP capacitors formed by M1 and M4 can be implemented on the two sides of the proof mass for vertical sensing, as shown in Figure 8.65b. Based on both z-sensitive VPP electrodes and conventional xy-sensitive comb electrodes, the full bridge configuration of differential capacitors was realized in all three axes in the tri-axis accelerometer with a single proof mass

Figure 8.64 (a) Conventional interdigital comb fingers for in-plane sensing, (b) vertical sensing comb electrodes, (c) equivalent model for vertical sensing combs, and (d) z-axis accelerometer employing vertical sensing comb. Source: Xie and Fedder. [167]. © 2002, Elsevier.

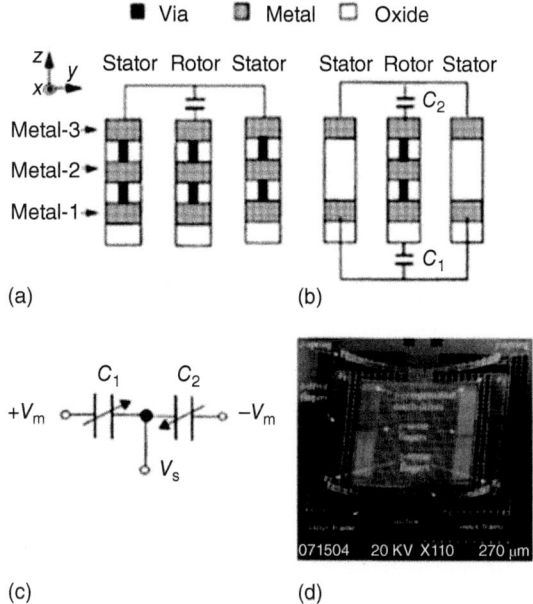

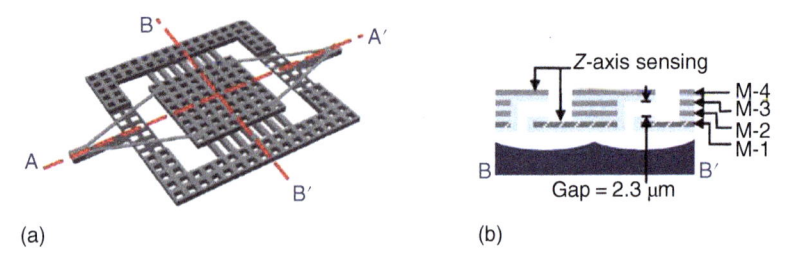

Figure 8.65 (a) Schematic of z-axis sensing unit showing proof mass and sensing electrodes and (b) BB' cross section of differential vertical sensing electrodes. Source: Sun et al. [106]. © 2010, IEEE.

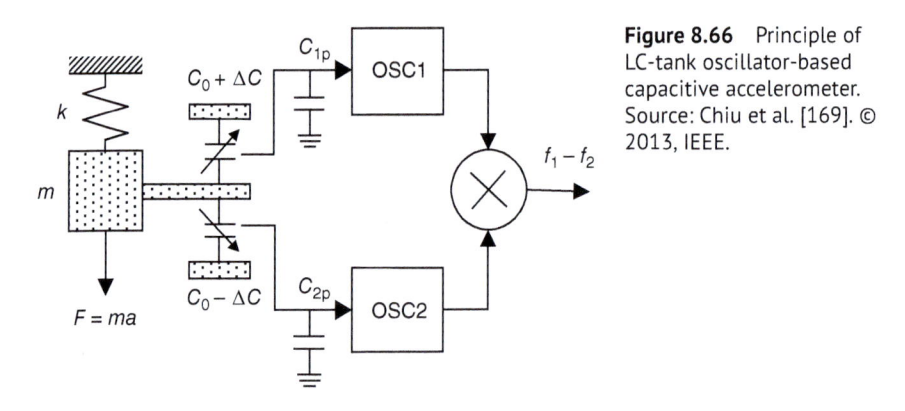

Figure 8.66 Principle of LC-tank oscillator-based capacitive accelerometer. Source: Chiu et al. [169]. © 2013, IEEE.

[106]. The resolutions in three directions were between 120 and 357 mg rtHz^{-1}, whereas the cross sensitivity ranged from 1% to 8.3%.

Oscillator-based readout circuits can also be applied to capacitive sensors. For example, simple RC ring oscillators were used in a capacitive tactile sensor [168] and various gas/humidity sensors [113, 116, 118, 122, 125, 126]. A more elaborate LC-tank oscillator was demonstrated in a capacitive accelerometer by Chiu et al. [169]. As shown in Figure 8.66, on-chip CMOS inductors are integrated with differential interdigital sensing capacitors to build two differential oscillators with oscillation frequencies $f_{1,2} = f_0 \pm \Delta f$. After mixing and low-pass filtering, the output difference frequency $2\Delta f$ is proportional to the capacitance change ΔC and thus the external acceleration. The CMOS inductors are embedded in the proof mass, as shown in Figure 8.67, to save the chip area as well as to improve the inductor quality factor by removing the underneath substrate. The absolute and relative sensitivities were 3.62 MHz g^{-1} and 1.9×10^{-3} $\Delta f/f_0$ g^{-1}, respectively. The noise floor was about 0.2 mg rtHz^{-1}.

8.6.2.3 Inductive Sensors

Inductive sensors are less common than resistive and capacitive sensors mainly because micro-CMOS or MEMS inductors usually have poor quality factors $Q = \omega L/R_s$ due to the series resistance R_s and the substrate loss caused by eddy currents. However, in the CMOS MEMS technology, the released structures made of BOEL

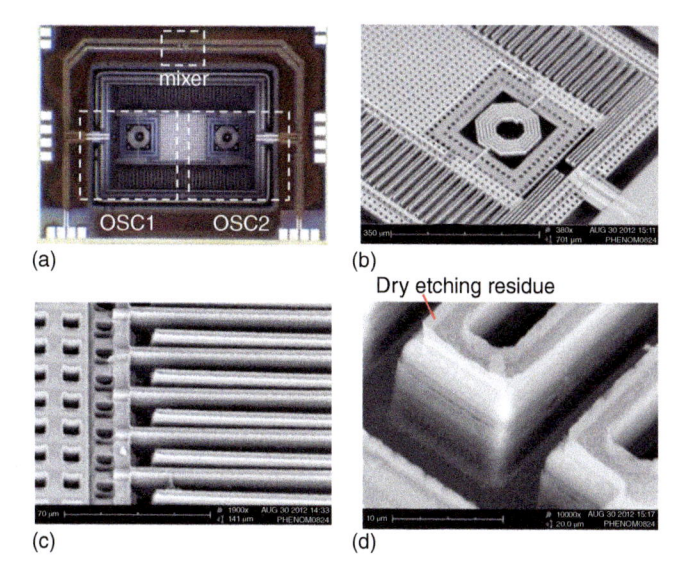

Figure 8.67 Fabricated LC-tank oscillator-based capacitive accelerometer. (a) Chip photograph, (b) released sensor, (c) sensing comb fingers, and (d) zoom-in of spring. Source: Chiu et al. [169]. © 2013, IEEE.

metal and oxide layers are prone to residual stress-induced curling, as shown in Figure 8.64d. In this case, the electrodes attached to the moveable proof mass and those attached to the fixed substrate will have relative displacement and tilt, resulting in reduction and uncertainty of capacitance values and sensor sensitivity. Even though curl matching frames were proposed to alleviate this problem [162], it is still difficult to precisely control the capacitance in every fabrication batch. If suspended piezoresistors are used for sensing, the residual stress will cause an initial offset of resistance values and affect the sensor performance. In contrast, the inductance of an inductor depends on the shape and area of the inductor winding. Therefore the inductance of a released sensing inductor is less sensitive to structural curling caused by residual stress.

In Ref. [170], Chiu et al. proposed and demonstrated an in-plane inductive accelerometer with on-chip sensing inductor and readout circuits. As shown in Figure 8.68, the mechanical spring-mass structure is similar to that of a conventional MEMS accelerometer. However, metal lines embedded in the springs are designed to form a closed coil winding. Therefore the spring and the inductor share the same physical structure which is termed "sprinductor." In Figure 8.68a, the sprinductors on both sides of the mass at rest have equal shape, area, and thus inductance. In Figure 8.68b, the mass is displaced by the external acceleration. Therefore the sprinductor on one side has a larger area and inductance, while the sprinductor on the other side has a smaller area and inductance. Such a differential sprinductor sensing pair was used in two LC tank oscillators with variable inductors and fixed capacitors, and the frequency-output detection scheme was similar to that in Figure 8.66. In [171], the similar design was extended and applied to a three-axis

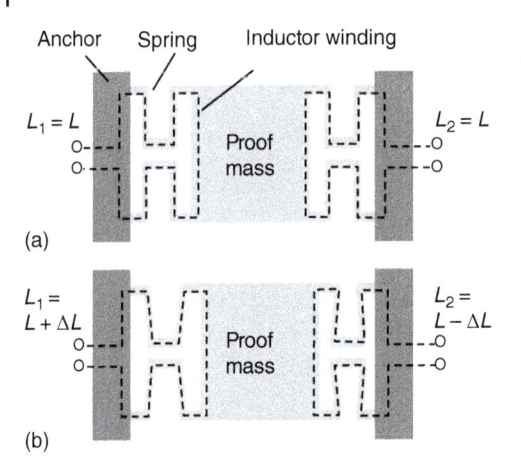

Figure 8.68 Principle of inductive CMOS MEMS accelerometer. (a) At rest and (b) under external acceleration. Source: Y. Chiu et al. [1/0]. © 2016, IEEE.

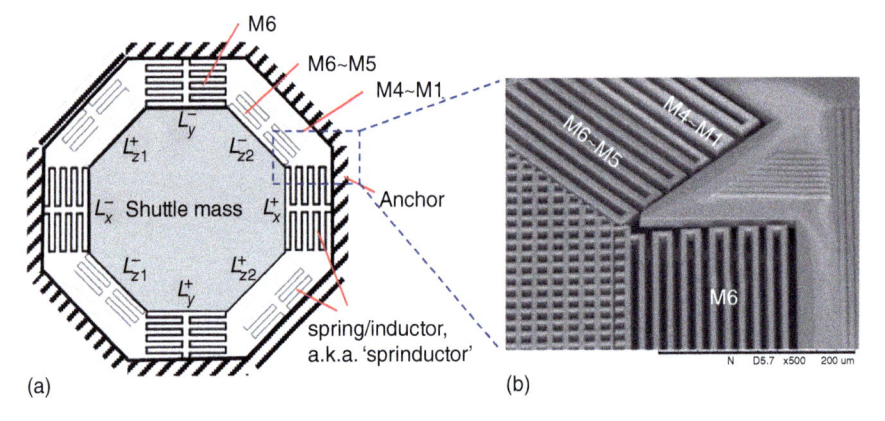

Figure 8.69 Three-axis inductive accelerometer with single proof mass. (a) Schematic and (b) z-axis sensing sprinductor. Source: Chiu et al. [171]. © 2017, IEEE.

accelerometer with only single proof mass. As shown in Figure 8.69, the octagonal proof mass is surrounded by eight sprinductors forming four differential sensing pairs. The x and y sensing sprinductors are similar to that shown in Figure 8.68. To sense the z-axis displacement, the effective coil planes of the sensing sprinductors are tilted from the chip surface (xy plane) by employing different metal layer stacking along the winding, as shown in Figure 8.69b. Thus when the proof mass moves in the z-direction, the tilt angles of the coil planes are changed, resulting in a change of coil areas and inductances. Three sets of differential LC oscillators were used to detect displacement and acceleration in three axes and reject cross-axis signals.

8.6.2.4 Resonant Sensors

Ring and LC-tank oscillators are electrical oscillators and suffer from low quality factors. The stability of oscillation frequency and the sensing resolution are thus limited. This issue can be addressed by replacing the electrical LC resonators with mechanical resonators which have quality factors Q in the range of 300–500 in air

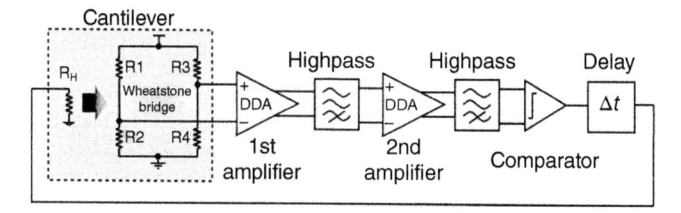

Figure 8.70 Oscillator circuit of mass sensing resonant cantilever beam. Source: Hagleitner et al. [151]. © 2002, IEEE.

and 1500–2000 in low-pressure environment. The frequency stability and sensing resolution, which are approximately proportional to $1/Q$, of such mechanical resonator-based oscillators can thus be improved.

To implement a MEMS oscillator, a mechanical resonant structure such as a cantilever beam or a mass-spring system is enclosed in a closed loop system. The displacement of the resonator is detected by a sensing mechanism and used to drive the resonator via a force transducer until oscillation is self-sustained. In consideration of material and process compatibility, the most common sensing mechanisms in CMOS MEMS devices are capacitive and piezoresistive sensing, while the driving forces are electrostatic, electrothermal, and possibly electromagnetic forces.

In the gas sensor in [150, 151], the adsorbed gas was detected by a mass-sensitive resonant-beam oscillator, as shown in Figure 8.70. The displacement was measured with a fully differential resistive Wheaton bridge, and the beam was driven by electrothermal actuation via resistive heating. The Q factor was 950 in air, the nominal oscillation frequency was 380 kHz, the short-term frequency stability was 0.03 Hz, and the limit of detection of toluene was less than 1 ppm. In Ref. [152], the electrothermal actuation was replaced with electromagnetic actuation to reduce the power consumption. In Ref. [128], a CMOS MEMS oscillator employing capacitive displacement sensing and electrostatic actuation was demonstrated for gas sensing (Figure 8.42a). The nominal oscillation frequency was 5475 kHz with 40 dB SNR; the calculated sensitivity was 23.6 pg Hz^{-1}. In Ref. [129], a similar device was demonstrated for chemical weapon detection with resistive displacement sensing and electrostatic actuation (Figure 8.42b). The oscillation frequency was about 90 kHz and a detection limit of 20 ppb for a nerve agent stimulant was demonstrated. In Ref. [172], an oscillator with capacitive displacement sensing and electrostatic actuation was demonstrated to operate in liquid to sense the elastic damping effect of liquid. A wide-range phase locked loop (PLL) was used to sustain the oscillation in the presence of potential large resonant frequency shift. During testing, the oscillation was sustained in air and various liquids with frequency ranging from 46 to 211 kHz.

8.7 Issues and Concerns

The CMOS MEMS process has the advantage of monolithic integration of the IC and micro-mechanical structures. Despite the merits, such as easy integration of various devices, mature fabrication processes, and availability in many IC foundry services, many issues cannot be ignored while implementing suspended mechanical devices using CMOS MEMS processes. In this section, several generic problems such as residual stresses, CTE mismatch, and creep of thin films, which may occur in many MEMS devices, are discussed first. After that, specific problems which may occur in the resonators and oscillators are also explained to show the detail design concerns for the application of CMOS MEMS.

8.7.1 Residual Stresses, CTE Mismatch, and Creep of Thin Films

The residual stresses of thin films, the mismatch of CTE between thin films, etc. will introduce problems for the suspended CMOS MEMS structures [140, 173]. The initial deformation of suspended MEMS structures, which resulted from thin film residual stresses, is a critical concern in device design and implementation. Furthermore, due to the mismatch of CTE between thin films, the suspended CMOS MEMS structures formed by the metal and dielectric composite films will be deformed by the variation of ambient (or operation) temperature. The shape of suspended CMOS MEMS structures may vary with time, which is another concern for the reliability of MEMS devices. Therefore, three major challenges of CMOS MEMS structures will be discussed in this subsection, including (i) initial deformation: due to the occurrence of thin film residual stresses right after the processes, (ii) thermal deformation: due to the occurrence of mismatch of thin film CTEs during operation, and (iii) long-time stability.

8.7.1.1 Initial Deformation – Residual Stress

In addition to the fundamental mechanical properties of thin films, such as the elastic modulus, Poisson's ratio, and CTE, the residual stresses are also important parameters that influence the performances of MEMS devices. Moreover, the variation of residual stresses could be exploited to monitor the status of thin film. Thus, many approaches and test structures have been reported to characterize the thin film residual stresses [174–177]. The suspended mechanical structures fabricated by the CMOS MEMS platform are stacked by multilayer metal and dielectric films. According to the characteristics of thin film processes, each layer has different residual stress. As a result, a net bending moment is applied on the suspended structure which is consisted of multilayer thin films. As shown in the SEM micrograph of Figure 8.71a, the suspended CMOS MEMS test cantilevers are deformed by thin film residual stresses. Furthermore, due to the thin film residual stresses, cantilevers in Figure 8.71b are even curled with a small radius of curvature. These unwanted deformations, which occurred right after the fabrication processes, would affect the performances of CMOS MEMS devices.

There are many reasons that cause residual stresses of deposited films. For example, the existence of defects in thin film frequently occur in low-temperature deposition processes [178], and the CTE mismatch between thin film and based layer could be the primary reason to cause residual stresses for high-temperature deposition processes [179]. Moreover, various steps after the

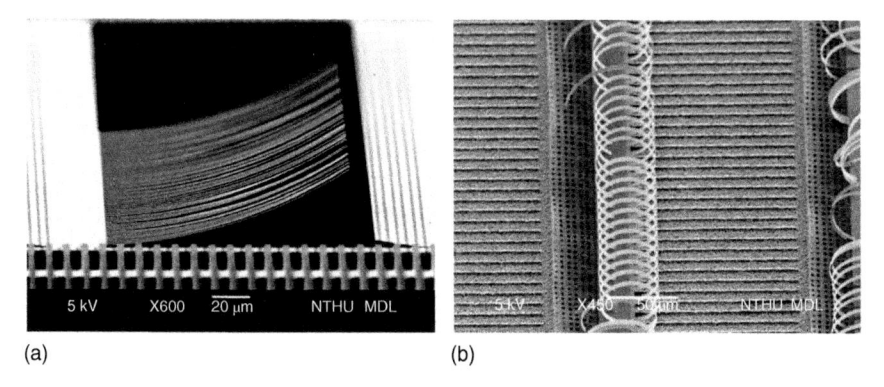

Figure 8.71 Test cantilevers fabricated by CMOS MEMS technology. (a) Cantilevers with same stacking bent upward after releasing from substrate and (b) extremely larger initial bending due to thin film residual stresses.

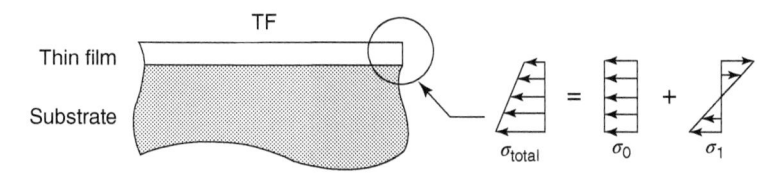

Figure 8.72 Thin film residual stress domination terms include uniform stress and gradient stress, which lead to the suspended structures with an out-of-plane deformation after thin film releasing from substrate. Source: Fang and Wickert [174]. © 1996, IOP Publishing.

deposition process may also introduce residual stresses to a thin film, for instance, the chemical mechanical planarization (CMP). Thus, the formation and state of thin film residual stresses for CMOS processes is complicated. As shown in Figure 8.72, the thin film residual stress can be expressed by the simplified equation [174]

$$\sigma_{total} = \sigma_0 + \sigma_1 \left(\frac{y}{h/2} \right) \tag{8.2}$$

where y is the coordinate across the film thickness h. The origin of coordinate is chosen at the mid-plane of film thickness. The parameter σ_0 represents the uniform residual stress and σ_1 indicates the gradient residual stress. The uniform and gradient residual stresses would be relieved after structures are released from the substrate, which would further lead to the initial deformation of structures. For a single-layer thin film structure (i.e. cantilever), it will be bent by the bending moment resulted from the gradient stress [174, 180]. For a multilayer thin film structure such as the CMOS cantilever formed by the metal and dielectric composite layers, it will be bent by the bending moment resulted from not only the gradient residual stresses but also the different uniform residual stresses of each layers [181].

Since the standard CMOS processes consist of many stacking layers and strict fabrication rules are also required by foundries, it is not straightforward to characterize residual stresses of each CMOS layers. As discussed in [181–183], designers rely on some test structures to predict the deformation of MEMS structures. Figure 8.73

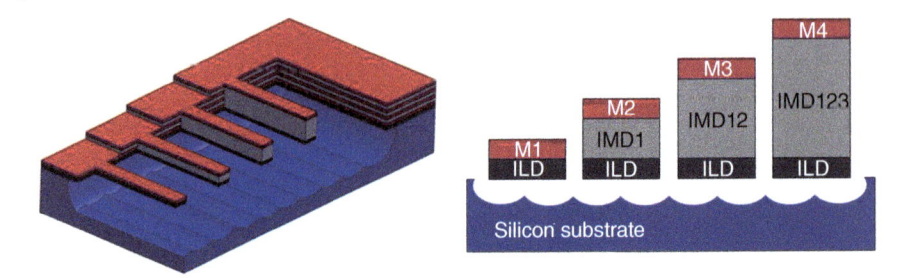

Figure 8.73 Test cantilevers with four stacking types by metal and dielectric composition are fabricated by TSMC 0.35 μm 2P4M CMOS process. Source: Cheng et al. [181]. © 2015, IOP Publishing.

shows the typical design of test cantilevers with different metal and dielectric layer stacking based on the TSMC 0.35 μm 2P4M CMOS process discussed in Section 8.3 [181]. The micrograph in Figure 8.74a shows the fabrication results of the test cantilevers depicted in Figure 8.73 [181]. Figure 8.74b further displays four arrays (as marked with I, II, III, and IV) with cantilevers of different metal and dielectric layer stacking. The suspended test cantilevers are bent by the residual stresses of metal and dielectric films. Typical deflection profiles of test cantilevers measured by the optical white light interferometer are shown in Figure 8.75. The dashed arrow in Figure 8.75a indicates the scanning direction while the results are plotted in Figure 8.75b. It indicates that the bending curvature of cantilevers will vary with the stacking layers. Thus, based on the results from test beams (cantilevers or bridges), a better layer stacking for structure design is determined. Many design concepts to reduce the effect of residual stresses for CMOS MEMS structures have also been proposed. For example, the design of stress compensation ring in [17] has been extensively exploited to increase the overlapping area of capacitive sensing electrodes for many CMOS MEMS sensors [184, 185]. As a second example,

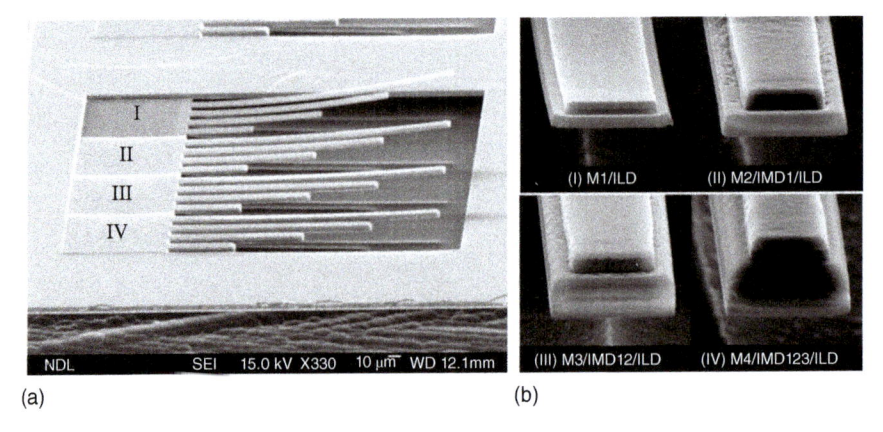

Figure 8.74 Fabrication results. (a) Four cantilever arrays as marked with I, II, III, and IV are associated with those shown in Figure 8.73, for example, the layers stacking of cantilevers in array I and II are M1/ILD and M2/IMD1/ILD, respectively, and (b) cross-sectional view of the test cantilevers depicted in I–IV, respectively. Source: Cheng et al. [181]. © 2015, IOP Publishing.

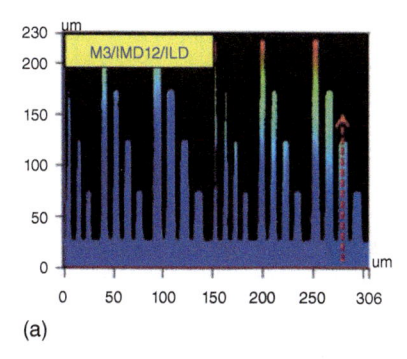

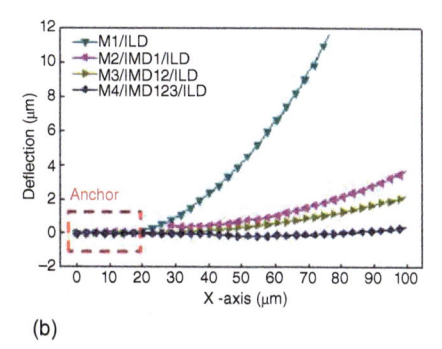

Figure 8.75 Typical measured deflection profiles by commercial optical interferometer for four-layer stacking as shown in Figure 8.73. (a) Top view with scanning direction labeled. (b) Deflection profile for various cantilevers.

symmetric layer stacking designs have been proposed in [173–181, 183–186] to eliminate the bending deflection due to the balance of moment. Moreover, the stress relief design for the CMOS MEMS microphone has been presented by Akustica [19] (acquired by Bosch in 2008). The serpentine spring array with shorter equivalent beam length is designed to reduce the deformation due to residual stresses of CMOS layers [187, 188].

8.7.1.2 Thermal Deformation – Thermal Expansion Coefficient Mismatch

Due to the mismatch of CTE between metal and dielectric films, suspended CMOS MEMS structures consisting of metal/dielectric composite layers frequently suffer from unwanted out-of-plane bending deformation, which resulted from the temperature variation during operation [140]. Such out-of-plane deformation will further influence the performances of CMOS MEMS devices, for example, causing changes to the sensing gap between two electrodes [140] or reducing the overlap area of comb electrodes [17]. Figure 8.76 illustrates a simple bilayer cantilever model to show the out-of-plane deformation due to the CTE mismatch of metal/dielectric layers. The cantilever consists of two thin films (denoted as Layer 1 and Layer 2) with the length of L, the elastic modulus of E_1 and E_2, the CTE of α_1 and α_2, and the thickness of h_1 and h_2. According to the mismatch of CTE between two films, the bilayer cantilevers will be bent out-of-plane with a constant radius of curvature R during temperature changes. The relationship between the radius of curvature R of the bilayer cantilever and the temperature change ΔT can be expressed as [189]

$$\frac{1}{R} = \frac{6(\alpha_2 - \alpha_1)\Delta T(1+m)^2}{h\left[3(1+m)^2 + (1+mn)\left(m^2 + \frac{1}{mn}\right)\right]} \tag{8.3}$$

where h is the total film thickness of the bilayer cantilever ($h = h_1 + h_2$) and the parameters m and n represent $m = h_1/h_2$ and $n = E_1/E_2$, respectively. Measurements in Figure 8.77 depict the bilayer (metal and dielectric layers) cantilever deflections varying with the ambient temperature. Since the top metal layer has a larger CTE than the bottom dielectric layer, the bilayer cantilever was bent downward as the

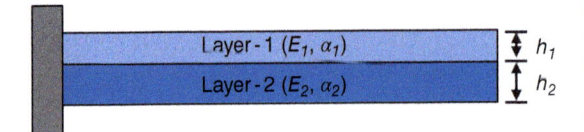

Figure 8.76 The bilayer cantilever model to determine the CTE of thin films. Source: Timoshenko [189]. (Courtesy of Prof. Weileun Fang, NTHU, TW).

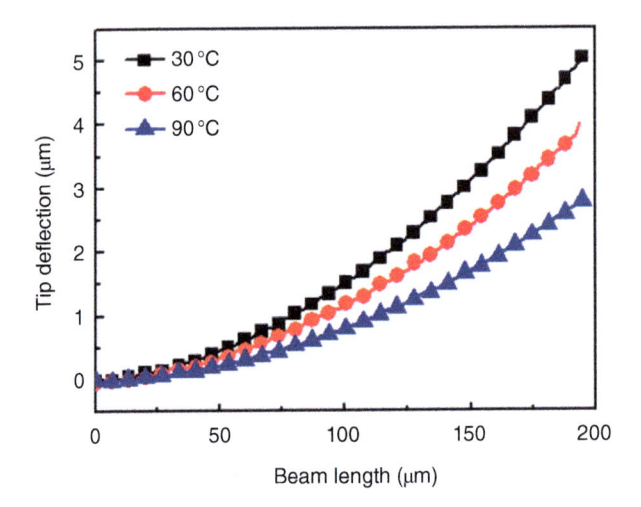

Figure 8.77 Typical measured deflection profiles of the bilayer (metal and dielectric layers) cantilever at three different ambient temperatures. Source: (Courtesy of Prof. Weileun Fang, NTHU, TW).

Table 8.2 List measured radius of curvature for various stacking types of bilayer cantilevers at two different environment temperatures (30–90°C).

CMOS films	Radius of curvature (30 °C)	Radius of curvature (90 °C)
M4/IMD123/ILD	2.09 ± 0.26	3.07 ± 0.58
M3/IMD12/ILD	1.89 ± 0.23	2.73 ± 0.52
M2/IMD1/ILD	0.97 ± 0.03	1.30 ± 0.06
M1/ILD	0.22 ± 0.02	0.26 ± 0.03
Unit	mm	mm

Source: Cheng et al. [181]. © 2015, IOP Publishing.

ambient temperature increased. Table 8.2 further shows the measured radius of curvatures for cantilevers of four different types of stacking layers (as displayed in Figure 8.73) at two different ambient temperatures. The initial bending deformation of cantilevers under 30 °C (room temperature) is resulted from thin film residual stresses. When the temperature is elevated from 30 to 90 °C, the in-use thermal deformation comes from the CTE mismatch between metal and dielectric layers.

In order to predict the thermal deformation and stresses, the technology to characterize the CTE of each CMOS layers has been reported [181]. The CTEs of thin films for the TSMC 0.35 μm 2P4M CMOS processes are also established. Moreover, several approaches have been developed to reduce the deformation due to CTE mismatch for CMOS MEMS devices. For example, the concept to fabricate CMOS MEMS structures using pure oxide stacking layers is presented in [184]. A tri-axis accelerometer with transparent pure oxide proof-mass is fabricated by CMOS MEMS platform. The metal layers are only used for sensing electrodes as well as electrical routings. Thus, the unwanted thermal deformation due to the mismatch of CTE can be reduced.

8.7.1.3 Long-time Stability – Creep

Since the metal layers of CMOS process are aluminum alloy, the creep issue is an important concern for suspension structures [190]. It may influence the mechanical properties of aluminum film and further change the performances of CMOS MEMS devices. As displayed in Figure 8.73, test cantilevers with three different layer stacking ($M2/IMD1/ILD$, $M3/IMD12/ILD$, and $M4/IMD123/ILD$) are fabricated by CMOS MEMS processes to evaluate the long-time stability of these films. As indicated in Figure 8.74, these three test cantilevers of different stacking layers are denoted as II, III, and IV, respectively. To avoid the influence of temperature and humidity fluctuation, test chips are stored in an environment chamber with temperature and humidity control. The cantilevers are deformed by residual stresses of thin films, and the change of deflection amplitude (tip deflection) could indicate the variation of thin film residual stresses. Measurements in Figure 8.78 depict the maximum tip deformations of three different test cantilevers varying with time (recording for 18 months long). Measurements on four samples (marked as #1–#4) are performed for cantilevers of the same stacking layers. The results show that tip deflections of cantilevers (i.e. residual stresses of the films) are unstable for the first 6 months. After that, the tip deflections of cantilevers have relatively small variations. Such time-varying deflection profile of structure is a critical design concern for CMOS MEMS devices.

In conclusion, the residual stresses and CTE mismatch of thin films are critical challenges for CMOS MEMS devices. The CMOS MEMS structures composed of metal/dielectric layers generally have initial deflection introduced by the thin film residual stresses right after process and also have deflection during operation due to the CTE mismatch of thin films. Presently, the available CMOS MEMS commercial products are generally no-moving structures. For example, the suspended CMOS MEMS structures for thermal isolation are adopted to realize the tri-axial thermal accelerometer by MEMSIC [35, 191]. The suspended CMOS MEMS structure with embedded heater and temperature sensors are exploited to detect the acceleration-induced temperature variation. As a second example, the metal-oxide (MOx)-based gas sensor is provided by Sensirion (Sensirion, https://www.sensirion.com/cn/) [192]. The BCD processes (Bipolar–CMOS–DMOS) together with the DRIE backside silicon substrate etching are employed to fabricate the device. The reaction of MOx and target gas will lead to the resistance change for detection, and hence the Si cavity etched by DRIE could enhance thermal isolation. For these two successful commercial products, no static deformation or dynamic responses are required for the suspended MEMS structures.

198 | 8 MEMS Using CMOS Wafer

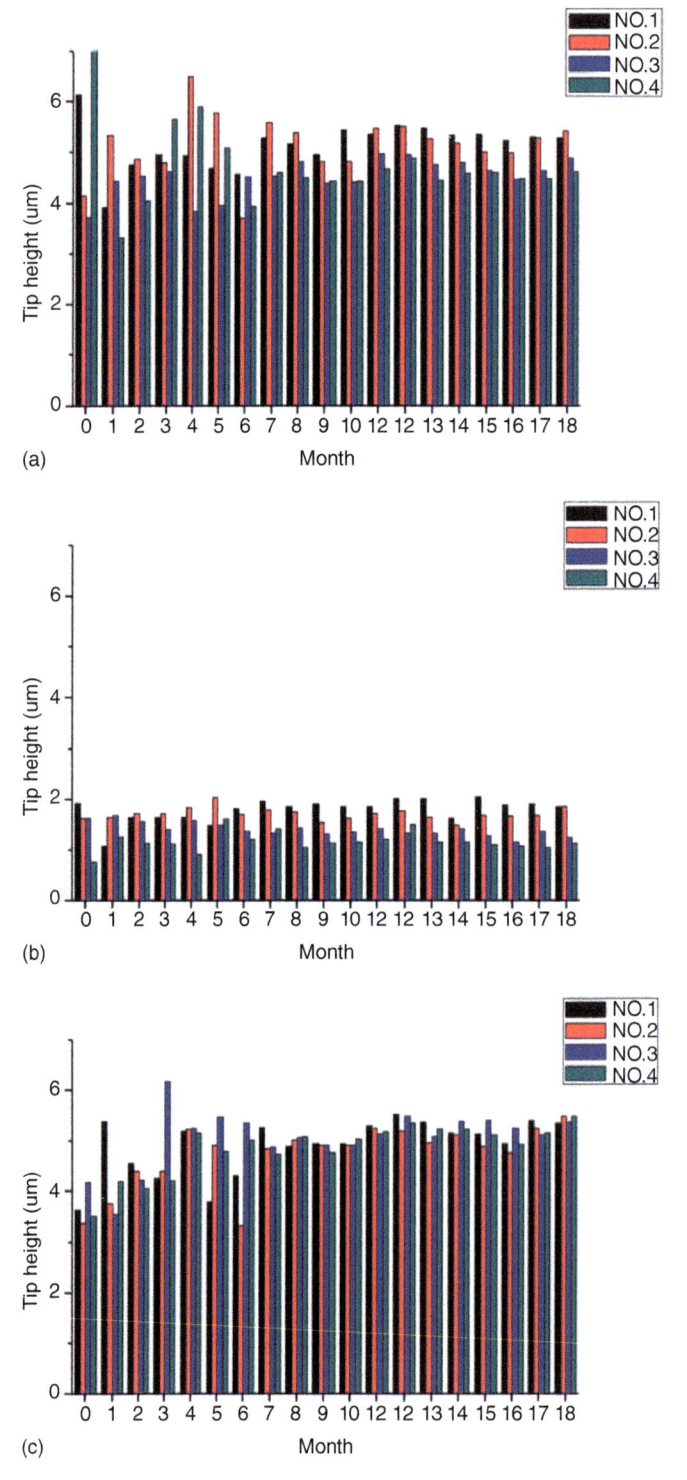

Figure 8.78 Measurements during 18 months to observe the stability of tip deflections for bilayer cantilevers with different metal and dielectric. (a)–(c) are respectively associated with those shown in Figure 8.73 (M2/IMD1/ILD, M3/IMD12/ILD, and M4/IMD123/ILD).

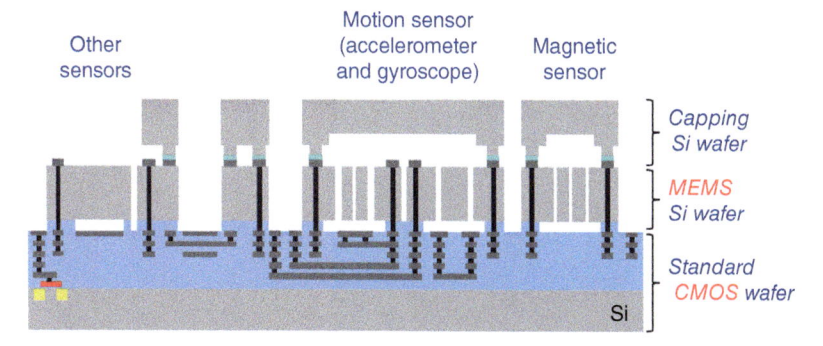

Figure 8.79 Vertical integration of Si mechanical structures on top of CMOS chip. Source: Fang [27]. (Courtesy of Prof. Weileun Fang, NTHU, TW).

As a future perspective, the commercial applications realized by CMOS MEMS processes would be significantly increased if the CMOS foundries could dedicate to the investigation and modification of thin film mechanical properties. Moreover, the post-CMOS wafer bonding and thinning processes have been established in foundries to vertically integrate Si mechanical structures on top of the CMOS chip [27, 193], as shown in Figure 8.79. As compared with the metal-dielectric composite thin films, the SCS has better and stable mechanical properties [194]. This is another option for the integration of MEMS structures and CMOS circuits. In fact, the vertical integration of MEMS mirrors and CMOS memory cell through the CMOS and post-CMOS processes has already been demonstrated by the well-known DLP products of TI (http://www.dlp.com) [195]. Efforts of TI show that the reliability and stress issues of thin films can be solved by the structure design and fabrication process. Similarly, the vertical integration of MEMS structures and CMOS circuits can also be achieved by the chemical vapor deposition (CVD) of poly SiGe thin film [196]. The CVD poly SiGe could meet the thermal budget of CMOS chip and hence can be deposited on the CMOS chip to implement MEMS structures. The vertical integration process scheme could even enable the implementation of MEMS devices on the 12 inch wafer [197].

8.7.2 Quality Factor, Materials Loss, and Temperature Stability

There are several important indexes to define the performance of a MEMS resonant transducer, which underlies many applications such as frequency domain signal processing (i.e. oscillators [198, 199] and filters [45]), low-power and low-noise resonant sensors [200], and actuators [201]. The two most important properties, namely, quality factor (Q) and TC_f, are comprehensively studied in silicon-based MEMS resonators [202, 203]. However, unlike single- and poly-crystalline silicon, which possesses ultra-high Q more than 10^6 (f-Q product $>10^{13}$) [204], the typical Q for *temperature-compensated* CMOS MEMS resonators only lies in a range of 10^3–10^5 [205]. To investigate the reason behind, the energy loss from different mechanisms in a temperature-compensated composite-material-based resonator should be carefully considered.

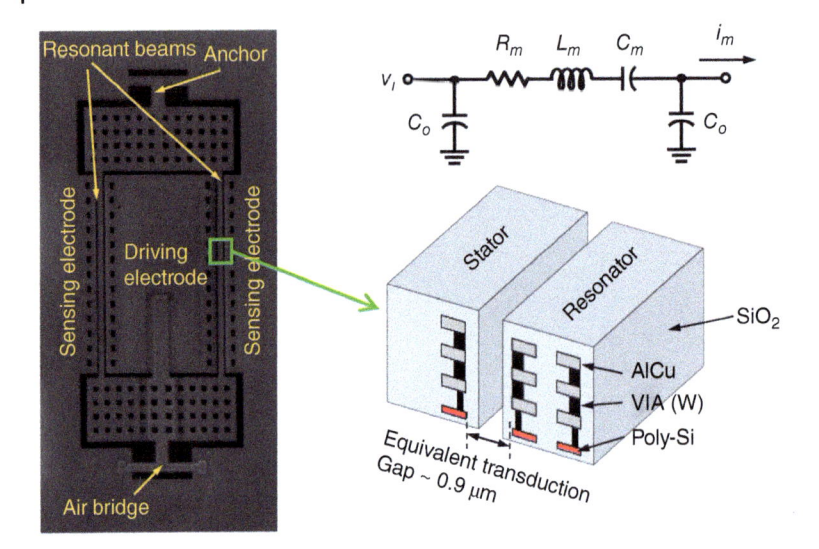

Figure 8.80 Temperature-compensated CMOS MEMS DETF resonator. Source: Prof. Ming-Huang Li, NTHU, TW.

Figure 8.80 illustrates an example of a temperature-compensated CMOS MEMS resonator, which is composed of resonant beams, forcing and pick-up electrodes, and supporting tethers (i.e. anchor). To improve the temperature stability, several materials with both positive and negative temperature coefficient of elastic modulus (TC_E) are well organized to form a composite-material resonator. Assuming the frequency variation from the effect of thermal expansion can be neglected, the following equation describes the relation between the TC_E and TC_f of a laterally vibrating beam resonator with composite material:

$$f(T) = (\beta l)^2 \sqrt{\frac{\sum E_i(T) \cdot I_i}{\sum \rho_i(T) \cdot A_i} \frac{1}{l^2}}, \quad TC_f = \frac{1}{f(T_o)}\left(\frac{\partial f_o}{\partial T}\right)_{T_o} \tag{8.4}$$

where l is the length of the resonator, βl is the mode constant, and E_i, I_i, ρ_i, and A_i are the elastic modulus, area moment of inertia, density, and cross-sectional area of the ith material, respectively. In a CMOS MEMS platform, the material constants (elastic modulus, density, and TC_E) of the metal layer (AlCu/TiN composite), tungsten (VIA), and silicon dioxide (IMD) are summarized in Table 8.3. Apparently, the zero-TC_f designs can be targeted in this platform because both positive and negative TC_E's are offered by the constituent materials. For example, the DETF CMOS MEMS resonator in [199] shows a passively compensated first-order $TC_f < 1$ ppm K^{-1}.

To understand the relation between Q and TC_f, we firstly consider the individual Qs contributed by distinct energy loss mechanisms. The total Q-factor of a resonator can be generally approximated as

$$\frac{1}{Q_{Total}} \approx \frac{1}{Q_{Anchor}} + \frac{1}{Q_{TED}} + \frac{1}{Q_{MAT}} \tag{8.5}$$

where Q_{Anchor}, Q_{TED}, and Q_{MAT} are the Q-factors from anchor loss, thermoelastic damping (TED), and material loss, respectively. The damping from the air is

Table 8.3 List of material properties of CMOS MEMS FEOL/BEOL materials.

	E (GPa)	ρ (kg m^{-3})	TC_E (ppm)	$Q_{Material}$
Metal (AlCu)	70	2700	−620	>100 k
VIA (W)	411	19500	−6	~2000
Poly-Si	160	2300	−30	>100 k
IMD (SiO$_2$)	70	2200	+180	~3000

Source: Li et al. [205]. © 2015, IEEE.

ignored for resonators operated in vacuum. Let's review the Qs in Equation (8.5) one by one:

8.7.2.1 Anchor Loss

The anchor loss comes from the outgoing elastic energy from the resonator body to the environment through its supporting tethers [202, 205]. Typically, it is considered to be one of the most significant Q-limiting factors for MEMS resonators. Therefore, there are tons of research papers discussing the optimal design for maximizing the Q_{Anchor}. For resonators with in-plane motion, balanced dual-resonator design with a common anchor is an effective solution to obtain high Q_{Anchor}. For instance, the dual-ring silicon resonator in [204] attains very high Q by maximizing the acoustic energy confinement in the system. However, the anchor loss would not be the dominant loss mechanism for CMOS MEMS since the simulated Q_{Anchor} are well above the measured Qs for common designs. Instead, the TED and material loss are the major energy loss contributors for a general case.

8.7.2.2 Thermoelastic Damping (TED)

TED is from the coupling between mechanical strain and temperature. As the thermal expansion represents the change in dimension in response to a change in temperature, this effect also works in the other direction. For most of the materials, compressive strain will increase the temperature of the solid and vice versa. Therefore, heat flows might be generated to relax the temperature gradient (caused by the strain gradient) in a deformed resonator during vibration, and then the energy is permanently lost. For transversely vibrating flexural beams, the Q_{TED} can be well predicted by Zener's formula [206]. For resonators formed by composite materials such as CMOS MEMS resonators, finite-element analysis is widely used for Q_{TED} estimation. However, for a temperature-compensated DETF CMOS MEMS resonator [205], the Q_{TED} is still one order above measured Q.

8.7.2.3 Material and Interface Loss

For composite CMOS MEMS resonators, materials contacting with each other may dissipate energy in numerous ways, from direct strain coupling to lossy metals to hysteresis movement of metal-oxide interface. It is pretty difficult to develop an analytical model to predict it. Therefore, to deduce the Q_{MAT} for a complex resonator,

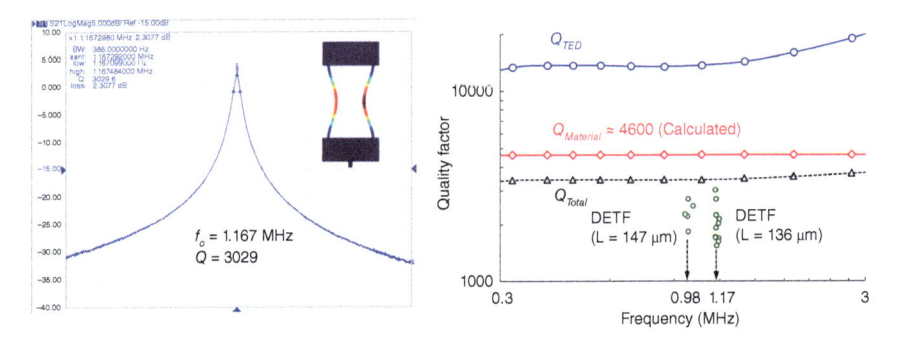

Figure 8.81 Estimated Qs using the method proposed in [205]. The circle legends indicate the measurement results. The highest Q-factor measured is 3029. Source: Li et al. © 2015, IEEE.

a simple model can be applied to extract the individual material Q from the experiment data:

$$\frac{1}{Q_{MAT}} = \frac{SE_{AlCu}}{Q_{AlCu}} + \frac{SE_{SiO2}}{Q_{SiO2}} + \frac{SE_W}{Q_W} + \frac{SE_{PolySi}}{Q_{PolySi}} \quad (8.6)$$

where the Q_j's are the Q for the individual material and the SE_j is the fractional strain energy of the j-th structural material. With many developed CMOS MEMS resonators, the extracted Q_j's are summarized in Table 8.3 by using the method in [205]. In conclusion, for a resonator made by composite material, the interface loss between metal and oxide is recognized as the dominant loss mechanism.

As a result, we applied this model to a CMOS MEMS tuning fork resonator composed of AlCu, W, and SiO_2 in a 0.35 μm 2P4M CMOS process; the upper-limit of the Q can be well predicted by the model [205], as shown in Figure 8.81. It also suggests that the material losses place a trade-off between high Q and low TC_f. According to [37], *mere-oxide* resonators with minimum metal for electrical routing (to reduce the impact of interface loss) typically exhibit very high Q and humongous TC_f. For example, the Q of a 48MHz CMOS MEMS Lamé mode resonator is more than 10 000, showing a very promising f–Q product of 5.4×10^{11} but with $TC_f \sim 80$ ppm K^{-1} [207] (Figure 8.82). Similar concept is also adopted

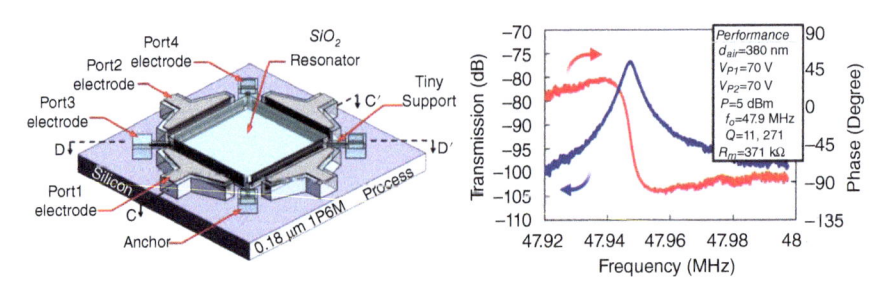

Figure 8.82 Mere-oxide CMOS MEMS Lamé mode resonator. Source: Chen et al. [37]. © 2012, IEEE.

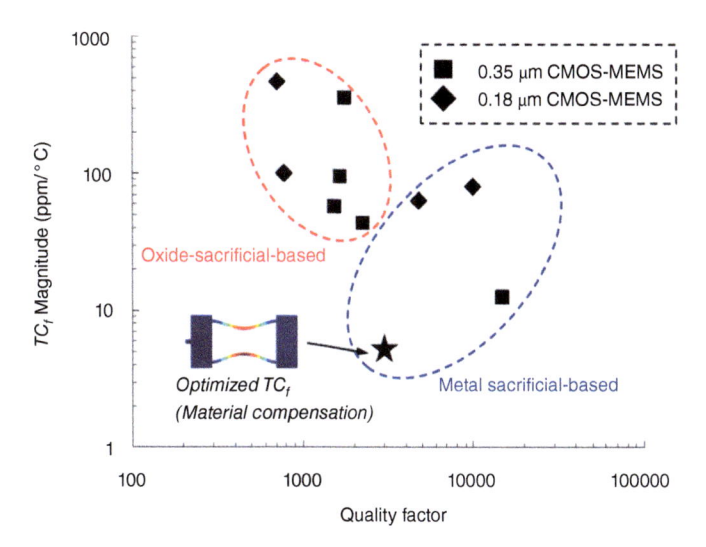

Figure 8.83 Summary of TC_f and Q of published CMOS MEMS resonators. Source: Li et al. [205]. © 2015, IEEE.

for a "dog-bone" II-Bar resonator with embedded piezoresistive detectors [40] with $Q > 12\,000$. Finally, Figure 8.83 summarizes the performance of temperature coefficients and Qs of the CMOS MEMS resonators [205].

8.7.3 Dielectric Charging

Electrostatic transduction is widely used in CMOS MEMS devices to excite and detect the motion of the mechanical structure. However, in the metal-sacrificial process introduced in Section 8.2.1, the dielectric sidewalls create charging phenomena in electrostatic actuators and capacitive sensors. Charging in dielectrics can lead to resonator frequency drift and decrease the lifetime of the device. The charge-induced frequency drift arises from the electron traps in the sidewall dielectric layers, which generates a built-in voltage $\Delta V(\tau)$. Please also note that $\Delta V(\tau)$ contains a time-dependent variable τ to describe the charge relaxation time. As a result, the presence of dielectric charging causes the resonant frequency shifts over time, which limits the applications of CMOS MEMS devices. For example, the resonant frequency of the CMOS MEMS DETF resonator drifted for more than 2500 ppm over 40 minutes at high temperature [52].

Fortunately, this issue can be mitigated by removing the dielectric sidewalls in the electrostatic transducer. This can be done by oxide-sacrificial process and TiN-C composites (Section 8.2.1). Moreover, coating some conductive media on the dielectric sidewalls can effectively eliminate the charge-related frequency drifts. In [208], the lift-off patterned atomic layer deposited (ALD) conductive titania (TiO_2) selectively coats CMOS MEMS sidewalls. The thin-TiO_2 layer drastically lowers the charging time constant to less than 1 second, thus greatly improving the long-term frequency stability, as shown in Figure 8.84.

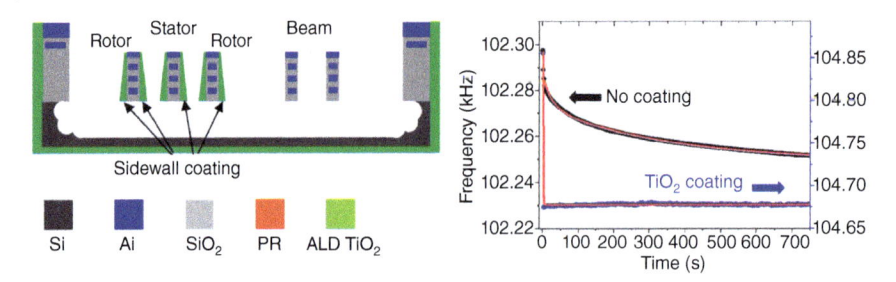

Figure 8.84 TiO$_2$-coated CMOS MEMS resonator for charge-drift elimination. Source: Lin et al. [208]. © 2019, IEEE.

8.7.4 Nonlinearity and Phase Noise in Oscillators

Nonlinearity in a mechanical resonator arises from the natural coupling between the oscillating amplitude and actuation force. In CMOS MEMS devices, the electrostatic transduction contributes significant bias-dependent nonlinearities to the capacitive resonators. Considering a DETF resonator in Figure 8.80 with a DC bias voltage V_P, the nonlinear equation of motion (EOM) can be expressed as

$$m_{re}\ddot{x} + c_{re}\dot{x} + (k_{m1} + k_{e1})x + (k_{m3} + k_{e3})x^3 = \frac{2\varepsilon_o A}{d^2} V_P V_{drive} \tag{8.7}$$

where k_{e1} and k_{e3} are the bias-dependent negative spring constants and k_{m1} and k_{m3} are the linear and the cubic mechanical nonlinear spring, respectively. In this equation, the resultant cubic nonlinear stiffness parameters can be further written as

$$k_3 = k_{m3} + k_{e3} = 0.767 \frac{k_{m1}}{W_B^2} - \frac{2\varepsilon_o A}{d^5} V_P^2 \tag{8.8}$$

where W_B is the width of the vibrating beam and A is the total surface area of the electrostatic transducers. For capacitive resonators with sub-μm transduction gaps (d), the mechanical spring constants are much smaller than the electrical spring constants under high V_P. Figure 8.85 illustrates the steady-state solution with normalized magnitude to show the nonlinear amplitude–frequency (A–f) response

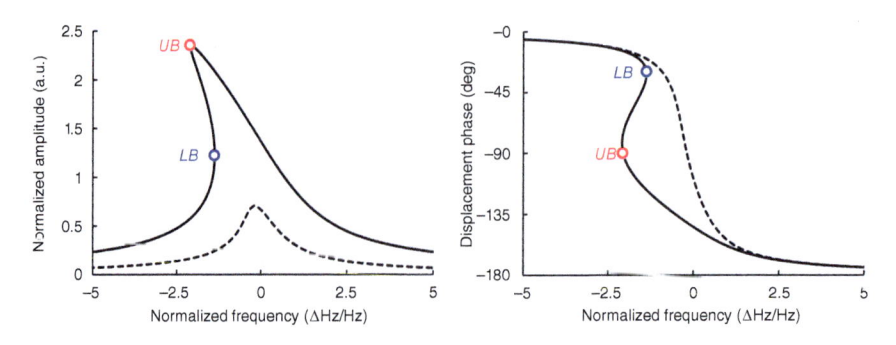

Figure 8.85 Normalized vibrating magnitude of a nonlinear resonator. Assume the cubic nonlinearity is negative. The UB and LB denotes the upper and lower bifurcation points. Source: Li et al. [209]. © 2018, IEEE.

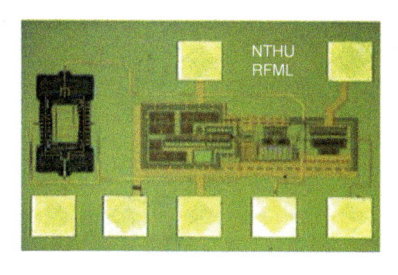

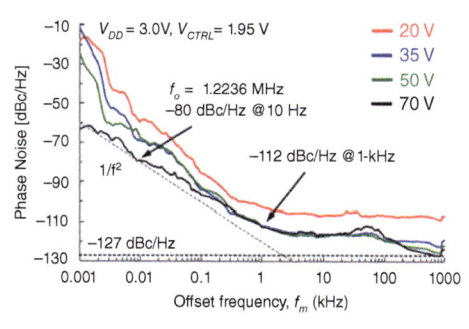

Figure 8.86 Phase noise cancellation effect in a nonlinear CMOS MEMS oscillator. Source: Li et al. [209]. © 2018, IEEE.

[209]. Please note that the peak bends to the left owing to the negative cubic spring.

Typically, the nonlinear A–f response induces additional amplitude-to-phase modulation (AM-PM) phase noise conversion when the resonator is driven at large amplitude [210]. This effect would generally convert the amplitude noise component (i.e. the voltage and current noise) to phase noise with a steeper slope (typically -30 dB dec^{-1}). Recently, researchers have demonstrated that in operating the oscillator at the lower bifurcation point (LB in Figure 8.85), the phase variations on the oscillation frequency is minimized [211–213]. Figure 8.86 shows the phase noise cancellation effect demonstrated in a CMOS MEMS oscillator, showing a very competitive FoM of 190 dB [209].

8.8 Concluding Remarks

Process development is one of the bottlenecks for the commercialization of MEMS products. There are many CMOS foundries to provide mature and standard fabrication processes to date. It could be a cost-effective solution to design and implement MEMS devices by using these existing CMOS fabrication resources. Thus, the CMOS MEMS approach plays an important role to bridge the CMOS and MEMS technologies. In this section, various post-CMOS process modules to etch the thin films as well as the silicon substrate are introduced first. After that, the integration of these process modules to implement different MEMS sensors and actuators on two standard CMOS processes are demonstrated. This section also shows the possibility to integrate many processes, such as the deposition, molding, assembly, etc., to fabricate MEMS structures with materials other than those for CMOS process. Finally, the fabrication and integration of different MEMS sensors and actuators and the sensing and control circuits are demonstrated. These examples show the future goals for the CMOS MEMS technology: to respectivly enable the MPW service for foundries and combo sensor platform for design houses. Nevertheless, many issues regarding processes and materials such as, the residual stresses, CTE mismatch, reliability, etc., still need to be solved before the commercialization of CMOS MEMS devices.

8 MEMS Using CMOS Wafer

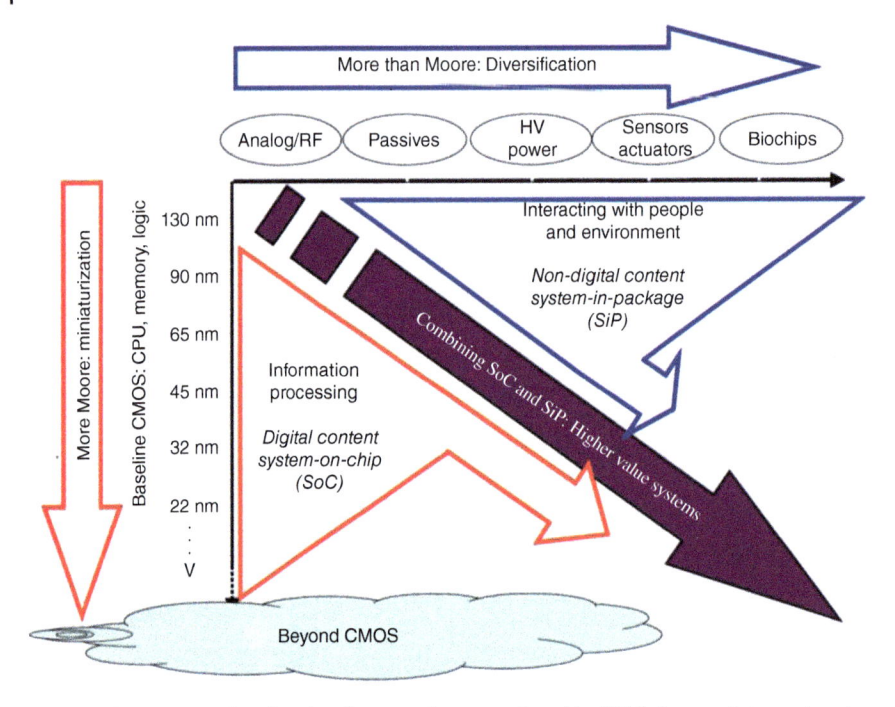

Figure 8.87 International technology roadmap predicted by ITRS. Source: International Technology Roadmap for Semiconductors. 2005 Edition [214].

According to the report from the International Technology Roadmap for Semiconductors (ITRS) in 2005, the developments of semiconductor-related technologies, as shown in Figure 8.87, have two different trends: (i) Miniaturization (in the vertical axis) following the Moore's law to continuously reduce the size of components, named "More Moore," and (ii) Diversification (in the horizontal axis), increasing semiconductor process technologies to many different applications, named "More than Moore" [214]. However, the "More Moore" trend to miniaturize the size of devices will eventually reach the physical limitation. Moreover, the investment of very expensive equipment is required for the advanced process technologies. In this regard, leveraging the existing CMOS process technologies to develop the MEMS devices such as sensors, actuators, biochip, etc. through the trend of "More than Moore" could be a cost-effective approach to extend the applications of semiconductor industries. Thus, the CMOS MEMS could be a promising approach to bridge the gap between the mature CMOS technology and emerging MEMS and also provide additional value to the semiconductor industries. Presently, commercial CMOS MEMS products using the suspended metal and dielectric composite layers in CMOS process are available in (Sensirion, https://www.sensirion.com/cn/; MEMSIC, http://www.memsic.com/memsic/), and the suspended poly SiGe and Si mechanical structures on top of CMOS chip are available in (TDK InvenSense, https://www.invensense.com/technology/; IMEC. http://www.imec.be/). Moreover, since diversification, and not miniaturization, is the major focus for microsystems developed toward the roadmap of "More than Moore," the metal and dielectric thin film materials available for the CMOS processes could not fulfill the requirements for diverse applications. Therefore, many different approaches, in addition to the patterned metal-dielectric layers of standard CMOS process, have been established to realize the suspended mechanical structures on standard CMOS chip, for instance, the bonding and thinning of thick Si and the deposition of additional poly Si or poly SiGe layer. Despite that the main focus of this section is the first approach, the other two approaches could offer different characteristics of CMOS MEMS structures. The applications of CMOS MEMS devices can be further enhanced if other functional materials, such as the piezoelectric film, porous material, etc., are available and compatible with the CMOS and post-CMOS processes.

References

1 Bustillo, J.M., Howe, R.T., and Muller, R.S. (1998). Surface micromachining for microelectromechanical systems. *Proc. IEEE* 86 (8): 1552–1574.
2 Zyung, T., Kim, S.H., Chu, H.Y. et al. (2005). Flexible organic LED and organic thin-film transistor. *Proc. IEEE* 93 (7): 1265–1272.
3 Lee, J.B., Chen, Z., Allen, M.G. et al. (1995). A miniaturized high-voltage solar cell array as an electrostatic MEMS power supply. *IEEE J. Microelectromech. Syst.* 4 (3): 102–108.
4 Huang, H., Winchester, K.J., Suvorova, A. et al. (2006). Effect of deposition conditions on mechanical properties of low-temperature PECVD silicon nitride films. *Mater. Sci. Eng. A* 435 (5): 453–459.
5 Fan, L.-S., Tai, Y.-C., and Muller, R.S. (1988). Integrated movable micromechanical structures for sensors and actuators. *IEEE Trans. Electron Dev.* 35 (6): 724–730.
6 Fan, L.-S., Tai, Y.-C., and Muller, R.S. (1988). *IC-Processed Electrostatic Micro-Motors*, 666–669. San Francisco, CA: IEEE IEDM.
7 Tang, W.C., Nguyen, T.-C.H., Judy, M.W., and Howe, R.T. (1990). Electrostatic-comb drive of lateral polysilicon resonators. *Sens. Actuators A* 21: 328–331.
8 Pister, K.S.J., Judy, M.W., Burgett, S.R., and Fearing, R.S. (1992). Microfabricated hinges. *Sens. Actuators A: Phys.* 33: 249–256.
9 Wu, M.C. (1997). Micromachining for optical and optoelectronic systems. *Proc. IEEE* 85 (11): 1833–1856.
10 Comtois, J.H. and Bright, V.M. (1997). Applications for surface-micromachined polysilicon thermal actuators and arrays. *Sens. Actuators A: Phys.* 58: 19–25.
11 Muller, R.S. and Lau, K.Y. (1998). Surface-micromachined microoptical elements and systems. *Proc. IEEE* 86 (8): 1705–1720.
12 Langfelder, G., Dellea, S., Zaraga, F. et al. (2012). The dependence of fatigue in microelectromechanical systems on the environment and the industrial packaging. *IEEE Trans. Ind. Electron.* 59 (12): 4938–4948.

13 Renard, S. (2000). Industrial MEMS on SOI. *J. Micromech. Microeng.* 10: 245–249.

14 Perlmutter, M. and Robin, L. (2012). High-performance, low cost Inertial MEMS: a market in motion!. Proceedings of the 2012 IEEE/ION Position, Location and Navigation Symposium, Myrtle Beach, SC, 225–229.

15 Schaller, R.R. (1997). Moore's law: past, present and future. *IEEE Spectr.* 34 (6): 52–59.

16 Fedder, G.K. (2005). CMOS-based sensors. In: *IEEE Sensors 2005*, 125–128. Irvine, CA, USA.

17 Zhang, G., Xie, H., de Rosset, L.E., and Fedder, G.K. (1999). A lateral capacitive CMOS accelerometer with structural curl compensation. 12th IEEE Micro Electro Mechanical Systems (MEMS), Orlando, FL, USA, 606–611.

18 Xie, H., Erdmann, L., Zhu, X. et al. (2002). Post-CMOS processing for high-aspect-ratio integrated silicon microstructures. *IEEE J. Microelectromech. Syst.* 11 (2): 93–101.

19 Neumann, J.J. and Gabriel, K.J. (2002). CMOS-MEMS membrane for audio-frequency acoustic actuation. *Sens. Actuators A: Phys.* 95: 175–182.

20 Baltes, H., Paul, O., and Brand, O. (1998). Micromachined thermally based CMOS microsensors. *Proc. IEEE* 86 (8): 1660–1678.

21 Chavan, A.V. and Wise, K.D. (2002). A monolithic fully-integrated vacuum-sealed CMOS pressure sensor. *IEEE Trans. Electron Dev.* 49 (1): 164–169.

22 Hierlemann, A., Brand, O., Hagleitner, C., and Baltes, H. (2003). Microfabrication techniques for chemical/biosensors. *Proc. IEEE* 91 (6): 839–863.

23 Gu, L., Huang, Q.-A., and Qin, M. (2004). A novel capacitive-type humidity sensor using CMOS fabrication technology. *Sensors Actuators B Chem.* 99 (2-3): 491–498.

24 Reinke, J., Fedder, G.K., and Mukherjee, T. (2010). CMOS-MEMS variable capacitors using electrothermal actuation. *IEEE J. Microelectromech. Syst.* 19 (5): 1105–1115.

25 Chen, W.-C., Fang, W., and Li, S.-S. (2011). A generalized CMOS-MEMS platform for micromechanical resonators monolithically integrated with circuits. *J. Micromech. Microeng.* 21 (6): 065012.

26 Dai, C.-L., Xiao, F.-Y., Juang, Y.-Z., and Chiu, C.-F. (2005). An approach to fabricating microstructures that incorporate circuits using a post-CMOS process. *J. Micromech. Microeng.* 15 (1): 98–103.

27 Fang, W., Li, S.-S., Cheng, C.-L. et al. (2013). "CMOS MEMS: a key technology towards the "more than Moore" era," in the 17th International Conference on Solid-State Sensors, Actuators and Microsystems (TRANSDUCERS'13), Barcelona, Spain, pp. 2513-2518.

28 Sun, C.-M., Wang, C., Tsai, M.-H. et al. (2009). Monolithic integration of capacitive sensors using a double-side CMOS MEMS post process. *J. Micromech. Microeng.* 19 (1): 015023.

29 Tsai, M.-H., Sun, C.-M., Liu, Y.-C. et al. (2009). Design and application of a metal wet-etching post-process for the improvement of CMOS-MEMS capacitive sensors. *J. Micromech. Microeng.* 19 (10): 105017.

30 Sun, C.-M., Tsai, M.-H., Wang, C. et al. (2009). "Implementation of a monolithic TPMS using CMOS-MEMS technique," in the 15th International Conference on Solid-State Sensors, Actuators and Microsystems (TRANSDUCERS'09), Denver, CO, pp. 1730–1733.

31 Qu, H. (2016). CMOS MEMS fabrication technologies and devices. *Micromachines (Basel)*. 7 (1): 14.

32 Mansour, R.R. (2013). RF MEMS-CMOS device integration: An overview of the potential for RF researchers. *IEEE Microw. Mag.* 14 (1): 39–56.

33 Chen, C.-Y., Li, M.-H., and Li, S.-S. (2018). CMOS-MEMS resonators and oscillators: A review. *Sens. Mater.* 30: 733–756.

34 Chen, C.-Y., Li, M.-H., Zope, A.A., and Li, S.-S. (Oct. 2019). A CMOS-integrated MEMS platform for frequency stable resonators - Part I: Fabrication, implementation and characterization. *IEEE/ASME J. Microelectromech. Syst. (JMEMS)* 28 (5): 744–754.

35 Liu, Y., Tsai, M., Chen, W. et al. (2013). Temperature-compensated CMOS-MEMS oxide resonators. *J. Microelectromech. Syst.* 22 (5): 1054–1065.

36 Chen, W., Li, M., Liu, Y. et al. (2012). A fully differential CMOS–MEMS DETF oxide resonator with Q >4800 and positive TCF. *IEEE Electron Dev. Lett.* 33 (5): 721–723.

37 Chen, W., Fang, W., and Li, S. (2012). "VHF CMOS-MEMS oxide resonators with Q > 10,000," presented at the 2012 IEEE International Frequency Control Symposium Proceedings, pp. 1–4.

38 Li, M.-H., Chen, C.-Y., Li, C.-S. et al. (2015). Design and characterization of a dual-mode CMOS-MEMS resonator for TCF manipulation. *J. Microelectromech. Syst.* 24 (2): 446–457.

39 Chen, W., Fang, W., and Li, S. (2012). High-Q integrated CMOS-MEMS resonators with deep-submicrometer gaps and quasi-linear frequency tuning. *J. Microelectromech. Syst.* 21 (3): 688–701.

40 Li, C.-S., Li, M.-H., Chin, C.-H. et al. (2003). A piezoresistive CMOS-MEMS resonator with high Q and low TCF. Presented at the 2013 Joint European Frequency and Time Forum & International Frequency Control Symposium (EFTF/IFC), 425–428.

41 Chang, J., Li, C., Chen, C., and Li, S. (2015). Performance evaluation of CMOS-MEMS thermal-piezoresistive resonators in ambient pressure for sensor applications. Presented at the 2015 Joint Conference of the IEEE International Frequency Control Symposium & the European Frequency and Time Forum, 202–204.

42 Liu, T.-Y., Chu, C.-C., Li, M.-H. et al. (2017). CMOS-MEMS thermal-piezoresistive oscillators with high transduction efficiency for mass sensing applications. 19th International Conference on Solid-State Sensors, Actuators and Microsystems (TRANSDUCERS'17), 452–455.

43 Lin, F., Tian, W., and Li, P. (2013). CMOS-based capacitive micromachined ultrasonic transducers operating without external DC bias. Presented at the 2013 IEEE International Ultrasonics Symposium (IUS), 1420–1423.

44 Chin, C.-H., Li, M.-H., Chen, C.-Y. et al. (2015). A CMOS–MEMS arrayed resonant-gate field effect transistor (RGFET) oscillator. *J. Micromech. Microeng.* 25 (11): 115025.

45 Chen, C.-Y., Li, M.-H., Chin, C.-H., and Li, S.-S. (2016). Implementation of a CMOS MEMS filter through a mixed electrical and mechanical coupling scheme. *J. Microelectromech. Syst.* 25 (2): 262–274.

46 Chin, C.-H., Li, C.-S., Li, M.-H. et al. (2014). Fabrication and characterization of a charge-biased CMOS-MEMS resonant gate field effect transistor. *J. Micromech. Microeng.* 24 (9): 095005.

47 Su, H.-C., Li, M.-H., Chen, C.-Y., and Li, S.-S. (2015). A single-chip oscillator based on a deep-submicron gap CMOS-MEMS resonator array with a high-stiffness driving scheme. The 18th International Conference on Solid-State Sensors, Actuators and Microsystems (TRANSDUCERS'15), 133–136.

48 Lopez, J.L., Verd, J., Uranga, A. et al. (2009). A CMOS–MEMS RF-tunable bandpass filter based on two high-Q 22-MHz polysilicon clamped-clamped beam resonators. *IEEE Electron Dev. Lett.* 30 (7): 718–720.

49 Sarkar, N., Lee, G., and Mansour, R.R. (2013). CMOS-MEMS dynamic FM atomic force microscope. The 17th International Conference on Solid-State Sensors, Actuators and Microsystems (Transducers'13 & Eurosensors XXVII), 916–919.

50 Xie, H. and Fedder, G.K. (2003). Fabrication, characterization, and analysis of a DRIE CMOS-MEMS gyroscope. *IEEE Sens. J.* 3 (5): 622–631.

51 Li, S.-S. (2014). CMOS-MEMS resonators. In: *Encyclopedia of Nanotechnology* (ed. B. Bhushan), 1–19. Dordrecht: Springer Netherlands.

52 Li, M.-H., Chen, C.-Y., and Li, S.-S. (2015). A reliable CMOS-MEMS platform for titanium nitride composite (TiN-C) resonant transducers with enhanced electrostatic transduction and frequency stability. Presented at the 2015 IEEE International Electron Devices Meeting (IEDM), 18.4.1–18.4.4.

53 Li, S.-S. (2013). CMOS-MEMS resonators and their applications. Presented at the 2013 Joint European Frequency and Time Forum & International Frequency Control Symposium (EFTF/IFC), 915–921.

54 Bahl, G., Melamud, R., Kim, B. et al. (2010). Model and observations of dielectric charge in thermally oxidized silicon resonators. *J. Microelectromech. Syst.* 19 (1): 162–174.

55 Dorsey, K.L. and Fedder, G.K. (2010). Dielectric charging effects in electrostatically actuated CMOS MEMS resonators. Presented at the SENSORS, 2010 IEEE, 197–200.

56 Li, M.-H., Chen, C.-Y., Li, C.-S. et al. (2015). A monolithic CMOS-MEMS oscillator based on an ultra-low-power ovenized micromechanical resonator. *J. Microelectromech. Syst.* 24 (2): 360–372.

57 Dai, C.-L., Kuo, C.-H., and Chiang, M.-C. (2007). Microelectromechanical resonator manufactured using CMOS-MEMS technique. *Microelectron. J.* 38 (6): 672–677.

58 Kovacs, G.T.A., Maluf, N.I., and Petersen, K.E. (1998). Bulk micromachining of silicon. *Proc. IEEE* 86 (8): 1536–1551.

59 Tsai, M., Liu, Y., and Fang, W. (2012). A three-axis CMOS-MEMS accelerometer structure with vertically integrated fully differential sensing electrodes. *J. Microelectromech. Syst.* 21 (6): 1329–1337.

60 Xie, H. and Fedder, G.K. (2001) A CMOS-MEMS lateral-axis gyroscope. 14th IEEE Micro Electro Mechanical Systems (MEMS) (Cat. No.01CH37090), 162–165.

61 Münch, U. and Baltes, H. (2000). *Industrial CMOS Technology for Thermal Imagers*. Hartung-Gorre Verlag.

62 Tabata, O. (1996). pH-Controlled TMAH etchants for silicon micromachining. *Sens. Actuators A Phys.* 53 (1): 335–339.

63 Chen, C.-Y., Li, M.-H., Li, C.-S., and Li, S.-S. (2014). Design and characterization of mechanically coupled CMOS-MEMS filters for channel-select applications. *Sens. Actuators A Phys.* 216: 394–404.

64 Kloeck, B., Collins, S.D., de Rooij, N.F., and Smith, R.L. (1989). Study of electrochemical etch-stop for high-precision thickness control of silicon membranes. *IEEE Trans. Electron Dev.* 36 (4): 663–669.

65 Jain, A., Qu, H., Todd, S., and Xie, H. (2005). A thermal bimorph micromirror with large bi-directional and vertical actuation. *Sens. Actuators A Phys.* 122 (1): 9–15.

66 Qu, H. and Xie, H. (2007). Process development for CMOS-MEMS sensors with robust electrically isolated bulk silicon microstructures. *J. Microelectromech. Syst.* 16 (5): 1152–1161.

67 Chen, C.-Y., Li, M.-H., Li, C.-S., and Li, S.-S. (2019). A CMOS-integrated MEMS platform for frequency stable resonators - Part II: Design and analysis. *IEEE/ASME J. Microelectromech. Syst. (JMEMS)* 28 (5): 755–765.

68 Yeh, S. and Fang, W. (2019). Inductive micro tri-axial tactile sensor using a CMOS chip with a coil array. *IEEE Electron Dev. Lett.* 40 (4): 620–623.

69 Chang, K., Lee, Y., Sun, C., and Fang, W. (2017). Novel absorber membrane and thermocouple designs for CMOS-MEMS thermoelectric infrared sensor. 30th IEEE Micro Electro Mechanical Systems (MEMS), 1228–1231.

70 Liu, T.-Y., C.-A. Sung, C.-H. Weng, C.-C. Chu, A. A. Zope, G. Pillai, and S.-S. Li. (2018). Gated CMOS-MEMS thermal-piezoresistive oscillator-based PM2.5 sensor with enhanced particle collection efficiency. 31th IEEE Micro Electro Mechanical Systems (MEMS), 75–78.

71 Li, C.-S., Li, M.-H., Chen, C.-Y. et al. (2015). A low-voltage CMOS-microelectromechanical systems thermal-piezoresistive resonator With $Q > 10,000$. *IEEE Electron Dev. Lett.* 36 (2): 192–194.

72 Chu, C.-C., Dey, S., Liu, T.-Y. et al. (2018). Thermal-piezoresistive SOI-MEMS oscillators based on a fully differential mechanically coupled resonator array for mass sensing applications. *J. Microelectromech. Syst.* 27 (1): 59–72.

73 Bhattacharya, S. and Li, S. (2019). A fully differential SOI-MEMS thermal piezoresistive ring oscillator in liquid environment intended for mass sensing. *IEEE Sensors J.* 19 (17): 7261–7268.

74 Dahiya, R.S., Metta, G., Valle, M., and Sandini, G. (2009). Tactile sensing–from humans to humanoids. *IEEE Trans. Robot.* 26 (1): 1–20.

75 Lin, Y.-C., Hsieh, C.-J., Sun, C.-T. et al. (2013). CMOS-based tactile sensors using oxide as sacrificial layer. The 17th International Conference on Solid-State Sensors, Actuators and Microsystems (TRANSDUCERS & EUROSENSORS XXVII), 1895–1898.

76 Liu, Y., Sun, C., Lin, L. et al. (2011). Development of a CMOS-based capacitive tactile sensor with adjustable sensing range and sensitivity using polymer fill-in. *J. Microelectromech. Syst.* 20 (1): 119–127.

77 Lai, W. and Fang, W. (2015). Novel two-stage CMOS-MEMS capacitive-type tactile-sensor with ER-fluid fill-in for sensitivity and sensing range enhancement. 18th International Conference on Solid-State Sensors, Actuators and Microsystems (TRANSDUCERS'15), 1175–1178.

78 Li, C., Wu, P., Lee, S. et al. (2008). Flexible dome and bump shape piezoelectric tactile sensors using PVDF-TrFE copolymer. *J. Microelectromech. Syst.* 17 (2): 334–341.

79 Sedaghati, R., Dargahi, J., and Singh, H. (2005). Design and modeling of an endoscopic piezoelectric tactile sensor. *Int. J. Solids Struct.* 42 (21): 5872–5886.

80 Wen, C.-C. and Fang, W. (2008). Tuning the sensing range and sensitivity of three axes tactile sensors using the polymer composite membrane. *Sens. Actuators A Phys.* 145–146: 14–22.

81 Tu, S., Lai, W., and Fang, W. (2017). Vertical integration of capacitive and piezo-resistive sensing units to enlarge the sensing range of CMOS-MEMS tactile sensor. 30th IEEE Micro Electro Mechanical Systems (MEMS), 1048–1051.

82 Wang, H., Kow, J., Raske, N. et al. (2018). Robust and high-performance soft inductive tactile sensors based on the Eddy-current effect. *Sens. Actuators A Phys.* 271: 44–52.

83 Yeh, S.-K., Chang, H.-C., and Fang, W. (2018). Development of CMOS MEMS inductive type tactile sensor with the integration of chrome steel ball force interface. *J. Micromech. Microeng.* 28 (4): 044005.

84 Yeh, S.-K., Chang, H., Lu, C., and Fang, W. (2018). A CMOS-MEMS electromagnetic-type tactile sensor with polymer-filler and chrome-steel ball sensing interface. 2018 IEEE Sensors, 1–4.

85 Yeh, S.-K., Lee, J.-H., and Fang, W. (2019). On the detection interfaces for inductive type tactile sensors. *Sensors Actuators A Phys.* 297: 111545.

86 Hudson, R.D. and Hudson, J.W. (Jan. 1975). The military applications of remote sensing by infrared. *Proc. IEEE* 63 (1): 104–128.

87 Stephens, E.R. (1961). Long-path infrared spectroscopy for air pollution research. *Infrared Phys.* 1 (3): 187–196.

88 Gitelman, L., Stolyarova, S., Bar-Lev, S. et al. (2009). CMOS-SOI-MEMS transistor for uncooled IR imaging. *IEEE Trans. Electron Dev.* 56 (9): 1935–1942.

89 Eminoglu, S., Tanrikulu, M.Y., and Akin, T. (2003). A low-cost 128 x128 uncooled infrared detector array in CMOS process. *J. Microelectromech. Syst.* 17 (1): 20–30.

90 Dillner, U., Kessler, E., and Meyer, H.-G. (2013). Figures of merit of thermoelectric and bolometric thermal radiation sensors. *J. Sens. Sens. Syst.* 2 (1): 85–94.

91 Hyseni, G., Caka, N., Hyseni, K., and Teknik, F. (2010). Infrared thermal detectors parameters: semiconductor bolometers versus pyroelectrics. *WSEAS Transducers Circuits Syst.* 9: 238–247.

92 Völklein, F., Wiegand, A., and Baier, V. (1991). High-sensitivity radiation thermopiles made of Bi-Sb-Te films. *Sens. Actuators A Phys.* 29 (2): 87–91.

93 Socher, E., Bochobza-Degani, O., and Nemirovsky, Y. (2002). Novel CMOS compatible frontside micromachining of integrated thermoelectric sensors. Proceedings of the 21st IEEE Convention of the Electrical and Electronic Engineers in Israel (Cat. No.00EX377), 417–420.

94 Modarres-Zadeh, M. and Abdolvand, R. (2014). High-responsivity thermoelectric infrared detectors with stand-alone sub-micrometer polysilicon wires. *J. Micromech. Microeng.* 24 (12): 125013.

95 Chen, C. and Huang, W. (2011). A CMOS-MEMS thermopile with low thermal conductance and a near-perfect emissivity in the 8-14-µm wavelength range. *IEEE Electron Dev. Lett.* 32 (1): 96–98.

96 Völklein, F. and Wiegand, A. (1990). High sensitivity and detectivity radiation thermopiles made by multi-layer technology. *Sensors Actuators A Phys.* 24 (1): 1–4.

97 Shen, T.-W., Chang, K.-C., Sun, C.-M., and Fang, W. (2019). Performance enhance of CMOS-MEMS thermoelectric infrared sensor by using sensing material and structure design. *J. Micromech. Microeng.* 29: 024001.

98 Discera Inc. (2011). *Low-Power Precision CMOS Oscillator Discera DSC1001 Datasheet*. San Jose, CA: Discera Inc.

99 SiTime Corporation (2013). *SiT8208 Ultra Performance Oscillator Datasheet*. Sunnyvale, CA: SiTime Corporation.

100 Verd, J., Uranga, A., Teva, J. et al. (2006). Integrated CMOS-MEMS with on-chip readout electronics for high-frequency applications. *IEEE Electron Dev. Lett.* 27 (6): 495–497.

101 Lopez, J.L., Verd, J., Teva, J. et al. (2008). Integration of RF-MEMS resonators on submicrometric commercial CMOS technologies. *J. Micromech. Microeng.* 19: 015002.

102 Li, C.-S., Hou, L.-J., and Li, S.-S. (2012). Advanced CMOS-MEMS resonator platform. *IEEE Electron Dev. Lett.* 33: 272–274.

103 Liu, Y.-S. and Wen, K.-A. (2019). Implementation of a CMOS/MEMS accelerometer with ASIC processes. *Micromachines* 10 (30642025): 50.

104 Narducci, M., Yu-Chia, L., Fang, W., and Tsai, J. (2013). CMOS MEMS capacitive absolute pressure sensor. *J. Micromech. Microeng.* 23: 055007.

105 Li, M.-H., Chen, W.-C., and Li, S.-S. (2012). Mechanically coupled CMOS-MEMS free-free beam resonator arrays with enhanced power handling capability. *IEEE Trans. Ultrason. Ferroelectr. Freq. Control* 59 (3): 346–357.

106 Sun, C.-M., Tsai, M.-H., Liu, Y.-C., and Fang, W. (2010). Implementation of a monolithic single proof-mass tri-axis accelerometer using CMOS-MEMS technique. *IEEE Trans. Electron Dev.* 57 (7): 1670–1679.

107 Chen, J.-H. and Huang, C.-W. (2018). 0.35 µm CMOS–MEMS low-mechanical-noise micro accelerometer. *Microsyst. Technol.* 24 (1): 299–304.

108 Chiang, C. (2018). Design of a CMOS MEMS accelerometer used in IoT devices for seismic detection. *IEEE J. Emerg. Select. Top. Circuits Syst.* 8 (3): 566–577.

109 Eranna, G., Joshi, B.C., Runthala, D.P., and Gupta, R.P. (2004). Oxide materials for development of integrated gas sensors—a comprehensive review. *Crit. Rev. Solid State Mater. Sci.* 29 (3-4): 111–188.

110 Graf, M., Barrettino, D., Zimmermann, M. et al. (2004). CMOS monolithic metal-oxide sensor system comprising a microhotplate and associated circuitry. *IEEE Sensors J.* 4 (1): 9–16.

111 Kappler, J., Bârsan, N., Weimar, U. et al. (1998). Correlation between XPS, Raman and TEM measurements and the gas sensitivity of Pt and Pd doped SnO_2 based gas sensors. *Fresenius J. Anal. Chem.* 361 (2): 110–114.

112 Afridi, M., Suehle, J.S., Zaghloul, M.E. et al. (2002). A monolithic CMOS microhotplate-based gas sensor system. *IEEE Sensors J.* 2 (6): 644–655.

113 Yang, M.-Z., Dai, C.-L., Shih, P.-J., and Chen, Y.-C. (2011). Cobalt oxide nanosheet humidity sensor integrated with circuit on chip. *Microelectron. Eng.* 88 (8): 1742–1744.

114 Dai, C.-L., Chen, Y.-C., Wu, C.-C., and Kuo, C.-F. (2010). Cobalt oxide nanosheet and CNT micro carbon monoxide sensor integrated with readout circuit on chip. *Sensors* 10 (3): 1753–1764.

115 Yang, M.-Z., Dai, C.-L., and Wu, C.-C. (2011). A zinc oxide nanorod ammonia microsensor integrated with a readout circuit on-a-chip. *Sensors* 11 (12): 11112–11121.

116 Yang, M.-Z., Dai, C.-L., and Wu, C.-C. (2014). Sol-gel zinc oxide humidity sensors integrated with a ring oscillator circuit on-a-chip. *Sensors* 14 (11): 20360–20371.

117 Liao, W.-Z., Dai, C.-L., and Yang, M.-Z. (2013). Micro ethanol sensors with a heater fabricated using the commercial 0.18 μm CMOS process. *Sensors* 13 (10): 12760–12770.

118 Yang, M.-Z. and Dai, C.-L. (2015). Ethanol microsensors with a readout circuit manufactured using the CMOS-MEMS technique. *Sensors* 15 (1): 1623–1634.

119 Fong, C.-F., Dai, C.-L., and Wu, C.-C. (2015). Fabrication and characterization of a micro methanol sensor using the CMOS-MEMS technique. *Sensors* 15 (10): 27047–27059.

120 Dai, C.-L., Liu, M.-C., Chen, F.-S. et al. (2007). A nanowire WO_3 humidity sensor integrated with micro-heater and inverting amplifier circuit on chip manufactured using CMOS-MEMS technique. *Sensors Actuators B Chem.* 123 (2): 896–901.

121 Hu, Y.-C., Dai, C.-L., and Hsu, C.-C. (2014). Titanium dioxide nanoparticle humidity microsensors integrated with circuitry on-a-chip. *Sensors* 14 (3): 4177–4188.

122 Yang, M.-Z., Dai, C.-L., and Shih, P.-J. (2014). An acetone microsensor with a ring oscillator circuit fabricated using the commercial 0.18 μm CMOS Process. *Sensors* 14 (7): 12735–12747.

123 Lazarus, N., Bedair, S.S., Lo, C.-C., and Fedder, G.K. (2010). CMOS-MEMS capacitive humidity sensor. *J. Microelectromech. Syst.* 19 (1): 183–191.

124 Lazarus, N. and Fedder, G.K. (2011). Integrated vertical parallel-plate capacitive humidity sensor. *J. Micromech. Microeng.* 21 (6): 065028.

125 Dai, C.-L. (2007). A capacitive humidity sensor integrated with micro heater and ring oscillator circuit fabricated by CMOS–MEMS technique. *Sens. Actuators B Chem.* 122 (2): 375–380.

126 Yang, M.-Z., Dai, C.-L., and Lu, D.-H. (2010). Polypyrrole porous micro humidity sensor integrated with a ring oscillator circuit on chip. *Sensors* 10 (11): 10095–10104.

127 Chung, V., Yip, M.-C., and Fang, W. (2015). Resorcinol–formaldehyde aerogels for cmos-mems capacitive humidity sensor. *Sens. Actuators B Chem.* 214: 181–188.

128 Bedair, S.S. and Fedder, G.K. (2004). CMOS MEMS oscillator for gas chemical detection. IEEE SENSORS 2004, Vienna, 955–958.

129 Voiculescu, I., Zaghloul, M.E., McGill, R.A. et al. (2005). Electrostatically actuated resonant microcantilever beam in CMOS technology for the detection of chemical weapons. *IEEE Sensors J.* 5 (4): 641–647.

130 Tsai, H.-H., Lin, C.-F., Juang, Y.-Z. et al. (2010). Multiple type biosensors fabricated using the CMOS Bio MEMS platform. *Sens. Actuators B Chem.* 144 (2): 407–412.

131 Li, D.-C., Yang, P.-H., and Lu, M.S.-C. (2010). CMOS open-gate ion-sensitive field-effect transistors for ultrasensitive dopamine detection. *IEEE Trans. Electron Dev.* 57 (10): 2761–2767.

132 Lai, W.-A., C.-H. Lin, Y.-S. Yang, and M. S.-C. Lu (2012). Ultrasensitive detection of avian influenza virus by using CMOS impedimetric sensor arrays. 25th IEEE Micro Electro Mechanical Systems (MEMS) (February 2012), 894–897.

133 Yin, T.-I., Zhao, Y., Lin, C.-F. et al. (2011). The application of capillary force to a cantilever as a sensor for molecular recognition. *Appl. Phys. Lett.* 98 (10): 104102.

134 Yin, T.-I., Zhao, Y., Horak, J. et al. (2013). A micro-cantilever sensor chip based on contact angle analysis for a label-free troponin I immunoassay. *Lab Chip* 13 (5): 834–842.

135 Huang, Y.-J., Huang, C.-W., Lin, T.-H. et al. (2013). A CMOS cantilever-based label-free DNA SoC with improved sensitivity for hepatitis B virus detection. *IEEE Trans. Biomed. Circuits Syst.* 7 (6): 820–831.

136 Huang, C.-W., Hsueh, H.-T., Huang, Y.-J. et al. (2013). A fully integrated wireless CMOS microcantilever lab chip for detection of DNA from Hepatitis B virus (HBV). *Sensors Actuators B Chem.* 181: 867–873.

137 Neumann, J.J. and Gabriel, K.J. (2003). A fully-integrated CMOS-MEMS audio microphone. 12th International Conference on Solid-State Sensors, Actuators and Microsystems (TRANSDUCERS'03), vol. 1, 230–233.

138 Tang, P.-K., Wang, P.-H., Li, M.-L., and Lu, M.S.-C. (2011). Design and characterization of the immersion-type capacitive ultrasonic sensors fabricated in a CMOS process. *J. Micromech. Microeng.* 21 (2): 025013.

139 Li, M.-L., Wang, P.-H., Liao, P.-L., and Lu, M.S.-C. (2011). Three-dimensional photoacoustic imaging by a CMOS micromachined capacitive ultrasonic sensor. *IEEE Electron Dev. Lett.* 32 (8): 1149–1151.

140 Cheng, C.-L., Chang, H.-C., Chang, C.-I., and Fang, W. (2015). Development of a CMOS MEMS pressure sensor with a mechanical force-displacement transduction structure. *J. Micromech. Microeng.* 25 (12): 125024.

141 Lin, W.-C., C.-L. Cheng, C.-L. Wu, and W. Fang (2017). Sensitivity improvement for CMOS-MEMS capacitive pressure sensor using double deformarle diaphragms with trenches. 19th International Conference on Solid-State Sensors, Actuators and Microsystems (TRANSDUCERS'17), 782–785.

142 Dai, C.-L., Lu, P.-W., Wu, C.-C., and Chang, C. (2009). Fabrication of wireless micro pressure sensor using the CMOS process. *Sensors* 9 (11): 8748–8760.

143 Uddin, A., Milaninia, K., Chen, C., and Theogarajan, L. (2011). Wafer scale integration of CMOS chips for biomedical applications via self-aligned masking. *IEEE Trans. Compon. Packag. Manuf. Technol.* 1 (12): 1996–2004.

144 Huang, Y. and Mason, A.J. (2013). Lab-on-CMOS integration of microfluidics and electrochemical sensors. *Lab Chip* 13 (19): 3929–3934.

145 Uddin, A., Yemenicioglu, S., Chen, C.-H. et al. (2013). Integration of solid-state nanopores in a 0.5 μm CMOS foundry process. *Nanotechnology* 24 (15): 155501.

146 Lindsay, M., Bishop, K., Sengupta, S. et al. (2018). Heterogeneous integration of CMOS sensors and fluidic networks using wafer-level molding. *IEEE Trans. Biomed. Circuits Syst.* 12 (5): 1046–1055.

147 Ghafar-Zadeh, E., Sawan, M., and Therriault, D. (2007). Novel direct-write CMOS-based laboratory-on-chip: Design, assembly and experimental results. *Sens. Actuators A Phys.* 134 (1): 27–36.

148 Chien, J.-C., Ameri, A., Yeh, E.-C. et al. (2018). A high-throughput flow cytometry-on-a-CMOS platform for single-cell dielectric spectroscopy at microwave frequencies. *Lab Chip* 18 (14): 2065–2076.

149 Chiu, Y., Chen, B.-T., and Hong, H.-C. (2015). Integrated CMOS MEMS liquid capacitive inclinometer. 18th International Conference on Solid-State Sensors, Actuators and Microsystems (TRANSDUCERS'15), 1152–1155.

150 Hagleitner, C., Hierlemann, A., Lange, D. et al. (2001). Smart single-chip gas sensor microsystem. *Nature* 414: 293–296.

151 Hagleitner, C., Lange, D., Hierlemann, A. et al. (2002). CMOS single-chip gas detection system comprising capacitive, calorimetric and mass-sensitive microsensors. *IEEE J. Solid State Circuits* 37 (12): 1867–1878.

152 Li, Y., Vancura, C., Barrettino, D. et al. (2007). Monolithic CMOS multi-transducer gas sensor microsystem for organic and inorganic analytes. *Sens. Actuators B Chem.* 126 (2): 431–440.

153 Eddy, D.S. and Sparks, D.R. (1998). Application of MEMS technology in automotive sensors and actuators. *Proc. IEEE* 86 (8): 1747–1755.

154 Ishihara, T., Suzuki, K., Suwazono, S. et al. (1987). CMOS integrated silicon pressure sensor. *IEEE J. Solid State Circuits* 22 (2): 151–156.

155 Baskett, I., Frank, R., and Ramsland, E. (1991). The design of a monolithic, signal conditioned pressure sensor. Proceedings of the IEEE 1991 Custom Integrated Circuits Conference IEEE, 27.331–27.3.4.

156 Sugiyama, S., Takigawa, M., and Igarashi, I. (1983). Integrated piezoresistive pressure sensor with both voltage and frequency output. *Sens. Actuators* 4: 113–120.

157 Barrettino, D., Graf, M., Taschini, S. et al. (2006). CMOS monolithic metal–oxide gas sensor microsystems. *IEEE Sensors J.* 6 (2): 276–286.

158 Chiu, Y., Huang, T.-C., and Hong, H.-C. (2014). A three-axis single-proof-mass CMOS-MEMS piezoresistive accelerometer with frequency output. *Sens. Mater.* 26 (2): 95–108.

159 Lu, C., Lemkin, M., and Boser, B.E. (1995). A monolithic surface micromachined accelerometer with digital output. *IEEE J. Solid State Circuits* 30 (12): 1367–1373.

160 Boser, B.E. and Howe, R.T. (1996). Surface micromachined accelerometers. *IEEE J. Solid State Circuits* 31 (3): 366–375.

161 Lemkin, M. and Boser, B.E. (1999). A three-axis micromachined accelerometer with a CMOS position-sense interface and digital offset-trim electronics. *IEEE J. Solid State Circuits* 34 (4): 456–468.

162 Luo, H., Zhang, G., Carley, L.R., and Fedder, G.K. (2002). A post-CMOS micromachined lateral accelerometer. *J. Microelectromech. Syst.* 11 (3): 188–195.

163 Wu, J., Fedder, G.K., and Carley, L.R. (2004). A low-noise low-offset capacitive sensing amplifier for a 50-μg/rtHz monolithic CMOS MEMS accelerometer. *IEEE J. Solid State Circuits* 39 (5): 722–730.

164 Tan, S.-S., Liu, C.-Y., Yeh, L.-K. et al. (2011). An integrated low-noise sensing circuit with efficient bias stabilization for CMOS MEMS capacitive accelerometers. *IEEE Trans. Circuits Syst. I: Regular Papers* 58 (11): 2661–2672.

165 Michalik, P., J. M. Sánchez-Chiva, D. Fernández, and J. Madrenas, (2015). CMOS BEOL-embedded lateral accelerometer. 2015 IEEE Sensors Conference, Busan, South Korea (November 2015).

166 Liao, S.-H., Chen, W.-J., and Lu, M.S.-C. (2013). A CMOS MEMS capacitive flow sensor for respiratory monitoring. *IEEE Sensors J.* 13 (5): 1401–1402.

167 Xie, H. and Fedder, G.K. (2002). Vertical comb-finger capacitive actuation and sensing for CMOS-MEMS. *Sens. Actuators A Phys.* 95 (2-3): 212–221.

168 Ko, C.-T., Tseng, S.-H., and Lu, M.S.-C. (2006). A CMOS micromachined capacitive tactile sensor with high-frequency output. *J. Microelectromech. Syst.* 15 (6): 1708–1714.

169 Chiu, Y., Hong, H.-C., and Wu, P.-C. (2013). Development and characterization of a CMOS-MEMS accelerometer with differential LC-tank oscillators. *J. Microelectromech. Syst.* 22 (6): 1285–1295.

170 Chiu, Y., Hong, H.-C., and Lin, C.-W. (2016). Inductive CMOS MEMS accelerometer with integrated variable inductors. 29th IEEE Micro Electro Mechanical Systems (MEMS), 974–977.

171 Chiu, Y., Hong, H.-C., and Chang, C.-M. (2017). Three-axis CMOS MEMS inductive accelerometer with novel Z-axis sensing scheme. 19th International Conference on Solid-State Sensors, Actuators and Microsystems (TRANSDUCERS'17), 410–413.

172 Chiang, C.-H., Chou, M.-C., Hsieh, P.-H., and Lu, M.S.-C. (2016). Design and characterization of a CMOS MEMS capacitive oscillator for resonant sensing in liquids. *IEEE Sens. J.* 16 (5): 1136–1142.

173 Huang, Y.-J., Chang, T.-L., and Chou, H.-P. (2009). Study of symmetric microstructures for CMOS multilayer residual stress. *Sens. Actuators A Phys.* 150: 237–242.

174 Fang, W. and Wickert, J.A. (1996). Determining mean and gradient residual stresses in thin films using micromachined cantilevers. *J. Micromech. Microeng.* 6: 301–309.

175 Fang, W. and Wickert, J.A. (1994). Post buckling of micromachined beams. *J. Micromech. Microeng.* 4: 116–122.

176 Mehregany, M., Howe, R.T., and Senturia, S.D. (1987). Novel microstructures for the in situ measurement of the mechanical properties of thin films. *J. Appl. Phys.* 62: 3579–3584.

177 Guckel, H., Randazzo, T., and Burns, D.W. (1985). A simple technique for the determination of mechanical strain in thin films with applications to polysilicon. *J. Appl. Phys.* 57: 1671–1675.

178 Ceiler, M.F. Jr.,, Kohl, P.A., and Bidstrup, S.A. (1995). Plasma-enhanced chemical vapor deposition of silicon dioxide deposited at low temperatures. *J. Electrochem. Soc.* 142 (6): 2067–2071.

179 EerNisse, E.P. (1979). Stress in thermal SiO2 during growth. *Appl. Phys. Lett.* 35: 8–10.

180 Greek, S. and Chitica, N. (1999). Deflection of surface micromachined devices due to internal, homogeneous or gradient stresses. *Sensors Actuators A Phys.* 78: 1–7.

181 Cheng, C.-L., Tsai, M.-H., and Fang, W. (2015). Determining the thermal expansion coefficient of thin films for a CMOS MEMS process using test cantilevers. *J. Micromech. Microeng.* 25: 025014.

182 Valle, J., Fernández, D., Madrenas, J., and Barrachina, L. (2017). Curvature of BEOL cantilevers in CMOS-MEMS processes. *J. Microelectromech. Syst.* 26 (4): 895–909.

183 Lakdawala, H. and Fedder, G.K. (1999). Analysis of temperature-dependent residual stress gradients in CMOS micromachined structures. Proceedings of the 15th International Conference on Solid-State Sensors, Actuators and Microsystems (TRANSDUCERS'99), Sendai, Japan (June 1999), 526–529.

184 Tsai, M.-H., Liu, Y.-C., Liang, K.-C., and Fang, W. (2015). Monolithic CMOS-MEMS pure oxide tri-axis accelerometers for temperature stabilization and performance enhancement. *J. Microelectromech. Syst.* 24 (6): 1916–1927.

185 Chang, C.-I., Tsai, M.-H., Sun, C.-M., and Fang, W. (2014). Development of CMOS-MEMS in-plane magnetic coils for application as a three-axis resonant magnetic sensor. *J. Micromech. Microeng.* 24: 035016.

186 Yen, T.-H., Tsai, M.-H., Chang, C.-I. et al. (2011). Improvement of CMOS-MEMS accelerometer using the symmetric layers stacking design. IEEE Sensors Conference, Limerick, Ireland (October 2011), 145–148.

187 Fedder, G.K., Howe, R.T., Liu, T.-J.K., and Quevy, E.P. (2008). Technologies for cofabricating MEMS and electronics. *Proc. IEEE* 96 (2): 306–322.

188 Neumann, J.J. and Gabriel, K.J. (2005). *CMOS-MEMS Acoustic Devices, CMOS-MEMS*, 1e, 193–224. Wiley-VCH.

189 Timoshenko, S. (1925). Analysis of bi-metal thermostats. *J. Opt. Soc. Am.* 11: 233–255.

190 Modlinski, R., Witvrouw, A., Ratchev, P. et al. (2004). Creep characterization of Al alloy thin films for use in MEMS applications. *Microelectron. Eng.* 76: 272–278.

191 Jiang, L., Cai, Y., Liu, H., and Zhao, Y. (2013). A micromachined monolithic 3 axis accelerometer based on convection heat transfer. IEEE NEMS Conference, Suzhou, China (April 2013), 248–251.

192 Rüffer, D., Hoehne, F., and Bühler, J. (2018). New digital metal-oxide (MOx) sensor platform. *Sensors* 18: 1052.

193 Cheng, C.-W., Liang, K.-C., Chu, C.-H. et al. (2013). Single chip process for sensors implementation, integration, and condition monitoring. Proceedings of the Seventh International Conference on Solid-State Sensors, Actuators and Microsystems (TRANSDUCERS'13), Barcelona, Spain (June 2013), 730–733.

194 Petersen, K.E. (1982). Silicon as a mechanical material. *Proc. IEEE* 70 (5): 420–457.

195 Hornbeck, L.J. (1997). Digital light processing for high-brightness, high-resolution applications. *Proc. SPIE* 3013: 27–40.

196 Ruiz, P.G., Meyer, K.D., and Witvrouw, A. (2013). *Poly-SiGe for MEMS-Above-CMOS Sensors*, 1e. Springer.

197 Cheng, C.-W., Chu, C.-H., Hung, L.-M., and Fang, W. (2017). 12 inch MEMS process for sensors implementation and integration. Proceedings of the 19th International Conference on Solid-State Sensors, Actuators and Microsystems (TRANSDUCERS'17), Kaohsiung, Taiwan (June 2017), 402–405.

198 Li, M.-H., Chen, C.-Y., Liu, C.-Y., and Li, S.-S. (2016). A sub-150μW BEOL-embedded CMOS-MEMS oscillator with a 138dBΩ ultra-low-noise TIA. *IEEE Electron Dev. Lett.* 37 (5): 648–651.

199 Liu, C.-Y., Li, M.-H., Ranjith, H.G., and Li, S.-S. (2016). A 1 MHz 4 ppm CMOS-MEMS oscillator with built-in self-test and sub-mW ovenization power. Proceedings of the, IEEE International Electron Devices Meeting (IEDM'16), San Francisco, USA (3–7 December 2016), 26.7.1–26.7.4.

200 Chiu, W.-C., Li, M.-H., Chou, C., and Li, S.-S. (2016). A ring-down technique implemented in CMOS-MEMS resonator circuits for wide-range pressure sensing applications. 2016 IEEE International Frequency Control Symposium (IFCS'16), New Orleans, Louisiana, USA (9–12 May 2016), 1–3.

201 Sarkar, N., Mansour, R.R., Patange, O., and Trainor, K. (2011). CMOS-MEMS atomic force microscope 2011 16th International Solid-State Sensors, Actuators and Microsystems Conference, Beijing, 2610–2613.

202 Ghaffari, S., Ng, E., Ahn, C.H. et al. (2015). Accurate modeling of quality factor behavior of complex silicon MEMS resonators. *J. Microelectromech. Syst.* 24 (2): 276–288.

203 Melamud, R., Chandorkar, S.A., Kim, B. et al. (2009). Temperature-Insensitive Composite Micromechanical Resonators. *J. Microelectromech. Syst.* 18 (6): 1409–1419.

204 Ghaffari, S., Chandorkar, S.A., Wang, S. et al. (2013). Quantum limit of quality factor in silicon micro and nano mechanical resonators. *Scientific Rep.* 3: 3244.

205 Li, M.-H., Li, C.-S., and Li, S.-S. (2015). Exploring the Q-factor limit of temperature compensated CMOS-MEMS resonators. Proceedings of the 28th IEEE International Conference on Micro Electro Mechanical Systems (MEMS'15), Estoril, Portugal (18–22 January 2015), 853–856.

206 Kim, B., Hopcroft, M.A., Candler, R.N. et al. (2008). Temperature dependence of quality factor in MEMS resonators. *J. Microelectromech. Syst.* 17 (3): 755–766.

207 Wang, S., Bahr, B., Chen, W.-C. et al. (2015). Temperature coefficient of frequency modeling for CMOS-MEMS bulk mode composite resonator. *IEEE Trans. Ultrason. Ferroelectr. Freq. Control (T-UFFC)* 62 (6): 1166–1178.

208 Lin, Y.-C., Guney, M.G., and Fedder, G.K. (2019). ALD titania sidewalls on a CMOS-MEMS resonator oscillator and effects on resonant frequency drift. 32nd IEEE Micro Electro Mechanical Systems (MEMS), Seoul, Korea (27–31 January 2019), 640–643.

209 Li, M.-H., Chen, C.-Y., and Li, S.-S. (2018). A study on the design parameters for MEMS oscillators incorporating nonlinearities. *IEEE Trans. Circuits Syst. I Reg. Papers* 65 (10): 3424–3434.

210 H. K. Lee, P. A. Ward, A. E. Duwel, J. C. Salvia, Qu, Y.Q., Melamud, R., Chandorkar, S.A. et al. (2011). Verification of the phase-noise model for MEMS oscillators operating in the nonlinear regime. 2011 16th International Solid-State Sensors, Actuators and Microsystems Conference (June 2011), 510–513.

211 Villanueva, L.G., Kenig, E., Karabalin, R.B. et al. (2013). Surpassing fundamental limits of oscillators using nonlinear resonators. *Phys. Rev. Lett.* 110: 177208.

212 Kenig, E., Cross, M.C., Villanueva, L.G. et al. (2012). Optimal operating points of oscillators using nonlinear resonators. *Phys. Rev. E* 86: 056207.

213 Li, M.-H., Chen, C.-Y., Chin, C.-H. et al. (2014). Optimizing the close-to-carrier phase noise of monolithic CMOS-MEMS oscillators using bias-dependent nonlinearity, Technical Digest IEEE International Electron Devices Meeting (IEDM'14), San Francisco, CA (December 2014), 22.3.1–22.3.4.

214 International Technology Roadmap for Semiconductors (2005). Edition.

9

Wafer Transfer

Masayoshi Esashi

Tohoku University, Micro System Integration Center (μSIC), 519-1176 Aramaki-Aza-Aoba, Aoba-ku, Sendai 980-0845, Japan

9.1 Introduction

Micro-electro mechanical systems (MEMS) are used being combined with integrated circuit (IC) or large scale integration (LSI) in many cases. This heterogeneous integration (hetero-integration) requires process flexibility to fabricate high-performance MEMS on LSI without damage to the LSI [1–3]. Multiple elements and multiple interconnections with reduced stray capacitance and stray inductance can be achieved by the MEMS on LSI. The heterogeneous integration can be performed by wafer transfer of multiple MEMS on a carrier wafer to an LSI wafer. The concept of heterogeneous integration combined with wafer level packaging (WLP) is shown in Figure 9.1. LSI wafer is prepared (1 in the figure). MEMS or thin films of functional material are formed on a Si carrier wafer (1' in the figure) and transferred on an LSI wafer by bonding with adhesive polymer (adhesive bonding) (2 in the figure) or bump bonding. The carrier wafer is removed (3 in the figure). MEMS can be formed on the LSI wafer with some additional process steps and removal of the adhesive polymer (4 in the figure). The heterogeneous integration devices should be protected from environment and hence WLP [4] is required. Lid wafer made of glass or low-temperature co-fired ceramics (LTCC) with through substrate vias (TSV) is prepared (4' in the figure), and it is anodically bonded to the MEMS on LSI wafer (5 in the figure). The bonded wafer is finally diced to chips and packaged heterogeneous integration chips can be obtained (6 in the figure).

The wafer transfer methods for the MEMS on LSI can be categorized as (a) film transfer, (b) device transfer (via-last), and (c) device transfer (via-first) as shown in Figure 9.2.

Figure 9.2a is the process sequence of the film transfer method. A film of functional material such as diamond or PZT (lead zirconate titanate) is formed on a carrier wafer by chemical vapor deposition (CVD), sputter deposition, or other methods. The film is transferred on the LSI wafer by the adhesive bonding and then the carrier wafer is removed. This removal can be done by etching out the carrier

3D and Circuit Integration of MEMS, First Edition. Edited by Masayoshi Esashi.
© 2021 WILEY-VCH GmbH. Published 2021 by WILEY-VCH GmbH.

9 Wafer Transfer

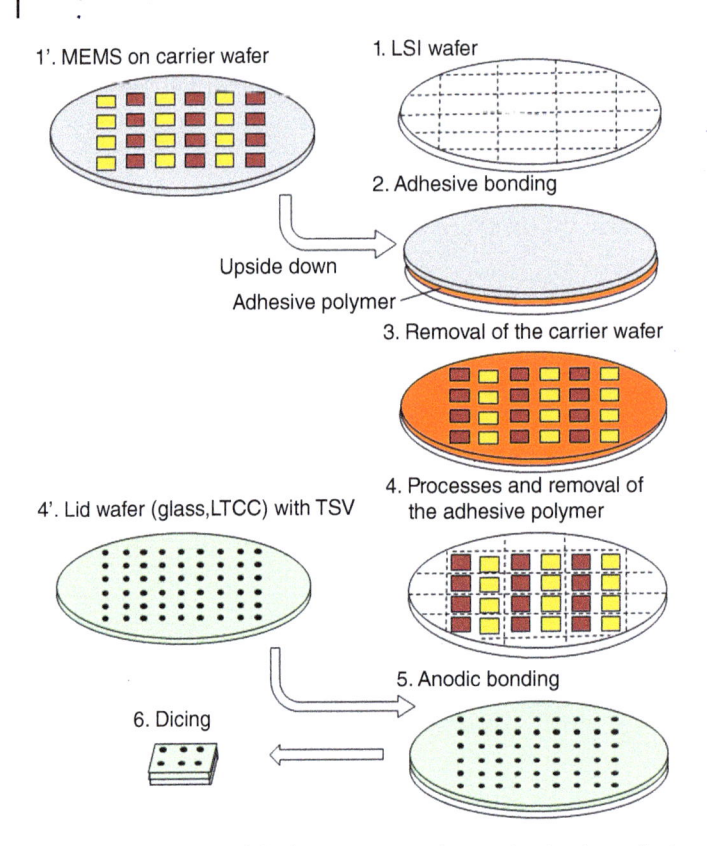

Figure 9.1 Concept of the heterogeneous integration by the wafer transfer and the wafer level packaging (WLP).

wafer or by etching out a sacrificial layer between the film and the carrier wafer. MEMS are fabricated using the film, and its electrical and mechanical connections to the LSI wafer are made by using an electroplating or other deposition methods of metal combined with patterning. The adhesive polymer is etched out and we can get the MEMS on LSI. Since the MEMS fabrication is carried out on the LSI wafer, the process is limited not to damage the LSI.

Figure 9.2b shows the process sequence of the device transfer (via-last) method. The MEMS are fabricated on a carrier wafer and the wafer is bonded to the LSI wafer using the adhesive bonding. After removing the carrier wafer, electrical and mechanical connections to the LSI wafer are made through vias by electroplating or other deposition methods of metal and by patterning. The adhesive polymer is etched out finally. This process (b) is called via-last because vias are made after bonding.

Figure 9.2c is the process sequence of the device transfer (via-first). The MEMS are fabricated on a carrier wafer having the adhesive polymer layer on it. Bumps for electrical interconnection (via) are formed on the LSI wafer. These wafers are bonded using the bumps and finally the carrier wafer is removed by etching it out or by removing the adhesive polymer. In case we can use the MEMS formed directly on the carrier wafer, the adhesive polymer is not necessary. This process (c) is called via-first because vias are made before bonding.

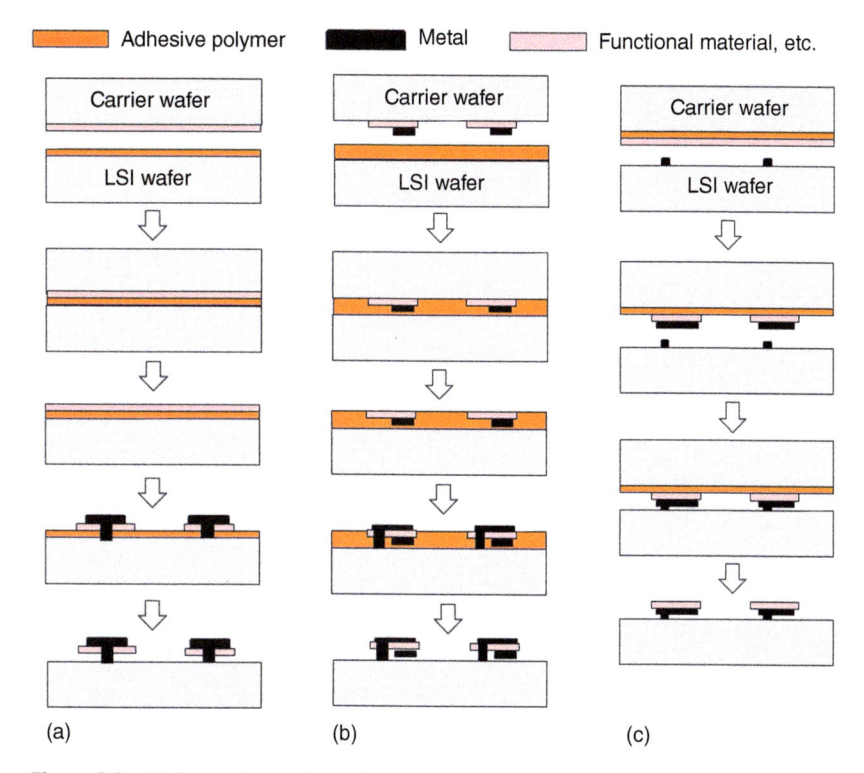

Figure 9.2 Wafer level transfer of heterogeneous MEMS to LSI. (a) Film transfer, (b) device transfer (via-last), and (c) device transfer (via-first).

Examples of each method will be shown in Sections 9.2, 9.3 and, 9.4 respectively.

Contrary to the wafer level transfer mentioned earlier, chip level wafer transfer can be used as will be explained in Section 9.5.

Figure 9.3 shows a concept of multiband system for cognitive wireless communication. The heterogeneous integration using the wafer level transfer or the chip level wafer transfer can be used to fabricate key components for such systems as explained in the following.

9.2 Film Transfer

A wide-tuning range complementary metal oxide semiconductor (CMOS)-film bulk acoustic resonator (FBAR) for voltage-controlled oscillator (VCO) has been developed on 0.18 µm CMOS LSI [5]. Figure 9.4 shows its structure and photograph (a), VCO circuit (b), and fabrication process (c). The FBAR uses aluminum nitride (AlN) as its piezoelectric layer, and it has an air gap under it to prevent energy loss from the resonator as shown in (a). The circuit (b) is a Pierce oscillator in which electrical frequency tuning can be made by connecting capacitors digitally. The fabrication process (c) explained in the following corresponds to the film transfer shown Figure 9.2a. Silicon-on-insulator (SOI) wafer and CMOS LSI wafer are used (1 in the figure). The SOI wafer is flipped and bonded with adhesive polymer BCB (benzocyclobutene) on the CMOS wafer (2 in the figure). Thin Si layer is formed on

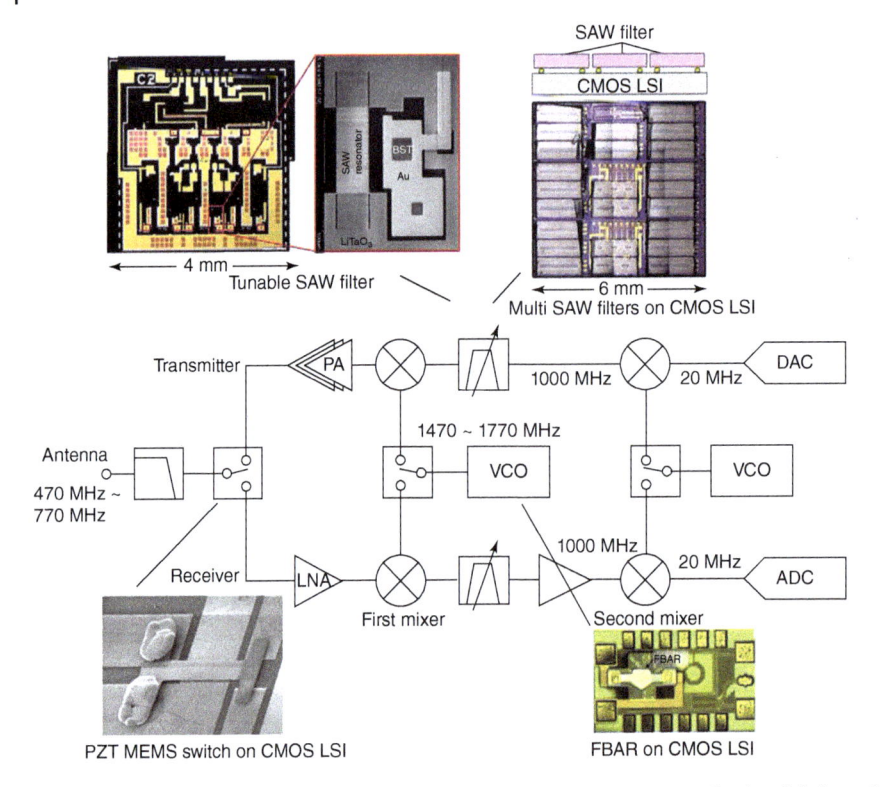

Figure 9.3 Concept of multiband system for cognitive wireless communication fabricated using the heterogeneous integration.

the CMOS wafer by removal of the handle Si layer of the SOI wafer (3 in the figure). Ru and AlN are sputter deposited and patterned (4 in the figure). Since the AlN can be formed at 300 °C, the CMOS LSI is not damaged in this process step. Top Al electrode is fabricated by lift-off, and electrical interconnections with the CMOS LSI are made by Cr/Au (5 in the figure). Sacrificial etching is performed to remove the Si underneath the FBAR to make the air gap (6 in the figure).

A 20 × 20 array of boron-doped diamond (BDD) electrodes is formed on a CMOS LSI having 20 × 20 array of operational amplifier. This BDD on LSI was applied to simultaneous multipoint amperometric detection of biochemical substances [6]. The structure of the BDD electrodes on the CMOS LSI is shown in Figure 9.5a. Figure 9.5b is I–V converter circuit in which current is detected by integrating it in the feedback capacitor C. The integration time is determined by the current level. The voltage applied to the BDD electrode and the output voltage obtained by integrating the current are e_i and e_o in Figure 9.5b. The fabrication process is explained using Figure 9.5c. Metallization of Cr/Pt/Au/Pt/Cr is carried out on the LSI wafer (1 in the figure). BCB (Benzocyclobutene) is coated on the LSI. The BDD layer is prepared on a carrier wafer by microwave chemical vapor deposition (MW-CVD) at 800 °C. It is patterned in O_2 plasma using Al as a mask. The boron is doped for electrical conduction of the diamond (2 in the figure). The carrier wafer

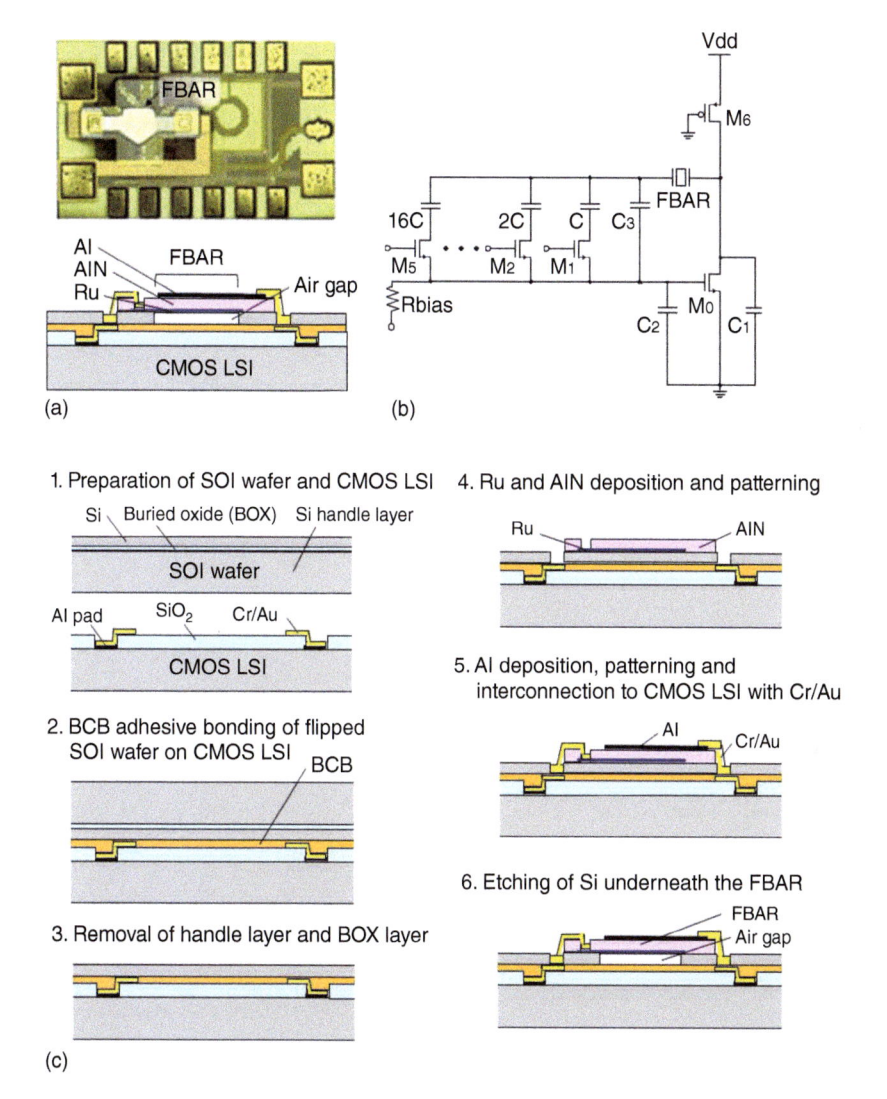

Figure 9.4 FBAR on CMOS VCO. (a) Structure, (b) oscillator circuit, and (c) fabrication process.

which has the BDD layer is flipped and bonded on the BCB-coated LSI wafer. The BDD film is transferred on the LSI by the film transfer explained in Figure 9.2a (3 in the figure). The BDD is exposed by dry etching of the Si carrier wafer in SF_6 plasma (4 in the figure). The BCB is selectively etched by plasma ($SF_6 + O_2$) using patterned Al as a mask (5 in the figure). Cr and Au are deposited and patterned for electrical interconnection between the LSI pad and the BDD (6 in the figure). It is covered with thick photoresist (SU-8) and windows are opened to expose the BDD (7 in the figure).

The cyclic voltammogram (CV) of the BDD compared with that of Au is shown in Figure 9.6a. Oxygen gas is generated by oxidation of hydroxide ion at positive voltage. Hydrogen gas is generated by reduction of hydrogen ion at negative voltage.

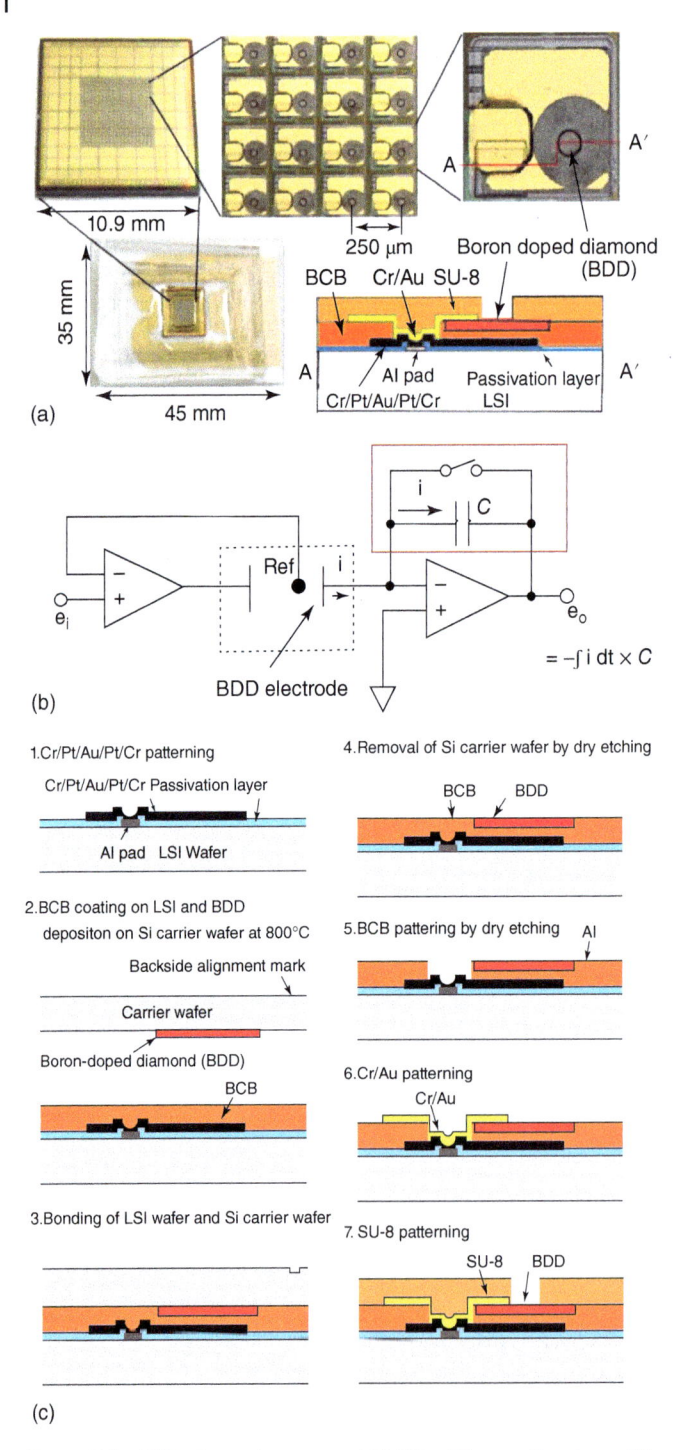

Figure 9.5 BDD electrode array on LSI (20 × 20 operational amplifier array) for electrochemical bio-sensing. (a) Structure, (b) current detection circuit, and (c) fabrication process.

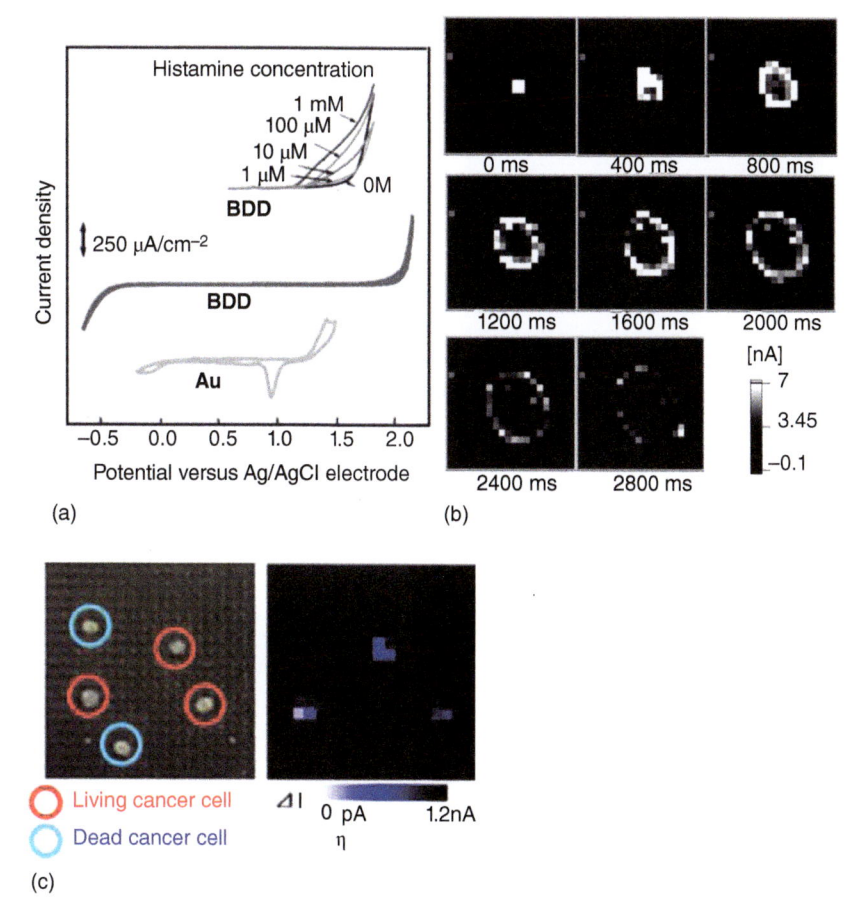

Figure 9.6 Electrochemical detection of biochemical substances. (a) Cyclic voltammogram (CV) on Au and BDD in 0.5 M H_2SO_4 and CV on BDD in 0.1 mM histamine in Dulbecco's phosphate buffer solution, (b) diffusion of dropped histamine, and (c) distinction between living cancer cell and dead cancer cell using the difference of oxygen reduction current.

These voltages are large in the case of BDD compared with Au, because the diamond has not catalytic property. The CV of the BDD in histamine solution is also shown in Figure 9.6a. The histamine could be detected by amperometry using the BDD with a high oxidation potential around 1.5 V. This voltage is too high for a conventional Au electrode to detect. The diffusion behavior of the histamine in a solution was imaged in real time by parallel measurement of the oxidation current at 20×20 points (Figure 9.6b). This CMOS-based amperometric sensor array has been successfully applied to drug screening for cancer cells on the array (Figure 9.6c). The living cancer cells are identified by electrochemical measurement of the consumed oxygen surrounding the living cells.

9.3 Device Transfer (via-last)

PZT (lead zirconate titanate) actuated MEMS switches were fabricated on an LSI wafer as shown in Figure 9.7 [7]. The photographs of the PZT MEMS switch are shown in Figure 9.7a. The piezoelectric MEMS switch works at lower driving voltage and occupies smaller area than electrostatic MEMS switches. The fabrication process is shown in Figure 9.7b. For the MEMS switch, the PZT is deposited on a Si carrier wafer by a sol–gel method. Symmetrical structure made of two stacked PZT layers is formed in order to prevent bending and then patterned into device structures (1 in the figure). They are transferred on the LSI wafer using the adhesive polymer on the flipped carrier wafer (2 and 3 in the figure). The carrier wafer is etched out (4 in the figure). Holes are made for electrical interconnection (5 in the figure). After connecting the MEMS and the LSI using electroplated Au (6 in the figure), the polymer is removed by O_2 plasma to release the MEMS switches (7 in the figure). The PZT cantilever bended downward by 6 μm by applying 10 V to the PZT layer. The piezoelectric MEMS also enable wide-range variable MEMS capacitors comparing to electrostatic MEMS capacitors, which have pull-in phenomena caused by large electrostatic force at narrow gap.

Distributed tactile sensors (tactile sensor network) are needed on the skin of nursing care robots, rehabilitation robots, etc. to ensure their collision safety and enable body communication. The tactile sensor network acquires sensing data from tactile sensors on the skin by autonomous data transmission (event driven) [8]. The

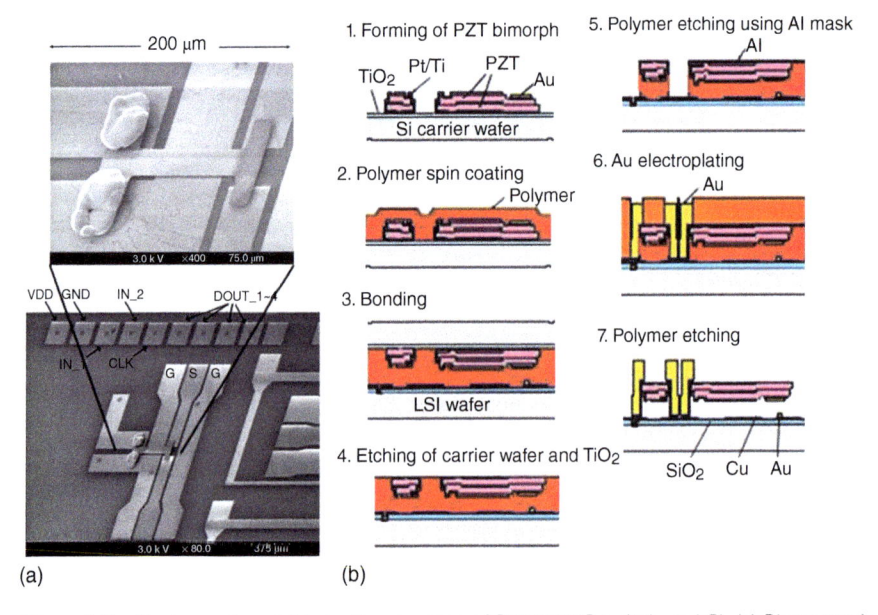

Figure 9.7 Photograph and fabrication process of PZT MEMS switch on LSI. (a) Photograph and (b) fabrication process.

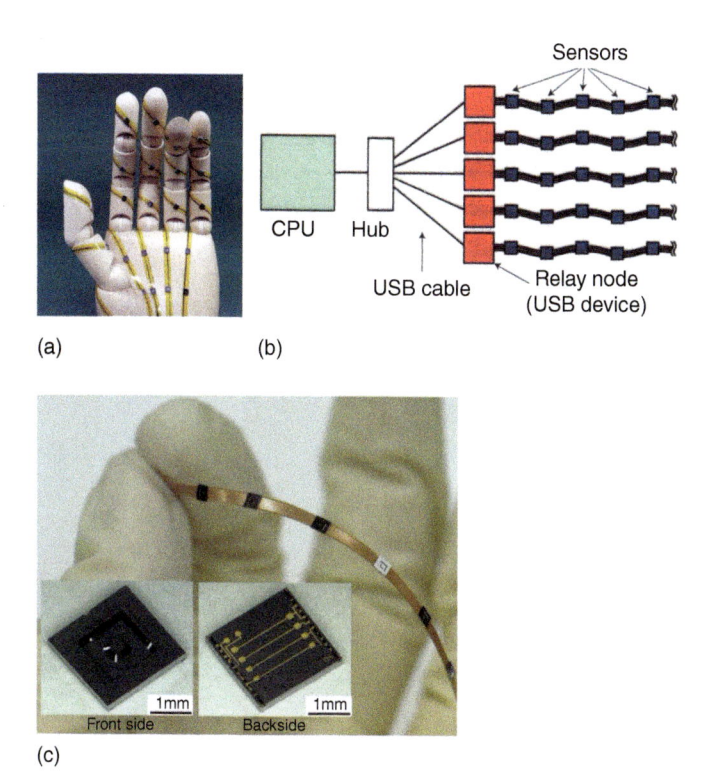

Figure 9.8 Tactile sensor network. (a) Application to robot skin, (b) concept of network, and (c) photographs of tactile sensors on flexible circuit and sensor chip.

photograph and the concept are shown in Figure 9.8a,b, respectively. The tactile sensor chips are connected to a flexible cable with common four wire bus for power supply, ground, and two signals as shown in Figure 9.8c. The capacitive tactile force sensor is formed on a communication LSI by adhesive bonding of a MEMS wafer using BCB (benzocyclobutene) (Figure 9.9a). The interconnection to the flexible cable are made at the backside of the sensor chip through the tapered via. The photograph and the block diagram of the LSI chip are shown in Figure 9.9b,c.

The fabrication process of the tactile sensor is shown in Figure 9.10. V grooves are formed on the LSI wafer using a dicing saw (1 in the figure). The grooves are insulated by depositing SiO_2, and then metal interconnections are made between bonding pads and the grooves by Ti and Au (2 in the figure). Polymer (BCB) is coated (3 in the figure), and after making via in the polymer Al pattern is made for the sensing capacitor (4 in the figure). Si MEMS wafer having diaphragms for sensing capacitors are bonded to the polymer (5 in the figure). The LSI wafer is thinned from the backside by grinding and polishing. This makes the bottom of the V groove exposed (6 in the figure). BCB is coated on the backside (7 in the figure), and the

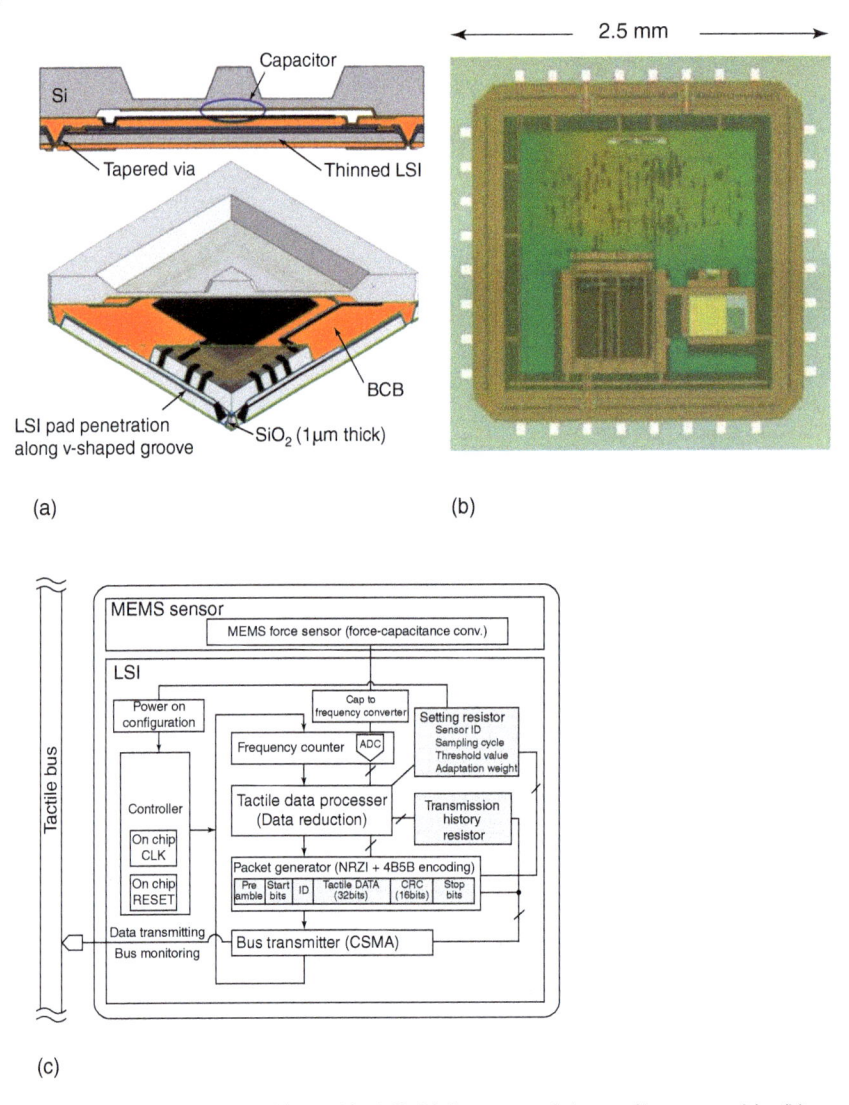

Figure 9.9 Tactile sensor chip and its LSI. (a) Structure of the tactile sensor chip, (b) photograph of the LSI, and (c) block diagram of the LSI.

polymer and the SiO$_2$ are etched to expose the metal connected from the front side (8 in the figure). Finally pads on the backside are formed by Ti and Au (9 in the figure).

The function of the communication LSI for event-driven data transmission at a clock rate of 45 MHz was confirmed. An example of packet communication of the tactile sensor network is shown in Figure 9.11a. The host computer recognizes the sensor position and the applied force from the sensor ID and the force data in the packet signal. The force is detected as a capacitance change, which is converted to a digital output, and the linear relationship between the force and the digital output was observed as shown in Figure 9.11b.

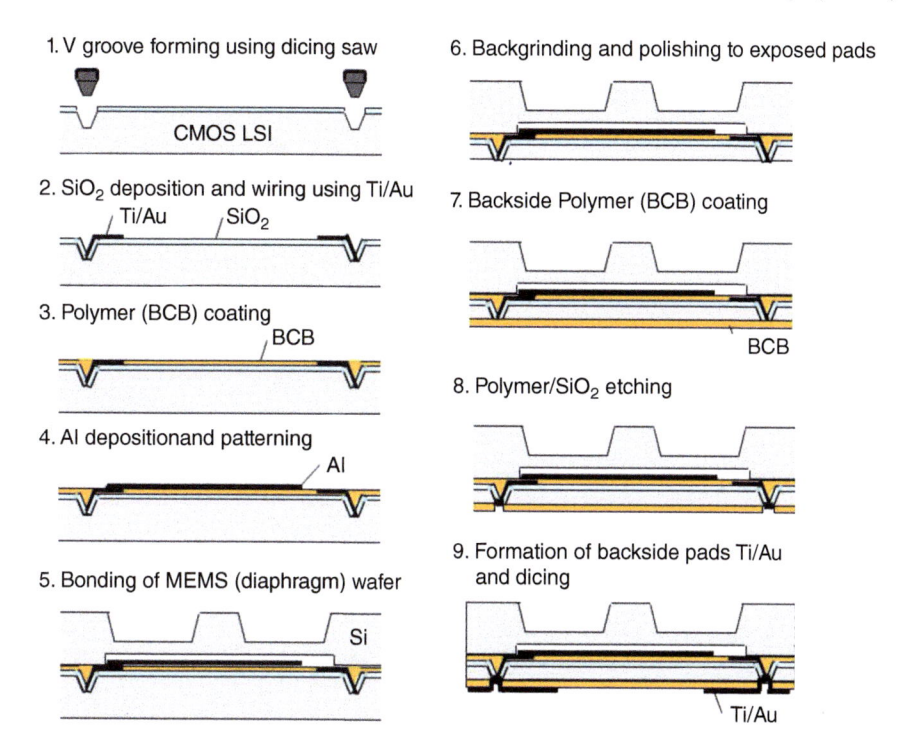

Figure 9.10 Fabrication process of the tactile sensor.

Steps shown:
1. V groove forming using dicing saw (CMOS LSI)
2. SiO$_2$ deposition and wiring using Ti/Au
3. Polymer (BCB) coating
4. Al deposition and patterning
5. Bonding of MEMS (diaphragm) wafer (Si)
6. Backgrinding and polishing to exposed pads
7. Backside Polymer (BCB) coating
8. Polymer/SiO$_2$ etching
9. Formation of backside pads Ti/Au and dicing

9.4 Device Transfer (Via-First)

Digital fabrication of LSI based on maskless lithography is expected for cost-effective small-volume production and short-term development. Extremely high throughput direct electron beam (EB) lithography is required because the latest LSI wafer has as many as 1 trillion (10^{12}) nanoscale transistors on it. In order to meet the demand, a massive parallel EB exposure system with an active matrix nc-Si (nanocrystal silicon) emitter array has been developed using the device transfer (via-first) method shown in Figure 9.2c. The nc-Si emitter has cascaded tunnel junctions, and accelerated ballistic electrons are emitted through a thin (10 nm thick) Au layer. The emitted current by the ballistic electrons is obtained at low voltage (10 V). The nc-Si is formed by anodizing Si in a HF (hydrofluoric acid) solution and is oxidized by following electrochemical oxidation (ECO). The low-voltage electron emission is needed for making high-density array of active matrix emitters. This is because the size of the transistor and hence the integration density depends on the voltage required. The active matrix LSI is developed for 100 × 100 array on a 10 mm square chip, which means the size of the cell for one emitter is 100 × 100 µm. The nc-Si emitter array with through Si via (TSiV) is connected to the LSI for active matrix drive as shown in Figure 9.12. This has been developed for the Massively Parallel Electron Beam Direct Write (MPEBDW) [9, 10].

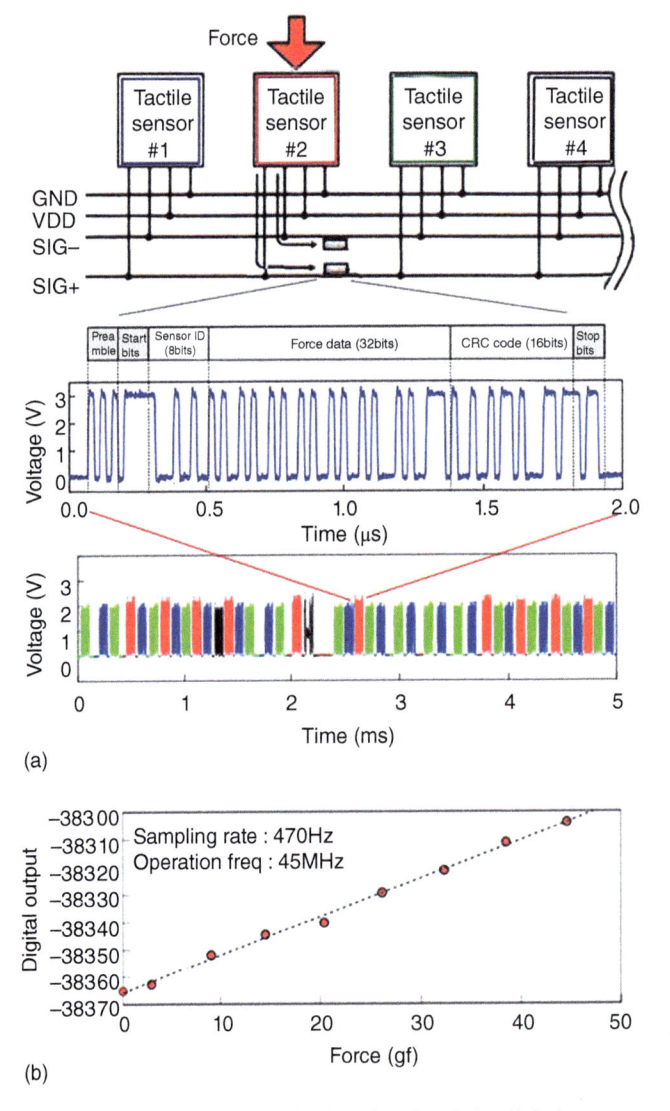

Figure 9.11 Packet communication signal and the digital output versus force. (a) Packet communication signal and (b) digital output versus force.

The nc-Si emitter array with the TSiV was fabricated using the device transfer (via-first) method as shown in Figure 9.13. A 200 μm thick Si wafer is prepared (1 in the figure). After making through vias by deep reactive ion etching (DRIE), the wafer is thermally oxidized, poly Si is deposited, and phosphorous is diffused for n$^+$ doing (2 in the figure). The vias are filled with poly Si deposition, both surfaces are polished, and phosphorous is diffused for n$^+$ doing (3 in the figure). Columnar poly Si is deposited for the purpose of the nc-Si and the poly Si is patterned (4 in the figure). Si_3N_4 is deposited and patterned (5 in the figure). The next process

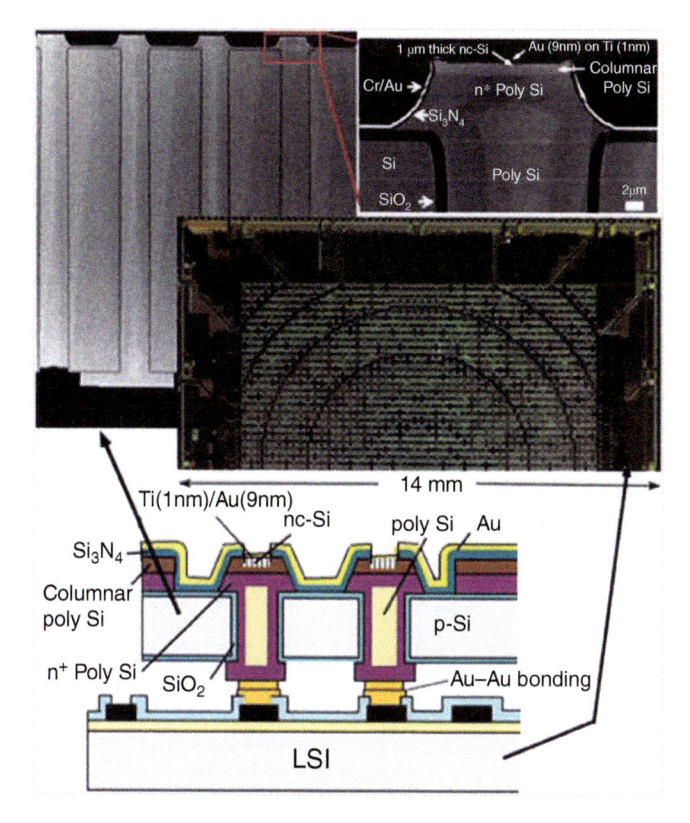

Figure 9.12 Cross-sectional photographs of the nc-Si electron emitter with TSiV and the structure of active matrix emitter array.

step is the formation of the nc-Si. The poly Si is anodized in HF (55%) + C$_2$H$_5$OH (1 : 1) solution, followed by ECO in ethylene glycol and potassium nitrate, HWA (high-pressure water vapor annealing), super-critical rinsing and drying (SCRD), and annealing in H$_2$ (6 in the figure). Backside electrode for the bump bonding and the thin surface electrode (Ti (1 nm) + Au (9 nm)) are formed (7 in the figure). The wafer is bonded to the LSI wafer with bumps. Finally bonding pads of the LSI are exposed by half-cut dicing of the emitter array chip (8 in the figure).

A preliminary experiment was conducted. The 1 : 1 exposure system with a magnetic focusing shown in Figure 9.14a was used for this experiment. Photographs of the nc-Si emitter array (top view) and a resist pattern exposed to the electron from the nc-Si emitter array are shown in Figure 9.14b. The 12 μm square exposed pattern corresponding to the nc-Si emitter size was formed on the resist successfully.

LSI for the active matrix emitter driver was developed and its basic operation was confirmed. Circuit of one cell in 100 × 100 matrix array for the active matrix driving LSI (a) and an example of functional operation (b) are shown in Figure 9.15. The LSI receives external writing bitmap data and switches 100 × 100 electron beamlets on and off.

9 Wafer Transfer

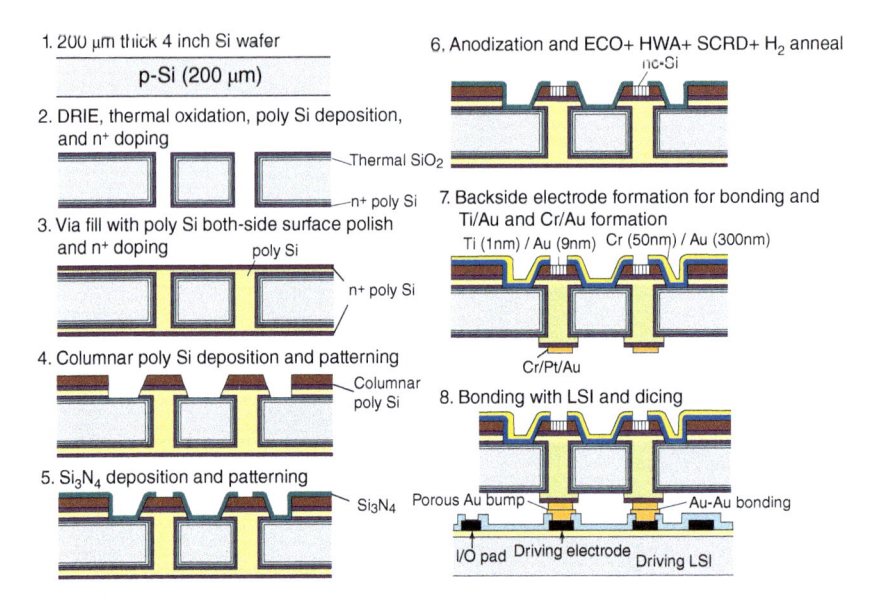

Figure 9.13 Fabrication process of active matrix nc-Si emitter array.

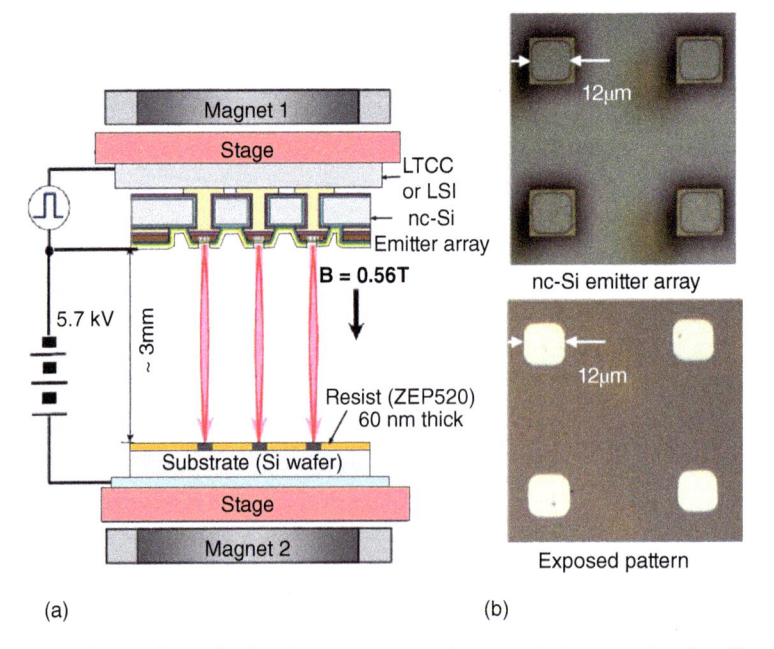

1. 200 μm thick 4 inch Si wafer
   p-Si (200 μm)
2. DRIE, thermal oxidation, poly Si deposition, and n+ doping
   Thermal SiO$_2$
   n+ poly Si
3. Via fill with poly Si both-side surface polish and n+ doping
   poly Si
   n+ poly Si
4. Columnar poly Si deposition and patterning
   Columnar poly Si
5. Si$_3$N$_4$ deposition and patterning
   Si$_3$N$_4$
6. Anodization and ECO+ HWA+ SCRD+ H$_2$ anneal
   nc-Si
7. Backside electrode formation for bonding and Ti/Au and Cr/Au formation
   Ti (1nm) / Au (9nm)   Cr (50nm) / Au (300nm)
   Cr/Pt/Au
8. Bonding with LSI and dicing
   Porous Au bump   Au-Au bonding
   I/O pad Driving electrode   Driving LSI

Figure 9.14 Setup for 1 to 1 exposure experiment and photographs of nc-Si emitter array and exposed pattern on photoresist. (a) Setup and (b) photographs of emitter array and exposed pattern.

(a) Magnet 1 / Stage / LTCC or LSI / nc-Si Emitter array / B = 0.56T / 5.7 kV / ~3mm / Resist (ZEP520) 60 nm thick / Substrate (Si wafer) / Stage / Magnet 2

(b) nc-Si emitter array (12μm)
Exposed pattern (12μm)

(a)

(b)

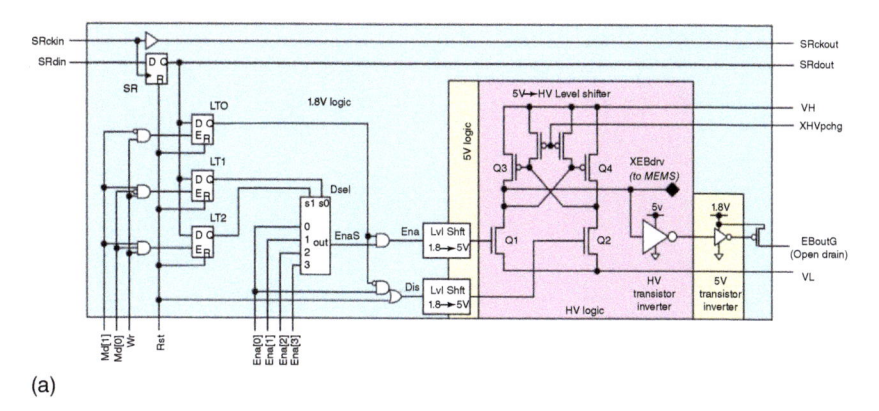

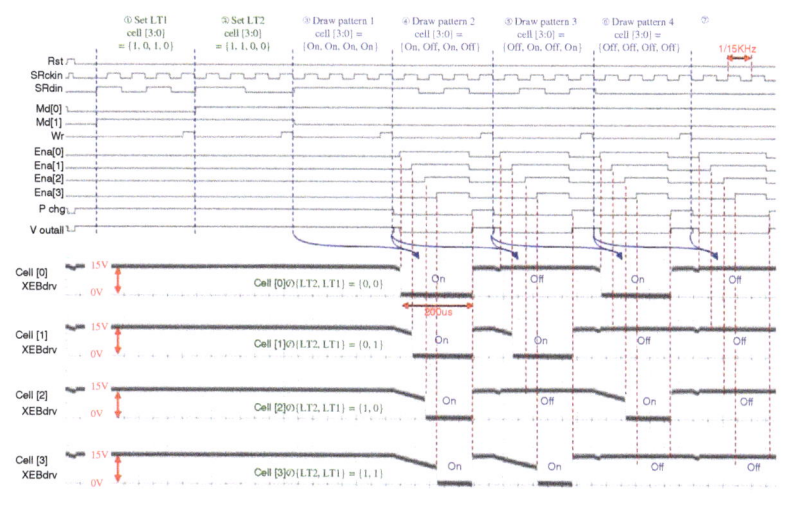

Figure 9.15 Circuit of one cell in 100 × 100 matrix array for the active matrix driving LSI and example of functional operation. (a) Circuit of one cell and (b) example of functional operation.

This EB exposure system has a function of electronic aberration compensation. The acceleration voltage is controlled to compensate the refraction angle of objective lens. The focal length of the objective (reduction) lens is controlled by adjusting the voltage for the condenser lens in front of the electron emitter array. For the purpose of the former compensation, the active matrix control LSI has concentric rings, which are electrically isolated in order for different ring offset voltages to be applied. This is performed using electrical isolation and TSiV shown in Figure 9.16. The fabrication process of the TSiV will be explained later in Figure 17.26. The isolation and the TSiV have to be made by post process as shown in Figure 9.17.

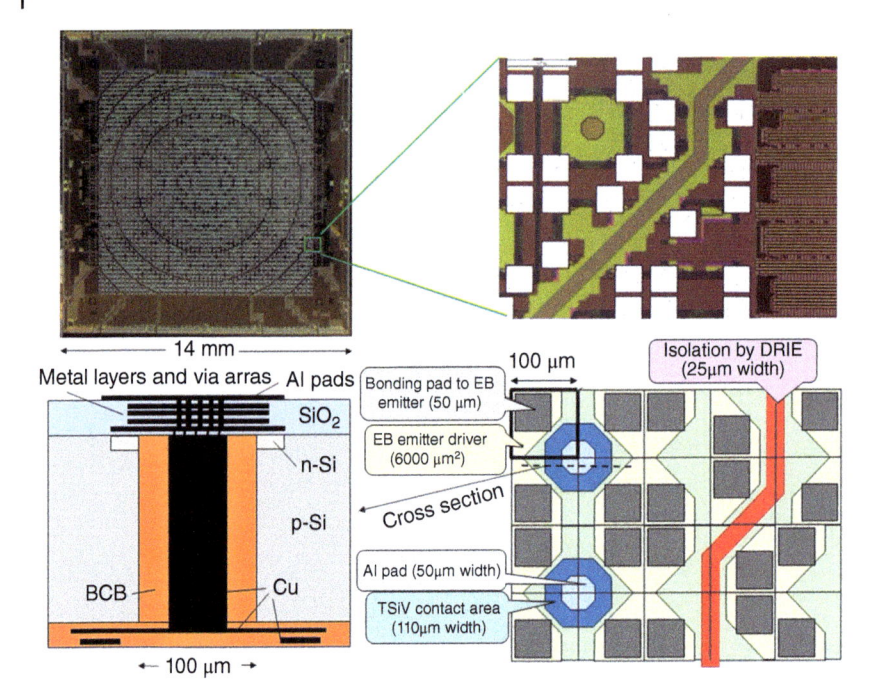

Figure 9.16 Layout and cross section of isolation and TSiV interconnection in the active matrix control LSI.

The functional test of the LSI is carried out at first (1 in the figure). Individual rings are isolated by DRIE of the Si (2 in the figure). Isolated trenches are filled with polymer (BCB (Benzocyclobutene)), and the TSiVs are made for electrical interconnection between the front side and the backside (3 in the figure). Finally wirings on the reverse side are made (4 in the figure).

A reduction electron optics was designed to reduce the electron beamlets by an objective lens with a factor of 100 at an acceleration voltage of 5 k eV. Photograph and construction of prototype 1/100 exposure system using the 100 × 100 active matrix electron emitter array is shown in Figure 9.18.

9.5 Chip Level Transfer

Contrary to the batch wafer transfer explained in Figure 9.1 [11], we can use chip level transfer, which is required when the size of the MEMS chip is different from that of the LSI chip [12, 13]. Selective transfer process using laser lift-off (debonding) is used as shown in Figure 9.19. Au pads are formed on the MEMS wafer for Au–Au bump bonding. The MEMS wafer is bonded to a glass carrier wafer using bonding interlayer (acrylic resin), and grooves are made on the MEMS wafer by dicing (1 in the figure). Au metalized silicone bumps are formed on the LSI wafer (2 in the figure). The MEMS wafer on the carrier wafer is aligned with the LSI wafer, and these wafers are bonded by pressing them uniformly at 180 °C after Ar plasma activation

1. Before reprocess (functional test of LSI)

```
Wire bond pads    Input buffers    Top metal layer        Internal wire bond pads for test
                                                                            LSI wafer
                                R2            R1              R0
```

2. Isolation of individual rings for aberration correction
   Disconnect

3. Filling isolation trench with polymer and TSV formation

4. Wiring on reverse side

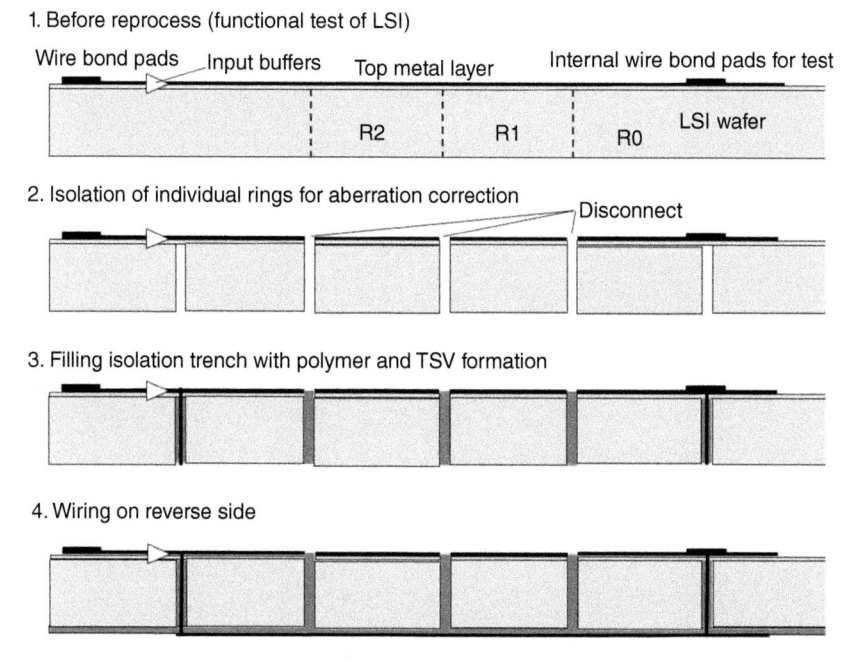

Figure 9.17 Process sequence to fabricate isolation and TSiV interconnection.

(3 in the figure). Selective transfer by laser lift-off (debonding) is made by irradiating the interfacial acrylic resin using Nd:YVO$_4$ third harmonic laser (λ = 355 nm) through the glass carrier wafer (4 in the figure). The acrylic resin is carbonized to lose adhesion, and the MEMS device is transferred on the surface of the LSI wafer (5 in the figure). The bump bonding is reinforced using an underfill polymer if necessary. The MEMS dies remaining on the glass carrier wafer can be transferred to another LSI wafer as shown in Figure 9.20. In the opposite way MEMS chips from different carrier wafers can be transferred on the same chip in LSI wafer. The selective transfer technology was applied to the multi-surface acoustic wave (SAW) filters on LSI [12]. The photograph of the multi-SAW filters transferred on an LSI chip and frequency characteristics of three different SAW filters on the LSI chip are shown in Figure 9.21.

SAW filter plays an important role in our communication systems. A one-chip bandwidth-tunable SAW filter was developed by the chip level wafer transfer of BaSrTiO$_3$ (BST) film to a LiTaO$_3$ wafer [14]. The BST film is needed for variable capacitors (varactors). The fabrication process is shown in Figure 9.22. Pt, BST, Pt, and Au are successively deposited on a sapphire wafer (1 in the figure). Au, Pt, and BST are etched using photoresist as a mask (2 in the figure). The Pt layer is irradiated with Nd:YVO$_4$ third harmonic laser (wavelength 355 nm) through the sapphire wafer. The purpose of this process step is to weaken the adherence between the Pt and the BST, and this is called laser pre-irradiation (3 in the figure). Using Au–Au bonding the sapphire wafer is bonded to the LiTaO$_3$ wafer, which has interdigital transducer (IDT) electrodes for SAW filter (4 in the figure). The patterned BST, Pt, and Au layers are transferred to the LiTaO$_3$ wafer from the sapphire wafer to the

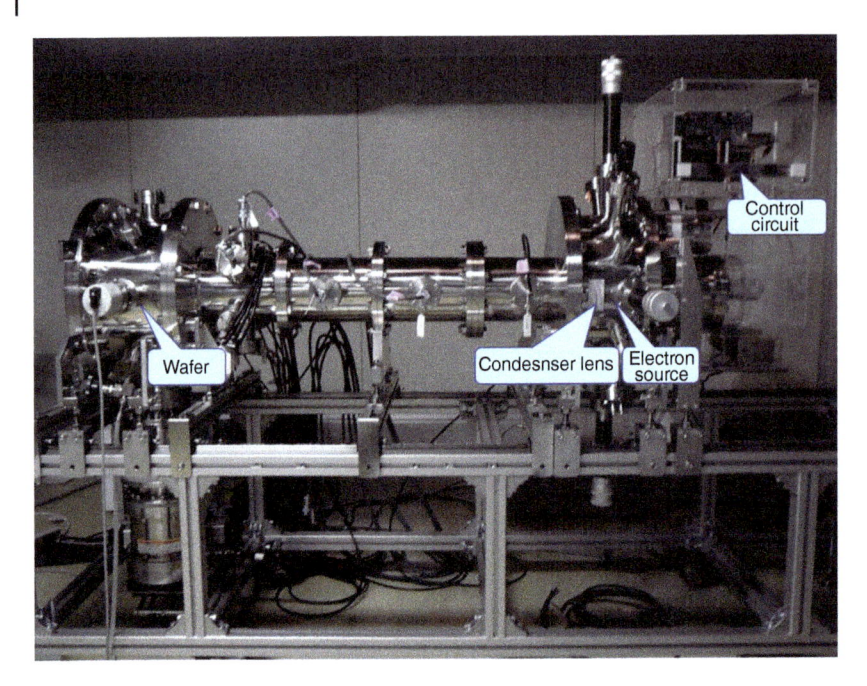

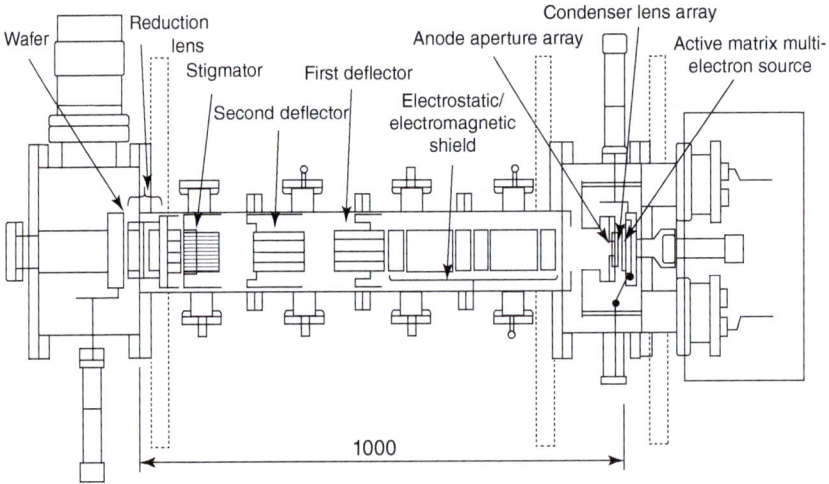

Figure 9.18 Photograph and construction of prototype 1/100 exposure system using 100 × 100 active matrix electron emitter array.

LiTaO$_3$ wafer (5 in the figure). The side of the BST is insulated with polyimide and metalized with Ti, Pt, and Au for interconnections (6 in the figure).

Figure 9.23 shows the circuit diagram (a), photographs (b), and characteristics (c) of the bandwidth-tunable SAW filter [15]. The Yp and the Ys are the SAW filters and the bandwidth is tunable by connecting variable capacitors Cp and Cs made of

9.5 Chip Level Transfer

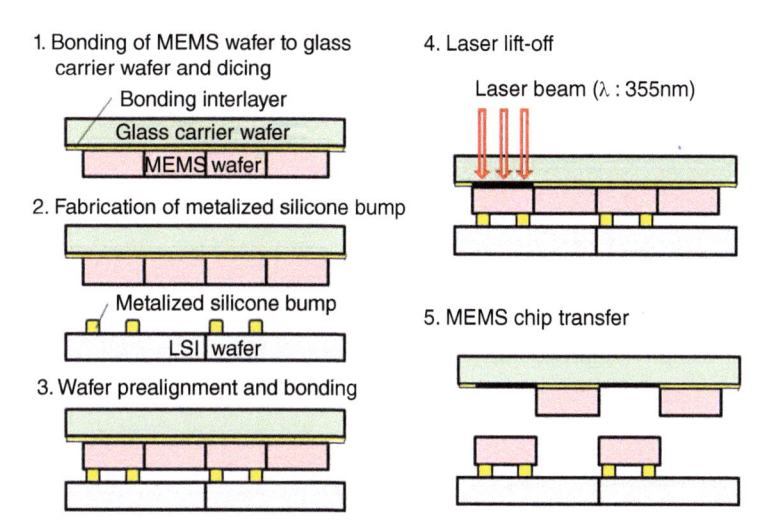

1. Bonding of MEMS wafer to glass carrier wafer and dicing
2. Fabrication of metalized silicone bump
3. Wafer prealignment and bonding
4. Laser lift-off
5. MEMS chip transfer

Figure 9.19 Chip level transfer process.

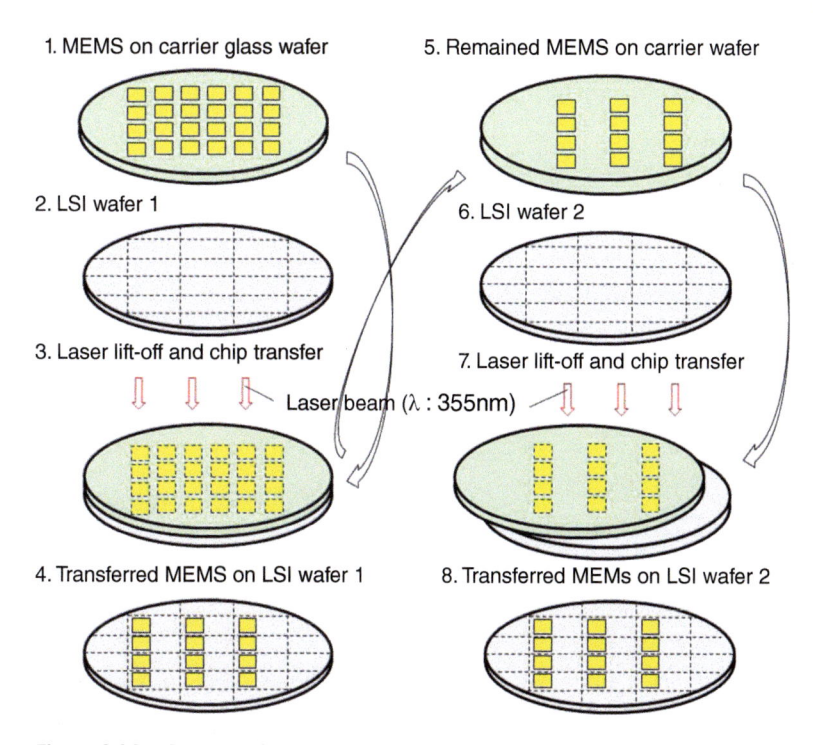

1. MEMS on carrier glass wafer
2. LSI wafer 1
3. Laser lift-off and chip transfer
4. Transferred MEMS on LSI wafer 1
5. Remained MEMS on carrier wafer
6. LSI wafer 2
7. Laser lift-off and chip transfer
8. Transferred MEMs on LSI wafer 2

Figure 9.20 Concept of the chip level transfer to multiple LSI wafers.

Figure 9.21 Multi-SAW filters on LSI chip by the chip level transfer.

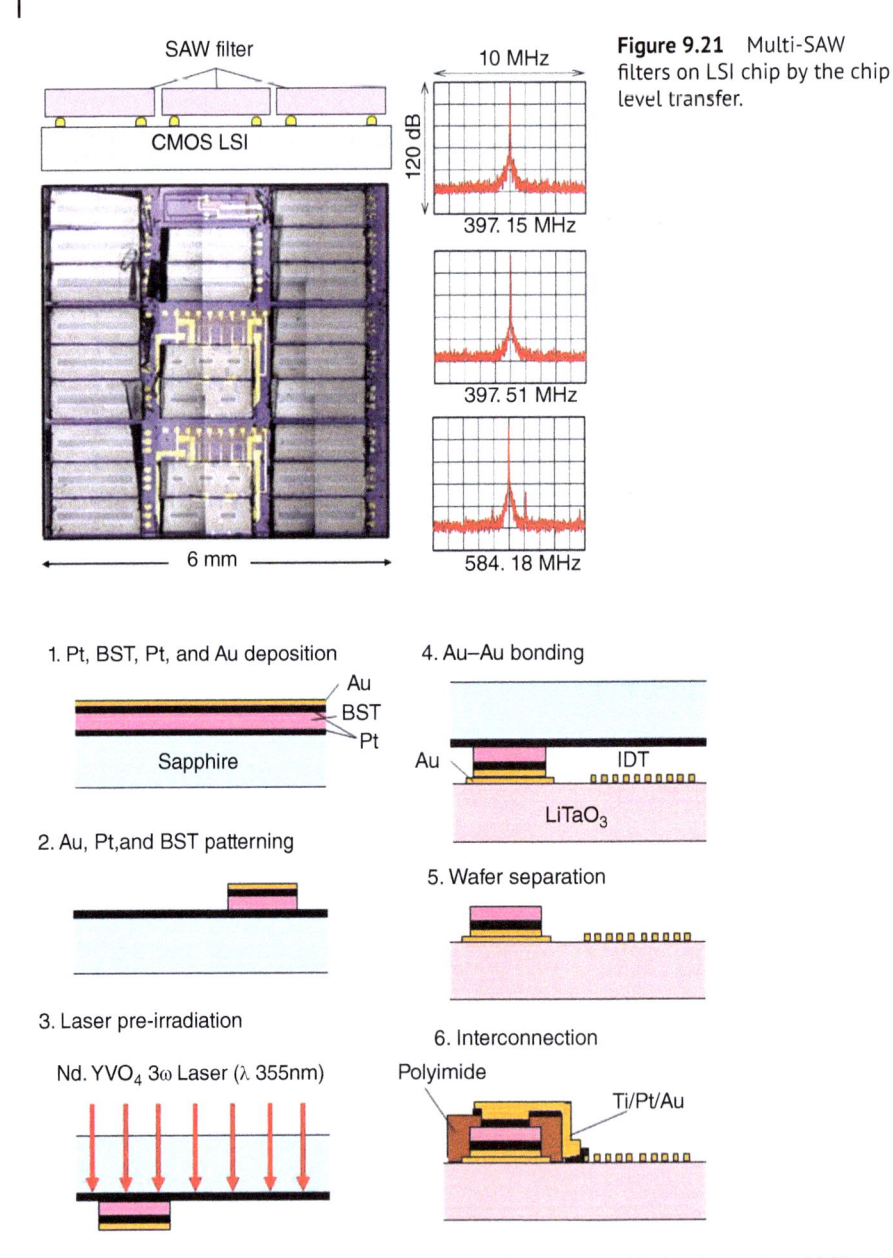

Figure 9.22 Fabrication process of tunable SAW filter by the chip level transfer of BST varactor.

the BST as shown in the circuit diagram. Owing to the nonlinear dielectric constant of the ferroelectric BST, the capacitance decreases to approximately 50% by applying 5 V. The 3-dB bandwidth of the filter is tuned between 3.25 and 6.25 MHz by applying DC voltage to the BST varactors, while the center frequency is constant at 1.004 GHz.

Figure 9.23 Tunable SAW filter using BST varactor. (a) Circuit (resistors for biasing are not shown), (b) photograph, and (c) characteristics.

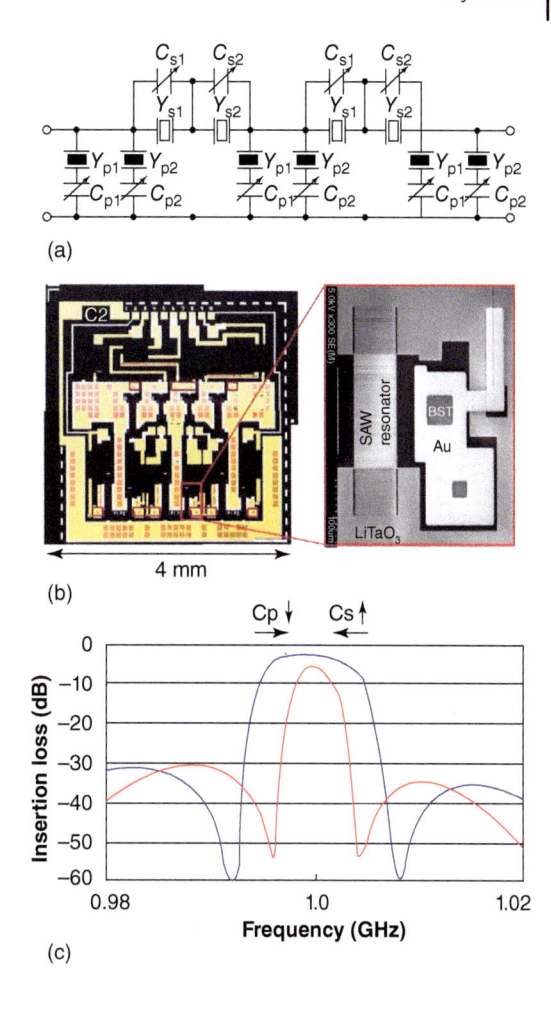

References

1 Esashi, M. and Tanaka, S. (2013). Heterogeneous integration by adhesive bonding. *Micro and Nano Systems Letters* 1: 3.
2 Esashi, M. and Tanaka, S. (2016). Stacked integration of MEMS on LSI. *Micromachines* 7: 137.
3 Lapisa, A., Stemme, G., and Niklaus, F. (2011). Wafer-level heterogeneous integration for MOEMS, MEMS and NEMS. *IEEE Journal of Selected Topics in Quantum Electronics* 17 (3): 629–644.
4 Esashi, M. (2008). Wafer level packaging of MEMS. *Journal of Micromechanics and Microengineering* 18 (7): 073001(13pp).
5 Kochhar, A., Matsumura, T., Zhang, G. et al. (2012). Monolithic fabrication of film bulk acoustic resonators above integrated circuit by adhesive-bonding-based film transfer. 2012 IEEE International Ultrasonics Symposium, Dresden, Germany (7–10 October 2012), 295–298.

6 Hayasaka, T., Yoshida, S., Inoue, K.Y. et al. (2015). Integration of boron-doped diamond microelectrode on CMOS-based amperometric sensor array by film transfer technology. *Journal of Microelectromechanical Systems* 24 (4): 958–967.

7 Matsuo, K., Moriyama, M., Esashi, M., and Tanaka, S. (2012). Low-voltage PZT-actuated MEMS switch monolithically integrated with CMOS circuit. Technical Digest 25th IEEE International Conference on Micro Electro Mechanical Systems (MEMS 2012), Paris, France (29 January – 2 February 2012), 1153–1156.

8 Makihata, M., Muroyama, M., Nakano, Y. et al. (2013). A 1.7 mm^3 MEMS-on-CMOS tactile sensor using human-inspired autonomous common bus communication. Technical Digest 17th International Conference on Solid-State Sensors, Actuators and Microsystems (Transducers 2013 & Eurosensors XXVII), Barcelona, Spain (16–20 June 2013), 2729–2732.

9 Esashi, M., Kojima, A., Ikegami, N. et al. (2015). Development of massively parallel electron beam direct write lithography using active-matrix nanocrystalline-silicon electron emitter arrays. *Microsystems & Nanoengineering* 1: 15029.

10 Esashi, M., Miyaguchi, H., Kojima, A. et al. (2018). *Development of Massive Parallel Electron Beam Write System: Aiming at Digital Fabrication of Integrated Circuits*. Sendai: Tohoku University Press (in Japanese).

11 Dospont, M., Drechsler, U., Yu, R. et al. (2004). Wafer-scale microdevice transfer/interconnect: its application in an AFM-based data-storage system. *IEEE Journal of Microelectromechanical Systems* 13: 895–901.

12 Guerre, R., Drechsler, U., Jubin, D., and Despont, M. (2008). Selective transfer technology for microdevice distribution. *Journal of Microelectromechanical Systems* 17 (1): 157–165.

13 Hikichi, K., Seiyama, K., Ueda, M. et al. (2014). Wafer-level selective transfer method for FBAR-LSI integration. Proceedings 2014 IEEE International Frequency Control Symposium, Taipei, Taiwan (19–22 May 2014), 246–249.

14 Samoto, T., Hirano, H., Somekawa, T. et al. (2013). Wafer-to-wafer transfer process of barium strontium titanate metal-insulator-metal structures by laser pre-irradiation and gold-gold bonding for frequency tuning applications. Technical Digest 17th International Conference on Solid-State Sensors, Actuators and Microsystems (Transducers 2013 & Eurosensors XXVII), Barcelona, Spain (16–20 June 2013), 171–174

15 Hirano, H., Samoto, T., Kimura, T. et al. (2014). Bandwidth-tunable SAW filter based on wafer-level transfer-integration of BaSrTiO$_3$ film for wireless LAN system using TV white space. Proceedings IEEE Ultrasonic Symposium, Chicago, USA (3–6 September 2014), 803–806

10

Piezoelectric MEMS

T Takeshi Kobayashi (AIST)

National Institute of Advanced Industrial Science and Technology

10.1 Introduction

10.1.1 Fundamental

Piezoelectric materials generate an electric charge by applying force or generate a force by applying a voltage. Micro electro-mechanical systems (MEMS) devices that use piezoelectric materials as sensors and actuators are called as piezoelectric MEMS devices. Piezoelectric MEMS devices are superior to electrostatic MEMS ones in that they have a simpler structure (10 μm or more in dimensional accuracy), higher power generation, and low-voltage driving. Among dielectrics, materials that have no central symmetry and have spontaneous polarization are called as piezoelectrics, and aluminum nitride (AlN) is classified into this. Materials whose spontaneous polarization can be reversed by applying voltage are called as ferroelectrics. PZT is classified as a ferroelectric.

Piezoelectric performance index includes piezoelectric d (m V^{-1}), e (C m^{-2}), and g (Vm N^{-1}) constants. The piezoelectric d constant is called the piezoelectric strain constant, which represents the strain that occurs when a voltage is applied with no stress. The piezoelectric e constant is called the piezoelectric stress constant, which represents the stress that occurs when a voltage is applied while the strain is constrained to be zero. The piezoelectric g constant is called the piezoelectric output constant and represents the voltage generated when pressure is applied with no electrical displacement. Piezoelectric d and e constants are often used as an index as actuator devices, and piezoelectric g constants are often used as an index as sensor devices.

The following relationship exists between these piezoelectric constants.

$$d = se = \varepsilon g \tag{10.1}$$

where s and ε are compliance and dielectric constant, respectively.

AlN thin films, which have high electromechanical coupling constants, are used in film bulk acoustic resonators (FBARs). On the other hand, lead zirconate titanate (PZT) thin films, which have large piezoelectric constants, are suitable for actuators,

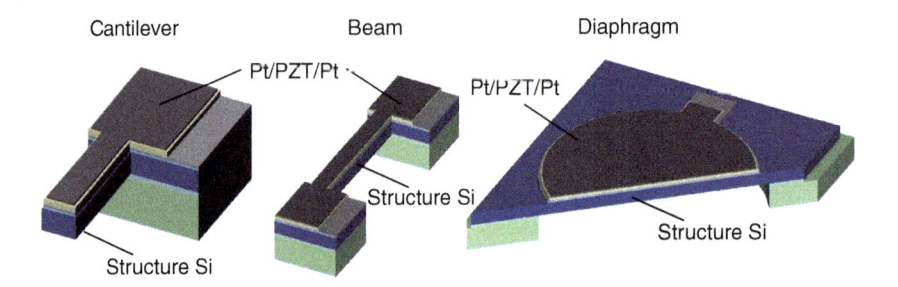

Figure 10.1 Schematics of a cantilever, beam, and diaphragm in which PZT thin films are integrated on the MEMS structure.

such as inkjet heads. This chapter focuses on piezoelectric MEMS actuator devices using PZT thin films.

10.1.2 PZT Thin Films Property as an Actuator

Figure 10.1 shows schematics of a cantilever, beam, and diaphragm in which PZT thin films are integrated on a MEMS structure. This section describes the characteristics required for PZT thin films, taking the simplest of these, for example, as a cantilever. As shown in the figure, in PZT–MEMS, PZT thin films sandwiched between upper and lower electrodes are formed on the elastic layer. When a voltage is applied, the PZT thin films expands and the cantilever contracts in the horizontal direction. Since one side is constrained by the elastic layer, the cantilever bends up and down.

Smit et al. [1] show that the displacement of the tip of a PZT cantilever is expressed by the following equation:

$$\delta = -\frac{3 s_s s_f h_s (h_s + h_f)}{s_s^2 h_f^4 + s_f^2 h_s^4 + 2 s_s s_f h_s h_f (2 h_s^2 + 2 h_f^2 + 3 h_s h_f)} \cdot d_{31} \cdot L^2 \cdot V \tag{10.2}$$

The first term on the right side of Equation (10.2) includes the compliance (reciprocal of Young's modulus, s_s and s_f, respectively) and the film thickness (h_s and h_f) of the elastic layer and the PZT thin film, which represents the mechanical properties of the PZT cantilever. d_{31}, L, and V are transverse piezoelectric constant, length of the cantilever, and applied voltage, respectively. Equation (10.2) considers only the PZT thin film and the elastic layer. If the thickness of the elastic layer is less than 10 times than that of the PZT thin film, the electrode and insulating film are also taken into account [2]. On the other hand, if the elastic layer is thick enough than PZT thin film, Equation (10.2) is approximated to $h_s \gg h_f$ to obtain Equation (10.3) [3].

$$\delta = -\frac{3 s_s}{h_s^2} \cdot e_{31} \cdot L^2 \cdot V \tag{10.3}$$

Here, e_{31} is also a transverse piezoelectric constant as well as d_{31}, and the relationship between the two constants is expressed as follows:

$$d_{31} = s_f \cdot e_{31} \tag{10.4}$$

Equation (10.3) can estimate piezoelectric constants without using compliance of PZT thin films, which are difficult to determine. When using Equation (10.3), PZT

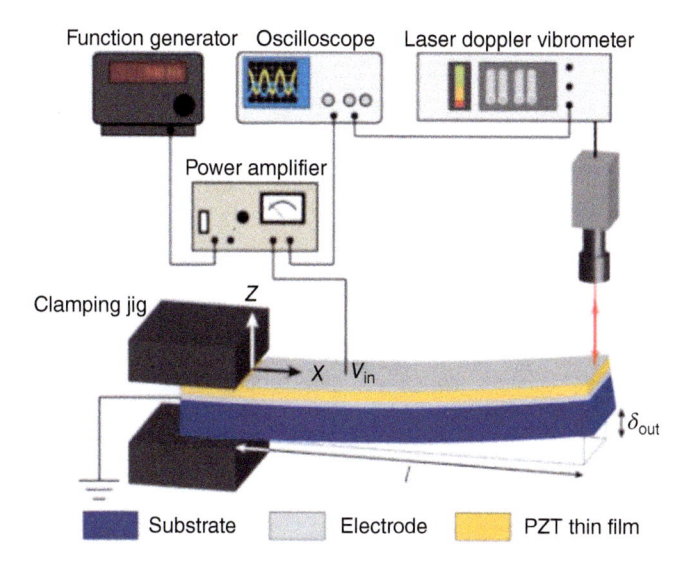

Figure 10.2 Measurement setup for tip displacement of piezoelectric cantilevers. Source: Kanno [4]. Copyright (2018) The Japan Society of Applied Physics.

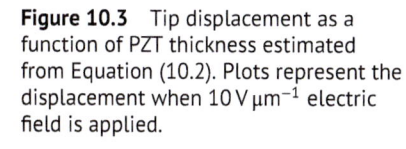

Figure 10.3 Tip displacement as a function of PZT thickness estimated from Equation (10.2). Plots represent the displacement when $10\,V\,\mu m^{-1}$ electric field is applied.

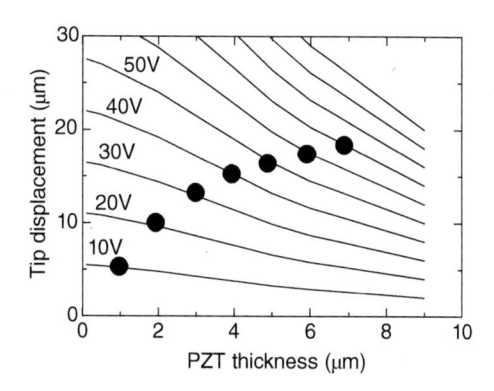

thin films are formed on silicon wafers, to create a macrocantilever by cutting to an appropriate size. Transverse piezoelectric constant e_{31} is determined by measuring the tip displacement when the voltage is applied as shown in Figure 10.2 [4]. In the case of a PZT thin film, it can be said that it has excellent piezoelectric characteristics if the value of d_{31} is $-100\,pm\,V^{-1}$ and the value of e_{31} is $-15\,C\,m^{-2}$ or more.

Piezoelectric constant d_{33} is also referred to evaluate the piezoelectric thin films [5], because measurement of d_{33} is easier compared to d_{31}. d_{33} is usually two to three times of d_{31}.

The film thickness h_f is also an important parameter. Calculations from Equation (10.2) show that when the applied electric field is constant, the displacement increases as the film thickness increases, as shown in Figure 10.3. However, in actual devices, the film thickness is often 1–5 μm because of limitations on the film thickness, which can be formed, electric field resistance, and voltage, which can be applied on the circuit side.

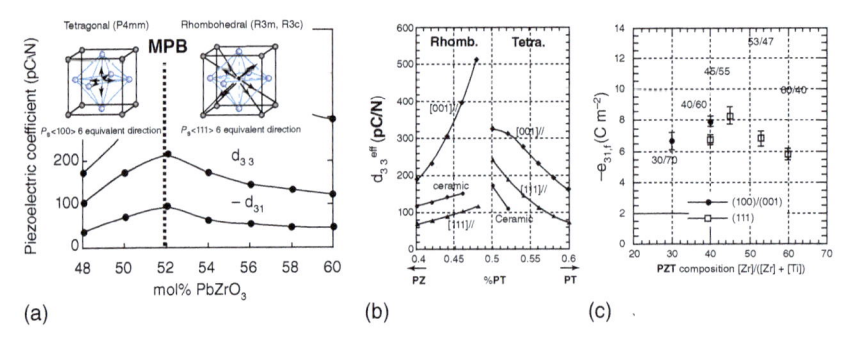

Figure 10.4 Relationship between the piezoelectric constant and Zr/Ti ratio for (a) PZT ceramics. Source: Reproduced from Damajanovic [6] with the permission of AIP Publishing, (b) (100)- and (111)-oriented PZT thin films (calculation). Source: Reproduced from Du et al. [7] with the permission of AIP Publishing, and (c) (100)/(001)- and (111)-oriented PZT thin films (experimental). Source: Ledermann et al. [8]. © 2003, Elsevier.

10.1.3 PZT Thin Film Composition and Orientation

The piezoelectric constant of PZT depends on the Zr/Ti ratio. The origin is that the crystal structure of PZT depends on the Zr/Ti ratio. PZT has rhombohedral structure in a Zr-rich composition and cubic structure in a Ti-rich composition, and the boundaries are called morphotropic phase boundary (MPB). It is well known that the piezoelectric constant of PZT shows a maximum near MPB composition such as Zr/Ti = 52/48. Most of the PZT ceramics practically used are MPB composition. It is interpreted that the maximum of the piezoelectric constant is because of the crystal structure, which becomes unstable between rhombohedral and cubic structure, and the polarization switching becomes easy to occur. Figure 10.4a shows the relationship between the piezoelectric constant of PZT ceramics and the Zr/Ti ratio [6]. It can be seen that all of the piezoelectric constants of d_{31}, d_{33}, and d_{15} show the maximum at the composition of Zr/Ti = 52/48.

The piezoelectric constant of a PZT thin film is greatly affected not only by its composition but also by its orientation. Uchino used thermodynamic-based phenomenological calculations to compare the longitudinal piezoelectric constant d_{33} of PZT near the MPB composition for the (001) and (111) orientations [7]. They have shown that (001)-oriented PZT thin films have four times larger d_{33} than that of (111)-oriented PZT thin films as shown in Figure 10.4b. Muralt et al. fabricated (100)/(001)-oriented and (111)-oriented PZT thin films of various compositions. As shown in Figure 10.4c, for most Zr/Ti composition, it has been shown that the (100)/(001)-oriented PZT thin films have a higher piezoelectric constant [8]. Other groups have reported similar results [9–12].

10.2 PZT Thin Film Deposition

10.2.1 Sputtering

Most of the PZT thin films used in commercialized PZT-MEMS devices are formed by sol–gel or sputtering. Several companies including EPSON, ROHM, RICOH,

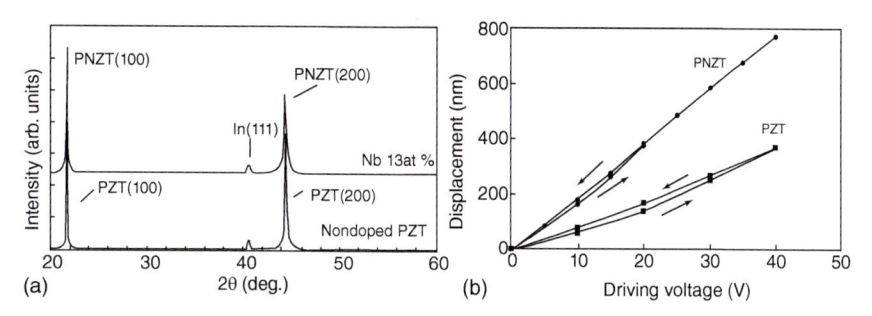

Figure 10.5 (a) XRD patterns of sputtered PZT and PNZT thin films and (b) displacement of the tip of the silicon cantilevers with PZT thin films. Source: Fujii [13]. © 2009, Elsevier.

STMicroelectronics, and TSMC employ sol–gel, while Panasonic, FUJIFILM, SAE Magnetics, Silicon Sensing Systems, Ulvac, and Robert Bosch employ sputtering. The standard sputtering use PZT ceramics as a target and perform sputtering with Ar and O_2 gas on a substrate heated over than 500 °C. Since PbO tends to evaporate during sputtering, the PZT ceramic target contains an excessive amount of Pb. PZT thin films formed by sputtering often has a (100)/(001) orientation.

FUJIFILM has reported PZT thin film deposition by sputtering [13]. $Pb_{1.3}(Zr_{0.52}Ti_{0.48})O_3$ ceramic target was sputtered with 0.5 Pa of Ar + 2.5% O_2 gas to form a film on Ir/Ti/SiO$_2$/Si substrates heated to 525 °C. In addition, Nb-doped PZT (PNZT) thin films were also prepared by using an Nb-doped PZT ceramic target. The deposited PZT thin films have (100)/(001) orientation as shown in Figure 10.5. The piezoelectric constant d_{31} of the Nb-doped PZT thin film reaches −259 pm V^{-1}, and e_{31} calculated assuming a Young's modulus of 49 GPa is −13 C m^{-2}, which is twice of that of a normal PZT thin film [14].

As shown in Figure 10.6, Yoshida et al. formed an $SrRuO_3/LaSrCoO_3/CeO_2/YSZ$ buffer layer on Si substrates to form epitaxially grown PZT thin films having not

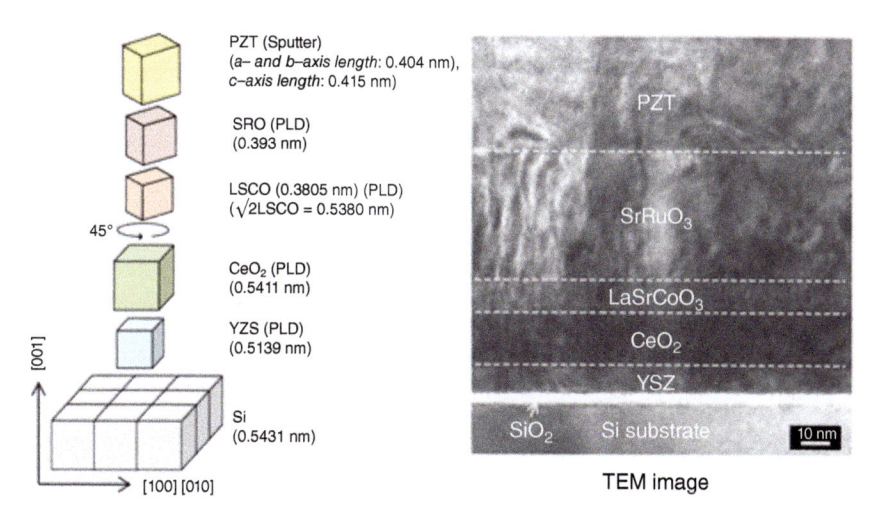

Figure 10.6 Schematic of epitaxially grown PZT thin films using $SrRuO_3/LaSrCoO_3/CeO_2/$ YSZ buffer layer. Source: Yoshida et al. [15]. © 2014, IEEE.

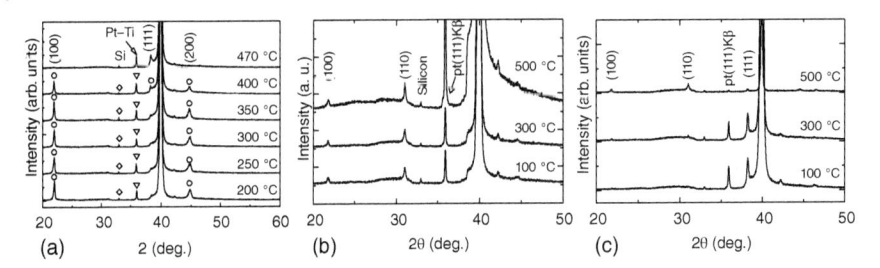

Figure 10.7 XRD patterns of sol–gel-derived PZT thin films deposited on (a) Pt/Ti. Source: (a) Kobayashi et al. [16] © 2005, Elsevier, (b) Pt, and (c) pre-heated Pt/Ti. The temperatures shown in the figures represent thermal decomposition temperatures. Source: (b, c) Kobayashi et al. [17]. © 2007, Taylor & Francis Ltd.

only a (100)/(001) orientation but also an in-plane orientation. The e_{31} of the PZT thin films is $-11\,\text{C}\,\text{m}^{-2}$. Since the dielectric constant is as small as 200, the figure of merit $e_{31}{}^2/\varepsilon_0\varepsilon_\text{r}$ proportional to the electromechanical coupling constant K is about five times larger than that of the oriented PZT thin film [15].

10.2.2 Sol–Gel

PZT thin films can also be prepared by spin coating a solution in which a metal organic compound of Pb, Zr, and Ti is dissolved and crystallizing the solution by heat treatment. Although it is classified into MOD, CSD, etc. according to the type of a solution, it is unified with sol–gel.

10.2.2.1 Orientation Control

In sol–gel method, (100)/(001)-oriented PZT thin films with a thickness of 1 μm or more suitable for PZT-MEMS is prepared by repeating spin coating of precursor solutions, thermal decomposition heat treatment, and crystallization heat treatment. The crystal orientation of PZT thin films is determined by the first film deposition. Since the PZT thin films formed by the sol–gel method can have various orientations such as a random, (111), and (110) orientation in addition to (001) orientation, it is necessary to optimize film deposition conditions.

It is known that the thermal decomposition temperature and the bottom electrode materials affect the orientation of PZT thin films. As shown in Figure 10.7a, Kobayashi et al. reported that when Pt/Ti is used as the bottom electrode, PZT thin films have (100)/(001) and (111) orientation at the thermal decomposition temperature of 400 and 470 °C, respectively [16]. They also reported that Pt lower electrode leads to randomly oriented PZT thin films regardless of the thermal decomposition temperature (Figure 10.7b) and the annealed Pt/Ti lower electrode dose (111)-oriented PZT thin films as shown in Figure 10.7c [17].

It is considered that the (001)-oriented PbO layer formed at the initial stage of crystallization promotes nucleation of (100)/(001)-oriented PZT thin films. Kobayashi et al. report that (100)/(001)-oriented PZT thin films are formed in the case of a Pt/Ti lower electrode, and randomly oriented PZT thin films are formed in the case of a Pt lower electrode. From these results, it is supposed that the adhesion layer Ti

Sample	Sequence	Number of layer
A	(470 => 700) × 13	13
B	(250 => 700) × 13	13
C	(250 => 700) × 1 => (470 × 9 => 700) × 1	10
D	(470 × 10 => 700) × 13	13

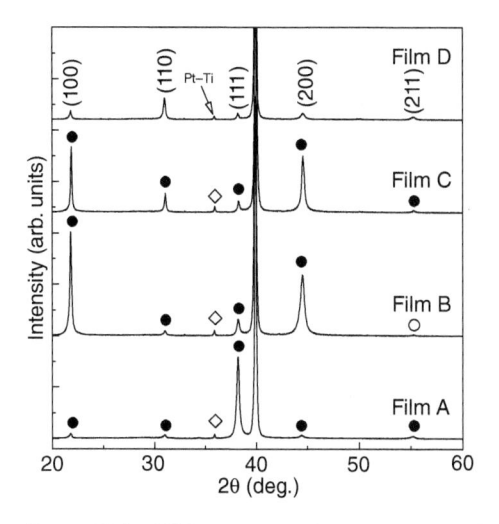

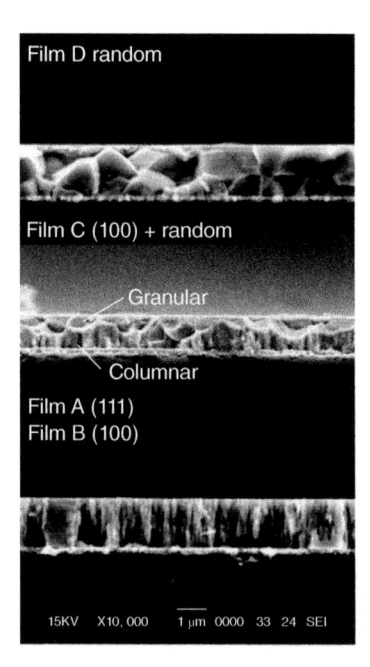

Figure 10.8 XRD patterns and cross-sectional SEM images of the sol–gel-derived 1 μm thick PZT thin films in which film thickness is increased by various multi-coating sequences. Source: Kobayashi et al. [16]. © 2005, Elsevier.

diffused in Pt during the thermal decomposition or crystallization annealing, reacted with the PZT precursor, and nucleated the (001)-oriented PbO layer. On the other hand, for the (111) orientation, nucleation of (111) orientation by the TiO_2 layer formed at the initial stage of crystallization is known as one of the mechanisms. In the case of Kobayashi et al., it is probable that TiO_2 was formed on the Pt surface by Pt/Ti annealing and promoted the nucleation of PZT-(111) orientation. Gong et al. [11] and Muralt et al. [18] reported an orientation control of PZT thin films by seed layer. According to Gong et al., when the PbO layer was used as the seed layer, (100)/(001)-oriented PZT thin films were formed, and when the TiO_2 layer was used as the seed layer, (111)-oriented PZT thin films were formed, which is in good agreement with the results of Kobayashi et al.

10.2.2.2 Thick Film Deposition

In order to form PZT thin films with a thickness of 1 μm or more by sol–gel method, spin coating, thermal decomposition heat treatment, and crystallization heat treatment are repeated. For 1 μm-thick PZT thin films, it has been reported that crystallizing after each film formation and thermal decomposition results in (100)/(001) or (111) orientation, while results in random orientation when crystallizing after repeating film formation and thermal decomposition are 10 times (Figure 10.8) [16]. It has also been reported that (100)/(001) and (111)-oriented PZT thin films have a columnar structure, and random oriented PZT has a granular structure. Kobayashi

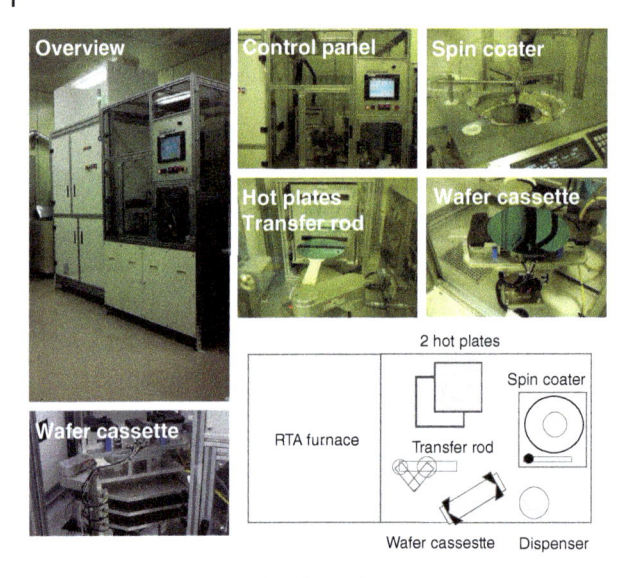

Figure 10.9 Automatic film formation system that can automatically perform spin coating, thermal decomposition heat treatment, and crystallization heat treatment.

et al. have developed an automatic film formation system, which can automatically perform spin coating, thermal decomposition heat treatment, and crystallization heat treatment (Figure 10.9).

When PZT thin films with a film thickness of 1 µm or more was prepared, (100)/(001)-oriented PZT thin film were formed under thermal decomposition temperature of 250 °C. However, more than 300 particles with a diameter of about 100 µm in which the PZT solution aggregated were generated on a 4-inch wafer. On the other hand, when the thermal decomposition temperature was set to 300 °C or higher, the number of particles was reduced to 10 or less, but the PZT film was randomly oriented. By setting the first layer at a pyrolysis temperature of 250 °C and the second and subsequent layers at a pyrolysis temperature of 300 °C, the number of particles on a 4-in. wafer was reduced to 10 or less as shown in Figure 10.10.

Muralt et al. repeated the film formation and thermal decomposition four times, then made the process of crystallization by rapid thermal annealing one cycle, and repeated several times to produce a (100)/(001)-oriented MPB–PZT thin film with a thickness of 4 µm. The prepared PZT thin films showed excellent piezoelectric properties of $-e_{31} = 10\,\text{C}\,\text{m}^{-2}$ [8]. The Zr/Ti ratio of the PZT thin film formed by one deposition was inclined in the cross-sectional direction in the range of 44/56–65/35 (Figure 10.11a). It was proposed that the piezoelectric PZT constant can be further improved by reducing this composition deviation. They realized the higher piezoelectric constant of $-e_{31} = 18\,\text{C}\,\text{m}^{-2}$ by sequentially reducing the Zr/Ti ratio deviation from 15% to 3% by depositing four types of PZT solutions with different compositions. (Figure 10.11b) [19].

10.2.3 Electrode Materials and Lifetime of PZT Thin Films

It is known that the lifetime of PZT thin films for a ferroelectric memory is determined by polarization reversal fatigue caused by repeated application of an

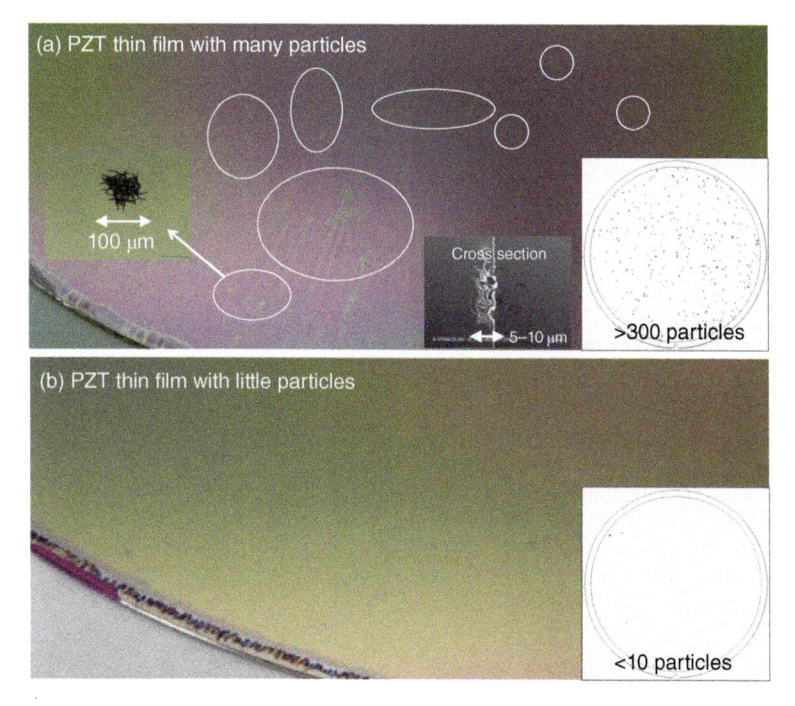

Figure 10.10 Surface images of the sol–gel-derived PZT thin films on 4 inch wafers. By setting the first layer at a pyrolysis temperature of 250 °C and the second and subsequent layers at a pyrolysis temperature of 300 °C, the number of particles on a 4-in. wafer was reduced to 10 or less.

AC voltage. In particular, it has been shown that polarization fatigue is remarkable when a Pt electrode is used [20]. As a cause of polarization fatigue, a model has been proposed in which oxygen vacancies grow with repeated polarization reversal [21]. It has been shown that this fatigue characteristic is greatly improved by compensating for oxygen vacancies using conductive oxide electrodes such as IrO_2 [22], RuO_2 [23], $LaNiO_3$ [20], and so on. On the other hand, in most PZT thin films for actuators, a unipolar voltage other than an AC voltage is usually applied. It has been pointed out that the polarization inversion fatigue does not occur under such conditions, but rather the life is determined by dielectric breakdown against DC voltage application, which is governed by the migration of oxygen vacancies through the PZT films [24].

10.3 PZT–MEMS Fabrication Process

10.3.1 Cantilever and Microscanner

This section describes the fabrication process for piezoelectric MEMS cantilevers and optical scanners from silicon-on-insulator (SOI) wafers on which PZT thin films are formed [25]. First, SOI wafers are annealed at 1100 °C in O_2 or H_2O/O_2 gas to form a thermal oxide film on the surface of the structure silicon. The thermal oxide film serves not only as an insulator between the structure silicon and the

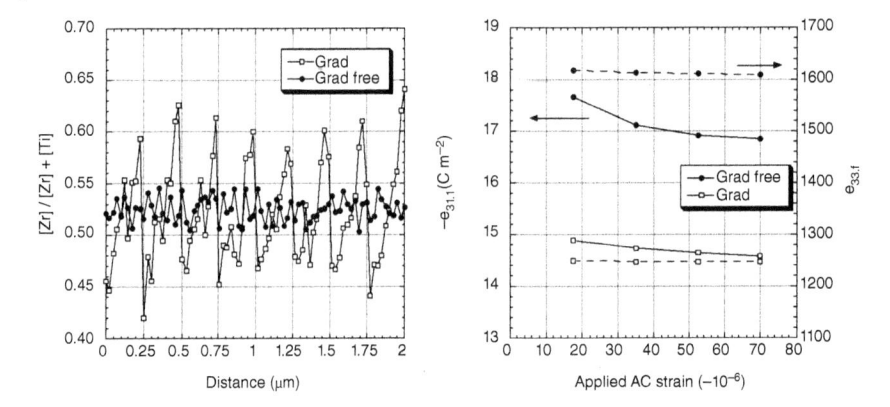

Figure 10.11 (a) Deviation of Zr/Ti ratio in the PZT thin films in the cross-sectional direction and (b) transverse piezoelectric constant e_{31}. Piezoelectric PZT constant are improved by reducing this composition deviation. Source: Reproduced from Calame and Muralt [19] with the permission of AIP Publishing.

lower electrode but also as a barrier against the diffusion of Pb into Si, which occurs in the formation of the PZT thin film, especially in the case of the sol–gel method. Instead of the thermal oxide film, alumina and zirconia can be also used. Subsequently, a lower electrode is formed. The most typical lower electrode is Pt/Ti, but an oxide electrode such as $SrRuO_3$ or $LaNiO_3$ may be formed directly above the lower electrode. After forming the lower electrode, PZT thin films are formed by sputtering or sol–gel. Finally, an upper electrode is formed. As a material of the upper electrode, Au, Ru, and Ir can be used.

Hereafter, a fabrication process of PZT–MEMS by etching a SOI wafer on which Pt/Ti/PZT/Pt/Ti/SiO$_2$ has been formed will be described. Figure 10.12 shows schematics of the fabrication process for the piezoelectric microcantilevers and optical microscanners. A photoresist mask is often used as an etching mask. First, the Pt/Ti top electrode is etched by Ar ion milling using mask 1. Subsequently, the PZT thin films are etched using mask 2. In the case of wet etching of PZT thin films, a strong acid mixed with HF, HNO_3, and HCl is used. The etch rate of wet etching is as fast as about 1 μm min^{-1}. However, since an undercut of about 5 μm occurs, wet etching is suitable for patterning large structures of 100 μm or more. Dry etching is more suitable for patterning smaller structures. In the case of dry etching, reactive ion etching (RIE) using SF_6 or CF_4 is used. However, it is necessary to note that the selectivity of PZT to the bottom electrode Pt/Ti is not so high because of the large contribution of physical sputtering.

The top electrode and the PZT thin films have almost the same shape. Since the undercut in the case of wet etching and the process damage of the sidewall in the case of dry etching occur, so the pattern of the PZT etching should be several μm larger compared to the pattern of the top electrode etching. Thereafter, the Pt/Ti bottom electrode and the thermal oxide films are etched with mask 3. RIE using CHF_3 and CF_4 is used for etching the thermal oxide film. Figure 10.13 shows the image of the wafer surface after the steps described earlier are completed. It can be confirmed

10.3 PZT–MEMS Fabrication Process

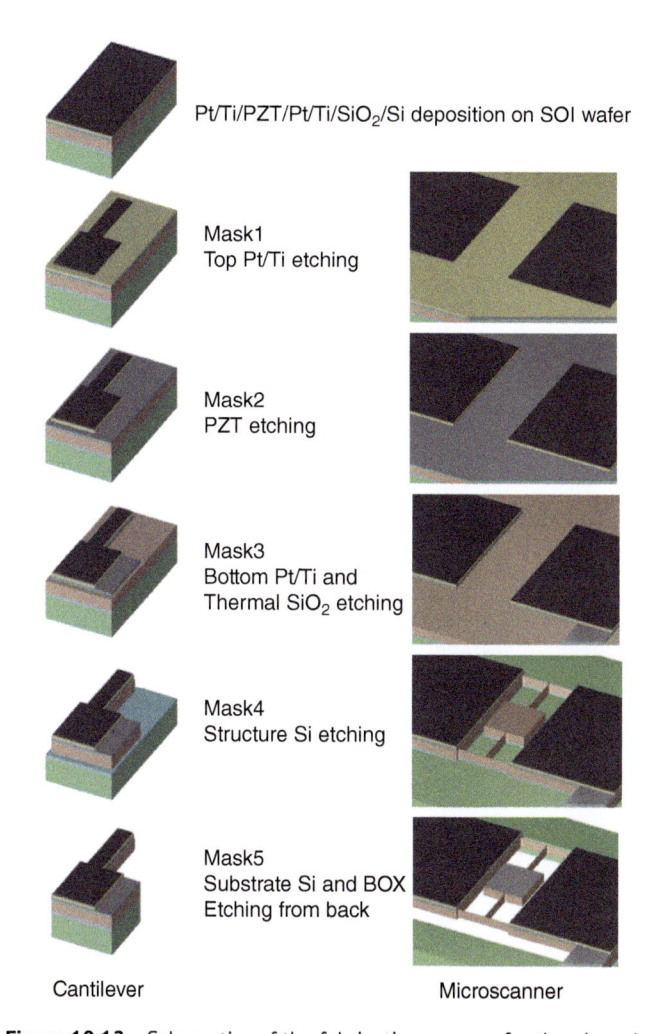

Figure 10.12 Schematics of the fabrication process for the piezoelectric microcantilevers and optical microscanners.

Figure 10.13 Surface image of the wafer after bottom Pt/Ti etching through mask 3.

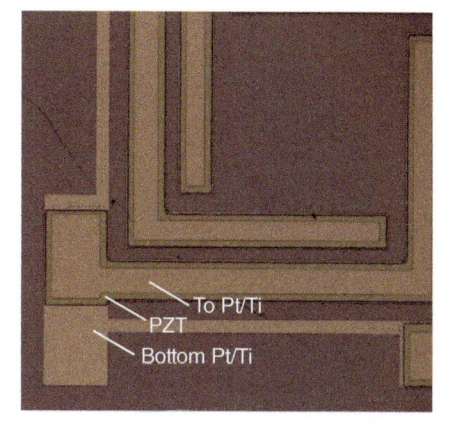

that the pattern width of PZT thin films etching is several μm wider than that of the top electrode and that the PZT thin films are undercut by several μm.

Subsequently, through mask 4, the structural silicon is etched by deep reactive ion etching (DRIE) in which silicon is deeply etched by repetitive SF_6 etching and side wall protection. Finally, the substrate is etched through from the backside of the wafer with mask 5 by DRIE to release the structural silicon. When etching the substrate from the back, a mask that can withstand etching for a long time is required. For a photoresist, a resist having a thickness of 10 μm or more is usually used. Alternatively, metal thin films such as aluminum which is hardly etched by SF_6 can be used as a metal mask.

Since the silicon wafers after through-etching of the silicon substrate are very fragile, the wafers are usually put into the etching equipment in a state where the wafer is attached to the dummy wafer. A photoresist, silicone oil, vacuum grease, or the like is used for bonding the wafers, but it is important to select a bonding material, which can be easily removed by ashing or chemical cleaning. The end point of the through-etching can be determined by confirming the appearance of buried oxide (BOX) layer on the entire surface of the wafer. After through-etching of the substrate is completed, the wafer bonded to the dummy wafer is immersed in an organic solvent such as acetone to separate the dummy wafers and the processed SOI wafers. After cleaning the SOI wafer, the BOX layer is removed by RIE using CHF_3 or CF_4 or wet etching using BHF to complete the process.

10.3.2 Poling

Most of the PZT thin films for MEMS are polycrystalline ones. PZT thin films deposited by sputtering often have the same direction of spontaneous polarization just by being deposited. On the other hand, the direction of spontaneous polarization of PZT thin films prepared by sol–gel method differs for each crystal grain. As shown in Figure 10.14, in the (100)/(001)-oriented PZT thin films, the direction of spontaneous polarization of (100)-oriented crystal grains is horizontal, and that of

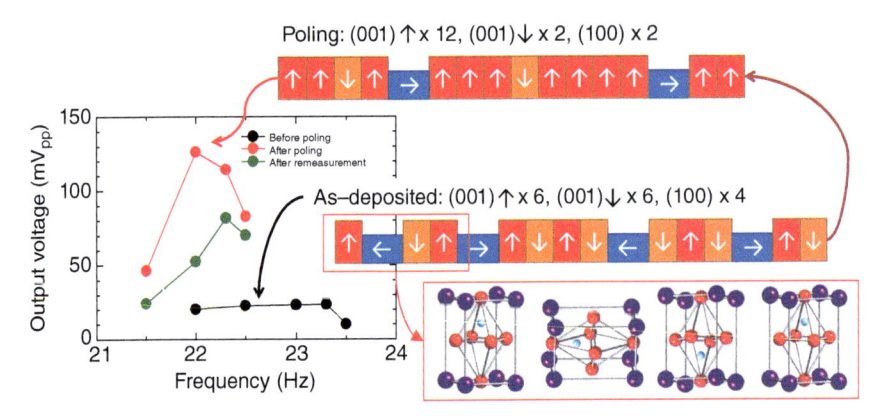

Figure 10.14 Schematics of polarization direction of as-deposited and poled PZT thin films.

(001)-oriented crystal grains is vertically upward or downward. Then, the direction of spontaneous polarization of the PZT thin films is aligned by applying a voltage, which is called as poling.

Poling of the PZT thin films is sufficient if a DC voltage of 10 V μm^{-1} or more is applied for several minutes at room temperature. However, if the poled PZT thin films are heated to 200°C or more by heat treatment such as reflow, the effect of poling disappears. By poling while heating, the effect of poling is maintained even when heated [26]. In addition, it is possible to reduce the poling time to several seconds or less by pulse poling in which a pulse voltage, which is twice or higher than the DC voltage, is applied. When a large number of MEMS chips need to be poled, pulse poling is effective for improving productivity [27].

References

1 Smits, J. and Choi, W. (1991). *IEEE Trans. Ultrason. Ferroelectron. Freq. Control* 38: 256.
2 Dekkers, M., Boschker, H., van Zalk, M. et al. (2013). *J. Micromech. Microeng.* 23: 025008.
3 Kanno, I., Kotera, H., and Wasa, K. (2003). *Sens. Actuators A* 107: 68.
4 Kannno, I. (2018). *Jpn. J. Appl. Phys.* 57: 040101.
5 Taylor, D.V. and Damjanovic, D. (2000). *Appl. Phys. Lett.* 76: 1615.
6 Damajanovic, D. (1998). *Rep. Prog. Phys.* 61: 1267.
7 Du, X.H., Zheng, J., Belegundu, U., and Uchino, K. (1998). *Appl. Phys. Lett.* 72: 2421.
8 Ledermann, N., Muralt, P., Baborowski, J. et al. (2003). *Sens. Actuators A* 105: 162.
9 Park, C.-S., Kim, S.-W., Park, G.-T. et al. (2000). *J. Mater. Res.* 20: 243.
10 Chen, H.D., Udayakumar, K.R., Gaskey, C.J., and Cross, L.E. (1995). *Appl. Phys. Lett.* 67: 3411.
11 Wen, G., Li, J.F., Chu, X. et al. (2004). *J. Appl. Phys.* 96: 590.
12 Hoffmann, M. et al. (2003). *IEEE Trans. Ultrason. Ferroelectron. Freq. Control* 50: 1240.
13 Fujii, T., Hishinuma, Y., Mita, T., and Arakawa, T. (2009). *Solid State Commun.* 149: 1799.
14 Fujii, T., Hishinuma, Y., Mitam, T., and Naono, T. (2010). *Sens. Actuators A* 163: 220.
15 Yoshida, S., Hanzawa, H., Wasa, K., and Tanaka, S. (2014). *IEEE Trans. Ultrason. Ferroelectr., Freq. Control* 61: 1552.
16 Kobayashi, T., Ichiki, M., Tsaur, J., and Maeda, R. (2005). *Thin Solid Films* 489: 74.
17 Kobayashi, T., Ichiki, M., and Maeda, R. (2007). *Ferroelectrics* 357: 233.
18 Muralt, P. et al. (1998). *J. Appl. Phys.* 83: 3835.
19 Calame, F. and Muralt, P. (2007). *Appl. Phys. Lett.* 90: 062907.
20 Chen, M.S., Wu, T.B., and Wu, J.M. (1996). *Appl. Phys. Lett.* 68: 1430.

21 Scott, J.F., Araujo, C.A., Melnick, B.M. et al. (1991). *J. Appl. Phys.* 70: 382.
22 Nakamura, T., Nakao, Y., Kamisawa, A., and Takasu, H. (1994). *Jpn. J. Appl. Phys.* 33: 5207.
23 Alshareef, H.N., Kingon, A.I., Chen, X. et al. (1994). *J. Mater. Res.* 9: 2968.
24 Akkopru-Akgunm, B., Zhu, W., Randall, C.A. et al. (2019). *APL Mater.* 7: 120901.
25 Kobayashi, T., Tsaur, J., and Maeda, R. (2005). *Jpn. J. Appl. Phys.* 44: 7078.
26 Nogami, H., Kobayashi, T., Okada, H. et al. (2012). *Jpn. J. Appl. Phys.* 51: 09LD11.
27 Kobayashi, T., Suzuki, Y., Makimoto, N. et al. (2014). *AIP Adv.* 4: 117116.

Part III

Bonding, Sealing and Interconnection

11

Anodic Bonding

Masayoshi Esashi

Tohoku University, Micro System Integration Center (μSIC), 519-1176 Aramaki-Aza-Aoba, Aoba-ku, Sendai 980-0845, Japan

11.1 Principle

Anodic bonding is used for bonding of glass to metals or semiconductors. The principle of the glass–Si anodic bonding is shown in Figure 11.1. A flat surfaces of glass and Si are faced and negative voltage (500–1000 V) is applied to the glass at elevated temperature around 400 °C. Silicide glass is formed by Si–O network as shown in Figure 11.2, and there are mobile positive ions as Na$^+$ and immobile negative ions as SiO$^-$. The Na$^+$ ions move and hence SiO$^-$ space charge layer is formed near the Si surface during the anodic bonding process. The negative space charge in the glass and induced positive charge in the Si cause electrostatic attraction for bonding. The current by the displacement of the Na$^+$ ions is observed as shown in Figure 11.1 [1]. The Na$^+$ appears as NaOH on the surface.

The anodic bonding was invented in Mallory & Co. Inc. in the United States in 1968 and this bonding method is called Mallory bonding, field-assisted bonding, and electrostatic bonding as well [2, 3].

Figure 11.3a,b shows the photographs of the bonded glass–Si wafer and an accelerometer chip obtained by dicing the bonded wafer. The color of the Si surface is changed by the bonding because of the oxidation of the Si surface. The glass–Si anodic bonding has been used for packaged MEMS devices [4, 5]. The MEMS packaging by the anodic bonding and that with vacuum or controlled pressure cavity will be described in Chapters 17 and 18 respectively.

Bonding at the glass–Si interface made by the electrostatic force will be explained in the following [6]. Figure 11.4 shows the distributions of the charge and the voltage schematically. The voltage across the polarized space charge layer Vp is given from the Poisson's equation by

$$Vp = \frac{\rho Xp^2}{2\varepsilon'\varepsilon_0} \tag{11.1}$$

where ρ is the charge density in the glass, Xp is the thickness of the polarized space charge layer, ε_0 is the permittivity of vacuum, and ε' is the relative permittivity of

3D and Circuit Integration of MEMS, First Edition. Edited by Masayoshi Esashi.
© 2021 WILEY-VCH GmbH. Published 2021 by WILEY-VCH GmbH.

11 Anodic Bonding

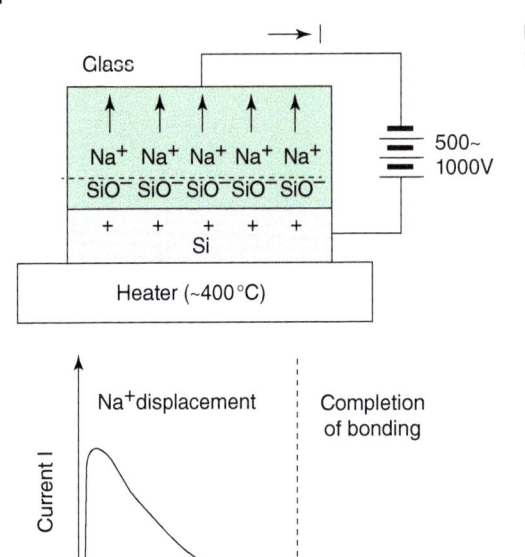

Figure 11.1 Principle of the anodic bonding.

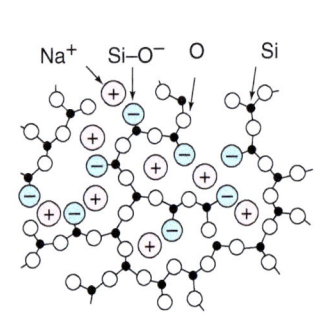

Figure 11.2 Structure of glass.

the glass. The charge per unit area of the polarized space charge layer σ_s is given by

$$\sigma_s = \rho X p = \varepsilon_0 E \tag{11.2}$$

where E is the electric field in the gap.

The electrostatic force per unit area P between the glass and the Si is given by

$$P = \frac{1}{2}\varepsilon_0 E^2 = \frac{1}{2}\frac{\sigma_s^2}{\varepsilon_0} \tag{11.3}$$

Combining Equations (11.1), (11.2), and (11.3), the electrostatic force per unit area P is given by

$$P = V p \rho \varepsilon' \tag{11.4}$$

An example of the anodic bonding tool is shown schematically in Figure 11.5. It is composed of a movable stage for the Si wafer, a holder for the glass wafer,

Figure 11.3 Photographs of bonded wafer and fabricated accelerometer chip. (a) Bonded glass–Si wafer and (b) accelerometer chip obtained by dicing the wafer.

Figure 11.4 Distributions of charge and voltage.

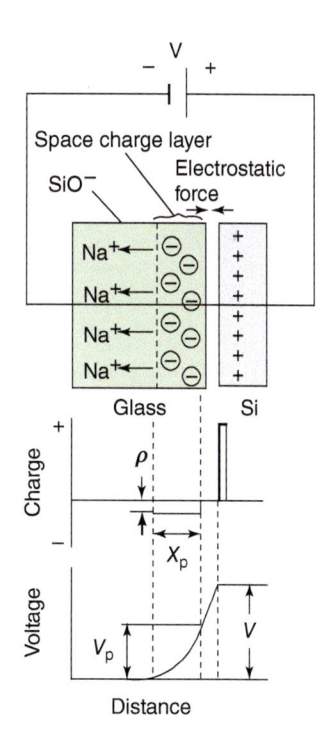

and microscopes. After alignment of the glass wafer and the Si wafer under the microscope, the glass wafer becomes in contact with the Si wafer. In the case of bonding in atmosphere, these wafers are heated using the heater on the stage and negative voltage is applied to the glass (left in Figure 11.5). Bonding in environment as vacuum is performed by transferring the contacted glass–Si into a chamber and by heating and applying the voltage (right in Figure 11.5).

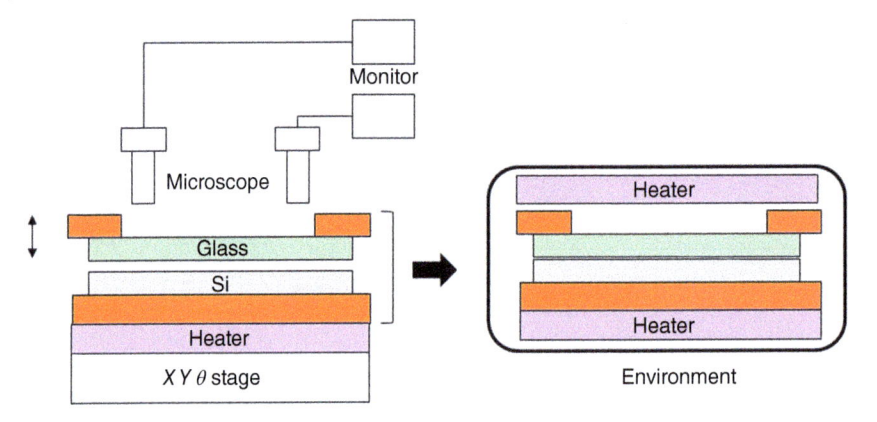

Figure 11.5 Anodic bonding tool.

11.2 Distortion

Thermal expansion of the glass should match with that of Si to prevent distortion after the bonding. The thermal expansion $L(T) - L(T_0)$ of glasses used for the anodic bonding is shown in Figure 11.6, where $L(T)$ is the length at temperature T and $L(T_0)$ is that at room temperature (operating temperature) T_0. The thermal expansion differences $L(T) - L_{Si}(T)$ between glasses and Si, which cause the distortion, are also shown in Figure 11.6. Glasses used for the anodic bonding are Pyrex glass (Corning 7740) (SiO_2 83%, B_2O_3 12%, Zn/MgO/CaO <1%, Na_2O 1%, and Al_2O_3 1%) (http://www.hoyaoptics.com/pdf/silicon_sensor.pdf), Schott Tempax Float (SiO_2 81%, B_2O_3 13%, Na_2O/K_2O 4%, and Al_2O_3 2%), Asahi glass SW-3 (SiO_2 60–65%, B_2O_3 5–10%, Zn/MgO/CaO 10–16%, Na_2O 2–4%, and Al_2O_3 12–20%) (http://www.hoyaoptics.com/pdf/silicon_sensor.pdf), and HOYA SD2 (SiO_2 58.5%, B_2O_3 1.8%, Al_2O_3 22.3%, Na_2O 2.5%) [7]. These glasses have coefficient of thermal expansion (CTE) of around 3.3 ppm/°C, which is close to that of Si. The thermal expansion curve of the glass is closely matched with that of the Si by replacing the B_2O_3 by Al_2O_3 as SW-3 and SD2. The bonding temperature of Asahi glass SW-YY is lowered down to 250 °C by replacing the Na with Li, which has smaller ion diameter than Na [7]. Anodic bonding at room temperature using thin glass film containing Li was reported [8].

Figure 11.7 shows distortions of Pyrex glass–Si structure after anodic bonding in different heating conditions [9]. When Si side is heated, the temperature of the Si is higher than that of the Pyrex glass. The Si shrinks more than the Pyrex glass after the bonding, and the Pyrex glass–Si bends as shown in Figure 11.7a. On the other hand when glass side is heated, the Pyrex glass–Si bends in opposite direction as shown in Figure 11.7c. When uniformly heated the Pyrex glass–Si bends as shown in Figure 11.7b. Curvature change of the anodically bonded glass–Si structure was studied [10].

Figure 11.8 shows examples of gap measurement between glass and Si in MEMS accelerometer [9]. Optical interference method ((a) in the figure) was used for this measurement. Photographs with and without fringes are shown in Figure 11.8b, in

Figure 11.6 Thermal expansions of glasses used for anodic bonding. Sources: Based on http://www.hoyaoptics.com/pdf/silicon_sensor.pdf, S. Takaki [7].

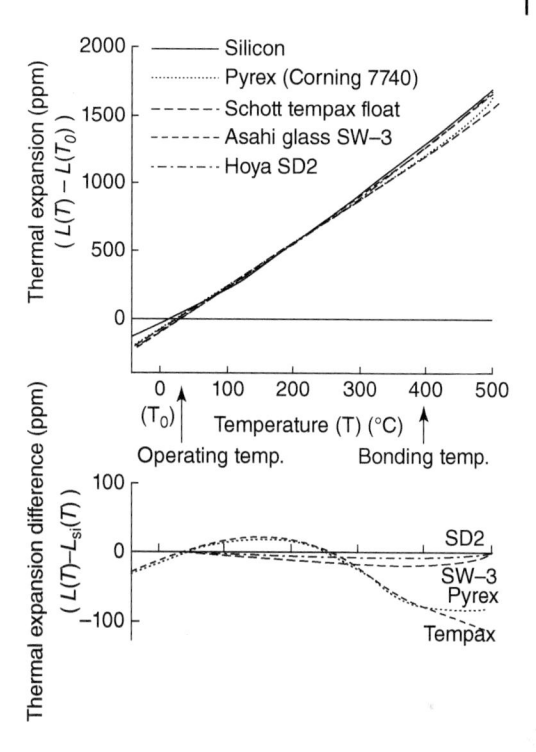

which the fringe means the gap variation caused by the distortion and the pitch of the fringe corresponds to half wavelength (273 nm) of the light.

Small distortion is caused by the space charge layer in the glass. Figure 11.9 shows a measurement of the bending of glass having Al electrodes on both sides [11]. The space charge layer in the glass is formed on the positive side by the Na^+ ion displacement and it causes the bending.

Thickness of the space charge layer can be estimated from the depth dependency of glass etching rate as shown in Figure 11.10a. The etching rate is higher in the space charge layer than its bulk [12]. The larger charge density by the current makes the thicker space charge layer and the thickness observed was around 1–2 µm. The thickness of the space charge layer depends on the size and the shape of the electrode on the glass as shown in Figure 11.10b. It is thick near the electrode. The thickness of the space charge layer was measured by electrostatic recoil detection analysis as well, and comparable thickness to that in Figure 11.10 was observed [13].

Mechanical strength of the glass–Si structure was studied and it was observed that tensile stress caused during the anodic bonding reduces the strength [14].

11.3 Influence of Anodic Bonding to Circuits

The influences of the anodic bonding to the integrated circuit (IC) were studied [15]. Figure 11.11 shows the experimental setup and test evaluation group (TEG) to study the influences of the high voltage (1 kV) at bonding temperature (400 °C) to the CMOS circuit. The n-Si substrate surface is doped with phosphorous for the

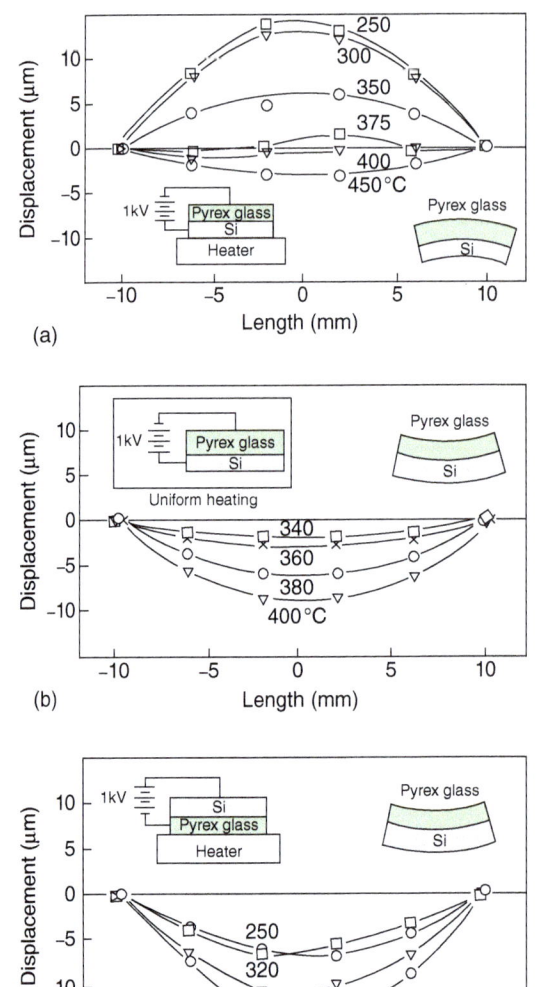

Figure 11.7 Distortion of Pyrex glass–Si structure after anodic bonding. (a) Si side heated, (b) uniformly heated, and (c) glass side heated. Source: Based on Shoji et al. [9].

purpose of channel stopper to prevent surface inversion by the electric field. Since MOS transistors are covered with gate poly Si, they are protected from the electric field. The surface of pn junction is exposed to the electric field, and hence leakage current of the pn junction increases as shown in Figure 11.12a. However this problem is solved by shielding the pn junction with metal as shown in Figure 11.12b. The method to use CMOS IC in the package made by the anodic bonding was applied to integrated capacitive sensors. Figures 11.13 and 11.14 are integrated capacitive pressure sensor [16] and integrated capacitive accelerometer [17], respectively. The circuit is used to detect small capacitance of the sensor. The CMOS circuits are located under the Ti/Pt layers on the glass to prevent the influences by light.

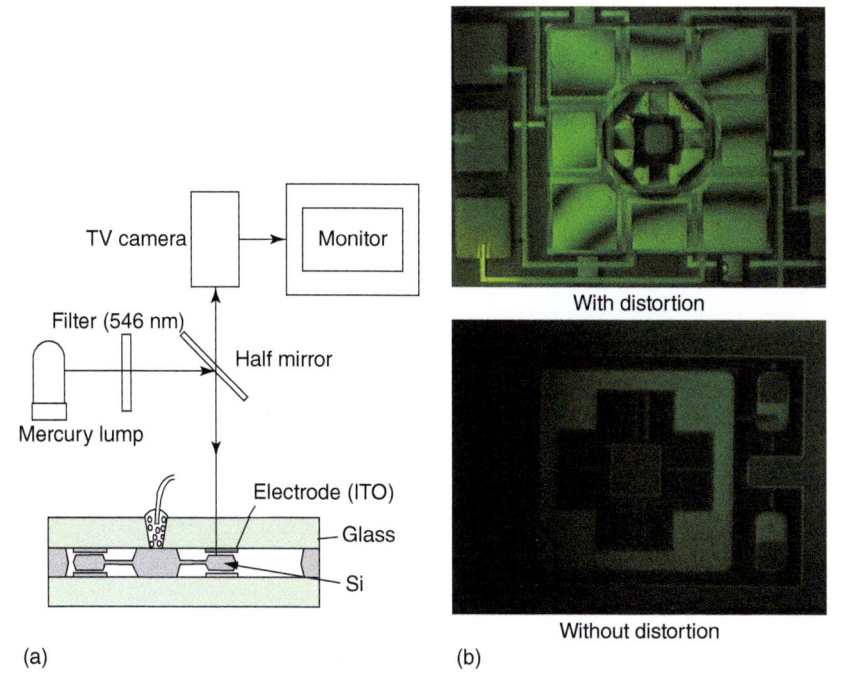

Figure 11.8 Measurement of the gap between glass and Si. (a) Gap measurement using optical interference and (b) photographs of interference image. Source: Modifed from Shoji et al. [9].

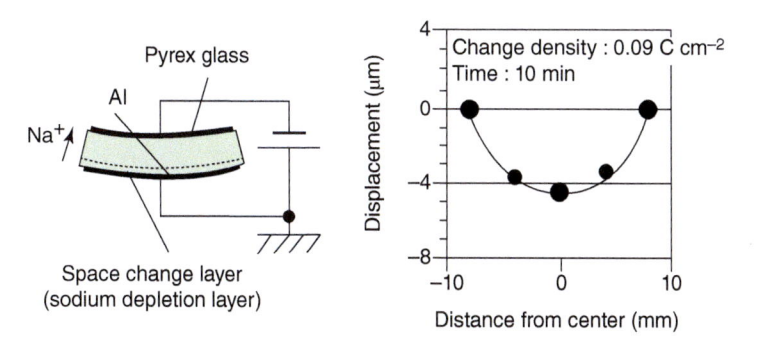

Figure 11.9 Bending of Pyrex glass caused by the space charge layer. Source: Shoji et al. [11].

11.4 Anodic Bonding with Various Materials, Structures and Conditions

11.4.1 Various Combinations

Anodic bonding can be used for bonding glass to materials as metal or semiconductor. The material should have close thermal expansion with the glass to prevent

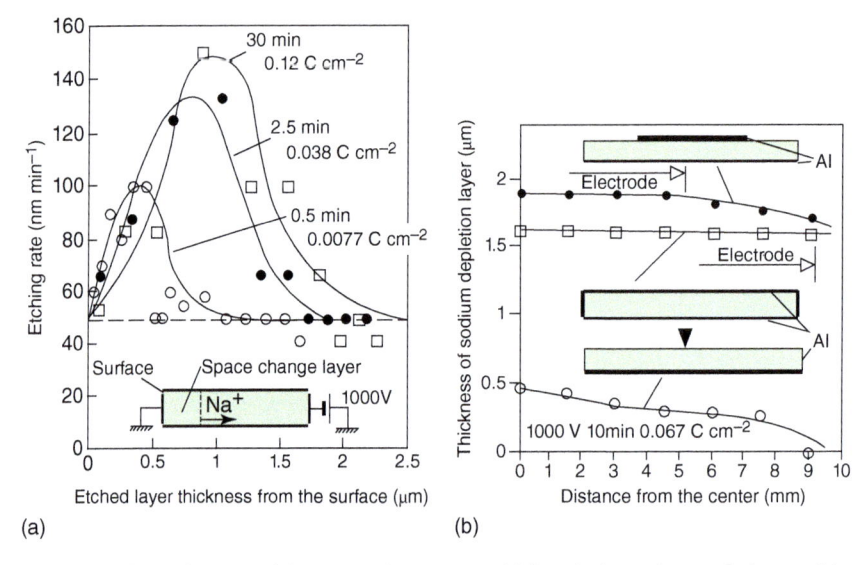

Figure 11.10 Thickness of the space charge layer. (a) Depth dependency of glass etching rate and (b) thickness variation depending on electrode rate. Source: Shoji et al. [11].

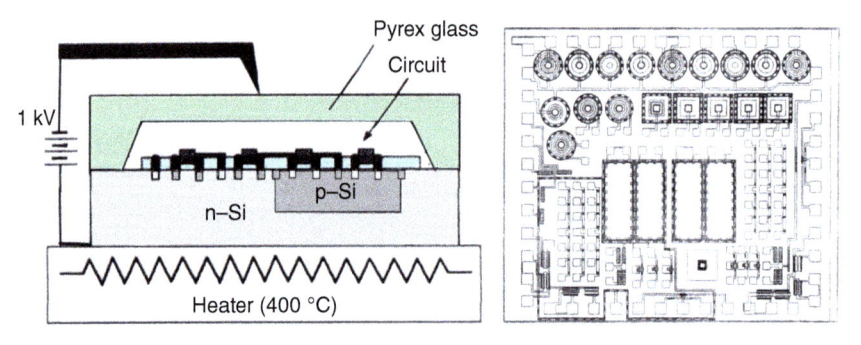

Figure 11.11 Anodic bonding of integrated circuit (IC) and test evaluation group (TEG). Source: Shirai and Esashi [15].

distortion after bonding. Au and Ag can't be used for the anodic bonding, because Au doesn't make chemical bond because it is not oxidized. On the other hand Ag diffuses into the glass. Figure 11.15 shows anodic bonding methods with various materials. The Si of which CTE is around 3.3 ppm/°C is bonded to glasses shown in Figure 11.6 (Figure 11.15a). The CTE of GaAs is 6.6 ppm/°C at 300 °C and it can be bonded to Corning 0211 glass of which the CTE is 7.5 ppm/°C (Figure 11.15b) [22]. Hydrogen plasma treatment is needed for the GaAs and glass surfaces [18].

Metal as Fe–Ni–Co alloy can be bonded to the Pyrex glass (Figure 11.15c) [19]. Kovar 10 (Ni 30%, Co 13%, and Fe 57%) and Fe–Ni alloy (Fe 59% and Ni 41%) have close thermal expansion with the Pyrex glass and Si as shown in Figure 11.16. These metals can be anodically bonded to make a metal–glass–Si (–glass) multilayer structure. Figure 11.17 shows a bakable microvalve as an example of this

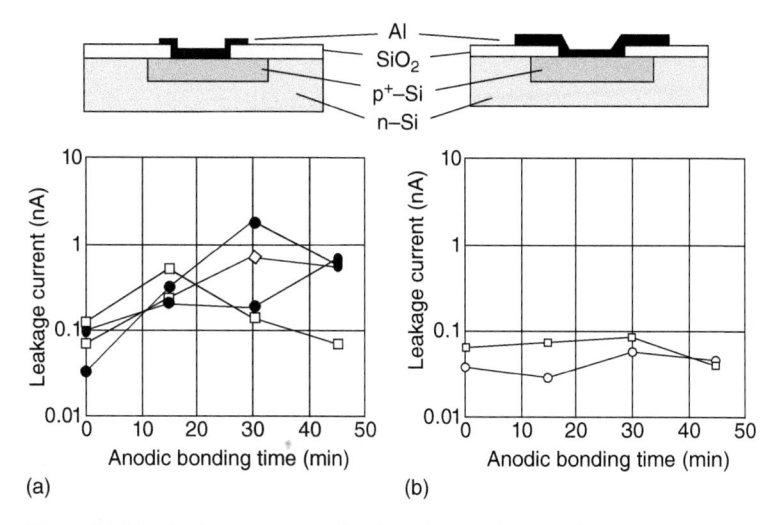

Figure 11.12 Leakage current of *pn* junctions without and with metal shield. (a) Without metal shield and (b) with metal shield. Source: Shirai and Esashi [15].

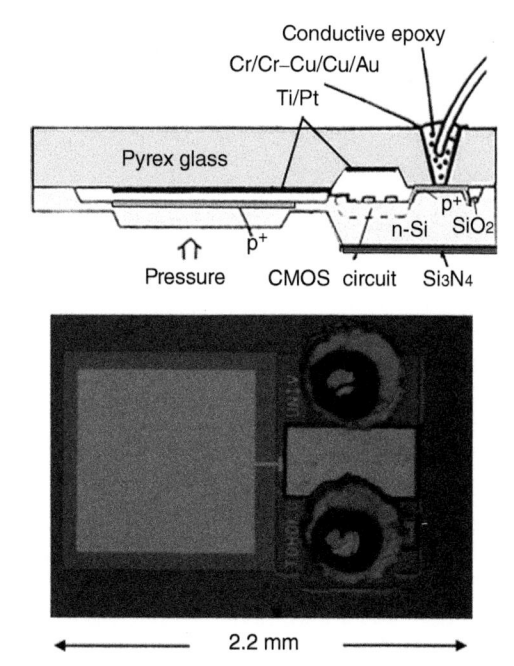

Figure 11.13 Integrated capacitive pressure sensor. Source: Matsumoto and Esashi [16].

multilayer structure [23]. The Kovar 10 block to which stainless steel pipes are welded is anodically bonded to the glass–Si–glass structure (Figure 11.17a). This is a pneumatic valve of which working principle is shown in Figure 11.17b. The valve is closed by pushing up the glass 2 by the control air as shown in the top figure. On the other hand it is opened by pushing down the glass 2 by the another control air as in the bottom figure. Figure 11.17c is the photograph of the pneumatic valve.

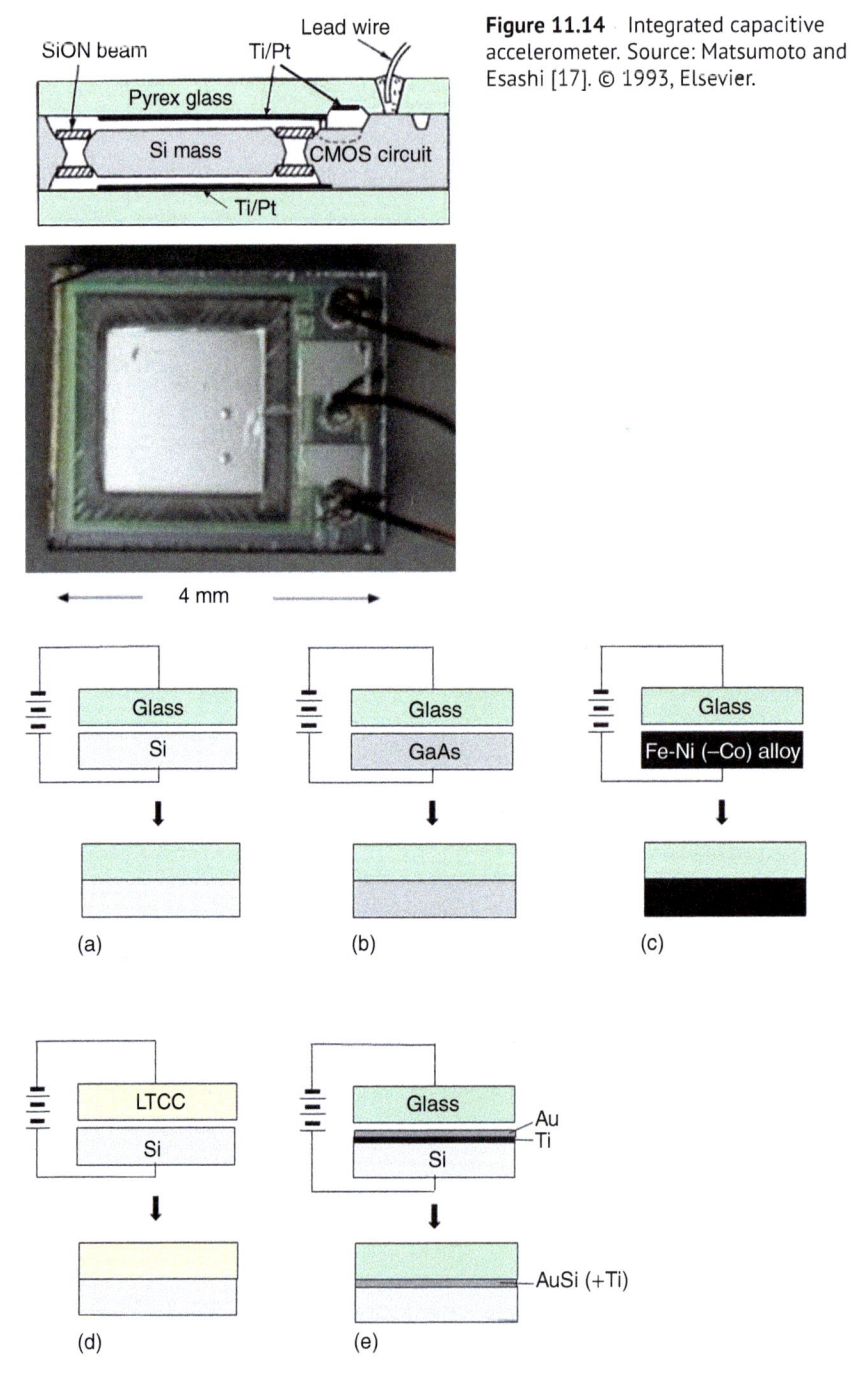

Figure 11.14 Integrated capacitive accelerometer. Source: Matsumoto and Esashi [17]. © 1993, Elsevier.

Figure 11.15 Anodic bonding with various materials. (a) Glass–Si. Source: Modified from Pomerrantz [2], (b) glass–GaAs. Source: Modifed from Huang [18], (c) glass–Fe–Ni (–Co) alloy. Source: Modified from Wallis [19], (d) LTCC–Si. Source: Modified from Tanaka et al. [20], and (e) glass–AuSi–Si. Source: Modified from Harpster and Najafi [21].

This can be used as bakable valves for reactive gasses because water vapor on the inner wall can be desorbed by heating up the valve, and hence clogging by reactive deposition can be prevented. The open–close characteristics of the valve at 120 °C is shown in Figure 11.17d [23].

Low-temperature co-fired ceramics (LTCC) can be anodically bonded to Si as shown in Figure 11.15d [20]. CTE of the LTCC is shown in Figure 11.18 being compared with those of Si and Pyrex glass. The CTE is adjusted to that of Si using the LTCC composed of ceramic (alumina (Al_2O_3) and cordierite (SiO_2, MgO, and Al_2O_3)) powder and glass (SiO_2, Al_2O_3, B_2O_3, and Na_2O) powder. Wafer level packaging process using the LTCC–Si anodic bonding is shown in Figure 11.19. Holes are made in a green sheet of the LTCC by punching (2 in the figure). The holes are plugged with Au paste and metallized patterns are made by screen printing of the Au paste (3 in the figure). LTCC wafers with the Au paste plugs are laminated (4 in the figure), and they are sintered to make ceramics (5 in the figure). During the sintering the LTCC wafers are fixed on a plate to prevent lateral dimensional change by allowing the shrinkage in thickness. The sintered LTCC is anodically bonded to Si (6 in the figure). The LTCC is more advantageous than the glass when vertical feedthroughs are required. Air leakage is caused at the metal feedthrough because of the different thermal expansion between the metals and the glass (ceramics). This problem will be discussed in Section 18.6. Problem of the air leakage through the metal feedthrough can be solved by using the laminated structure. Figure 11.20 shows a 4 inch laminated LTCC wafer, which has the Au feedthroughs.

Figure 11.15e is anodic bonding of glass to Au–Si eutectic solder [21]. Ti/Au (20 nm/1 µm) is deposited on a Si wafer, and the Au–Si eutectic solder is formed by heating the wafer to 410 °C. This wafer is brought in contact with a glass by force, and then anodic bonding is made between the melted Au–Si and the glass by applying −500 V to the glass. The bonding can be made even on a non-planar surface because the Au–Si is melted.

11.4.2 Anodic Bonding with Intermediate Thin Films

Anodic bonding methods using intermediate thin films are shown in Figure 11.21. Two Si wafers which have 1µm thick oxide on the surface are anodically bonded at 850–950 °C by applying 30–50 V for 1 hour (Figure 11.21a) [24]. Thin glass film formed on Si by sputter deposition is used as the intermediate layer for the anodic bonding (Figure 11.21b) [25]. However electrical breakdown occurs, and hence thin SiO_2 is used between the thin glass layer and the Si to prevent the breakdown (Figure 11.21c) [26]. Unpolarized PZT (lead titanate zirconate) has CTE of 3.4 ppm/°C which is close to the CTE of Si. Wafer of the PZT ceramics is anodically bonded to the Si at 400 °C using 2 µm thick intermediate Pyrex glass layer (Figure 11.21d) [27]. Two glass wafers can be successfully bonded anodically using thin layer of amorphous Si (a-Si) [28], Al [29], or Ti [30] (Figure 11.21e). Quartz wafer having a thin glass film on the surface is anodically bonded to another quartz wafer having a-Si layer on the surface at 400 °C (Figure 11.21f) [31].

270 | *11 Anodic Bonding*

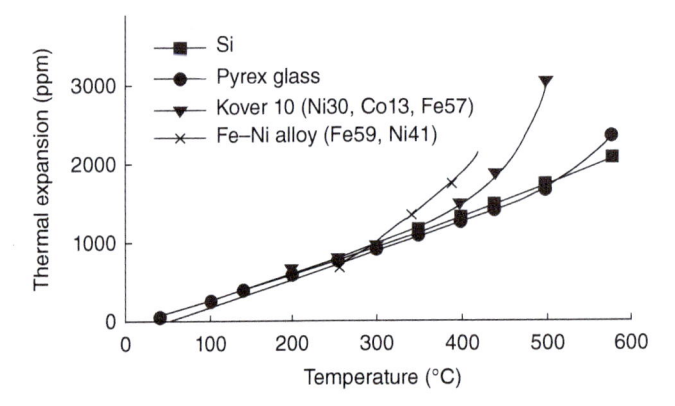

Figure 11.16 Thermal expansion of Kovar 10 and Fe–Ni alloy compared with silicon and Pyrex glass. Source: Sim et al. [23]. © 1996, IOP Publishing.

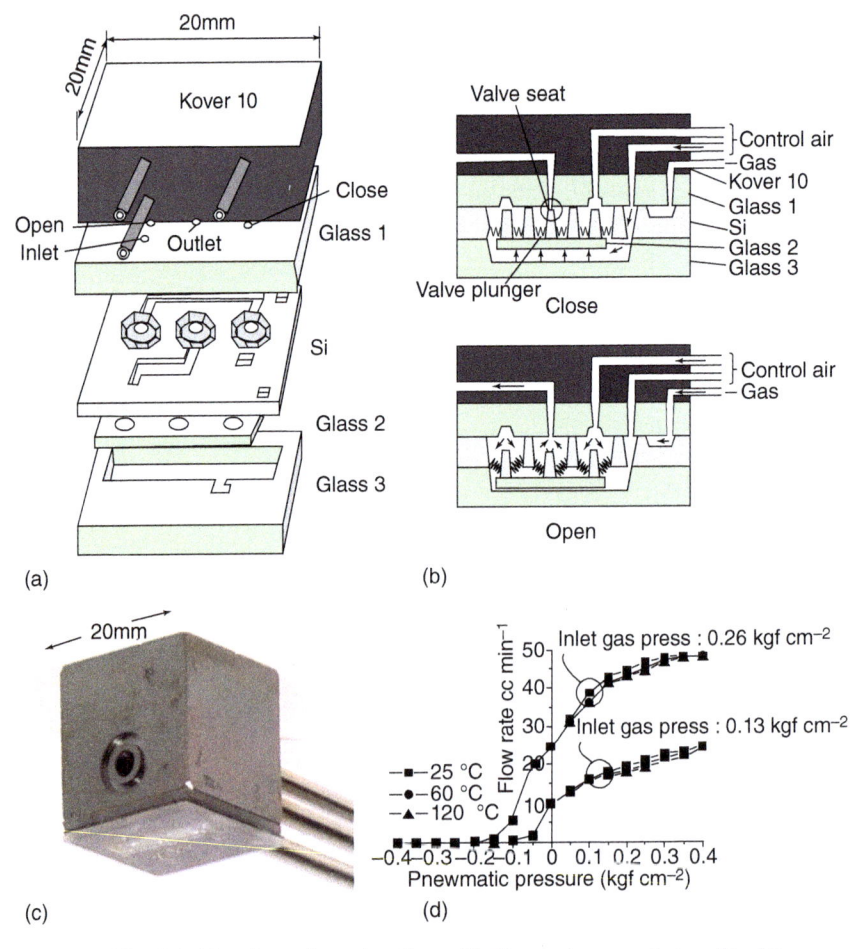

Figure 11.17 Bakable microvalve using Kovar 10–Pyrex glass anodic bonding (a) Structure, (b) principle, (c) photograph, and (d) characteristic. Source: Sim et al. [23]. © 1996, IOP Publishing.

Figure 11.18 Coefficient of thermal expansion (CTE) of LTCC compared with those of Si and Pyrex glass. Source: Tanaka et al. [20]. © 2011, IEEE.

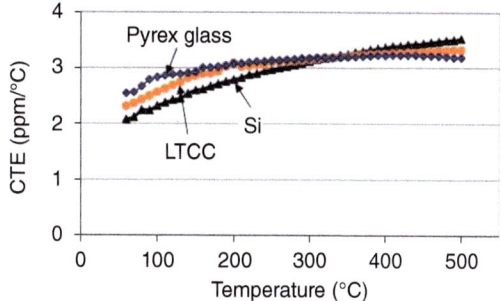

Figure 11.19 Wafer level packaging using LTCC–Si anodic bonding. Source: Modified from Tanaka et al. [20].

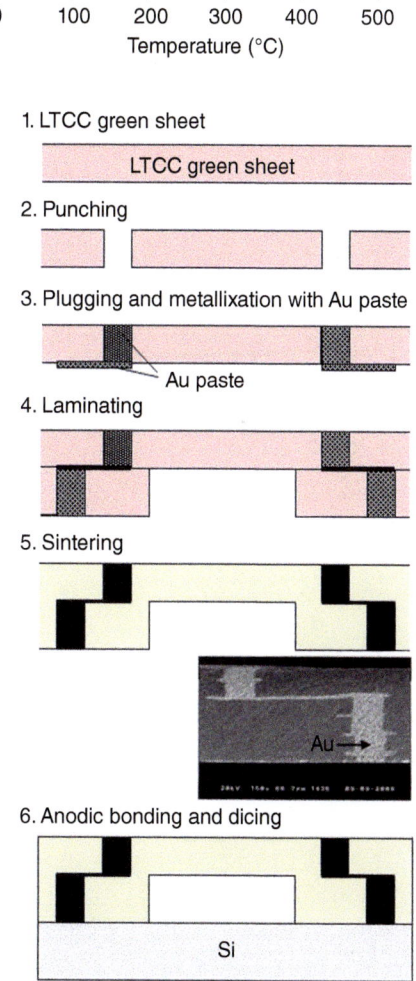

11.4.3 Variation of Anodic Bonding

Glass wafers can be anodically bonded successfully on both sides of a double-side polished Si wafer as shown in Figure 11.22. On the other hand it is not successful to bond Si wafers on both sides of a glass wafer. NaOH comes out from Na on the negative side of the glass because of the Na^+ ion moved by electric field. The NaOH makes it difficult to bond this side to another Si as shown in Figure 11.23a [32, 33].

Figure 11.20 4-inch laminated LTCC wafer with Au feedthrough and metallization.

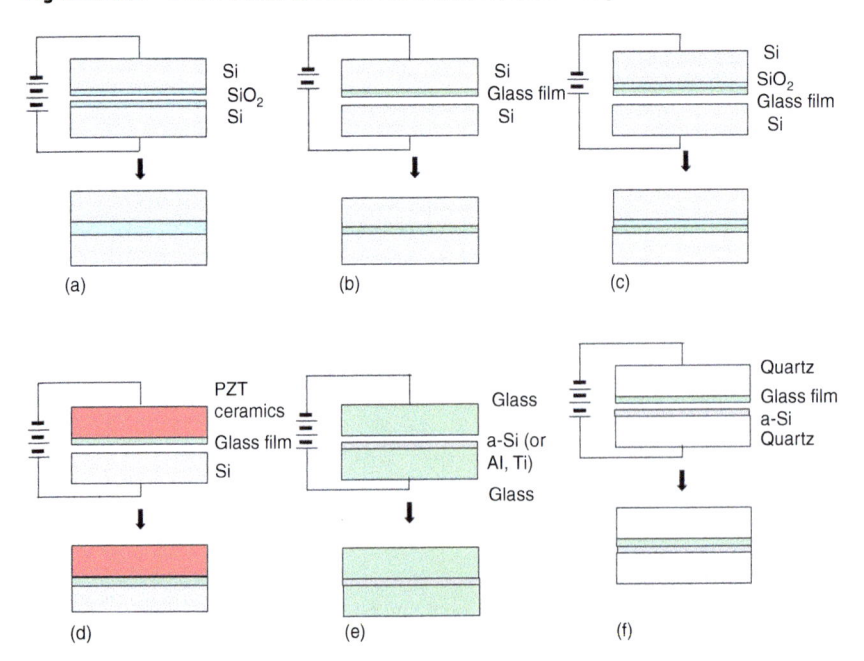

Figure 11.21 Anodic bonding methods using intermediate thin films. (a) Si–SiO$_2$ film–SiO$_2$ film–Si. Source: Modified from Anthony [24], (b) Si–glass film–Si. Source: Modified from Brooks and Donovan [25], (c) Si–SiO$_2$ film–glass film–Si. Source: Modified from Hanneborg et al. [26], (d) PZT ceramics–glass film–Si. Source: Modified from Tanaka et al. [27], (e) Glass–a-Si–glass. Sources: Sharon et al., Hu et al., Mrozek [28–30], and (f) Quartz–glass film–a-Si–quartz. Source: Modified from Danel and Delapierre [31].

Figure 11.22 Glass–Si–glass anodic bonding.

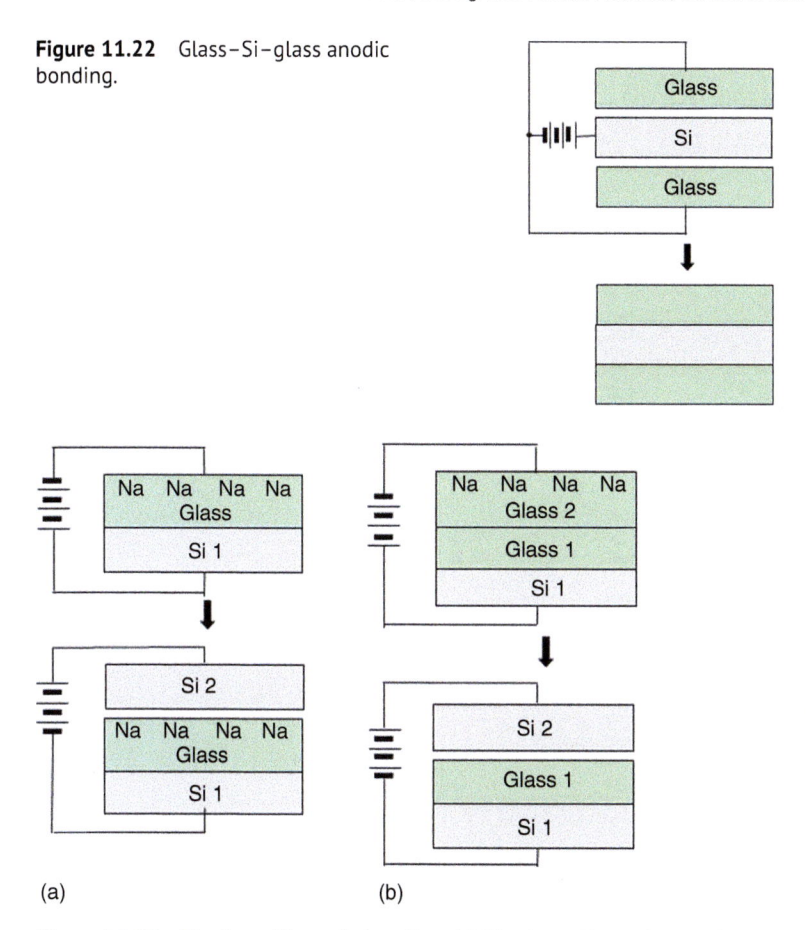

Figure 11.23 Si–glass–Si anodic bonding. (a) Si–glass–Si anodic bonding and (b) Si–glass–Si anodic bonding using glass electrode. [32].

This problem can be solved by using other glass as the electrode on the bonded glass as shown in Figure 11.23b. The Na+ ion moves to the other glass and the glass surface can be kept free from the NaOH. Si can be bonded to the glass by removing the other glass, and Si–glass–Si structure can be made by this anodic bonding.

Glass surface that was bonded to Si can't be bonded again to another Si. Figure 11.24 shows the fabrication process and cross-sectional photograph of a capacitive pressure sensor [34]. The through holes are sealed by thin p+ Si plates and electrical interconnections from the electrodes on the glass are made using the plates. The fabrication process is as follows. Glass wafer with through holes is prepared (1 in the figure). The Si wafer which has highly doped p+ Si (2 in the figure) is anodically bonded to the glass (3 in the figure). Si is etched out in alkaline solution selectively except the p+ Si (4 in the figure). Al is deposited and patterned on both sides of the glass wafer (5 in the figure). The glass wafer is anodically bonded to a Si wafer which has diaphragms for the pressure sensor (6 in the figure). The glass surfaces which were anodically bonded in step 3 of the process are not bonded in this process step. The bonded area is used for the device by dicing the wafer (7 in the figure).

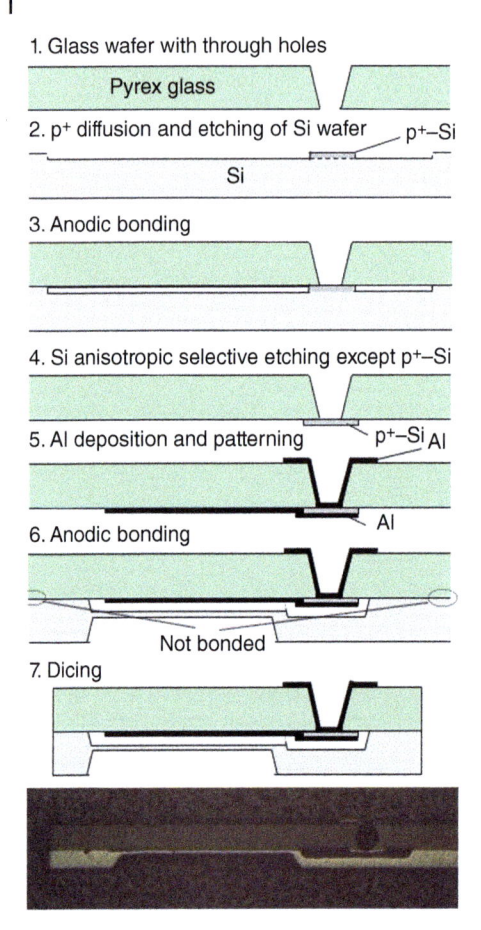

Figure 11.24 Fabrication process of capacitive pressure sensor. Source: Modified from Esashi [34].

1. Glass wafer with through holes
2. p+ diffusion and etching of Si wafer
3. Anodic bonding
4. Si anisotropic selective etching except p+–Si
5. Al deposition and patterning
6. Anodic bonding
7. Dicing

Cr is used as an etching mask for glass wet etching. Glass surface on which Cr is deposited before should be slightly etched in HF solution before the anodic bonding. This is because there remains Cr oxide on the glass surface even if the Cr is etched out.

Adhesion by anodic bonding can be prevented by partial electrical shielding. This is performed by electrically connecting the metal on the glass to the Si. Figure 11.25 shows assembly process to fabricate MEMS seismometer [35]. This is a highly sensitive accelerometer, which has a seismic mass suspended with thin beams. The glass has an Al pattern above the thin beams for the electrical shielding (3 in the figure). The interconnection of the Al to the Si is cut by laser through the glass after the anodic bonding (4 in the figure).

11.4.4 Glass Reflow Process

Etched part of Si wafer can be filled with glass by a glass reflow process at high temperature. The glass reflow process applied for the fabrication of through glass

Figure 11.25 Assembly process to prevent electrostatic adhesion. Source: Modified from Ko et al. [35].

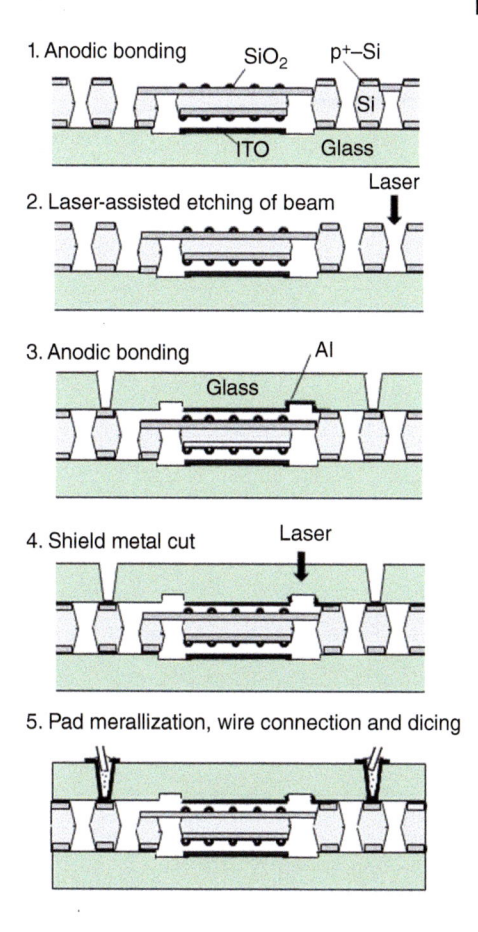

1. Anodic bonding

2. Laser-assisted etching of beam

3. Anodic bonding

4. Shield metal cut

5. Pad merallization, wire connection and dicing

via (TGV) for an intraocular pressure sensor is shown in Figure 11.26 [36]. The process sequence is as follows. Si mold is made by deep reactive ion etching (DRIE) of Si wafer (1 in the figure). The Si wafer is anodically bonded to a glass wafer in vacuum, which makes vacuum cavity between the glass wafer and the Si wafer (2 in the figure). The bonded wafers are annealed at 750 °C for 8 hours under the atmospheric pressure. This causes the glass reflow into the Si mold (3 in the figure). The top surface is planarized and polished using diamond pad (4 in the figure). The bottom surface is planarized and polished by chemical mechanical polishing (CMP). This makes the TGV (5 in the figure). The TGV wafer is metallized with Al (top surface) and Cr/Au (bottom surface) (6 in the figure). Si wafer is prepared for the diaphragm by etching, p^+ diffusion, and oxidation (6′ in the figure). The TGV wafer and the Si wafer are anodically bonded (7 in the figure). The Si is selectively etched out in alkaline solution, leaving the thin p^+ Si diaphragm (8 in the figure). Capacitive Si diaphragm pressure sensor for the intraocular pressure monitoring was fabricated by the glass reflow process [36].

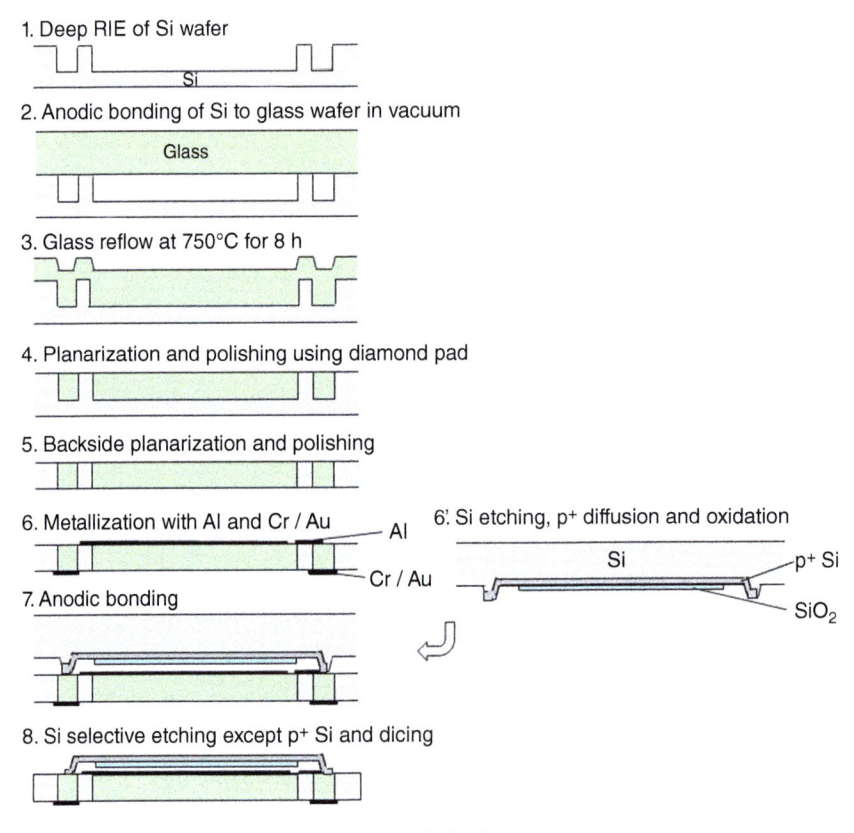

Figure 11.26 Glass reflow process for p$^+$ Si diaphragm pressure sensor.

References

1 Ko, W.H., Suminto, J.T., and Yeh, G.J. (1985). Bonding techniques for microsensors. In: *Micromachining and Micropackaging of Transducers* (eds. C.D. Fung, P.W. Cheung, W.H. Ko and D.G. Fleming), 41–61. Elsevier Science Publishers.
2 Pomerrantz, D.I. (1968). U.S. Patent No. 3, 397, 278.
3 Wallis, G. and Pomerantz, D.I. (1969). Field assisted glass-metal sealing. *J. Appl. Phys.* 40 (10): 3946–3949.
4 Esashi, M. (1994). Encapsulated micro mechanical sensors. *Microsyst. Technol.* 1 (1): 2–9.
5 Esashi, M. (2008). Wafer level packaging of MEMS. *J. Micromech. Microeng.* 18 (7): 073001 (13).
6 Esashi, M., Nakano, A., Shoji, S., and Hebiguchi, H. (1990). Low-temperature silicon-to-silicon anodic bonding with intermediate low melting point glass. *Sens. Actuators* A21–A23: 931–934.
7 Takaki, S. (2009). Glass wafer and bonding materials for MEMS encapsulating. *Mater. Integrat.* 22 (4): 8–13. (in Japanese).

8 Woetzel, S., Kessler, E., Diegel, M. et al. (2014). Low-temperature anodic bonding using thin film of lithium-niobate-phosphate glass. *J. Micromech. Microeng.* 24 (9): 095001 (6).

9 Shoji, Y., Yoshida, M., Minami, K., and Esashi, M. (1995). Diode integrated capacitive accelerometer with reduced structural distortion. The Eighth International Conference on Solid State Sensors and Actuators, and Eurosensors IX (Transducers'95·Eurosensors IX), Stockholm, Sweden (25–29 June 1995), 581–584,

10 Harz, M. and Engelke, H. (1996). Curvature changing or flattening of anodically bonded silicon and borosilicate glass. *Sens. Actuators A* 55: 201–209.

11 Shoji, Y., Minami, K., and Esashi, M. (1995). Glass-silicon anodic bonding for the reduction of structural distortion. *Trans. IEEJ* 115-A (12): 1208–1213. (in Japanese).

12 Wallis, G. (1969). Direct-current polarization during field-assisted glass-metal sealing. *J. Am. Ceram. Soc.* 53 (10): 563–567.

13 Nitzsche, P., Lange, K., Schmidt, B. et al. (1998). Ion drift processes in Pyrex-type alkali-borosilicate glass during anodic bonding. *J. Electrochem. Soc.* 145 (5): 1755–1762.

14 Johansson, S., Gustafsson, K., and Schweitz, J.-Å. (1988). Influence of bonded area ratio on the strength of FAB seals between silicon microstructures and glass. *Sens. Mater.* 1 (4): 209–221.

15 T. Shirai and M. Esashi, Damage to circuit during anodic bonding, Japan Institute of Electrical Engineering (JIEE) Technical Report, ST-92-7 (1992) 9-17 (in Japanese)

16 Matsumoto, Y. and Esashi, M. (1992). An integrated capacitive absolute pressure sensor. *Electron. Commun. Jpn., Part 2* 76 (1): 93–106.

17 Matsumoto, Y. and Esashi, M. (1993). Integrated silicon capacitive accelerometer with PLL servo technique. *Sens. Actuators A* 39: 209–217.

18 Huang, Q.-A., Lu, S.-J., and Tong, Q.-Y. (1990). A novel bonding Technology for GaAs sensors. *Sens. Actuators* A21–A23: 40–42.

19 Wallis, G., Dorsey, J., and Beckett, J. (1971). Field assisted seals of glass to Fe-Ni-Co alloy. *Ceram. Bull.* 50 (12): 958–961.

20 Tanaka, S., Matsuzaki, S., Mohri, M. et al. Wafer-level hermetic packaging technology for MEMS using anodically-bondable LTCC wafer. The 24th IEEE International Conference on Micro Electromechanical Systems (MEMS 2011), Cancun, Mexico (23–27 January 2011), 376–379.

21 Harpster, T.J. and Najafi, K. (2003). Field-assisted bonding of glass to Si-Au eutectic solder for packaging applications. IEEE The 16th Annual International Conference on Micro Electro Mechanical Systems (MEMS 2003), Kyoto, Japan (19-23 January 2003), 630–633

22 Hök, B., Dubon, C., and Ovrén, C. (1983). Anodic bonding of gallium arsenide to glass. *Appl. Phys. Lett.* 43 (3): 267–269.

23 Sim, D.Y., Kurabayashi, T., and Esashi, M. (1996). A bakable microvalve with a Kovar-glass-silicon-glass structure. *J. Micromech. Microeng.* 6 (2): 266–271.

24 Anthony, T.R. (1985). Dielectric isolation of silicon by anodic bonding. *J. Appl. Phys.* 53 (3): 1240–1247.
25 Brooks, A.D. and Donovan, R.P. (1971). Low-temperature electrostatic silicon-to-silicon seals using sputtered borosilicate glass. *J. Electrochem. Soc.* 119 (4): 545–546.
26 Hanneborg, A., Nese, M., and Øhlckers, P. (1991). Silicon-to-silicon anodic bonding with a borosilicate glass layer. *J. Micromech. Microeng.* 1 (3): 139–144.
27 Tanaka, K., Takata, E., and Ohwada, K. (1998). Anodic bonding of lead zirconate ceramics to silicon with intermediate glass layer. *Sens. Actuators A* 69: 199–203.
28 Sharon, J.W., Nai, M.L., Wong, C.K.S. et al. (2003). Low temperature glass-to-glass wafer bonding. *IEEE Trans. Adv. Packaging* 26 (3): 289–294.
29 Hu, L., Xue, Y., and Shi, F. (2017). Interfacial investigation and mechanical properties of glass-Al-glass anodic bonding process. *J. Micromech., Microeng.* 27 (10): 105004 (8).
30 Mrozek, P. (2009). Anodic bonding of glasses with interlayers for fully transparent device applications. *Sens. Actuators A* 151: 77–80.
31 Danel, J.S. and Delapierre, G. (1991). Quartz: a material for microdevices. *J. Micromech., Microeng.* 1 (4): 187–198.
32 Hu, L., Wang, H., Xue, Y. et al. (2018). Study on the mechanism of Si-glass-Si step anodic bonding process. *J. Micromech. Microeng.* 28 (4): 045003 (9).
33 Despont, M., Gross, H., Arrouy, F. et al. (1996). Fabrication of a silicon-Pyrex-silicon stack by a.c. anodic bonding. *Sens. Actuators A* 55: 219–224.
34 Esashi, M., Shoji, S., Wada, T., and Nagata, T. (1991). Capacitive absolute pressure sensors with hybrid structure. *Electron. Commun. Jpn. Part 2* 74 (4): 67–75.
35 Ko, S., Sim, D.Y., and Esashi, M. (1999). An electrostatic servo-accelerometer with mG resolution. *Trans. IEEJ* 119-E (7): 368–373.
36 R.M. Haque and Wise, K.D. (2010). An intraocular pressure sensor based on a glass reflow process. Solid-State Sens., Actuators and Microsyst. Workshop, Hilton Head, USA, 49–52.

12

Direct Bonding

Hideki Takagi

Device Technology Research Institute, National Institute of Advanced Industrial Science and Technology (AIST), Namiki 1-2-1, Tsukuba, Ibaraki 305-8564, Japan

12.1 Wafer Direct Bonding

Wafer direct bonding was first developed in order to fabricate silicon-on-insulator (SOI) wafers [1, 2]. Figure 12.1 shows two typical process flows for SOI fabrication. Bond and etch back (BESOI) process is suitable to fabricate relatively thick SOI wafers, which are widely used in various micro-electro mechanical systems (MEMS) processes. Because two wafers are necessary to fabricate a BESOI wafer and the fabrication process consists of many steps, the cost of SOI wafers inevitably increases.

On the other hand, Smart Cut process [3] is suitable for thinner SOI wafers. In the process, high-concentration H^+ ions are implanted into a Si wafer (wafer A) before bonding. During heat treatment after the bonding, a thin Si layer is exfoliated from the region where high concentration implanted H atoms exist and remains on the support wafer (wafer B). Finally the surface of the exfoliated layer is polished. The thickness of the exfoliated Si layer can be precisely controlled by ion energy. Very thin SOI wafers (<100 nm) are available by the method. Such thin SOI wafers are mainly used in micro-electronic devices, and they are also used to fabricate nanometer scale structures and sensing elements with extremely high sensitivity.

12.2 Hydrophilic Wafer Bonding

Hydrophilic wafer bonding is a method to bond wafers using hydrogen bonds between −OH groups on hydrophilic surface of wafers [1, 2, 4, 5]. This method is often called as "wafer direct bonding" or "wafer fusion bonding." In the method, the wafers to be bonded are cleaned by chemical solutions such as NH_3/H_2O_2 and H_2SO_4/H_2O_2 mixture. This treatment forms thin oxide layers on Si wafers and the surface becomes hydrophilic. Two wafers are then mated in air. After bonding initiation by applying slight load on a part of the mated two wafers, the bonding area spreads spontaneously as shown in Figure 12.2. This spontaneous bonding is

3D and Circuit Integration of MEMS, First Edition. Edited by Masayoshi Esashi.
© 2021 WILEY-VCH GmbH. Published 2021 by WILEY-VCH GmbH.

Figure 12.1 SOI wafer fabrication processes.

Bond and etch back

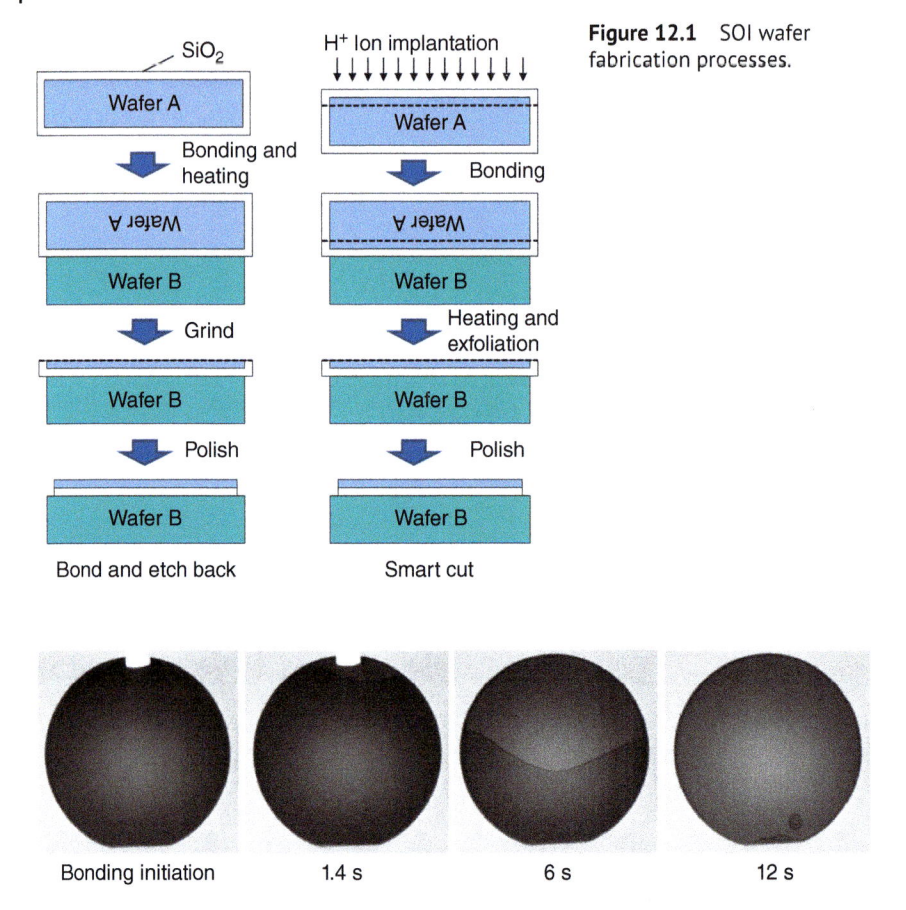

Smart cut

Figure 12.2 Infrared images of spontaneous propagation of bonding area.

assumed to be achieved by the attractive force between two surfaces. It should be noted that fringes near the orientation flat in the last picture are bonding defects caused by scratches and a particle. The surface roughness of the wafers is important to achieve the bonding. Although most of commercially available wafers are well polished enough for the spontaneous bonding, it is difficult to bond wafers with rough surface even with a large bonding load [4]. Therefore, it is important that the wafer processes for MEMS devices are designed considering surface roughness of the bonding areas before bonding.

Figure 12.3 illustrates the mechanism of the hydrophilic bonding. In the atmospheric condition, water molecules are adsorbed on the hydrophilic surface of SiO_2. When two wafers are mated, water molecules bridge surface of the two SiO_2 layers by hydrogen bonds. Because a layer of water molecules exists at the bonding interface, bonding strength is weak in this state as shown in Figure 12.4 (Chemical Treatment). Annealing processes are necessary to strengthen the bonding. It is supposed that the water molecules are diffused into SiO_2 layers during the annealing. Some of them react with Si and form oxide. Finally, Si—O—Si bonds are formed

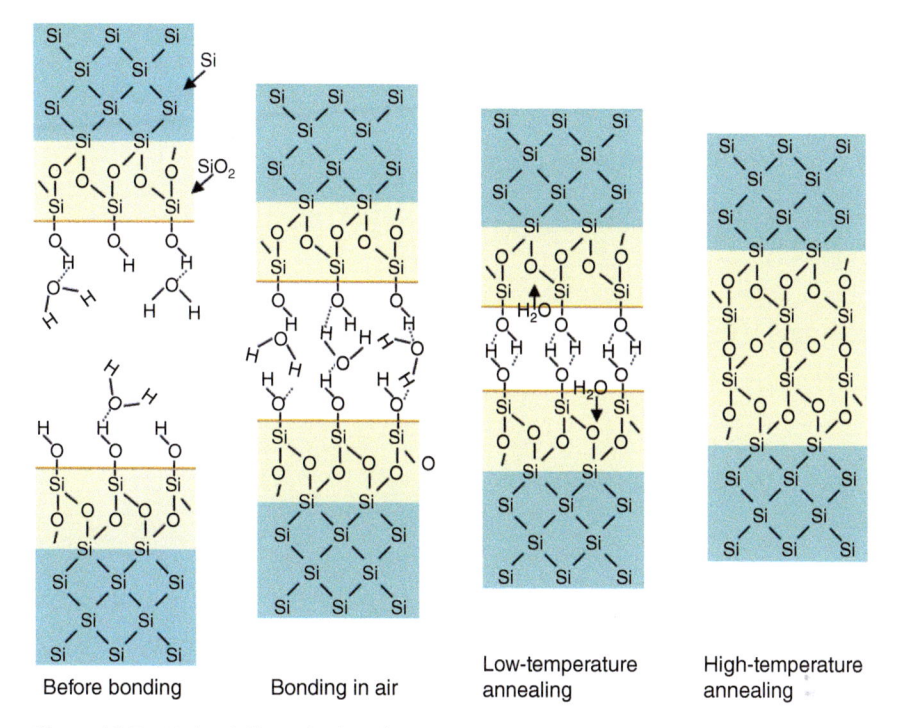

Figure 12.3 Hydrophilic wafer bonding process.

Figure 12.4 Change of bonding strength of hydrophilic wafer bonding by annealing.

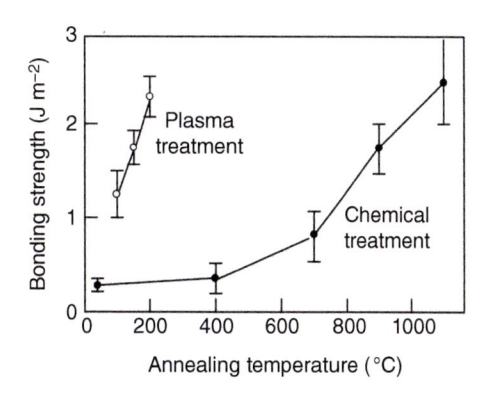

at the bonding interface. Figure 12.5 shows a high-resolution transmission electron microscope (HRTEM) image of the bonding interface prepared by the hydrophilic wafer bonding and annealing at 1100 °C. Although Si wafers only with native oxide were bonded, an about 3 nm thick oxide layer is observed at the bonding interface. Composition of the layer is analyzed by energy dispersive X-ray spectroscopy (EDS) in transmission electron microscope (TEM). Although the original bonding interface is supposed to exist at the center of the oxide layer, it is difficult to identify the bonding interface, and the oxide layer looks uniform.

The bonding strength shown in Figure 12.4 was measured by the method shown in Figure 12.6. This method was first proposed by Maszara et.al. [2], and it is often

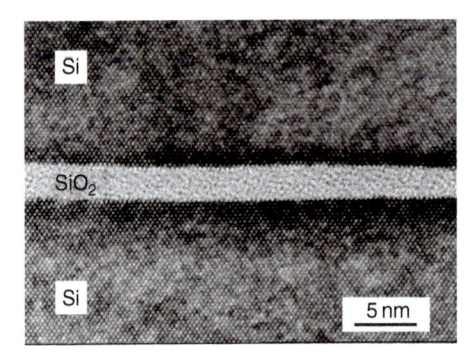

Figure 12.5 HRTEM image of Si/Si bonding interface prepared by the hydrophilic wafer bonding.

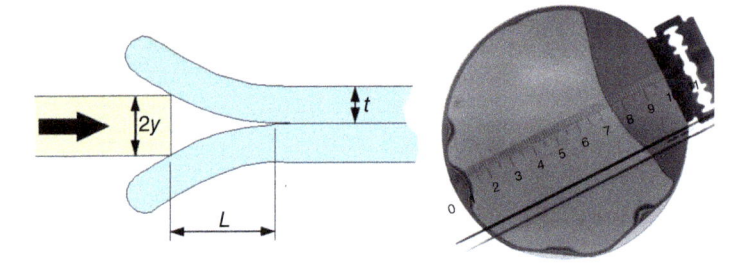

Figure 12.6 Bonding strength evaluation of wafer direct bonding by the "blade test.".

called as "Maszara method" or "blade test." The bonding strength is calculated as

$$\gamma = \frac{3}{8}\frac{Et^3 y^2}{L^4}$$

Here, γ is the specific surface energy in $J\,m^{-2}$, i.e. a half of bonding energy for unit area. Theoretical maximum value of γ for Si (100) is about $2.5\,J\,m^{-2}$. E is the modulus of elasticity: 1.66×10^{11} Pa for Si. $2y$ is the thickness of the blade and L is the crack length. This method is now widely used for the bonding strength evaluation in wafer-to-wafer bonding. It requires no special preparation procedure for the specimens and is, therefore, easy to perform. One drawback of the method is evaluation of strong bonding. If the bonding to be evaluated is too strong, the edges of wafers crack instead of opening a bonding interface when the blade is inserted.

The annealing temperature of >1000 °C used to strengthen the wafer direct bonding by chemical treatment is acceptable in SOI wafer fabrication. It is, however, too high for integration of processed wafers such as integrated circuits (ICs) and MEMS. Plasma treatments are often used to achieve the hydrophilic wafer bonding at lower temperatures [6, 7]. As shown in Figure 12.4 (Plasma Treatment), low-temperature annealing up to 200 °C is enough to achieve strong Si—Si bonding, which can be used in various integration processes of IC and MEMS wafers. Here, low-pressure capacitively coupled O_2 plasma in reactive ion etching (RIE) mode was used.

Hydrophilic wafer direct bonding is widely used to bond Si wafers with/without surface SiO_2 layers. It has been applied to MEMS fabrication process such as pressure sensors [8–10], optical devices [11], micro-fluidic devices [12], and force sensors

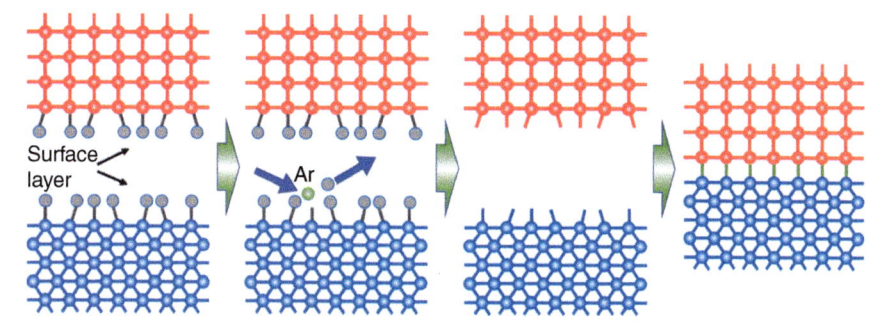

Figure 12.7 Principle of the surface activated bonding.

[13]. Sapphire wafers can be also bonded by the hydrophilic wafer bonding, and it is applied to pressure sensors for harsh environments [14, 15].

12.3 Surface Activated Bonding at Room Temperature

The surface activated bonding (SAB) is originally based on a simple idea that atoms on two clean surfaces can make strong interatomic bonds even at room temperature when they are mated. SAB originally uses cleaning of material surfaces in vacuum by sputter etching using high-energy ion/atom beam of inert gases, typically Ar, as shown in Figure 12.7. The cleaning process removes surface oxides, compound layers, adsorbed molecules, and so on, which stabilize the materials surface. Therefore, after the cleaning process, the surfaces become unstable, in other words active, states. Mating two such activated surfaces in vacuum enables strong bond formation at room temperature.

SAB was first applied to the bonding between soft metals [16]. Deformation of the soft metal by applying a large pressure was necessary to achieve atomic-level intimate contact at room temperature in case of bonding between relatively rough surfaces without chemical mechanical polishing (CMP). On the other hand, in wafer direct bonding process, intimate contact between two wafer surfaces is achieved spontaneously as shown in Figure 12.2. This property is quite advantageous for intimate contact formation at room temperature. Therefore, SAB is well suited to wafer direct bonding. Si wafers were successfully bonded by SAB without applying any heat treatments and mechanical loads [17, 18]. Strong bonding equivalent to bulk strength is achieved by the room-temperature process as shown in Figure 12.8.

In SAB, influence of surface sputter etching is an important issue. High-energy ion/atom beams used for sputter etching introduce crystal defects near the surface. In case of soft metals, the damage can be recovered even at room temperature, and a directly bonded interface between two crystals is formed as shown in Figure 12.9a [16]. On the other hand, Ar beam irradiation creates a damaged layer on Si surface, and an amorphous layer remains at the bonding interface as shown in Figure 12.9b [19]. EDS analysis revealed that the layer contains Si and implanted Ar. Surface sputter etching also influences to the surface roughness of the wafers. It has been

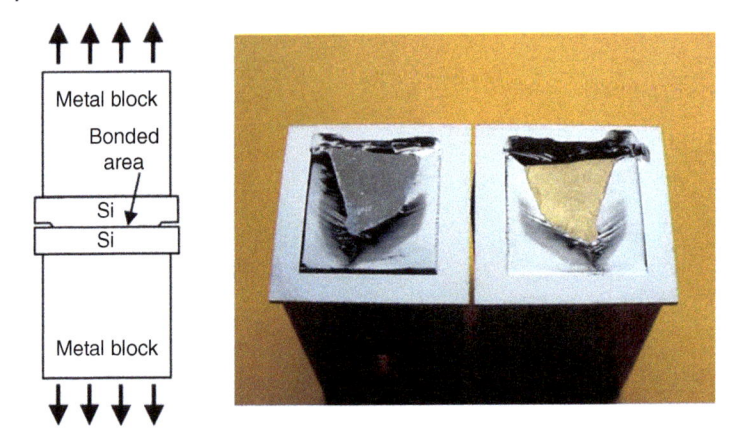

Figure 12.8 Fracture surface of Si/Si bonding prepared at room temperature by SAB.

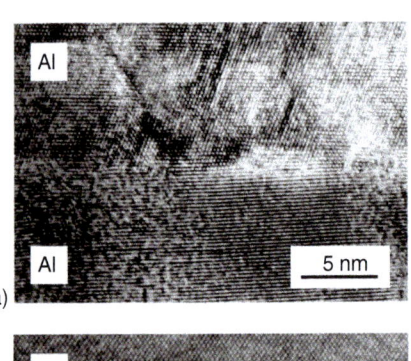

Figure 12.9 HRTEM images of (a) Al/Al and (b) Si/Si bonding interface prepared by SAB at room temperature.

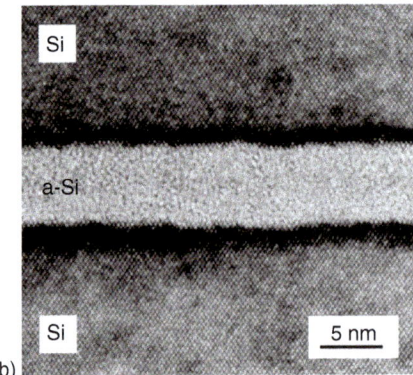

reported that Ar beam irradiation roughens the Si surface [20], whereas Ne beam irradiation smoothens the surface [21].

SAB has been successfully applied to various semiconductor wafers such as GaAs, InP, SiC, etc. [22, 23]. Atomically smooth surfaces of metals are also spontaneously bonded by SAB [24, 25]. In SAB, bonding is assumed to be achieved by interatomic bonds, which are the same as those in bulk material. For example, Si atoms bind to

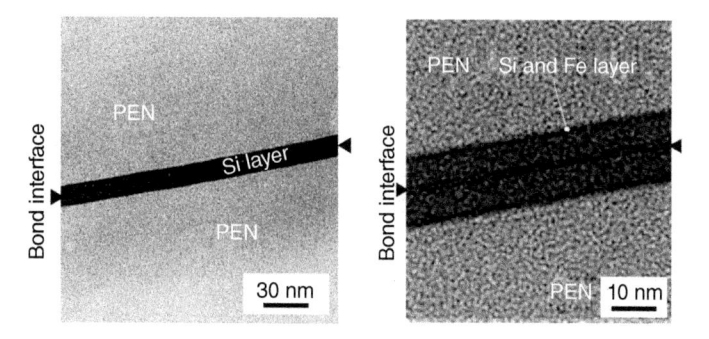

Figure 12.10 TEM images of bonding interface of PEN sheets prepared by SAB with Si intermediate layers.

each other by covalent bonds, and metal atoms bind by metallic bonds based on sharing of free electrons. Therefore, in SAB, bonding properties are largely influenced by interatomic bonds in materials to be bonded. Ionic compounds are also successfully bonded to metals and semiconductors. For example, Si_3N_4 ceramics strongly bonds to Al [16], and $LiNbO_3$, $LiTaO_3$, Al_2O_3, and $Gd_3Ga_5O_{12}$ single crystal wafers were successfully bonded to Si wafers [26, 27]. On the other hand, bonding between oxide materials is relatively weak, and in some cases annealing process is effective to improve the bonding strength [27].

The concept of SAB has been expanded in order to bond wide range of materials. Materials such as SiO_2 and polymers cannot be directly bonded by SAB. To bond such materials, deposition of very thin intermediate layers of metals or Si has been proposed [28–30]. The deposited atoms firmly adhere to the surface of SiO_2 and/or polymers and simultaneously form active surface. The thin film deposition can be regarded as a new process for surface activation. Figure 12.10 shows TEM images of bonding interface between PEN (poly-ethylene2,6-naphthalate) sheets using Si intermediate layers.

In the bonding processes using metal film deposition, Au film has a special advantage. Wafers with Au films are successfully bonded at room temperature in atmospheric air [25]. Because Au is not oxidized, it is assumed that Au surface remains in an active state even in ambient air. Room-temperature bonding in air is quite advantageous in various industrial applications. In addition, it is assumed that Au films can be bonded in various atmospheres such as low vacuum, pure gases, etc.

Various applications of the SAB have been developed in the field of wafer-level packaging and engineered substrates. In wafer-level packaging field, various MEMS devices have been already commercialized using SAB, and it is also applied to 3D integration [31]. In engineered substrate application, SAW filters for wireless communications using bonded substrates have been already commercialized [32], and still many efforts are paid to improve them [33]. One advantage of SAB against the hydrophilic bonding is that the bonding interface can be conductive. Engineered substrates for power electronics [23], multi-junction solar cells [34], etc. are being developed by the method. Special vacuum bonding apparatuses for SAB are now available from several suppliers [34–36].

References

1 Lasky, J.B. (1986). Wafer bonding for silicon-on-insulator technologies. *Appl. Phys. Lett.* 48 (1): 78–80.
2 Maszara, W.P., Goetz, G., Caviglia, A., and McKitterick, J.B. (1988). Bonding of silicon wafers for silicon-on-insulator. *J. Appl. Phys.* 64 (10): 4943–4950.
3 Bruel, M., Asper, B., and Auverton-Herve, A.-J. (1997). Smart-cut: a new silicon on insulator material technology based on hydrogen implantation and wafer bonding. *Jpn. J. Appl. Phys.* 36 (3B): 1636–1641.
4 Tong, Q.Y. and Goesele, U.M. (1999). Wafer bonding and layer splitting for microsystems. *Adv. Mater.* 11 (17): 1409–1425.
5 Haisma, J., Spierings, B.A.C.M., Biermann, U.K.P., and Gorkum, A.A.v. (1994). Diversity and feasibility of direct bonding—a survey of a dedicated optical technology. *Appl. Opt.* 33 (7): 1154–1169.
6 Suni, T., Henttinen, K., Suni, I., and Mäkinen, J. (2002). Effects of plasma activation on hydrophilic bonding of Si and SiO_2. *J. Electrochem. Soc.* 149 (6): G348–G351.
7 Pasquariello, D. and Hjort, K. (2002). Plasma-assisted InP-to-Si low temperature wafer bonding. *IEEE J. Sel. Top. Quant. Electron.* 8 (1): 118–131.
8 Christel, L., Petersen, K., Barth, P. et al. (1990). Single-crystal silicon pressure sensors with 500 × overpressure protection. *Sens. Actuators A* 21-23: 84–88.
9 Welham, C.W., Greenwood, J., and Bertioli, M.M. (1999). A high accuracy resonant pressure sensor by fusion bonding and trench etching. *Sens. Actuators A* 76: 298–304.
10 Pedersen, T., Fragiacomo, G., Hansen, O., and Thomsen, E.V. (2009). Highly sensitive micromachined capacitive pressure sensor with reduced hysteresis and low parasitic capacitance. *Sens. Actuators A* 157: 35–41.
11 Kwa, T.A. and Wolffenbuttel, R.F. (1992). Integrated grating/detector array fabricated in silicon using micromachining techniques. *Sens. Actuators A* 31: 259–266.
12 Enoksson, P., Stemme, G., and Stemme, E. (1996). Silicon tube structures for a fluid-density sensor. *Sens. Actuators A* 54: 558–562.
13 Brookhuis, R.A., Lammerink, T.S.J., Wiegerink, R.J. et al. (2012). 3D force sensors for biomedical applications. *Sens. Actuators A* 182: 28–33.
14 Ishihara, T., Sekine, M., Ishikura, Y. et al. (2005). Sapphire-based capacitance diaphragm gauge for high temperature applications. Digest of Technical Papers, The 13th International Conference on Solid State Sensors, Actuators and Microsystems (Transducers'05), Seoul, Korea (5–9 June 2005), 503–506.
15 Li, W., Liang, T., Chen, Y. et al. (2017). Interface characteristics of sapphire direct bonding for high-temperature applications. *Sensors* 17: 2080.
16 Suga, T., Takahashi, Y., Takagi, H. et al. (1992). Structure of Al-Al and $Al-Si_3N_4$ interfaces bonded at room temperature by means of the surface activation method. *Acta Metall. Mater.* 40: S1133–S1137.
17 Takagi, H., Kikuchi, K., Maeda, R. et al. (1996). Surface activated bonding of silicon wafers at room temperature. *Appl. Phys. Lett.* 68 (16): 2222–2224.

18 Takagi, H., Maeda, R., and Suga, T. (2003). Wafer-scale spontaneous bonding of silicon wafers by argon-beam surface activation at room-temperature. *Sens. Actuators A* 105: 98–102.

19 Takagi, H., Maeda, R., Hosoda, N., and Suga, T. (1999). Transmission electron microscope observations of Si/Si interface bonded at room temperature by Ar beam surface activation. *Jpn. J. Appl. Phys.* 38 (3A): 1589–1594.

20 Takagi, H., Maeda, R., Chung, T.R. et al. (1998). Effect of surface roughness on room-temperature wafer bonding by Ar beam surface activation. *Jpn. J. Appl. Phys.* 37 (7): 4197–4203.

21 Kurashima, Y., Maeda, A., and Takagi, H. (2013). Room temperature wafer direct bonding of smooth Si surfaces recovered by Ne beam surface treatments. *Appl. Phys. Lett.* 102: 251605.

22 Chung, T.R., Hosoda, N., Suga, T., and Takagi, H. (1998). 1.3 μm InGaAsP/InP lasers on GaAs substrate fabricated by the surface activated wafer bonding method at room temperature. *Appl. Phys. Lett.* 72 (13): 1565–1566.

23 Suda, J., Okuda, T., Uchida, H. et al. (2013). Characterization of 4H-SiC homoepitaxial layers grown on 100-mm-diameter 4H-SiC/poly-SiC bonded substrates. Technical Digest International Conference on Silicon Carbide Related Materials 2013, Miyazaki, Japan (29 September – 4 October 2013), 358.

24 Yakushiji, K., Takagi, H., Watanabe, N. et al. (2017). Three-dimensional integration of magnetic tunnel junctions for magnetoresistive random access memory application. *Appl. Phys. Express* 10: 063002.

25 Higurashi, E., Okumura, K., Kunimune, Y. et al. (2017). Room-temperature bonding of wafers with smooth Au thin films in ambient air using a surface-activated bonding method. *IEICE Trans. Electron.* E100 (2): 156–160.

26 Takagi, H., Maeda, R., Hosoda, N., and Suga, T. (1999). Room-temperature bonding of lithium niobite and silicon wafers by argon-beam surface activation. *Appl. Phys. Lett.* 74 (16): 2387–2389.

27 Takagi, H. and Maeda, R. (2006). Direct bonding of two crystal substrates at room temperature by Ar-beam surface activation. *J. Cryst. Growth* 292: 429–432.

28 Shimatsu, T. and Uomoto, M. (2010). Atomic diffusion bonding of wafers with thin nanocrystalline metal films. *J. Vac. Sci. Technol.* B28 (4): 706–714.

29 Kondou, R. and Suga, T. (2011). Si nonoadhesion layer for enhanced SiO_2-SiN wafer bonding. *Scr. Mater.* 65: 320–322.

30 Matsumae, T., Fujino, M., and Suga, T. (2015). Room-temperature bonding method for polymer substrate of flexible electronics by surface activation using nano-adhesion layers. *Jpn. J. Appl. Phys.* 54: 101602.

31 Shigetou, A., Ito, T., Sawada, K., and Suga, T. (2008). Bumpless interconnect of 6-μm pitch Cu electrodes at room temperature. Proceedings 58th Electronic Components Technology Conference 2008, Orlando, Florida USA (27–30 May 2008), 1405–1409.

32 Miura, M., Matsuda, T., Ueda, M. et al. (2005). Temperature compensated $LiTaO_3$/sapphire SAW substrate for high power applications. Proceedings IEEE International Ultrasonic Symposium 2005, Rotterdam, Netherlands (18–21 September 2005), 573–576.

33 Takai, T., Iwamoto, H., Takamine, Y. et al. (2019). High-performance SAW resonator with simplified LiTaO$_3$/SiO$_2$ double layer structure on Si substrate. Proceedings IEEE International Ultrasonic Symposium 2019, Glasgow, Scotland UK (6–9 October 2019), 1006–1013.

34 Dimroth, F., Tibbits, T.N.D., Niemeyer, M. et al. (2016). Four-junction wafer-bonded concentrator solar cells. *IEEE J. Photovoltaics* 6 (1): 343–349.

35 Utsumi, J., Ide, K., and Ichiyanagi, Y. (2016). Room temperature bonding of SiO$_2$ and SiO$_2$ by surface activated bonding method using Si ultrathin films. *Jpn. J. Appl. Phys.* 55: 026503.

36 Jung, A., Zhang, Y., Arroyo Rojas Dasilva, Y. et al. (2018). Electrical properties of Si-Si interfaces obtained by room temperature covalent wafer bonding. *J. Appl. Phys.* 123: 085701.

13

Metal Bonding

Joerg Froemel

Tohoku University, Advanced Institute for Materials Research, Aoba-ku, Katahira 2-1-1, Sendai 980-8577, Japan

Aluminum, copper, and gold are commonly used metallization materials for microelectronic and MEMS applications. Using those materials for wafer bonding technologies enable beside the mechanically stable encapsulation of sensitive sensors also an electrical interconnection between the sensor and integrated electronics within systems and devices. In this chapter, an overview of metal based wafer bonding as listed in Table 13.1 will be given.

Typically metal wafer bonding technologies are diffusion-based methods. They can be categorized into liquid and solid state technologies. A typical example of solid state bonding is thermos–compression bonding, whereas for liquid state it is eutectic bonding. Solid liquid interdiffusion bonding (SLID) bonding is partly liquid and partly solid state diffusion-based (Figure 13.1).

In general, metal-based wafer bonding uses relatively low temperatures (<500 °C). The low temperature processes are interesting for the use of heterogeneous materials in MEMS, the thermal budget limitations of integrated micro electronics in 3D-Integration and high precision sensors. Stress can still be an issue in devices using metal bonding frames. Metals have a relatively large thermal expansion coefficient, compared to semiconductor materials. Because of the difference between bonding temperature and application temperature thermo-mechanical stress is created. Therefore, metal bonding processes with lower temperature are preferable.

Wafer bonding using metal layers is also a method suitable for applications that require good electric and/or thermal conductance at the same time, such as 3D integration and TSV stacking. As a metal seal is very stable and in principle does not allow significant gas penetration, metal bonds can protect micro devices and structures from environmental particles and humidity. Therefore, they are good candidates for high reliability systems, as well as devices that need to operate under high vacuum conditions. Compared to other wafer bonding technologies, e.g. silicon direct bonding and anodic bonding, metal interlayer-based wafer bonding technologies exhibit more tolerance of particles. They can be pressed into the metal layer, or

Table 13.1 List and references of major metal wafer bond technologies.

Wafer bonding technology	Representative publications
Solid liquid Interdiffusion (SLID) bonding	[1] (L. Bernstein)
Eutectic bonding	[2] Au/Si (R. Wolffenbuettel)
	[3] Al/Ge (P. Zavracky)
	[4] Au/Sn (G. Matijasevic)
Metal thermocompression bonding	[5] Au/Au (C. Tsau)
	[6] Cu/Cu (A. Fan)
	[6] Al/Al (J. Martin)
Reactive Bonding	[7] (J. Braeuer)

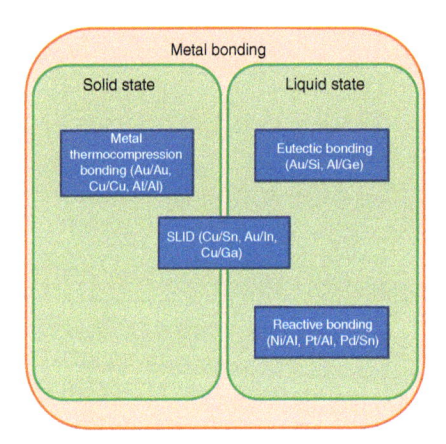

Figure 13.1 Overview of metal based wafer bonding methods.

embedded into a liquid phase. The major bond parameters, to be considered for high bond quality are bond time, temperature, force, and heating and cooling rates.

13.1 Solid Liquid Interdiffusion Bonding (SLID)

SLID is a possible way to bond at a low temperature with a resulting interface that is stable at temperatures higher than the bonding temperature. The resulting bonds can withstand higher temperatures than the process temperatures; it is an important advantage. At least two different metals are alloyed. Usually the two metals have widely different melting points, with one being quite low and defining the bonding temperature. The bonding process takes place at a temperature slightly higher that the lower melting point. At this temperature one metal is liquid and the liquid phase wets the surface of the metal that remains solid. Through the interface increased diffusion takes place from the solid phase into the liquid phase and a rapid increase of the high melting point metal inside the liquid phase occurs. After reaching saturation of the solubility, intermetallic phases with a higher melting points are formed.

When this melting points are higher than the bonding temperature solidification of this phase happens. This goes on until all of the liquid phase has been transformed into solid intermetallic phases. After that diffusion continues in the solid state until the equilibrium dictated by the phase diagram is reached. The second metal remains solid during the entire process (hence the name of the process). Sometimes the bonding process is stopped before the equilibrium has been reached and several phases remain in the interface. Table 13.2 shows the sequence of the SLID bonding process.

The required thickness t_{M1} and t_{M2} to form the intermetallic phase of $M1_xM2_y$ of the two metals M1, M2 can be calculated by:

$$\frac{t_{M1}}{t_{M2}} = \frac{\rho_{M1} M_{M1} x}{\rho_{M2} M_{M2} y} \tag{13.1}$$

In this formula, M_{M1} and M_{M2} are the respective molar masses and ρ_{M1} and ρ_{M2} the densities. Usual materials for the low melting point metal are Sn, In, or Ga, because of their very low melting point. Table 13.3 shows a selection of binary SLID material combinations. Bernstein made the first publication and coined the name SLID already 1966 [1]. Because the materials used in microfabrication become more and more heterogeneous, low temperature processes that enable the use of materials with some difference in thermal expansion coefficient are quite attractive. Another advantage is the tolerance of some surface roughness and profile.

Besides semiconductor wafer substrates, other materials can also be used with SLID bonding, e.g., ceramics or glasses. The requirement is that the substrate materials must be able to sustain the process temperatures, and they must be available in a wafer or similar shape to enable deposition and structuring processes.

13.1.1 Au/In and Cu/In

With a melting temperature of 157 °C indium is a good candidate for SLID bonding. The process temperature is reasonably low, but the material is solid at usual environment temperature. Copper [13, 14] and gold [15] are used with indium for bonding. The availability of electrolytes for the electrochemical deposition of indium is quite good. Therefore, the layers for Au/In and Cu/In SLID bonding are usually made by using electro-chemical deposition (ECD). Indium easily grows a natural oxide, that would prevent bonding. Strategies to offset this problem include the deposition of an additional thin layer of gold on top of indium, and indium oxide removal by chemical etching just before bonding. In and Au or Cu are in most cases deposited by a pattern-plating process. Figure 13.2 shows a typical example of electroplated indium on top of gold.

In many cases a symmetric layer design is used, where both wafers have a stack of Au/In or Cu/In respectively. Sometimes one side consists of only gold or copper. The bonding is done in a bonding chamber with inert or reducing gas atmosphere to prevent further oxidation of indium. Bonding temperature is usually between 180 and 200 °C for several 10 minutes.

Table 13.2 Sequence and steps of SLID bonding shown as cross section.

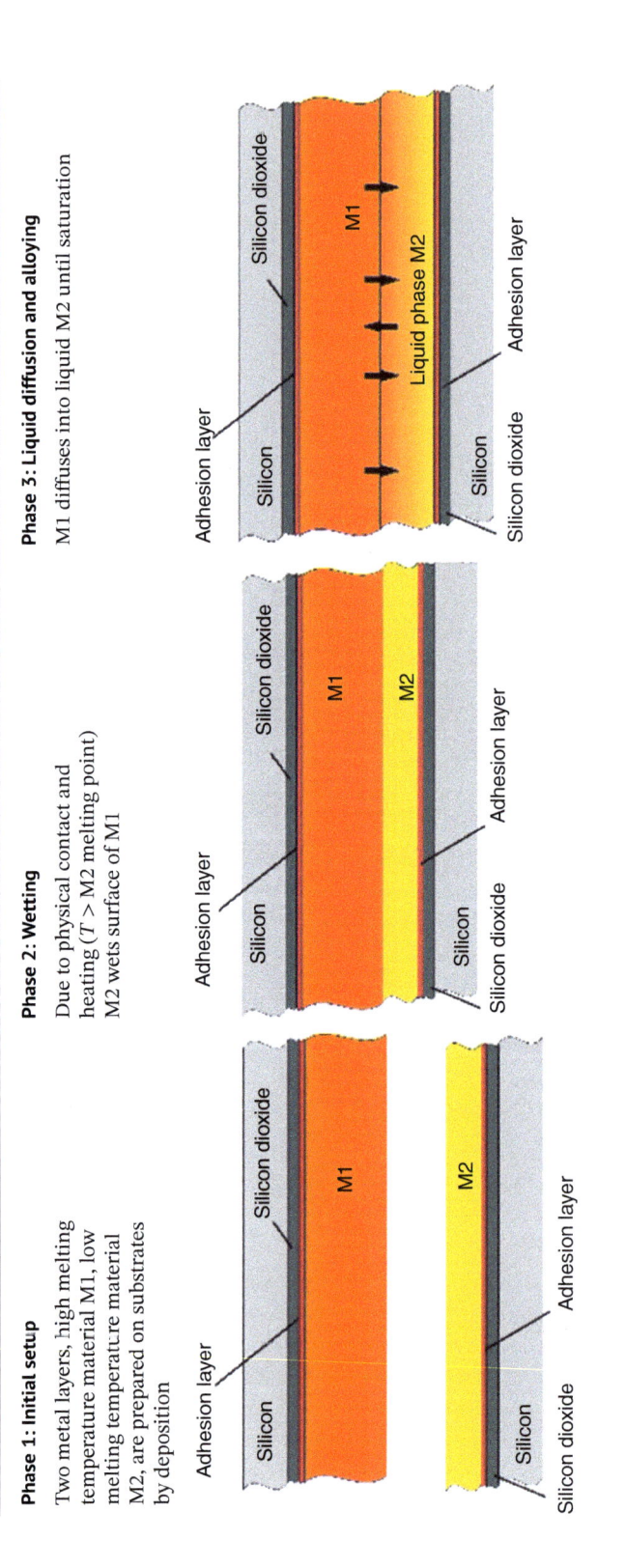

Phase 1: Initial setup

Two metal layers, high melting temperature material M1, low melting temperature material M2, are prepared on substrates by deposition

Phase 2: Wetting

Due to physical contact and heating ($T >$ M2 melting point) M2 wets surface of M1

Phase 3: Liquid diffusion and alloying

M1 diffuses into liquid M2 until saturation

Phase 4: Gradual solidification

At the interface between M1 and M2 intermetallic phases are solidifying

Phase 5: Solid-diffusion

After all M2 has been used up, the equilibrium intermetallic phase is expanding and consuming M1 and other phases

Phase 6: Equilibrium

Equilibrium is reached by complete formation of stable phase

Table 13.3 Selection of usual SLID bonding material combinations.

Material system	Low melting point (°C)	High melting point (°C)	Target intermetallic phases	Representative publication
Ag–In	157	962	Ag_2In, Ag_3In	[1] (L. Bernstein)
Au–In	157	1064	Au_7In	[1] (L. Bernstein)
Cu–In	157	1085	Cu_7In_3 (δ)	[1] (L. Bernstein)
Ag–Sn	232	962	Ag_3Sn, ζ	[8] (Y.-C. Chen)
Au–Sn	232	1064	Au_5Sn	[9] (C. C. Lee)
Cu–Sn	232	1085	Cu_3Sn	[10] (F. Bartels)
Au–Ga	30	1064	Au_7Ga_2	[11] (J. Froemel)
Cu–Ga	30	1085	Cu_9Ga_4	[12] (J. Froemel)

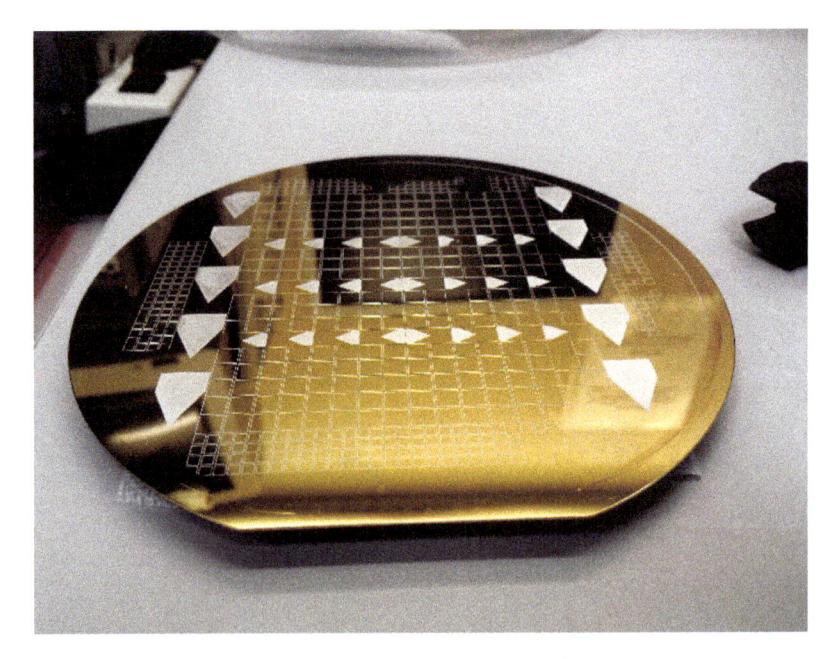

Figure 13.2 Electroplated indium on gold seed layer [12].

13.1.2 Au/Ga and Cu/Ga

Gallium has a very low melting point. It can be liquid at usual environmental temperatures. For practical purposes, it is important that all gallium is used up during the bonding process, given that liquid gallium in the interface would result in a very low bonding strength. MacKay [16] researched Ga/Ni, Ga/Ni/Cu, Ga/Cu, Ga/Ag/Cu amalgams. He identified as potential applications die attachment, flip chip bonding, heat sink attachment and via filling, among others. Intel has filed a patent to use such amalgams as a bonding material to hermetically seal MEMS devices [17]. They also apply the paste by screen printing in a very comparable way like glass

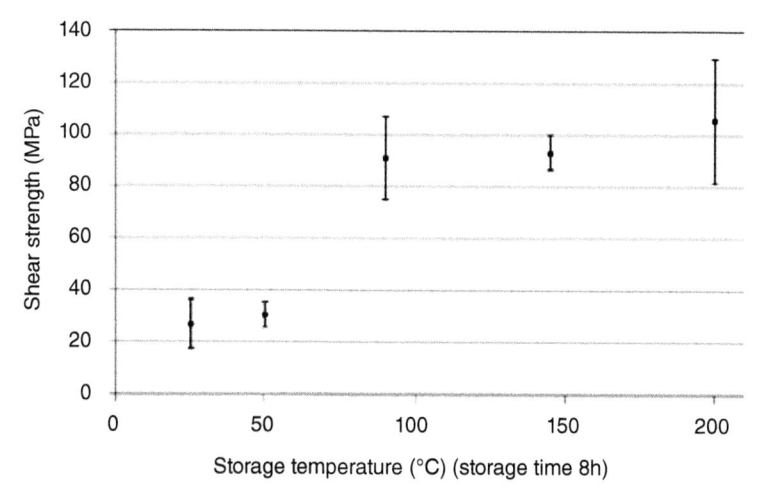

Figure 13.3 Shear strength of bonded gallium–copper samples annealed at different temperatures. Source: Froemel et al. [12]. © 2015, IEEE.

frit material is used for bonding in the field of MEMS packaging. A big problem is the very short shelf life of such paste material. Froemel et al. [11, 12] succeeded in gallium SLID bonding by using fully electroplated layers. In regards to Au/Ga samples with a thickness of 1.5 and 0.5 µm for gold and gallium, respectively have been used with one wafer having only gold and the other wafer having gallium on a seed layer of gold. A $GaCl_3$-based self-developed electroplating process on gold seed layer including structuring by pattern plating was used. Chromium with 20 nm thickness has been applied as an adhesion layer. For Cu/Ga 1.35 µm copper and 1 µm gallium was used. Although bonding can be done at 30 °C, subsequent annealing at higher temperatures will increase bonding strength and reduce electric contact resistance (Figures 13.3 and 13.4).

This characteristic is related to the formation of intermetallic phases in the interface (Figures 13.5–13.7).

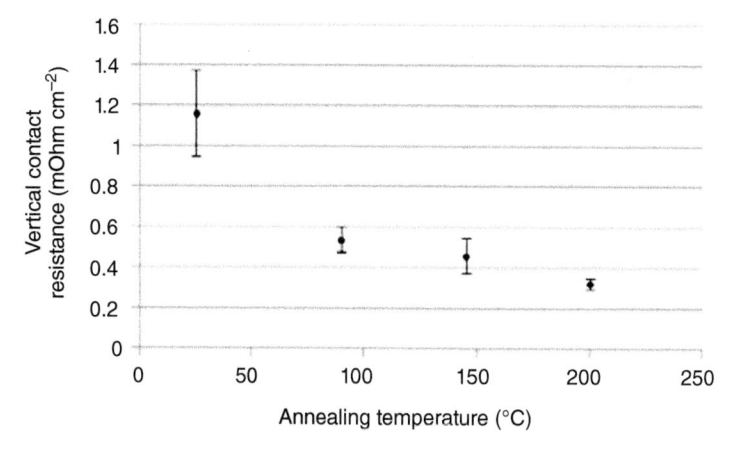

Figure 13.4 Electrical resistance of vertical contacts of bonded gallium–copper samples annealed at different temperatures. Source: Froemel et al. [12]. © 2015, IEEE.

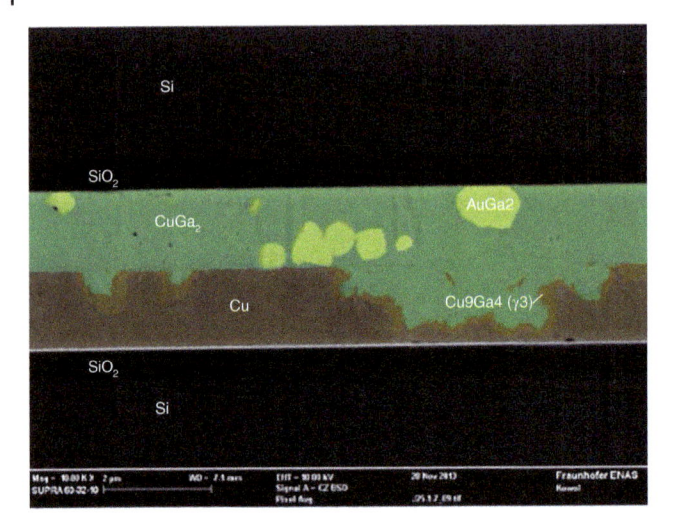

Figure 13.5 SEM cross section of gallium copper-bonded samples after annealing at 50 °C [12].

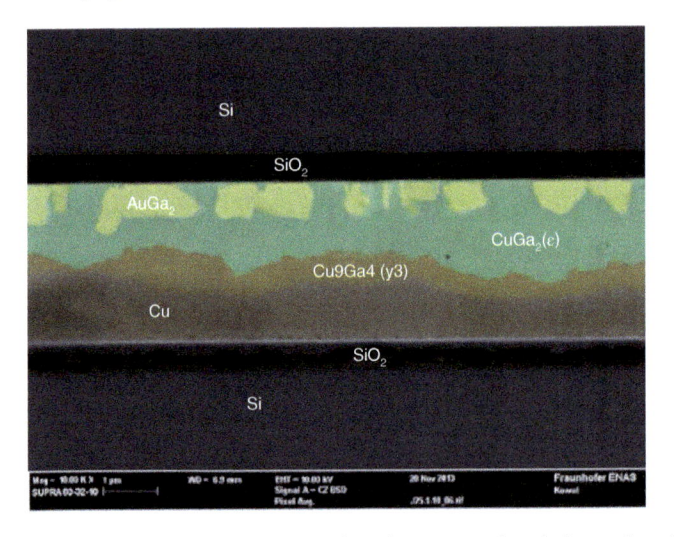

Figure 13.6 SEM cross section of gallium copper-bonded samples after annealing at 90 °C [12].

From 50 to 90 °C shear strength increases strongly and vertical contact resistance significantly becomes smaller. The structural difference between the interfaces annealed at 50 and 90 °C is the existence of an uninterrupted layer of Cu_9Ga_4 phase in between Cu intermetallic phase and $CuGa_2$ phase after annealing at 90 °C. The existence of this Cu_9Ga_4 layer is very important to achieve good bonding parameters. The visible $AuGa_2$ clusters are remnants of reaction between gallium and its seed layer of gold.

Gallium forms a natural oxide. But this oxide does not hinder the bonding, because gallium has a low viscosity during the process and any oxide is disrupted by mechanical force. On the other hand, it is necessary to take care of the natural oxide of copper. Removing the copper oxide with citric acid just before bonding

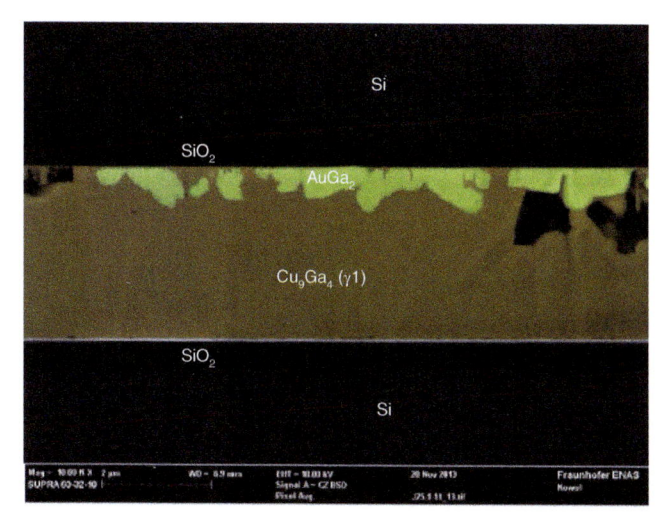

Figure 13.7 SEM cross section of gallium copper–bonded samples after annealing at 200 °C [12].

and covering the copper surface with a thin $CuGa_2$ protection layer can be used. Covering the copper with the $CuGa_2$ protection layer creates better results in regards to yield and mechanical strength.

13.1.3 Au/Sn and Cu/Sn

At first Cu/Sn SLID bonding was used in the field of 3D integration by using solder balls. It has the advantage of forming a stable intermetallic phase at a reasonable low temperature (<300 °C) [18]. Beside the application in stacking also recently it is used also as bonding material for wafer level packaging of MEMS [19]. In order to obtain the desired Cu_3Sn phase, the copper film thickness should be 1.3 times the thickness of the tin layer. The bonding process is typically at temperatures between 250 and 300 °C for several 10 minutes. In most cases tin is deposited onto copper layers on both substrates and the initial interface is formed by liquid tin. In this way the problem of removing the copper oxide can be avoided. Natural tin oxide is generally thinner than natural copper oxide and can be penetrated during the bond process due to the liquid interface. In case of bumping applications flux is used to deal with oxides, in MEMS applications this is not acceptable in most of the cases.

The Au/Sn binary phase diagram is more complicated than Cu/Sn. It also contains eutectic points and more intermetallic phases. Beside bumping, it is also used for MEMS applications [20, 21]. Although the bonding temperature is similar to Cu/Sn SLID bonding, the bonding time can be less than 10 minutes. This is because of the much higher interdiffusion coefficient. As diffusion barrier/adhesion layer Ni-based layers are often used. For both tin based methods shear strength of higher than 80 MPa can be achieved. [22]

13.1.4 Void Formation

Often observed at SLID bonds is void formation. Voids have a large potential influence on the bond quality, as they can reduce strength (because cracks can

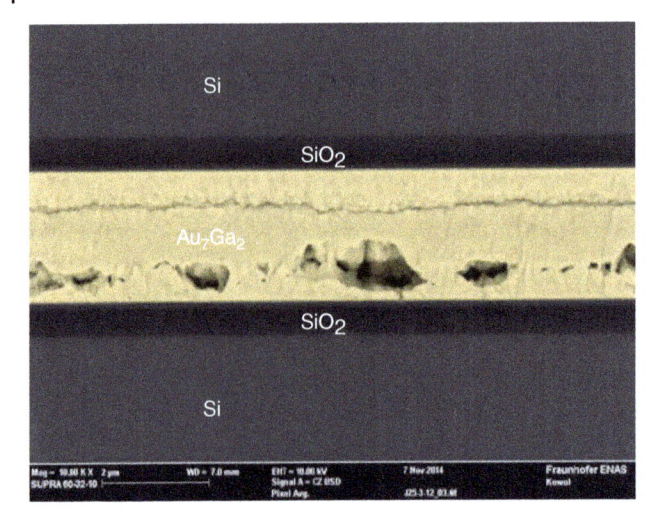

Figure 13.8 SEM picture of cross section of Au/Ga bond with clearly visible Kirkendall voids.

propagate easily along them), can reduce the electric conductance (because the effective conductive area is reduced), and can worsen hermetic properties (because the voids open paths for gas transport in the interface). The main reason for void formation is related to the diffusion process. Interdiffusion is described by the interdiffusion coefficient, which is a common property of both diffusion partners. It does not describe the individual behavior of each alloying partner. In the last century Kirkendall made experiments to understand the contribution of the individual elements to the overall diffusion [23]. Contrary to the state of the art at that time (It was thought, that diffusion happens only by exchange of atoms) he found that diffusion in solid state is mainly governed by vacancy movement and therefore real material transport through the solid happens. At an interface between two materials, there will be diffusion by both materials into each other. In almost all cases those diffusion coefficients will be different. If one of the coefficients is much larger than the other, voids can remain in the volume of the element with higher mobility. Those voids that are created by this effect are called Kirkendall Voids (Figure 13.8). Strategies to suppress Kirkendall voids include controlling the annealing temperature and alloying with additional elements [24, 25].

13.2 Metal Thermocompression Bonding

Metal thermocompression bonding is a form of solid state bonding, more specifically diffusion bonding. Available material combinations include Cu/Cu [5], Au/Au [26], Al/Al [27], Ag/Ag [28], and Ti/Ti [29]. Metal thermocompression bonding is important, as typically used metallization materials can be used without a liquid phase. The bonding happens in three phases: (i) interface formation, (ii) grain reorientation, and (iii) grain growth.

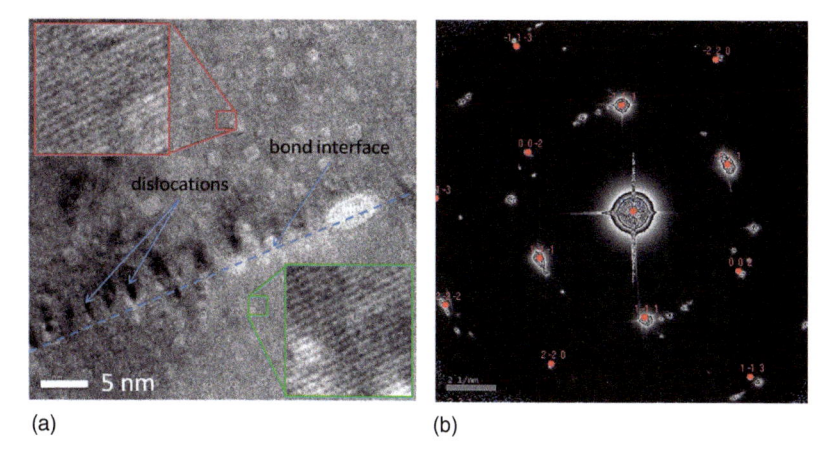

Figure 13.9 (a) TEM observation of a bond interface of gold–gold thermocompression bond with parallel aligned crystal planes [30]. (b) X-ray diffraction measurement of the interface revealing the plane as (111).

13.2.1.1 Interface Formation

A typical surface of a polycrystalline metal has a certain surface roughness. This roughness depends on the layer thickness, deposition parameters and grain size, among others. To establish a successful bond, both surfaces have to be plastically deformed until they are physically in contact as much as possible.

13.2.1.2 Grain Reorientation

In most cases a sputtered or electroplated metal has a polycrystalline matrix. It means the crystal orientation of each grain is randomly oriented. When the grain surfaces of the two metal layers come into contact a reorientation of the crystal direction happens to accommodate to the grain's interface. For metals with cubic face centered crystal, such as Au, Cu, Al, and Ag, structure the (111) layer aligns parallel to the bond interface. This implies that the closest area packing at the contacted interface seems to be a favorable condition for bonding. Figure 13.9 shows the this alignment at the example of an Au/Au interface.

This kind of reorientation of the grains crystal structure requires energy, for example in form of heat. It is known that recrystallization of metals can only happen above a certain temperature. This leads to the conclusion that a strong bonding, that not only relies on plastic deformation of the soft material but also on atomic bonding, can only be achieved at high enough temperatures. The required energy depends on the grain size. A larger grain needs more energy to change its orientation. Reducing the grain size is a way to significantly lower the required temperature for a successful bond. This can be achieved by mechanical force (cutting of the surface) [31], reducing the thickness of the metal layer to nanometer scale by sputtering [32], by using metal nanoparticles [33], or plasma pretreatment [34].

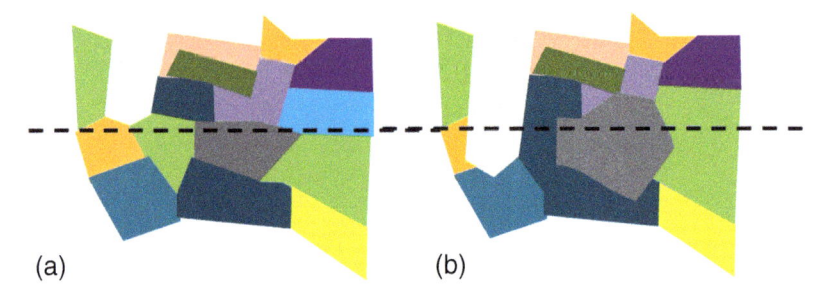

Figure 13.10 Schematic of bond interface (a) before annealing, (b) after annealing. The grains are growing trough the interface.

Table 13.4 Typical bond parameters for interface materials.

	Au/Au	Cu/Cu	Al/Al	Ag/Ag	Ti/Ti
Bond pressure on wafer (MPa)	~10	~10	~30	~60	3
Bond temperature (°C)	<300	400	400	300	400

Note: Those parameters can be significantly modified by the pretreatment methods described in the text.

13.2.1.3 Grain Growth

The final step of thermocompression bonding is growth of the grains in the interface. By further applying temperature diffusion leads to the growth of grains into the opposing metal layer (Figure 13.10). Also small voids in the interface can be closed in the process. If successful, the bond interface vanishes completely and the resulting bond strength is governed by the interface material intrinsic strength.

Typical bond parameters for various materials are shown in Table 13.4.

Beside flatness, homogeneity, and cleanliness also the existence of a natural oxide layer plays an important role in thermo-compression bonding. In case of Au/Au bonding this is usual of no concern, as gold does not form a stable natural oxide. But for other materials it must be addressed. For Cu/Cu bonding a number of established pre-treatments exist. One possibility is wet chemical etching of the oxide before the bonding process by HCl [35], or acetic acid [36]. Because native oxide may form again between the chemical pretreatment and the bonding process, also combination processes with additional treatment in formic acid vapour, forming gas, or Ar beam directly in the bonding chamber have been developed [37–39]. Furthermore methods to reduce the copper oxide with atomic hydrogen exists [40], as well as strategies to temporarily employ thiol-based self-assembled monolayer (SAM) to prevent reoxidation [41]. In the case of Al/Al bonding, as the natural aluminum oxide is chemically quite strong, usually high bonding force is used to break the surface oxide by deformation and enable diffusion. Another method is the use of a very thin tin layer on top of the aluminum to prevent oxidation [42]. The tin becomes liquid during the bonding and is finally diffused into the aluminum.

Figure 13.11 Example binary phase diagram of two elements E1 and E2, showing a eutectic point at 40at% of E2.

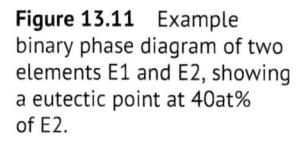

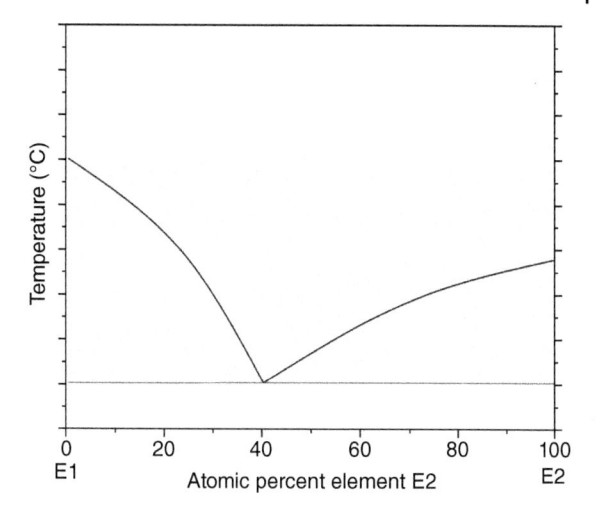

Table 13.5 Example for eutectic material combinations used in microfabrication.

Eutectic alloy	Eutectic temperature (°C)	Composition at the eutectic point (wt%)
Cu/Sn	231	5/95
Au/Sn	280	80/20
Al/Ge	419	49/51
Au/Ge	361	28/72
Au/Si	363	97/3

13.3 Eutectic Bonding

A eutectic is a mixture of two materials that solidifies or melts at a temperature that is lower than the melting temperature of each material. The eutectic point is the lowest melting temperature over the whole range of the two materials. At the eutectic point the liquid and two solid solutions coexists. Figure 13.11 shows a typical example for a binary phase diagram of two elements with an eutectic point.

After a eutectic reaction and solidification, the two materials exist in an intermixed macro or micro structure. This effect can be used to firmly bond these two materials. The bond strength is very high because of the intermixing. Table 13.5 shows some often used material combinations for wafer-level bonding. In most cases the bonding temperature is 20–40 °C higher than the actual eutectic temperature. Other eutectic material combinations are known (e.g. Pd/Au, Au/Al), but they are not used because of too high process temperatures. Eutectic wafer bonding is a good candidate for hermetic sealing applications. The liquid formed during the process can seal also slightly rough surfaces.

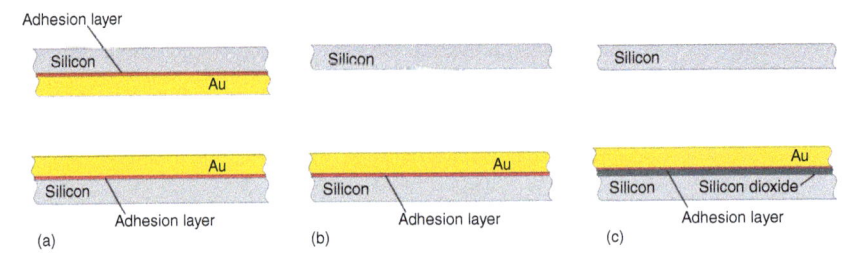

Figure 13.12 Typical set-ups for Au/Si eutectic bonds: (a) Gold sputtered on silicon (with an adhesion layer) on both sides, or (b) gold on one side, or (c) gold on silicon dioxide.

13.3.1 Au/Si

In case of Au/Si the eutectic point is at 363 °C with 19 at.% Si and 81 at.% Au [43]. Au/Si bonding is especially suitable for die attach, as well as wafer bonding, because the substrate material is silicon [2, 44]. If the silicon is supplied from the substrate the gold layer is often deposited by sputtering onto the silicon surface with an additional adhesion layer and the other side is just silicon. But also gold on both sides is possible, as well as gold on silicon dioxide, Figure 13.12.

In case an adhesion layer is used in between gold and silicon, the actual bonding temperature is higher than the eutectic temperature. The reason is the natural oxide of silicon below the adhesion layer. This oxide must be removed during the process by silicidation of the adhesion layer (often Ti, Cr, or Ta) first, to enable diffusion between silicon and gold [45]. If one side of the bond partners is just silicon, its surface oxide must be removed before bonding. Even if few nanometer thin, silicon dioxide is a very efficient diffusion barrier. This removal is often done by HF etching. Because of hydrogen termination after the etching, new oxide formation can be delayed, allowing the transfer of the wafer under atmosphere for a short period.

Figure 13.13 shows the interface of an Au/Si bond after the eutectic reaction. The bonding strength is very high. If the bond is destroyed usually the silicon material is breaking [46].

One potential problem is the uncontrolled flow of the eutectic liquid during the bonding process. Figure 13.14 shows a typical result with this problem.

To control this problem, rigorous temperature and mechanical pressure control during the bonding is needed. Other ways are the use of mechanical stoppers or trenches [47].

13.3.2 Al/Ge

The eutectic system aluminum–germanium exhibits an eutectic point at 419 °C, which can be seen from the binary phase diagram [47]. Al/Ge bonding is mainly used for MEMS packaging and 3D integration [48]. Despite needing a higher bond temperature, the Al/Ge system has the big advantage of using CMOS (complementary metal oxide semiconductor) compatible materials. It can be used to bond a MEMS with a germanium layer to an CMOS device with aluminum metallization, for example gyroscopes [49]. As the bonds are conductive, it can be also used

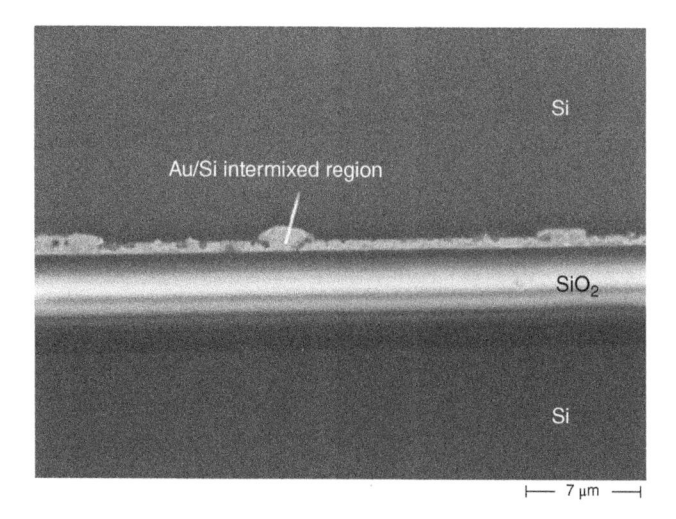

Figure 13.13 SEM image of a typical Au/Si bond interface after eutectic reaction, showing the Au/Si intermixed region.

Figure 13.14 Image of a part of an Au/Si bonded wafer, taken by infrared camera: uncontrolled outflow of the eutectic liquid in between bonding frames.

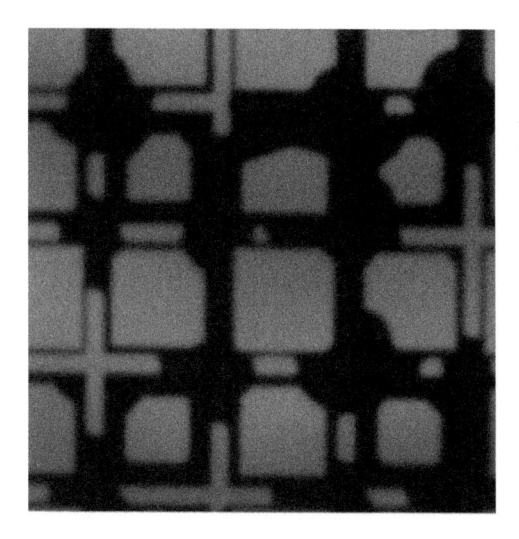

for electric contacts. Al/Ge bonding is also interesting for hybrid integration of (AlGaIn)N-based LEDs with CMOS technology [50]. Aluminum and germanium are deposited by sputtering as relatively thin layers. It is a cost advantage compared to bond technologies that require thicker, electroplated layers. The thinner layer is also better in regards to thermo-mechanical stress. On the other side Al/Ge bonding is less resistant to particles.

Care needs to be taken to make sure that the surfaces of both interface materials are not covered by natural oxide. One way is to remove the oxide by diluted hydrofluoric acid HF (e.g. 1 : 100). A better way to avoid trouble with the aluminum oxide is to use Al/Ge double layers on both sides, so that aluminum is not exposed [51]. In any case the bond process should be done in inert, or better reducing atmosphere, such as 95%N_2 and 5%H_2.

Al/Ge wafer bonding does not require application of high mechanical force. Because of the liquid phase is formed during the process, a high force would result in squeezing out of the metals. The role of the force is just to ensure good contact of the two wafers. A typical bond temperature is 430 °C for 5–10 minutes.

13.3.3 Au/Sn

Beside intermetallic phase formation the Au/Sn binary phase diagram also shows a eutectic point. To use this for bonding, an accurate stoichiometric composition is needed. Therefore, good control of the deposition process is essential. Au/Sn bonding has the disadvantage, that both wafers must be deposited with CMOS incompatible materials. After the two wafers have been brought into contact inside of a bonding chamber, it must be avoided to heat up the layer stack to temperatures higher than 120 °C to avoid premature intermetallic phase formation. Then the wafers should be heated as fast as possible to 300 °C. This temperature should be kept for less than five minutes.

References

1 Bernstein, L. (1966). Semiconductor joining by the solid-liquid-interdiffusion (SLID) process. *J. Electrochem. Soc.* 113 (12): 1282–1288.
2 Wolffenbuttel, R.F. and Wise, K.D. (1994). Low-temperature silicon wafer-to-wafer bonding using gold at eutectic temperature. *Sens. Actuators A. Phys.* 43 (1–3): 223–229.
3 Zavracky, P.M. and Vu, B. (1995). Patterned eutectic bonding with Al/Ge thin films for MEMS. *Micromach. Microfabricat. Process Technol.* 2639: 46–52.
4 Matijasevic, G.S., Lee, C.C., and Wang, C.Y. (1993). AuSn alloy phase diagram and properties related to its use as a bonding medium. *Thin Solid Films* 223 (2): 276–287.
5 Tsau, C.H., Schmidt, M.A., and Spearing, S.M. (2000). Characterization of low temperature, wafer-level gold-gold thermocompression bonds. *Mater. Res. Soc. Sympos. Proc.* 605: 171–176.
6 Fan, A., Rahman, A., and Reif, R. (Oct. 1999). Copper wafer bonding. *Electrochem. Solid-State Lett.* 2 (10): 534–536.
7 Braeuer, J., Besser, J., Wiemer, M., and Gessner, T. (2012). A novel technique for MEMS packaging: reactive bonding with integrated material systems. *Sensors Actuat. A: Phys.* 188: 212–219.
8 Chen, Y.-C., So, W.W., and Lee, C.C. (1997). A fluxless bonding technology using indium-silver multilayer composites. *IEEE Trans. Components, Packag. Manuf. Technol. Part A* 20 (1): 46–51.
9 Lee, C.C. and Wang, C.Y. (Feb. 1992). A low temperature bonding process using deposited gold-tin composites. *Thin Solid Films* 208 (2): 202–209.

10 Bartels, F., Morris, J.W., Dalke, G., and Gust, W. (1994). Intermetallic phase formation in thin solid-liquid diffusion couples. *J. Electron. Mater.* 23 (8): 787–790.

11 Froemel, J., Lin, Y.-C., Wiemer, M. et al. (2012). Low temperature metal inter-diffusion bonding for micro devices. 2012 3rd IEEE International Workshop on Low Temperature Bonding for 3D Integration, 163–163.

12 Froemel, J., Baum, M., Wiemer, M., and Gessner, T. (2015). Low-temperature wafer bonding using solid-liquid inter-diffusion mechanism. *J. Microelectromech. Syst.* 24 (6): 1973–1980.

13 Tian, Y., Wang, N., Li, Y., and Wang, C. (2012). Mechanism of low temperature Cu-In Solid-Liquid Interdiffusion bonding in 3D package. ICEPT-HDP 2012 Proceedings – 2012 13th International Conference on Electronic Packaging Technology and High Density Packaging, 216–218.

14 Panchenko, I., Bickel, S., Meyer, J. et al. (2017). Low temperature Cu/In bonding for 3D integration. Proceedings of 2017 5th International Workshop on Low Temperature Bonding for 3D Integration, LTB-3D 2017, 17.

15 Sohn, Y.C., Wang, Q., Ham, S.J. et al. (2007). Wafer-level low temperature bonding with Au-In system. Proceedings - Electronic Components and Technology Conference, 633–637.

16 MacKay, C.A. (1993). Amalgams for improved electronics interconnection. *IEEE Micro* 13 (2): 46–58.

17 Lu, D. and Heck, J. (2004). Microelectronic package having chamber sealed by material including one or more intermetallic compounds. US7061099B2.

18 Munding, A., Hübner, H., Kaiser, A. et al. (2008). Cu/Sn solid–liquid interdiffusion bonding. In: *Wafer Level 3-D ICs Process Technology. Integrated Circuits and Systems* (eds. C.S. Tan, R.J. Gutmann and L.R. Reif), 1–39. Boston, MA: Springer US.

19 Haubold, M., Baum, M., Schubert, I. et al. (2011). Low temperature wafer bonding technologies. 18th European Microelectronics & Packaging Conference, 1–8.

20 Heck, J.M., Arana, L.R., Read, B., and Dory, T.S. (2005). Ceramic via wafer-level packaging for mems. Proceedings of the ASME/Pacific Rim Technical Conference and Exhibition on Integration and Packaging of MEMS, NEMS, and Electronic Systems: Advances in Electronic Packaging 2005, vol. PART B, 1069–1074.

21 Belov, N., Chou, T.-K., Heck, J. et al. (2009). Thin-layer Au-Sn solder bonding process for wafer-level packaging, electrical interconnections and MEMS applications. Proceedings of the 2009 IEEE International Interconnect Technology Conference, IITC 2009, 128–130.

22 Aasmundtveit, K.E., Tollefsen, T.A., Luu, T.-T. et al. (2013). Solid-Liquid Interdiffusion (SLID) bonding—Intermetallic bonding for high temperature applications. EMPC 2013: European Microelectronics and Packaging Conference, 1–6.

23 Kirkendall, E.O. (1942). Diffusion of zinc in alpha brass. *Trans. AIME* 147: 104–110.

24 Cogan, S., Kwon, S., Klein, J., and Rose, R. (1983). Fabrication of large diameter external-diffusion processed Nb_3Sn composites. *IEEE Trans. Magn.* 19 (3): 1139–1142.

25 Yu, C., Yang, Y., Li, P. et al. (2012). Suppression of Cu_3Sn and Kirkendall voids at Cu/Sn-3.5Ag solder joints by adding a small amount of Ge. *J. Mater. Sci. Mater. Electron.* 23 (1): 56–60.

26 Chen, K.N., Fan, A., and Reif, R. (2002). Interfacial morphologies and possible mechanisms of copper wafer bonding. *J. Mater. Sci.* 37 (16): 3441–3446.

27 Yun, C.H., Martin, J.R., Tarvin, E.B., and Winbigler, J.T. (2008). Al to Al wafer bonding for MEMS encapsulation and 3-D interconnect. Proceedings of the IEEE International Conference on Micro Electro Mechanical Systems (MEMS), 810–813.

28 Liu, C., Hirano, H., Froemel, J., and Tanaka, S. (2017). Wafer-level vacuum sealing using AgAg thermocompression bonding after fly-cut planarization. *Sensors Actuat. A Phys.* 261: 210–218.

29 Takahata, T., Hirano, H., Froemel, J. et al. (2018). Wafer-level high vacuum packaging using titanium thin film as bonding and gettering material. *IEEJ Trans. Sensors Micromach. (in Japanese)* 138 (8): 387–391.

30 Froemel, J., Baum, M., Wiemer, M. et al. (2011). Investigations of thermocompression bonding with thin metal layers. 2011 16th International Solid-State Sensors, Actuators and Microsystems Conference, TRANSDUCERS'11, 990–993.

31 Al Farisi, M.S., Hirano, H., Frömel, J., and Tanaka, S. (2017). Wafer-level hermetic thermo-compression bonding using electroplated gold sealing frame planarized by fly-cutting. *J. Micromech. Microeng.* 27 (1): 015029.

32 Kon, H., Uomoto, M., and Shimatsu, T. (2014). Room temperature bonding of wafers in air using Au-Ag alloy films. Proceedings of 2014 4th IEEE International Workshop on Low Temperature Bonding for 3D Integration, LTB-3D 2014, 28.

33 Ishida, H., Ogashiwa, T., Kanehira, Y. et al. (2012). Low-temperature, surface-compliant wafer bonding using sub-micron gold particles for wafer-level MEMS packaging. Proceedings – Electronic Components and Technology Conference, 1140–1145.

34 Okada, H., Itoh, T., Froemel, J. et al. (2005). Room temperature vacuum sealing using surfaced activated bonding with Au thin films. Digest of Technical Papers - International Conference on Solid State Sensors and Actuators and Microsystems, TRANSDUCERS '05, vol. 1, 932–935.

35 Chen, K.N., Tan, C.S., Fan, A., and Reif, R. (2004). Morphology and bond strength of copper wafer bonding. *Electrochem. Solid-State Lett.* 7 (1): G14–G16.

36 Chen, K.N., Chang, S.M., Shen, L.C., and Reif, R. (2006). Investigations of strength of copper-bonded wafers with several quantitative and qualitative tests. *J. Electron. Mater.* 35 (5): 1082–1086.

37 Baum, M., Hofmann, L., Wiemer, M. et al. (2013). Development and characterisation of 3D integration technologies for MEMS based on copper filled TSV's and copper-to-copper metal thermo compression bonding. 2013 IEEE International Semiconductor Conference Dresden – Grenoble: Technology, Design, Packaging, Simulation and Test, ISCDG 2013.

38 Rebhan, B., Hesser, G., Duchoslav, J. et al. (2012). Low-temperature Cu-Cu wafer bonding. *ECS Trans.* 50 (7): 139–149.

39 Shigetou, A., Itoh, T., and Suga, T. (2005). Direct bonding of CMP-Cu films by surface activated bonding (SAB) method. *J. Mater. Sci.* 40 (12): 3149–3154.

40 Tanaka, K., Hirano, H., Kumano, M. et al. (2018). Bonding-based wafer-level vacuum packaging using atomic hydrogen pre-treated Cu bonding frames. *Micromachines* 9 (4): 181.

41 Tan, C.S., Lim, D.F., Singh, S.G. et al. (2009). Cu-Cu diffusion bonding enhancement at low temperature by surface passivation using self-assembled monolayer of alkane-thiol. *Appl. Phys. Lett.* 95 (19): 192108.

42 Chang, J. and Lin, L. (2010). MEMS packaging technologies & applications. Proceedings of 2010 International Symposium on VLSI Design, Automation and Test, VLSI-DAT 2010, 126–129.

43 Okamoto, H. and Massalski, T.B. (1983). The Au-Si (gold-silicon) system. *Bull. Alloy Phase Diagrams* 4 (2): 190–198.

44 Lani, S., Bosseboeuf, A., Belier, B. et al. (2006). Gold metallizations for eutectic bonding of silicon wafers. *Microsyst. Technol.* 12 (10–11): 1021–1025.

45 Wolffenbuttel, R.F. (1997). Low-temperature intermediate Au-Si wafer bonding; eutectic or silicide bond. *Sensors Actuat. A Phys.* 62 (1–3): 680–686.

46 Lin, Y.C., Baum, M., Haubold, M. et al. (2009). Development and evaluation of AuSi eutectic wafer bonding. TRANSDUCERS 2009 – 15th International Conference on Solid-State Sensors, Actuators and Microsystems, 244–247.

47 Gottfried, K., Wiemer, M., Franke, A. et al. (2012). Contact arrangement for establishing A spaced, electrically conducting connection between microstructured components, EP000002331455B1.

48 Perez-Quintana, I., Ottaviani, G., Tonini, R. et al. (2005). An aluminum-germanium eutectic structure for silicon wafer bonding technology. *Phys. Status Solidi* 2 (10): 3706–3709.

49 Nasiri, S. and Flannery, A. (2008). Method of fabrication of a Al/Ge bonding in a wafer packaging environment and a product produced therefrom. US7442570B2.

50 Goßler, C., Kunzer, M., Baum, M. et al. (2013). Aluminum-germanium wafer bonding of (AlGaIn)N thin-film light-emitting diodes. *Microsyst. Technol.* 19 (5): 655–659.

51 Chidambaram, V., Yeung, H.B., and Shan, G. (2012). Development of CMOS compatible bonding material and process for wafer level mems packaging application under harsh environment. *J. Electron. Mater.* 41: 136–141.

14

Reactive Bonding

Klaus Vogel[1], Silvia Braun[1], Christian Hofmann[1], Mathias Weiser[3], Maik Wiemer[1], Thomas Otto[2], and Harald Kuhn[2]

[1] Department System Packaging, Fraunhofer Institute for Electronic Nano Systems, Technologie-Campus 3, 09126 Chemnitz, Germany
[2] Fraunhofer Institute for Electronic Nano Systems, Technologie-Campus 3, 09126 Chemnitz, Germany
[3] Electrochemistry, Fraunhofer Institute for Ceramic Technologies and Systems, Winterbergstraße 28, 01277 Dresden, Germany

14.1 Motivation

Wafer bonding technologies such as glass frit bonding, thermal compression bonding, and transient liquid phase bonding require bonding temperatures of 200 °C and more to form a stable bond between two substrates. Reducing these temperatures is essential for the 3D integration of heterogeneous materials. Therefore, innovative bonding technologies utilizing local heating of the interface become more relevant. A new process is the bonding with integrated reactive material systems (iRMS). The multilayers generate the energy required for bonding by an exothermic reaction instead of an external heat source. Therefore, time-consuming annealing and cooling processes are not required, minimizing the thermal loading of the components.

14.2 Fundamentals of Reactive Bonding

Reactive material systems (RMS) are a new energetic material for wafer bonding application. They have a well-defined heterogeneous structure. The multilayer systems consist of multiple layers of at least two materials. These alternating thin layers have a total thickness up to 300 µm. Due to the low free energy of the reaction products and the chemical energy stored inside the multilayer stack, energy is released from the system, when reacting. The theoretical maximum energy provided by the system depends on the material combination. It is also affected by chemical reaction, reaction kinetics, thermal transport, and thermal energy dissipation, reducing the overall energy of the system.

In literature, two major reaction types have been documented. The first one is the thermal explosion. Upon ignition, a simultaneous reaction of all educts occurs,

3D and Circuit Integration of MEMS, First Edition. Edited by Masayoshi Esashi.
© 2021 WILEY-VCH GmbH. Published 2021 by WILEY-VCH GmbH.

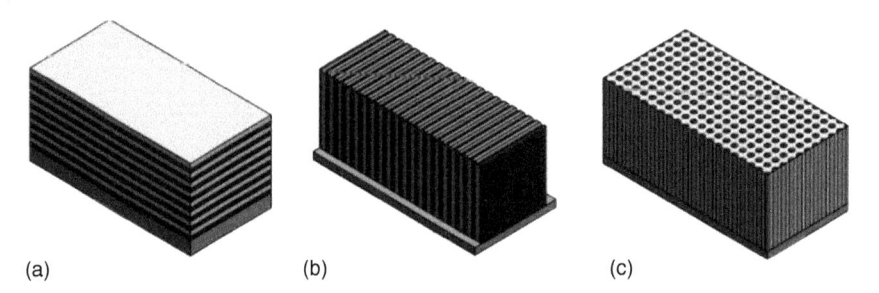

Figure 14.1 Schematic of reactive multilayer types: (a) planar multilayer stack, (b) vertical multilayer stack with 1D periodicity, and (c) vertical pillar systems with 2D periodicity.

requiring a fast annealing above the characteristic ignition temperature T_{IG}. The second and most commonly used reaction type is a self-propagating exothermic reaction (SER). Thereby, an external energy pulse is applied to a point of ignition. The energy released from the system causes neighboring multilayers to react as well, creating a SER within the multilayer stack. The reaction front propagates through the complete multilayer system with a typical speed between 0.01 and 80 m s^{-1}.

Depending on the material systems, the energy can be generated by two mechanisms. The first mechanism is based on the formation of a product composed of two educts. It can be observed during the alloying processes of two or more metals, reactant A and reactant B.

$$x \cdot A + y \cdot B \rightarrow A_x B_y$$

The second mechanism is based on metallo-thermic reactions such as thermite reactions. The energy is provided by a simultaneously occurring reduction of a metal oxide BO_n and the oxidation of the metal component A as second reactant.

$$x \cdot A + y \cdot BO_n \rightarrow yn \cdot B + A_x O_{ny}$$

As shown in Figure 14.1, the educts are deposited in alternating layers, enabling different reactive multilayer types. Horizontal multilayers are the most commonly used systems, stacking various layers of reactants A and B. More complex structures such as vertical multilayer stacks, vertical pillar systems, and particles have also been reported in recent years.

Planar multilayer systems utilizing a SER are most commonly used for microsystems. After starting the reaction at the point of ignition, the reaction releases thermal energy, causing the neighboring multilayer stack to react as well. The reaction therefore propagates through the multilayer stack. As shown in Figure 14.2, three zones can be identified in close proximity of the reaction front: the unreacted multilayer stack, the reaction zone itself, and the reacted products.

In addition to the material, the reactive properties of the multilayer stack strongly depend on the stack geometry. The overall thickness of the multilayer stack directly affects the overall released energy. It increases with thickness due to the availability of more material for the reaction. While the thickness of each single layer depends on the stoichiometry, the thickness of a bilayer can vary. With increasing bilayer thickness, the reaction velocity decreases. The increased average diffusion length for each reactant causes this behavior. The influence of the bilayer thickness on the reaction velocity for the Pd/Al integrated multilayer system is shown in Figure 14.3.

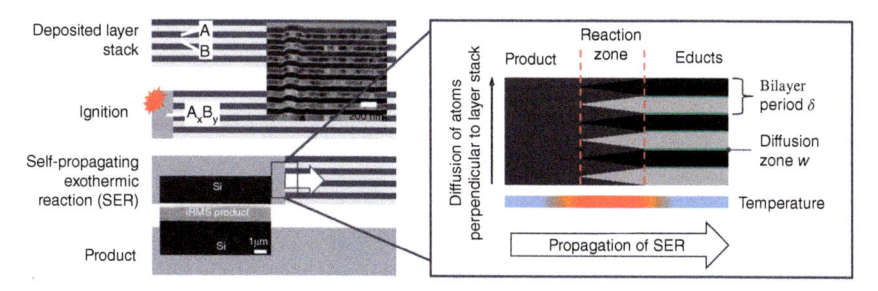

Figure 14.2 Schematic of SER in planar multilayer stack including temperature profile in close proximity of the reaction front.

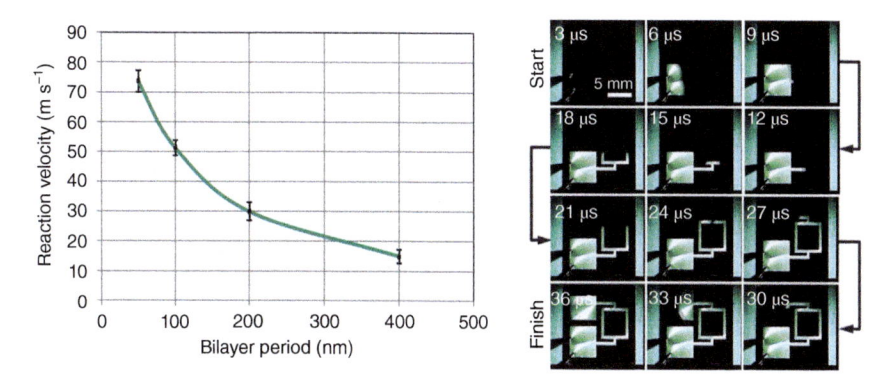

Figure 14.3 Influence of bilayer period on reaction velocity for Pd/Al iRMS measured with high-speed camera.

Due to diffusion of the reactants, the reaction velocity increases with decreasing bilayer thickness until reaching a maximum value. Decreasing the bilayer period further causes a fast reduction of the reaction velocity and converging towards zero. During deposition, intermixing already occurs at the interface between both reactants, forming a diffusion zone between both reactants. The already premixed reactants do not generate energy during the reaction process. With decreasing bilayer thickness δ, the diffusion zone w becomes predominant, causing a significant reduction of the heat of reaction ΔH_{Rx} in comparison to the theoretical heat of formation ΔH_0.

$$\Delta H_{Rx} = \Delta H_0 \left(1 - \frac{w}{0.5\,\delta}\right).$$

14.3 Material Systems

Many RMS had been published in recent years, focusing on particles, foils, and thin films. While particle and foil-based RMS only have limited relevance for microelectronics, thin film layers show a great potential as energy source for chip and wafer boning. Table 14.1 gives an overview of microsystems compatible RMS.

Table 14.1 Overview of binary reactive multilayer systems.

Material system	Type	Thickness (μm)	ΔH_{Rx} or ΔH_0 (kJ mol^{-1} atom)	Q (kJ cm^{-3})	Velocity (m s^{-1})	Reference
2Ni/Si	Foil	0.7…1.6	−17…−60	4.67[a]	22…27	[1–3]
3Ni/2Al	Foil	11	−51.5	n.a.	1.1…10.1	[4]
Zr/2Al	Foil	16…50	−51	4.63[a]	1.5…11.5	[2, 5]
Ti/Al	Foil	28	−35[a]	3.58[a]	0.08	[2, 6]
5Ti/3Si	Foil, film	1…3	−72	6.52[a]		[2, 7],
3Cu$_2$O/2Al	Foil	14	−55.5	n.a.	1	[8]
5Zr/3Si	Foil	n.a.	−72	5.37[a]	n.a.	[2, 9, 10]
CuO/Al	Foil/film	1…3	−42.6	n.a.	n.a.	[11]
3CuO/2Al	Film	1…2	−151.7	20.4[a]	n.a.	[2, 12]
Pt/Al	Film	1…3	−88…−90	10.29[a]	25…90	[2, 13]
Pt/2Al	Film	1…6	−56…−71	n.a.	16…56	[13]
Pd/Al	Film	1.8…2	−90	12.21[a]	12…75	[2, 12, 14]
2Pd/Sn	n.a.	n.a.	−51.2…−64.0	n.a.	n.a.	[15, 16]
Pd/2Sn	n.a.	n.a.	−37.9…−44.1	n.a.	n.a.	[15–17]
Pd/Sn	Film	n.a.	−55.1…−61.1	4.1[a]	n.a.	[2, 15, 16]
2Pd/Si	n.a.	n.a.	−64.2	n.a.	n.a.	[17]

a) Dimension converted.

The RMS can be categorized depending on their energetic level. The heat of formation for low, medium, and high energetic systems has to be smaller than −30 kJ mol^{-1}-atom. Low energetic systems such as Al/Co and Al/Ti have a heat of formation in the range of −30 kJ mol^{-1}-atom to −59 kJ mol^{-1}-atom. Medium energetic systems such as Ni/Si, Ti/Si, and Zr/Si generate up to −89 kJ mol^{-1}-atom. High energetic systems such as Pt/Al, Pd/Al, and CuO/Al release more than −90 kJ mol^{-1}-atom. Due to the high-energy level, high energetic systems enable reactive bonding with thin layers and are therefore the material of choice in microsystems technology.

14.4 State of the Art

Reactive material thin films generate energy based on a self-propagating high temperature reaction. At the end of the nineteenth century, the first thermite reactions had been reported by Goldschmidt [18, 19]. During the 1960s, Merzhanov et al. reported the high-temperature combustion synthesis for powder materials, focusing on various material systems, reaction kinetics, and modeling. Further research of the self-propagating high-temperature synthesis (SHS) led to the discovery of different reaction modes such as steady [20], spinning [21], chaotic [22], and repeated combustion [23].

With the emerging microsystems technology, the RMS were also deposited by physical vapor deposition (PVD). Anselmi-Tamburini reported the first SER in sputtered Ni/Al systems in 1989 [24]. Since the 1990s, many material systems have been deposited by both electron beam evaporation and magnetron sputtering [1, 4, 5, 7, 8, 11, 13, 14, 25, 26]. The RMSs have been intensively characterized with focus on reaction kinetics and phase changes during the reaction itself [4–6, 8, 11, 13, 25, 27]. In 2001, the first commercially available RMS foils for bonding applications were introduced to the market by Indium Corp. (formerly RNT) under the trademark NanoFoil®. These foils can be used as internal heat source for various bonding applications on chip and component level [9, 28], [29, 30], [31].

Further developments focus on the transfer of the reactive bonding process from component to wafer level. Foil type RMS have limitations regarding their minimum frame width and stability for wafer level application. Therefore, iRMS and oxide-based integrated reactive material systems (oiRMS) have been reported during recent years [12, 14, 26, 32].

14.5 Deposition Concepts of Reactive Material Systems

While macroscopic fabrication methods, such as milling, can be used to fabricate freestanding reactive foils, microsystems require a direct deposition of the reactive material on at least one substrate. Thereby, a homogeneous deposition of RMS is essential for the development of a bonding process on chip and wafer level. While PVD such as magnetron sputtering or electron beam evaporation are most commonly used, new concepts like electroplating or deposition of vertical systems also show great potential. Particle-based systems enable a deposition by printing. Figure 14.4 provides an overview of deposition concepts for integrated RMSs.

14.5.1 Physical Vapor Deposition

PVD can be used for the deposition of various RMS. While magnetron sputtering is the most common approach [4, 7, 9, 11–13, 33, 34], the reactive multilayers can also be deposited by electron beam evaporation [27].

For magnetron sputtering, sputter-up, sputter-down, and side-sputter arrangements are used. The metallic reactants are sputtered by direct current (DC) magnetron sputtering, while nonmetallic materials such as Si are deposited by radio frequency (RF) magnetron sputtering [9]. In the case of metal oxides, both RF and DC sputtering are used. These processes can be carried out, either by utilizing a metal oxide target or by reactive sputtering from a metal target [11, 12]. The deposition of reactive material systems by magnetron sputtering requires a uniform process with minimal thermal impact. An annealing of the substrates during the deposition process causes the reactants to diffuse. The resulting diffusion zone reduces the overall energy of the layer stack. To minimize this effect, an active

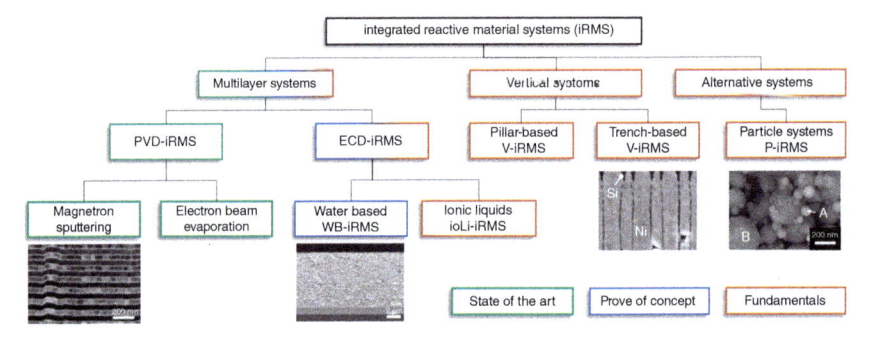

Figure 14.4 Overview of deposition technologies for iRMS.

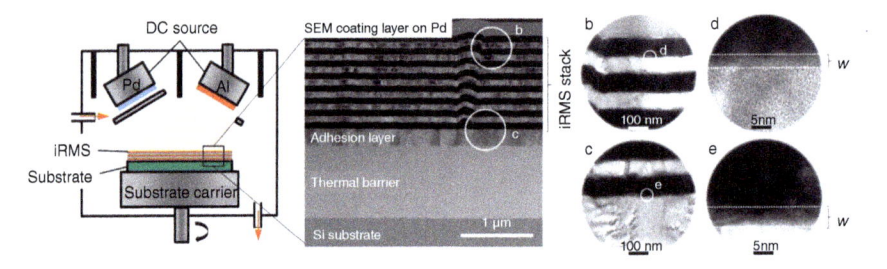

Figure 14.5 Schematic of process chamber for single chamber sputter-down iRMS deposition and as deposited Pd/Al iRMS including interdiffusion zones of first and last bilayer.

cooling of the substrates should be used during deposition. Figure 14.5 shows a cross section of a Pd/Al iRMS stack directly after deposition in a single-chamber DC magnetron sputter process.

Due to active cooling of the substrate holder, the diffusion zone does not exceed a total thickness of 5 nm. The diffusion zone for the first sputtered layer is thicker in comparison to the diffusion zone of the last layer due to prolonged thermal impact during sputtering. While an active cooling of the carrier indirectly cools the substrate as well, the thermal energy generated during sputtering dissipates from the surface through the iRMS and the substrate towards the carrier.

The iRMS on chip and wafer level can be patterned by both etching and lift-off process. Due to the number of bilayers, etching is a relatively complex process. The sequential etching of both materials leads to an extended exposure time of the top layers to the etchant. Therefore, each layer is etched differently. Furthermore, the etching rate of both components in different etchants varies as well, causing etching of material B during etching of material A as well. As shown in Figure 14.6, the sequential etching causes a significant under etching of the mask and a low resolution.

The resulting frame width of 6.9 and 37.9 µm, respectively, after wet etching is significantly smaller in comparison to the mask design with 20 and 50 µm. In comparison to wet etching, the patterning of the iRMS by lift-off creates much sharper structures with better resolution. The increased width of the test structure is caused

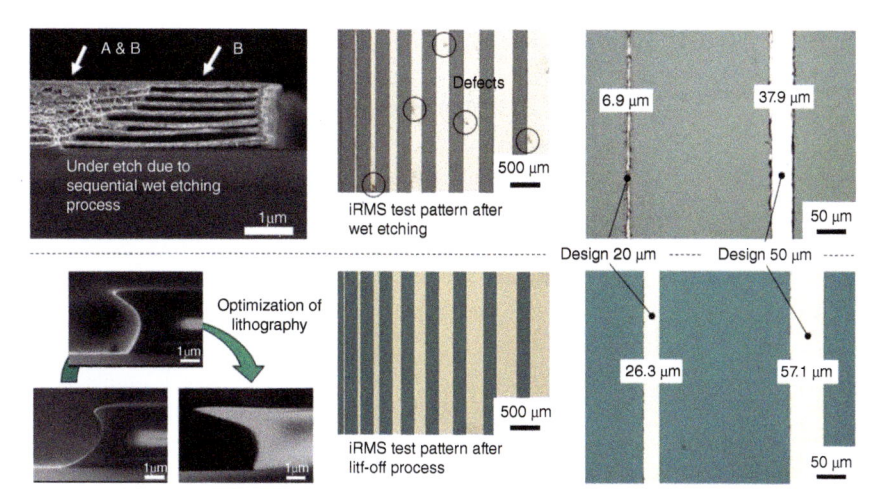

Figure 14.6 Influence of patterning on resolution of iRMS test structures for wet etching and lift-off process for Pd/Al iRMS [14].

by the undercut of the lift-off resist. It varies from 6.3 to 7.1 μm and is not dependent of the frame width itself. In combination with the capability of recycling the excess iRMS lifted from the wafer, the lift-off process is both more accurate and more cost-efficient in comparison to wet etching processes.

14.5.1.1 Conclusion Physical Vapor Deposition and Patterning

The deposition of reactive multilayers by PVD enables a homogeneous and uniform coating of chips and wafers with a large variety of material systems. While literature mainly focuses on magnetron sputtering processes, electron beam evaporated systems can also be found. The substrates can be coated with and without patterning. For patterning of iRMS, lift-off process is recommended due to its high resolution in comparison to sequential wet etching and its capability of material recycling after patterning.

In literature, the iRMS deposition is often carried out in a single-chamber magnetron sputter process. While the sputtering in one process camber enables easy access to the iRMS deposition, a dual-chamber process is more feasible for industrial application. Depending on the machine configuration, the iRMS are also deposited inside the chamber and on top of the sputter carriers. Based on the basic energy level of the RMS and the thickness of the bilayers, the sputtering chamber has to be cleaned for single-chamber processes. By using a dual-chamber process, the downtime of the deposition system can be reduced significantly.

14.5.2 Electrochemical Deposition of Reactive Material Systems

Another fabrication method for iRMS is the electrochemical deposition (ECD). The fabrication of iRMS using ECD is an attractive approach to extend the application field of this bonding technique to other industries, like printed circuit boards. In microsystem technology and microelectronics, the ECD is often used for through silicon vias (TSV), redistribution layers, micro bumps or bond frames. Thus, it is not an uncommon process technology. In comparison to PVD, the ECD is a cost-efficient process regarding equipment investment and maintenance. Furthermore, the iRMS are patterned directly during deposition using pattern-plating technique. This causes in a resource-efficient material consumption compared to the lift-off process in PVD. The material combinations for the fabrication of ECD-iRMS are very limited. From the theoretical point of view, considering the enthalpy of formation, only Pd-Sn, Pd-Zn, and Pd-In are possible to deposit from water-based electrolytes. If other electrolyte solvents are used, e.g. organic solvents or ionic liquids, the material combinations increase, e.g. Pd-Al from ionic liquids [9, 35].

There are two main approaches to deposit horizontal iRMS using ECD: dual bath technology (DBT) and single bath technology (SBT). In DBT, the substrates are transferred alternating between two metal electrolytes with a rinsing step in between. In SBT, both metal ions are dissolved in one solution. By switching the deposition current or potential, the noble or less-noble metal is deposited. The SBT does not need an additional rinsing step or bath transfer, which can lower the total process time and the possibility of contaminations in the electrolytes. Nevertheless, the bath chemistry is highly complicated in order to synthesize an electrolyte, which is able to deposit almost pure metal layers by switching the deposition potential.

The studies from Braeuer et al. show that the deposition of Pd on Zn results in dendritic growth of Pd, which is not applicable for multilayer deposition [36]. The same study shows that the stacking of Pd-Sn is successful but not reproducible with the used electrolytes [36]. In the following, new results in DBT and SBT for Pd-Sn multilayer are described.

14.5.2.1 Dual Bath Technology

Building on the results of Braeuer et al., the electrolyte selection was reviewed to achieve reproducible multilayer. Thus, the plating bath shall meet the following requirements:

- Good long-term stability
- Low layer stress
- Low surface roughness
  - Significantly smaller than layer thickness
  - Achievable at low current densities
- Deposition rates suitable for single layer process durations between 30 and 60 seconds
- Chemical compatibility between the two metal layer partners of the RMS (and to the substrate)

The aim in DBT is to use commercially available electrolytes, which could be modified or used as delivered. From the previous work, it is well known that the Pd layer introduces the stress in the RMS. Therefore, an extensive market study for Pd electrolytes has been carried out. In Table 14.2 an overview of the Sn and Pd electrolytes, which are finally chosen for experimental tests, is given. In the case of

14.5 Deposition Concepts of Reactive Material Systems

Table 14.2 Overview of Pd electrolytes used for deposition.

Name	c (g l^{-1})	pH	T (°C)	Characteristics
Pd1	8	7–9.5	30–65	Able to deposit 0.3–5 µm layer thickness
Pd2	3	6–7.5	35–50	Ammonia free
Pd3	10	7.5–8.5	38	Able to deposit 5 µm or more
Pd4	12	6–7	55–65	Ammonia and chloride-free, high current densities possible
Sn1	16 and 26	<2	15–30	Sulfuric acid based, bright deposit at low current densities, various brightener options

Table 14.3 Overview of applied bilayer combinations of Pd and Sn.

	j (mA cm^{-2})		t (s)		T (°C)		d (µm)	
Combination	Sn	Pd	Sn	Pd	Sn	Pd	Sn	Pd
Pd1/Sn1	7.5	10	30	25	RT	40	105	75
Pd2/Sn1	7.5	5	30	15	RT	40	105	75
Pd3/Sn1	15	5	15	48	RT	38	184	100
Pd4/Sn1	15	5	15	52	RT	55	184	100

Sn electrolytes, we have focused on sulfuric acid-based electrolytes because of the well-known results from Braeuer et al. with methane sulfonic acid [36].

The electrolytes were characterized using hull cell tests. The evaluation of layer quality and deposition rate single layer deposition were performed. After that the stacking of the different electrolytes was carried out. The deposition parameters are listed in Table 14.3.

The stacking behavior of Pd2/Sn1 and Pd4/Sn1 is not applicable for RMS. In the case of Pd2, the electrolyte is not able to smooth the prior deposited Sn layer, which results in high inhomogeneity. During the multilayer deposition of Pd4 and Sn1, it was observed that the Sn layers were dissolved in the Pd4 electrolyte. The dissolution of Sn in the Pd4 electrolyte cannot be explained without knowing the confidential additives of the electrolyte. The microstructures of those stacking trials are shown in Figure 14.7a,b.

The other two combinations Pd1/Sn1 and Pd3/Sn1 were stacked successfully. For Pd1/Sn1 a bilayer period of 180 nm with 40 bilayers was used. The sample was cut into pieces. One piece of the sample was ignited and did react completely. Another piece was analyzed via FIB, which is shown in Figure 14.7c. The layers show a high uniformity and thickness of 105 nm Sn (dark) and 75 nm Pd (light).

The bilayer period for Pd3/Sn1 was set to 284 nm and 25 bilayer were stacked. This sample was cut in pieces and analyzed via FIB. The ignition of the RMS was not successful, assumingly due to insufficient number of bilayers and disturbed bilayer

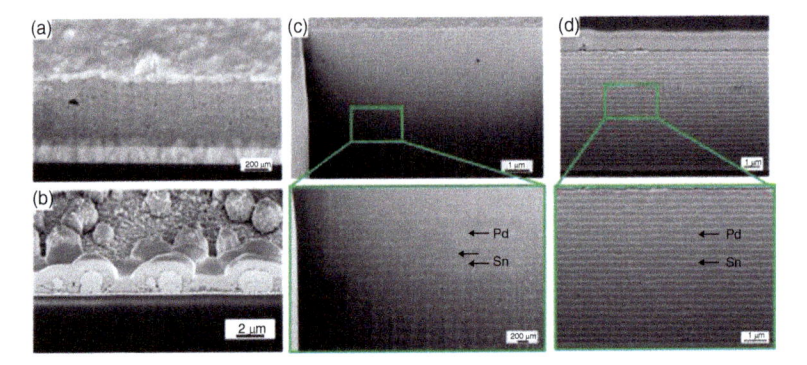

Figure 14.7 SEM images of FIB analysis of (a) Pd2/Sn1 stack, (b) Pd4/Sn1 stack, (c) Pd1/Sn1 40 Bilayer stack, and (d) Pd3/Sn1 25 bilayer stack.

Table 14.4 Overview of mixed electrolytes for SBT with Sn electrolyte base with 375 mM NaOH and 200 mM Na_2SnO_3.

Type	Amine1	Amine 2	Amine 3	pH
Sn	—	—	—	13.5
Sn, Pd	500 mM	—	—	14
Sn, Pd	—	500 mM	—	14
Sn, Pd	—	—	50 mM	14

ratio areas. Figure 14.7d, the FIB analysis is shown for 25 bilayers of the Pd3/Sn1 sample. The multilayer quality can be compared to Pd1/Sn1 in Figure 14.7. The porosity of the Sn layer could be also a reason for the ignition problems.

14.5.2.2 Single Bath Technology

As described earlier, in SBT the electrolyte contains both metal ions to be deposited. Currently, there is no electrolyte commercially available to produce multilayers of Pd/Zn, Pd/Sn, or Pd/In. Therefore, the synthesis of a stable Pd/Sn electrolyte is of high interest regarding the deposition of RMS. It should be able to deposit pure Pd and Sn layers on different potentials. Thus, the possible metal complexes were studied intensively regarding their chemical stability. For the mixed electrolyte, non-established amine Pd complexes were used in combination with alkaline Sn complexes. Table 14.4 shows the electrolytes for deposition tests.

The deposition was carried out using chronoamperometry for 600 s. The resulting current density-potential curves and photographs are shown in Figure 14.8. From the current density-potential curves, the Sn deposition in the pure Sn electrolyte starts at −1.2 V. The mixed electrolytes show starting metal deposition at −0.8 V, which can be assigned to Pd deposition. Furthermore, the hydrogen formation starts at lower potentials for the Pd/Sn electrolytes. The morphology of the layers is an important criterion for RMS fabrication. By comparing the photographs, the amine

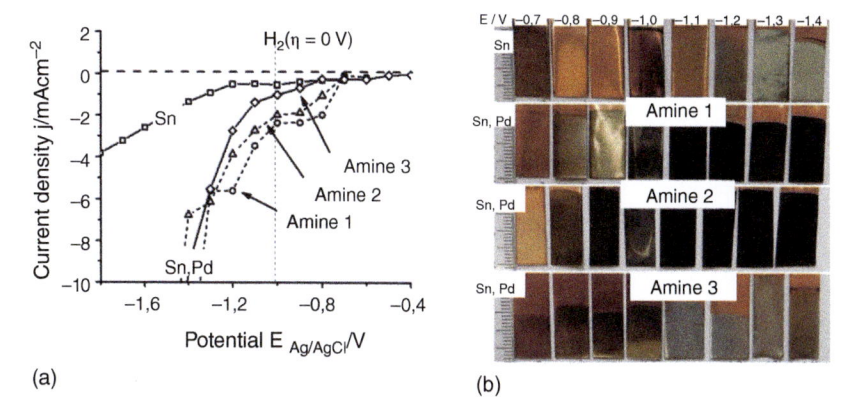

Figure 14.8 Current density-potential curves for a pure Sn electrolyte and three Pd/Sn mixed electrolytes with different complexing agents (a); photographs of deposited layers from chronoamperometry experiments (b).

1 and amine 2 show black and rough surfaces from −1.1 V and above. The electrolyte with amine 3 shows smooth and glossy surface up to the Sn range without using any kind of leveler or brightener additives. The analyzation of the layer composition shows that pure Pd deposition takes place at potentials between −0.8 and −1 V. In the Sn range from −1.2 to −1.4 V, an Pd-Sn alloy is deposited. That means that the deposition of a highly Sn-rich layer is not possible with the current electrolyte. The current electrolyte is a good base for further optimization regarding pH-value and Sn and Pd concentration.

14.5.2.3 Conclusion DBT and SBT

The ECD is still an attractive technique to produce RMS. The research is still ongoing regarding optimized electrolytes and process steps. In DBT, the realization of RMS has been shown with commercially available electrolytes. The process stability is still under investigation and thus not applicable for industrial scale until now.

The synthesis of a mixed electrolyte is under optimization. The base of the electrolyte is an alkaline stannate electrolyte with an amine Pd complex. It has been shown that the amine 3 could be suitable for this challenge. Even though, other amine derivative have to be investigated to achieve Sn-rich as well as Pd-rich layers.

14.5.3 Vertical Reactive Material Systems With 1D Periodicity

An alternative way of fabricating iRMS is the vertical approach. For these systems, the reactants needed for the SER are stacked vertically to each other (v-iRMS). Compared to the horizontal approach with time-consuming stacking of the reactants, this approach leads to a reduction of the necessary process steps. Furthermore, the formation of interdiffusion zones due to higher deposition temperatures, especially for PVD-iRMS, can be minimized. In addition, wet-chemical structuring processes of PVD-iRMS are complex due to different noble metals or are very time and resource consuming due to a lift-off process.

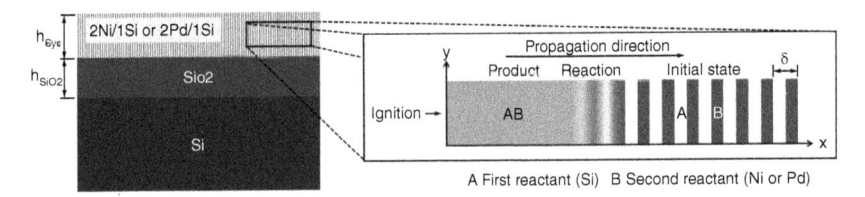

Figure 14.9 Schematic illustration of a self-propagating reaction along the vertically arranged RMS.

Table 14.5 Comparison of the systems 2Pd/1Si and 2Ni/1Si with $h_{Sys} = 10\,\mu m$ and $h_{SiO2} = 5\,\mu m$.

Thermodynamic property	Unit	2Pd/1Si	2Ni/1Si
Diffusion coefficient (D_0)	$cm^2\,s^{-1}$	0,07	2
Activation energy (E_A)	eV	1,4	1,5
Reaction enthalpy (ΔH_F)	$kJ\,mol^{-1}$	−43 [3]	−60 [1–3]
Ambient temperature $T_{ambient}$	°C	300	150
Maximum reaction temperature	°C	1000	1200

14.5.3.1 Dimensioning

In contrast, the material combinations for v-iRMS fabrication are very limited. Due to the good structuring ability of silicon in nanometer scale by using deep reactive ion etching (DRIE), a silicon-based layer system is necessary (reactant A = Si). Considering preferable thermodynamic properties (e.g. formation enthalpy ΔH_F) and the possibility of ECD from water-based electrolytes, only the metals palladium or nickel are suitable candidates for the second reactant (reactant B = Pd or Ni). This consideration leads to two possible systems: the low-energy system 2Pd/1Si and the medium-energy system 2Ni/1Si. Figure 14.9 shows a simplified configuration of the vertical iRMS. The SiO_2 serves as thermal decoupling to minimize heat input into the underlying silicon substrate during the reaction.

In order to determine the thermodynamic properties of the v-iRMS as a function of the vertical layer height h_{Sys}, the height of the thermal decoupling layer h_{SiO2}, and the bilayer period δ, FEM simulations can be used [3]. Table 14.5 shows the diffusion coefficient D_0, activation energy E_A, reaction enthalpy ΔH_F as well as the ambient and maximum reaction temperature for the systems 2Pd/1Si, and 2Ni/1Si with $h_{Sys} = 10\,\mu m$ and $h_{SiO2} = 5\,\mu m$.

The system 2Ni/1Si shows a higher maximum reaction temperature and reaction enthalpy (amount of released heat) in combination with a lower necessary ambient temperature. In particular, the ambient temperature has a wide influence on the dimensioning of the bilayer period. Figure 14.10 shows the necessary ambient temperature as a function of the bilayer period δ for $h_{Sys} = 10\,\mu m$.

Since the bilayer period affects the diffusion paths, a smaller period enhances the reaction. As a result, $T_{ambient}$ can be reduced. The selection of a suitable bilayer

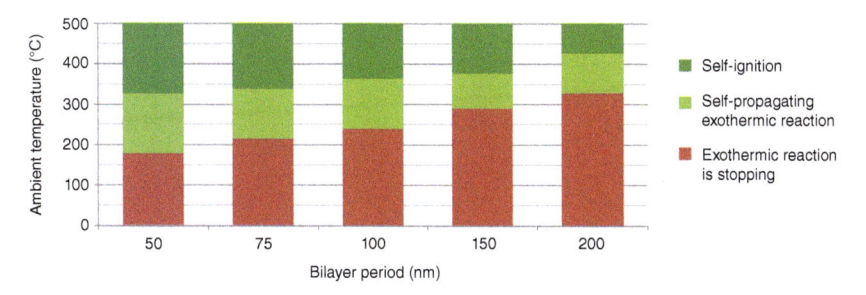

Figure 14.10 Ambient temperature $T_{ambient}$ for different bilayer periods δ in the vertical system 2Ni/1Si.

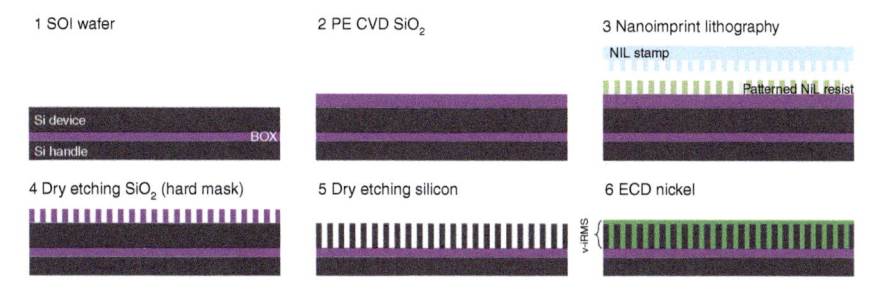

Figure 14.11 Process flow for fabrication of v-iRMS based on SOI substrate.

period strongly depends on the technological feasibility of fabricating the vertical silicon gratings in nanometer range. From the theoretical point of view, a bilayer period of $\delta = 200$ nm requires a silicon trench width of approximately 100 nm with an etching depth of 10 µm, which corresponds to an aspect ratio of AR = 100.

14.5.3.2 Fabrication

The fabrication of v-iRMS requires three basic process steps: nanoimprint lithography (NIL) based on ultraviolet light (UV-NIL), silicon dry etching, and ECD. The process flow is shown in Figure 14.11 by working with a silicon-on-insulator (SOI) substrate (Step 1). In order to use the silicon surface directly as an electrically conductive seed layer for electroplating, a silicon substrate with a low specific resistance in the range of 0.005 to 0.01 Ωcm is required. The primary purposes for the buried oxide (BOX) are the thermal decoupling as well as the barrier layer during the DRIE process of the Si device level. A silica layer serves as additional masking to enhance the selectivity during dry etching (Step 2). By means of NIL, the nano structures can be transferred into the NIL resist (Step 3). Step 5 shows the dry-etched silicon device (reactant A) by using the patterned silica masking (Step 4). ECD of nickel (reactant B) in the patterned silicon trenches generates the final vertical iRMS (Step 6).

The UV NIL requires a molding process of the corresponding nano grating from a master substrate onto a UV light-transmissive stamp ("daughter"). For preparation the master substrate, a separate fabrication process, e.g. e-beam lithography, is necessary. Figure 14.12a demonstrates the patterned NIL resist with the nano grating after UV NIL. The structured resist serves as a mask for the subsequent dry etching

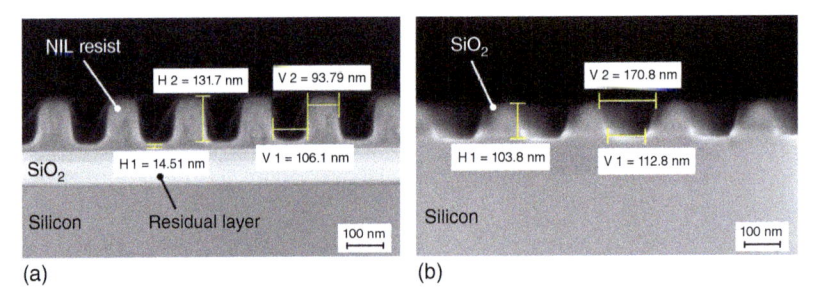

Figure 14.12 Cross-sectional SEM images of the patterned resist after NIL (a) and of the dry etched SiO_2 layer after ICP-RIE (b).

step of the SiO_2. Key points are the structure transfer, the residual resist layer in each trench bottom, the defect density, and the separation of the stamp from the resist (detachment). For this purpose, the SmartNIL™ Process, developed by the company EV Group (EVG), was used. The nano grating of the master wafer is molded onto an elastic foil by using a polymer layer, which works as the stamp for the NIL process. The transfer of the polymer structures into the NIL resist is done by means of a movable barrel. This allows a linear contact of the stamp with the resist, which significantly reduces defects. The dry etching of the SiO_2 masking by using reactive ion etching (RIE) with inductively coupled plasma (ICP) is shown in Figure 14.12b. In particular, the combination of bias and ICP power P_{Bias} and P_{ICP} has a decisive influence on the etching result.

Subsequently, the "High Aspect Ratio" (HAR) patterning of silicon (reactant A) by means of DRIE takes place. The target of the etching process is to generate very deep silicon trenches with high selectivity to the oxide mask and good lateral structure transfer (low sidewall tapering). Especially the frequency of the bias electrode f_{Bias}, the bias power P_{Bias}, and the cycle of the SF_6 etching and C_4F_8 passivation have an important function during the process. The structured silicon is shown for a 5 µm thick device layer in Figure 14.13a. These result in an aspect ratio for the trenches of 47 (:1). Due to the BOX the process stops directly on the oxide.

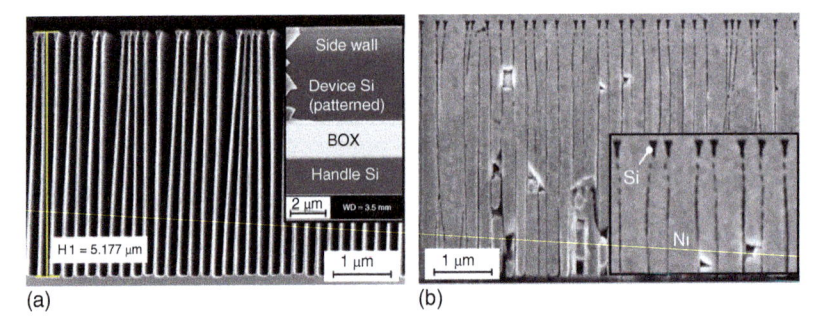

Figure 14.13 Cross-sectional SEM images of the patterned silicon device layer after DRIE (a) and the nickel-filled silicon trenches after ECD (b).

Table 14.6 Challenges for ECD in nano structured silicon trenches with high aspect ratio (HAR).

Process specification	Challenge	Approach
Electrical contact	Removal C_4F_8 polymer (DRIE)	O_2 plasma treatment
	Removal native SiO_2	Buffered oxide etch (BOE)
Pretreatment	Electrolyte wetting, entrapped air	Vacuum storage
Adhesion reactant B	Metal adhesion on silicon	DRIE scallops on sidewalls
ECD reactant B	Grain formation, grain sizes, homogeneity	Pulse plating (PP)

The ECD of the reactant B from water-based electrolytes in the etched silicon trenches poses a major challenge due to the hydrophobic properties of nano structured surfaces in combination with the very small lateral trench dimensions [37, 38]. The various problems with their approaches are listed in the following Table 14.6. Figure 14.13b shows a scanning electron microscopy (SEM) cross section of nickel-filled silicon trenches for the medium-energy system 2Ni/1Si. The deposition was carried out with a nickel sulfamate electrolyte in pulse plating mode at a mean current density of $J_{eff} = 0.4$ mA dm^{-2} and a bath temperature of $T = 60\,°C$.

14.5.3.3 Conclusion

The vertical iRMS approach, consisting of NIL, DRIE, and ECD, has been demonstrated for the medium-energy system 2Ni/1Si. It could be shown that this approach represents a promising alternative fabrication technology in comparison to the horizontal stacking of the reactants. However, further optimizations with consideration to stoichiometry, bilayer period, and defect-free metal deposition is necessary, since minimum technological deviations lead to a prevention of the exothermic reaction. Thus, the approach is currently still not suitable for industrial applications.

14.6 Bonding With RMS

Bonding based on RMS utilizes a self-propagating reaction to form a stable bond between two substrates. The energy of the exothermic reaction melts the surface of the bonding layer and forms a stable, mechanically strong bond interface between both substrates. The local heating of the interface by the exothermic reaction enables joining of the substrates without the requirement for external heating.

By applying an ignition impulse to the system, the reaction propagates through the complete stack. For ignition various methods have already been analyzed and reported. The technique has to provide a sufficient energy burst to the systems. Therefore, electrical ignition, laser irradiation, and mechanical discharge have been evaluated and characterized in recent years. Electrical ignition can be achieved by electrical spark initiation [27], electrostatic discharge, and DC heating. The

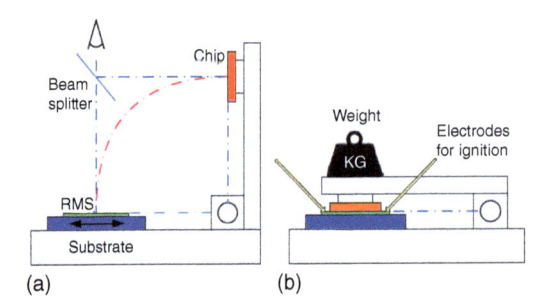

Figure 14.14 Schematic of chip-scale reactive bond tool: (a) alignment and (b) ignition and bonding. Source: Schumacher et al. [9].

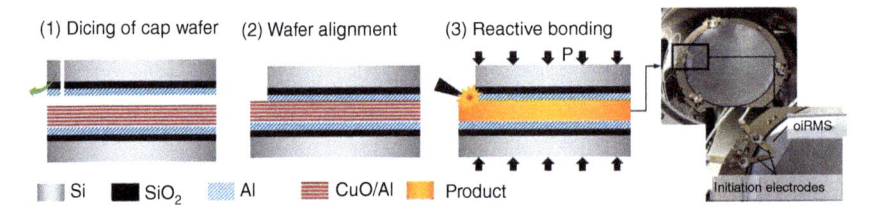

Figure 14.15 Process flow for bonding of eight inch silicon wafers with CuO/Al oiRMS and electrode setup inside process chamber. Source: Vogel et al. [12].

laser ignition provides a well-controlled energy transfer into the RMS, enabling a reproducible ignition of the reactive multilayers [39].

Based on the ignitable RMS, two or more substrates can be joined. While reactive foils are mainly used in combination with solder material, no additional solder is required for iRMS bonding. Instead, metallic bonding layers enable the formation of a stable connection between both substrates [12, 14]. Depending on the requirements of the application, different bonding setups have already been reported. For foil-based bonding on chip and component level, both specific bonding tools and adjusted commercial bonding tools can be used. As shown in Figure 14.14, Schumacher designed a specific bond tool to join chips and components with foil as well as iRMS [9].

All bonding tools require an ignition system to start the exothermic reaction. Schumacher therefore included two electrodes in the component level setup. An ignition system also has to be integrated for wafer level bonding as well. As reported by Vogel, the integration of two ignition electrodes to the bond chuck enables reactive bonding on wafer level up to eight inches. Due to the wafer flat, wafer substrates with a maximum diameter of six inches can easily be aligned within the setup. By rotating one of the wafers by 180°, the ignition electrodes can always access the reactive multilayer stack. For eight inch wafers, the flat is created artificially by dicing. Figure 14.15 gives an overview of both the preparation of the eight inch wafers as well as the electrode setup inside the adjusted SB8 bonding tool form SÜSS MicroTec AG [12].

In comparison to other metal-based bonding technologies, reactive bonding does not require external heating. In combination with the high reaction velocity, bonding can be achieved within seconds. Depending on the application, the substrate alignment and additional optional process steps, such as evacuating of bonding

Table 14.7 Comparison of Pd/Al-based reactive wafer level bonding process with other wafer level bonding technologies [26, 40, 41].

Parameter	Glass frit bonding	Copper thermal compression bonding	Reactive bonding with Pd/Al
Temperature	450 °C	400 °C	Room temperature
Process time	0.5…2 h	1…4 h	<1 min
Tool pressure	200…500 mbar	<40 MPa	0,5…5 MPa
Shear strength	32 N mm^{-2}	180 N mm^{-2}	235 N mm$^{-2a)}$

a) Dimension converted.
Source: Braeuer and Gessner [26].

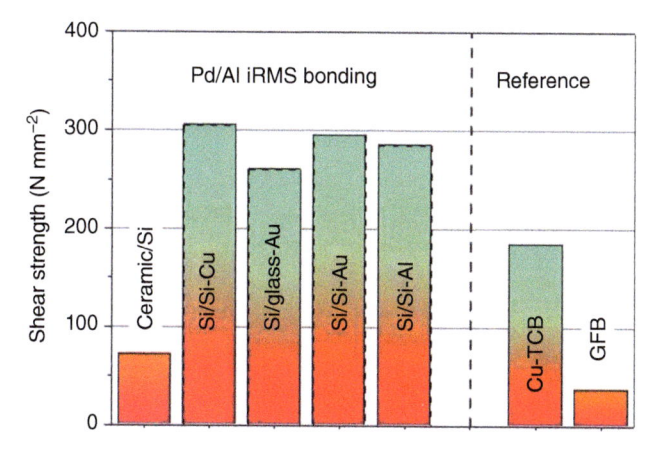

Figure 14.16 Shear strength of Pd/Al iRMS bonded wafers for the material combinations ceramic/Si, Si/Si, and Si/glass in comparison to the standard bonding technologies, copper thermal compression bonding (Cu-TCB) and Glass frit bonding (GFB) with two Si wafers.

chamber, are the most time-consuming parts of the reactive bonding process. Table 14.7 compares process parameters for state-of-the-art bonding technologies with reactive bonding.

The mechanical properties of the bond interface strongly depend on the interface material itself. While the strength of foil-based systems is mainly affected by the solder, the bonding strength of iRMS bonded systems depend on the RMS and bonding layer. As reported by Braeuer, the shear strength of Pd/Al iRMS bonded wafers is higher in comparison to glass frit or copper thermal compression bonded wafers [26, 41]. Depending on the bonding layers copper, aluminum, and gold as well as the substrate material, the shear strength exceeds 235 MPa [26]. The influence of the bonding layer and substrate material is shown in Figure 14.16.

Furthermore, reactive bonding with Pd/Al enables the hermetic encapsulation on wafer level. The leakage rate for bonded silicon–silicon substrates is in the range of 9,3 10^{-3} mbar·l/s. As reported by Spies, the shear strength of foil-based reactive bonds with Ni/Al foils strongly depends on the substrate materials. While reaching an average shear strength of 20 MPa for copper substrate, it exceeds 50 MPa for Al_2O_3 substrates [31].

14.7 Conclusion

RMS have been studied intensively throughout the last 30 years. While evaporated foil-based systems, such as Ni/Al NanoFoils®, are already in the market, the integration of RMS is still ongoing. The transfer of this innovative technology to chip and wafer level application enables the heterogeneous integration of new and sensitive materials as well as fully processed components. The bonding strength of Pd/Al-bonded substrates exceeds 235 N mm^{-2} and is therefore higher in comparison to many state-of-the-art technologies. The Pd/Al system furthermore enables the hermetic bonding of heterogeneous substrates within less than one minute. Commercially available reactive foils as well as new foil materials already enable the integration of components utilizing different solder materials. It is currently used to bond sputter targets and different joining applications.

New deposition techniques such as electroplating approaches show great potential for a more cost-efficient deposition of integrated reactive multilayers. While the first self-propagating reactions have already been reported for the DBT, further research is required to identify a stable process conditions for reproducible deposition. The fundamental research in the field of SBT is focusing on the development of suitable electrolytes. It shows great potential for further cost reduction in comparison to the DBT.

Vertical iRMS do not require a sequential layer deposition. Due to its complexity, further research is needed for the deposition of the second reactant and subsequent processing such as planarization and bonding.

References

1 Acker, J. and Bohmhammel, K. Optimization of thermodynamic data of the Ni-Si system. *Thermochimica Acta*. 337(1999): 187–193
2 Fischer, S.H. and Grubelich, N.C. (1998). Theoretical energy release of thermites, intermetallics, and combustible metals. *Int. Pyrotech. Semin.* 24: 1–56.
3 Masser, R., Braeuer, J., and Gessner, T. (2014). Modelling the reaction behavior in reactive multilayer systems on substrates used for wafer bonding. *J. Appl. Phys.* 115 (24): 1–7.
4 Gavens, A.J., Van Heerden, D., Mann, A.B. et al. (2000). Effect of intermixing on self-propagating exothermic reactions in Al/Ni nanolaminate foils. *J. Appl. Phys.* 87 (3): 1255–1263.
5 Barron, S.C., Kelly, S.T., and Kirchhoff, J. (2013). Self-propagating reactions in Al/Zr multilayers: anomalous dependence of reaction velocity on bilayer thickness. *J. Appl. Phys.* 114 (22): 1–10.

6 Gachon, J.C., Rogachev, A.S., Grigoryan, H.E. et al. (2005). On the mechanism of heterogeneous reaction and phase formation in Ti/Al multilayer nanofilms. *Acta Mater.* 53 (4): 1225–1231.

7 Boettge, B., Braeuer, J., Wiemer, M. et al. (2010). Fabrication and characterization of reactive nanoscale multilayer systems for low-temperature bonding in microsystem technology. *J. Micromech. Microeng.* 20 (6): 1–8.

8 Blobaum, K.J., Reiss, M.E., Plitzko Lawrence, J.M., and Weihs, T.P. (2003). Deposition and characterization of a self-propagating CuOx/Al thermite reaction in a multilayer foil geometry. *J. Appl. Phys.* 94 (5): 2915–2922.

9 Schumacher, A, U. Gaiß, S. Knappmann (2015). Assembly and packaging of micro systems by using reactive bonding processes. European Microelectronics and Packaging Conference, 1–6.

10 Meschel, S.V. and Kleppa, O.J. (2001). Thermochemistry of alloys of transition metals and lanthanide metals with some IIIB and IVB elements in the periodic table. *J. Alloys Compd.* 321 (2): 183–200.

11 Petrantoni, M., Rossi, C., and Salvagnac, L. (2010). Multilayered Al/CuO thermite formation by reactive magnetron sputtering: nano versus micro. *J. Appl. Phys.* 108 (8): 1–5.

12 Vogel, K., Roscher, F., Wiemer, M. et al. (2018). Reactive bonding with oxide based reactive multilayers. Smart Systems Integration 2018: International Conference and Exhibition on Integration Issues of Miniaturized Systems, Vol. 12, 71–78.

13 Adams, D.P. (2015). Reactive multilayers fabricated by vapor deposition: a critical review. *Thin Solid Films* 576: 98–128.

14 Braeuer, J., Besser, J., Wiemer, M., and Gessner, T. (2012). A novel technique for MEMS packaging: reactive bonding with integrated material systems. *Sensors Actuat. A Phys.* 188: 212–219.

15 Mathon, M., Gambino, M., Hayer, E. et al. (1999). [Pd2Sn] system: enthalpies of formation of the liquid [Pd1Sn] and heat capacities of PdSn, PdSn2, PdSn3 and PdSn4 compounds. *J. Alloys Compd.* 285: 123–132.

16 Amore, S., Delsante, S., Parodi, N., and Borzone, G. (2009). Thermochemistry of Pd-In, Pd-Sn and Pd-Zn alloy systems. *Thermochim. Acta* 481 (1–2): 1–6.

17 Meschel, S.V. and Kleppa, O.J. (1998). Standard enthalpies of formation of some 3d transition metal silicides by high temperature direct synthesis calorimetry. *J. Alloys Compd.* 267 (1–2): 128–135.

18 Goldschmidt, H. and Vautin, C. (1898). Aluminium as a heating and reducing agent. *J. Soc. Chem. Ind.* 6 (17): 543–545.

19 Goldschmidt, H. (1898). Ueber ein neues Verfahren zur Darstellung von Metallen und Legierungen mittelst Aluminium. *Justus Liebigs Ann. Chem.* 301 (1): 19–28.

20 Merzhanov, A.G. and Rumanov, E.N. (1999). Physics of reaction waves. *Rev. Mod. Phys.* 71 (4): 1173–1211.

21 Filonenko, A.K. and Barzykin, V.V. (1996). The effect of density on the limits and regularities of spin combustion of titanium in nitrogen. *Combust. Explos. Shock Waves* 32 (1): 45–49.

22 Mukasyan, A.S., Vadchenko, S.G., and Khomenko, I.O. (1997). Combustion modes in the titanium-nitrogen system at low nitrogen pressures. *Combust. Flame* 111 (1–2): 65–72.

23 Strunina, A.G., Dvoryankin, A.V., and Merzhanov, A.G. (1983). Unstable regimes of thermite system combustion. *Combust. Explos. Shock Waves* 19 (2): 158–163.

24 Anselmi-Tamburini, U. and Munir, Z.A. (1989). The propagation of a solid-state combustion wave in Ni-Al foils. *J. Appl. Phys.* 66 (10): 5039–5045.

25 Rogachev, A.S., Grigoryan, A.É., and Illarionova, E.V. (2004). Gasless combustion of Ti-Al bimetallic multilayer nanofoils. *Combust. Explos. Shock Waves* 40 (2): 166–171.

26 Braeuer, J. and Gessner, T. (2014). A hermetic and room-temperature wafer bonding technique based on integrated reactive multilayer systems. *J. Micromech. Microeng.* 24 (11): 1–9.

27 Ma, E., Thompson, C.V., Clevenger, L.A., and Tu, K.N. (1990). Self-propagating explosive reactions in Al/Ni multilayer thin films. *Appl. Phys. Lett.* 57 (12): 1262–1264.

28 Wang, J., Besnoin, E., Knio, O.M., and Weihs, T.P. (2004). Investigating the effect of applied pressure on reactive multilayer foil joining. *Acta Mater.* 52 (18): 5265–5274.

29 Wang, J., Besnoin, E., and Duckham, A. (2004). Joining of stainless-steel specimens with nanostructured Al/Ni foils. *J. Appl. Phys.* 95 (1): 248–256.

30 Long, Z., Dai, B., Tan, S. et al. (2017). Transient liquid phase bonding of copper and ceramic Al2O3 by Al/Ni nano multilayers. *Ceram. Int.* 43 (18): 17000–17004.

31 Spies, I., Schumacher, A., Knappmann, S. et al. (2018). Reactive joining of sensitive materials for MEMS devices: characterization of joint quality. Smart Systems Integration 2018: International Conference and Exhibition on Integration Issues of Miniaturized Systems, Vol. 12, 79–84.

32 Braeuer, J., Besser, J., and Tomoscheit, E. (2013). Investigation of different nano scale energetic material systems for reactive wafer bonding. *ECS Trans.* 50 (7): 241–251.

33 Braeuer, J., Besser, J., Wiemer, M., and Gessner, T. (2011). Room-temperature reactive bonding by using nano scale multilayer systeems. *Transducers* 11: 1332–1335.

34 Reiss, M.E., Esber, C.M., Van Heerden, D. et al. (2002). Self-propagating formation reactions in Nb/Si multilayers. *Mater. Sci. Eng. A* 261 (1–2): 217–222.

35 Hertel, S., Schröder, T.J., Wünsch, D. et al. (2016). Elektrochemische Abscheidung von Aluminium und Palladium aus ionischen Flüssigkeiten für das reaktive Waferbonden. *Eugen G. Leuze Verlag, Jahrb. Oberflächentechnik* 72: 119–131.

36 Braeuer, J., Besser, J., and Hertel, S. (2014). Reactive bonding with integrated reactive and nano scale energetic material systems (iRMS): state-of-the-art and future development trends. *ECS Trans.* 64 (5): 329–337.

37 Yan, Y.Y., Gao, N., and Barthlott, W. (2011). Mimicking natural superhydrophobic surfaces and grasping the wetting process: a review on recent progress in preparing superhydrophobic surfaces. *Adv. Colloid Interface Sci.* 169 (2): 80–105.

38 Barthlott, W., Schimmel, T., and Wiersch, S. (2010). The salvmia paradox: super-hydrophobic surfaces with hydrophilic pins for air retention under water. *Adv. Mater.* 22 (21): 2325–2328.

39 Picard, Y.N., Adams, D.P., Palmer, J.A., and Yalisove, S.M. (2006). Pulsed laser ignition of reactive multilayer films. *Appl. Phys. Lett.* 88 (14): 2004–2007.

40 Knechtel, R. (2005). Glass frit bonding: an universal technology for wafer level encapsulation and packaging. *Microsyst. Technol.* 12 (1-2): 63–68.

41 Vogel, K., Baum, M., Roscher, F. et al. (2014). Improvement of copper bonding by analyzing the mechanical properties of the bond interface. *Smart Syst. Integr. Micro- Nanotech.* XVIII: 529–536.

15

Polymer Bonding

Xiaojing Wang[1,2] and Frank Niklaus[2]

[1] Advanced Interdisciplinary Technology Research Center, National Innovation Institute of Defense Technology, 53 Dongda Street, 100071, Beijing, China
[2] Division of Micro and Nanosystems (MST), Department of Intelligent Systems (IS), School of Electrical Engineering and Computer Science (EECS), KTH Royal Institute of Technology, Malvinas väg 10, SE-100 44, Stockholm, Sweden

15.1 Introduction

Polymer bonding, also referred to as adhesive bonding, employs an intermediate polymer layer as the bonding material to join the surfaces of two substrates (e.g. wafers). Polymer bonding is being extensively employed in the semiconductor industry in applications such as wafer stacking to form three-dimensional (3D) integrated circuits (ICs) [1–3] and complementary metal oxide semiconductor (CMOS) imaging systems [4, 5] and in heterogeneous integration of microelectromechanical systems (MEMS) with ICs [3, 6–13]; in heterogeneous integration of photonics, e.g. III–V compound materials on wafers having silicon-based waveguides and ICs [14, 15]; and in manufacturing solar cells [16] and laser systems [17]. Polymer bonding is also extensively used for thin wafer handling, by temporarily bonding the thin wafers to handle wafers, thereby assisting grinding and etching processes [18], and manufacturing of through-substrate vias (TSVs) [19]. Furthermore, adhesive bonding is utilized for packaging of CMOS imaging sensors [20, 21]; manufacturing of radio frequency (RF) MEMS devices [22, 23]; fabrication of liquid crystal on silicon (LCoS) components [24]; packaging of MEMS, surface acoustic wave (SAW) filters, and CMOS imager devices [25–29]; and manufacturing of bio-MEMS and micro-total analysis systems (μTASs) [30]. Polymer adhesive bonding has also enabled recent progress in the integration of two-dimensional (2D) materials, including graphene, hexagonal boron nitride (hBN), and molybdenum disulfide (MoS2), from their specified growth substrates onto standard semiconductor substrates [31–33].

A typical polymer wafer bonding process starts with the coating of a well-defined polymer layer on one or both of the wafer surfaces to be bonded. After joining the two wafer surfaces, typically pressure and heat are applied to the wafer stack to induce intimate contact between the wafer surfaces. Finally, a polymer bond is

3D and Circuit Integration of MEMS, First Edition. Edited by Masayoshi Esashi.
© 2021 WILEY-VCH GmbH. Published 2021 by WILEY-VCH GmbH.

formed by diffusion or by fully or partly reflowing the polymer at the wafer interface and transforming the polymer layer from a liquid or viscoelastic state into a solid state. The mechanism for inducing the phase transitions of the polymer depends on the polymer type (e.g. thermoplastic or thermosetting) and can, for example, be initiated by melting and solidifying the polymer through heating and cooling (thermoplastic polymers) or as a result of thermal setting (cross-linking) of the polymer through heating (thermosetting polymers) [6, 34–36].

Polymer wafer bonding features many advantages, including insensitivity to the flatness and roughness of the wafer surfaces that are bonded, relatively low bonding temperatures (20–450 °C, depending on the polymer adhesive used), compatibility with standard CMOS wafers, and unlimited choice of the wafer materials that may be bonded. Since the polymer layer can to some extent tolerate and accommodate the topographic changes and particles at the wafer surfaces, no special wafer surface treatments such as planarization or excessive cleaning are required. Although being relatively simple, robust, and low cost, polymer wafer bonding also poses limitations on applications that require demanding temperature reliability, long-term stability, or hermeticity of the formed polymer bonds [6, 34–37].

15.2 Materials for Polymer Wafer Bonding

15.2.1 Polymer Adhesion Mechanisms

Similar to most bonding techniques, polymer wafer bonding makes use of the attraction between molecules when they are in sufficiently close contact. Different molecules can bond to each other by different possible inter-molecular interactions: (i) covalent bonds, (ii) ionic bonds, (iii) dipole–dipole interactions including hydrogen bonds, and (iv) van der Waals interactions. All of these interactions are based on electromagnetic forces but feature different energy contents as indicatively listed in Table 15.1 [35, 36]. The approximate energy of different bond types (covalent and typical dipole–dipole interactions) as a function of inter atomic distance is indicated in Figure 15.1. Covalent and van der Waals bonds can only be formed when the inter-atomic distance is roughly below 0.3–0.5 nm. The bond energy depend on the involved surface materials (type of molecules) and the interatomic distances, but none of the intermolecular bonds extend over distances longer than approximately 0.5 nm. Therefore, in order to realize bonding between two surfaces (e.g. a wafer surface and a polymer surface), the surfaces must be brought in sufficiently close contact. The typical surface roughness of macroscopically flat wafer surfaces, e.g. of a polished silicon (Si) wafer, is in the range of 0.3–1 nm (root mean square roughness (RMS)), and the peak-to-valley profile depth of these surfaces is in the order of several nanometers. Such dimensions induce large gaps between the wafer surfaces when they are brought in contact, which hinders the formation of interatomic bonds over large surface areas, as illustrated in Figure 15.2a.

To realize sufficiently close contact between two surfaces, at least one of the two surfaces must deform plastically or elastically to adapt to the surface topography of the second surface. This can be achieved, for example, by plastic or elastic surface

15.2 Materials for Polymer Wafer Bonding

Table 15.1 Comparison of the approximate energies of different bonds [35, 36].

Bond type	Energy content (kJ mol^{-1})
Covalent bonds	563–710
Ionic bonds	590–1050
Dipole–dipole interactions:	
Hydrogen bonds with fluorine	<42
Hydrogen bonds without fluorine	10–26
Other dipole–dipole bonds	4–21
Van der Waals interactions	2–4

Source: Nobel and Niklaus [35, 36].

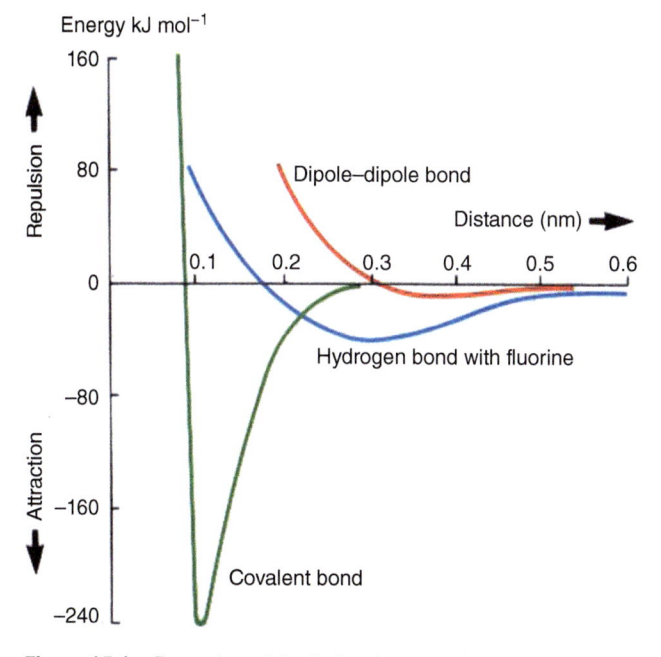

Figure 15.1 Examples of the indicative energies of different bonds as a function of the interatomic distance. Sources: Nobel and Niklaus [35, 36].

deformation resulting from applied pressure, by solid-state diffusion of polymer molecules induced by increased temperatures or by wetting of a surface with a liquid material (e.g. a polymer). In polymer wafer bonding, typically the intermediate polymer adhesive deforms and adapts to the surface topography of the wafers to be bonded. This may involve wetting of the wafer surfaces with a liquid/viscoelastic polymer (Figure 15.2b), and/or deformation of a solid/viscoelastic polymer layer due to a bond pressure, which is applied to the wafer stack.

Specifically, in most polymer wafer bonding approaches, a polymer adhesive is coated on one or both of the wafer surfaces to be bonded, thereby forming bonds

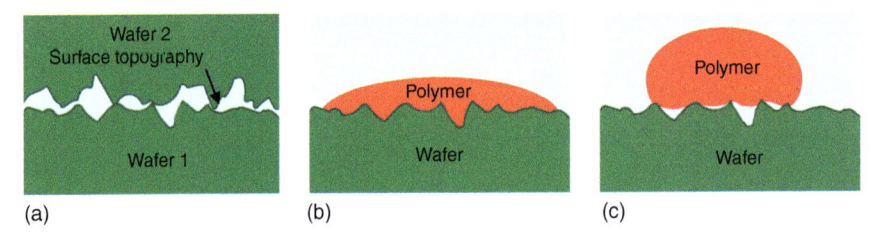

Figure 15.2 (a) Schematic close-up of a contact interface of two macroscopically flat solid surfaces. (b) Boundary layer of a solid surface and a liquid (e.g. a polymer) that is wetting the surface. (c) Boundary layer of a solid surface and a liquid (e.g. a polymer) that is not wetting the surface. Sources: Niklaus et al. [6], Ramm et al. [34], Niklaus [36].

between the molecules of the wafer surface and the polymer (given the polymer can wet the wafer surfaces). For the wafer bonding, the two wafer surfaces with the intermediate polymer layer(s) (coated on one or both wafer surfaces) are brought in intimate contact. During the polymer wafer bonding process, the polymer is then typically set in a liquid or semiliquid phase (e.g. by heating the wafer stack) and as a result is wetting both of the surfaces to be bonded by flowing (or being pressed) into the troughs of the matching surface. Then, the polymer adhesive is hardened (by removing the heat from the wafer stack) into a solid polymer to form permanent bonds between the molecules, thereby holding the surfaces together. The wetting of the surfaces by the polymer is critical to form a strong bond between the polymer and the surfaces. The wettability of a surface with a liquid or semiliquid polymer depends on the surface material and on the polymer involved, i.e. to ensure wetting of a surface with a liquid or semiliquid material as depicted in Figure 15.2b, and the solid surface must have a higher surface energy than the liquid. A counterexample is shown in Figure 15.2c, where the liquid does not wet the solid surface. A detailed discussion of surface energy and wettability can be found in Ref. [38]. The wettability of a surface by a liquid polymer can be strongly influenced by surface contaminants (e.g. weakly adsorbed organics), particles, moisture, and microscopic surface topography. Clean and contaminant-free surfaces can improve the wettability of the surface and can be realized by dedicated cleaning processes using solvents, oxidants, strong acids, or bases. In addition, surface pre-treatment with adhesion promoters can significantly increase the wettability of a surface. Adhesion promoters are typically coated on a surface to form a few monolayer thick surface functionalization, which bonds well to the wafer surface and enhances the wettability of the surface with the polymer and ultimately can result in increased bonding strength between the wafer surface and the polymer. Depending on the combination of the used wafer surface material and the polymer, specific adhesion promoters may be employed. High bond strengths and long-term bond stability are typically yielded when the filling of the troughs of the wafer surface topography with the polymer is more complete. Therefore, suitable polymers for wafer bonding typically feature good wettability of the wafer surface material(s) and low shrinkage during polymer hardening (or reflow) to minimize the amount of unfilled space at the bond interfaces. Unfilled space at the bond interfaces, which may be caused by polymer shrinkage during curing, can deteriorate the wafer bond quality or affect the long-term stability of the bond as water and gas molecules can diffuse in these nanoscale channels [38].

15.2.2 Properties of Polymers for Wafer Bonding

Polymers consist of large molecules (macromolecule), composed of many repeated subunits (monomers), which are covalently bonded to form polymer chains or networks. There exists a large number of different types of polymers with widely varying properties [39]. Polymers can be generally categorized into four broad types: (i) thermoplastic polymers, (ii) thermoset polymers, (iii) elastomeric polymers, and (iv) hybrid polymers. Thermoplastics can be reshaped by melting upon heating to specific temperatures and solidified again by cooling. Thermosetting polymers contain 3D cross-linked network of polymer chains and cannot be remelted or reshaped. However, before complete polymerization (cross-linking into polymer networks), they typically do undergo a semiliquid or liquid phase when heated the first time to high temperatures to initiate the cross-linking. Cross-linking reaction may also be initiated by other mechanisms than heating, such as mixing of two polymer components or exposure to ultraviolet (UV) light. Elastomers feature the capability to sustain large deformations with relatively low induced stresses. After the release of the external stress, they can recover to their original geometry without rupturing. Hybrid polymers are blends of the polymers belonging to the aforementioned three types, and they can have distinguishing properties compared to the individual constituents. In principle, all of these four types of polymers can be used as adhesives for joining wafers [6, 38].

As discussed earlier, there typically is a semiliquid or liquid phase of the intermediate polymer during wafer bonding in which the polymer reflows/deforms/diffuses and hence obtaining sufficiently close contact between the polymer and wafer surface molecules to form a bond. Then, the polymer hardens and transforms into a solid material to yield a permanent and reliable bond. Three common ways for such a polymer phase transformation to occur are the following [6, 35]:

- Polymers that are dissolved in solvents harden after the evaporation of the solvents. Such polymer adhesives are also called physical drying adhesives.
- Thermoplastic polymers can be melted by heating them. They solidify again upon cooling down to the melting temperature of the polymer, and the melting process can be repeated without changing the properties of the polymer. Such polymer adhesives are also called hot-melts.
- Thermosetting polymers can transition from a liquid or viscoelastic state before or during the polymer precursor is cross-linked (polymerized) to form polymer networks. The polymer precursor can be in a liquid phase (e.g. a resin) at room temperature or transform from a solid phase to a liquid/viscoelastic phase for a certain duration during the cross-linking process. For different thermosetting polymers, the cross-linking process can be initiated by various distinct mechanisms (i.e. ways of introducing the activation energy):
- Thermal curing (e.g. many thermosetting epoxies/polymers)

- Mixing of two or more components (e.g. two-component epoxies/polymers)
- Illumination with light (e.g. UV-curable adhesives/polymers)
- Presence of moisture (e.g. some polyurethanes and cyanoacrylates)
- Absence of oxygen (e.g. anaerobic adhesives/polymers)

There are also polymers that can harden and cure through a combination of two or several of the aforementioned mechanisms. For example, solvent-based thermosetting polymers (e.g. B-stage epoxies) involve evaporation of the solvent of the polymer precursor and subsequent cross-linking of the polymer precursor (e.g. by heating). Thus, the use of the solvent allows the adaptation of the viscosity of the polymer precursor to the specific reflow and coating requirements. Another example is the initiation of the cross-linking process of two-component polymers by illumination with UV light, with the cross-linking continuing even after removal of the UV light illumination.

In principle, most types of polymers can be used for wafer bonding, and there exist a variety of commercial polymer adhesives with distinct and specifically tuned material properties [6, 39]. When selecting a polymer for a wafer bonding application, the mechanical and environmental stability of the polymer has to be considered with respect to the application requirements. Polymers typically suffer to some extent from mechanical creep when exposed to an external load. The extent of creep is influenced by the ambient temperature, the period of time under the load, and the used polymer material [35, 39]. Polymers may also be affected by environmental stresses, such as chemicals, radiation (UV and gamma radiation), temperature, and biological deterioration, which can result in the changes of the properties of the polymers [35, 39]. It should also be noted that polymers are permeable to water and gasses. The permeability of polymers to gasses is typically several orders of magnitude higher than that of glasses and metals, as shown in Figure 15.3 [37]. Polymers exhibit a much higher permeability to gases and moisture due to the fact that gas molecules with small dimensions (on the order of 0.1–0.2 nm) can diffuse through the free space between the polymer molecules. For this reason it is not easily possible to form hermetic cavities or packages by using polymer wafer bonding.

The temperature stability of polymers is of particular importance when it comes to the choice of a polymer for a specific wafer bonding application. Generally, polymers transform from a hard (glassy) state into a viscous or rubber-like state at their polymer-specific glass transition temperature, and they are typically operated below their glass transition temperature. Typical thermoplastic polymers have operational temperatures of up to 200–300 °C, with their low-temperature end limited by brittleness. They can elongate and deform to a large extent when heated, until reaching a viscous state at higher temperatures. They typically have good peel strength, poor creep resistance, and varying chemical resistance depending on the specific polymer [35, 39]. Thermosetting polymers typically have a higher working temperature of up to 300–450 °C. They are generally more rigid than thermoplastic polymers and have good chemical resistance, creep resistance, and fair peel strength. When fully cross-linked, thermosetting polymers cannot reflow anymore but continue

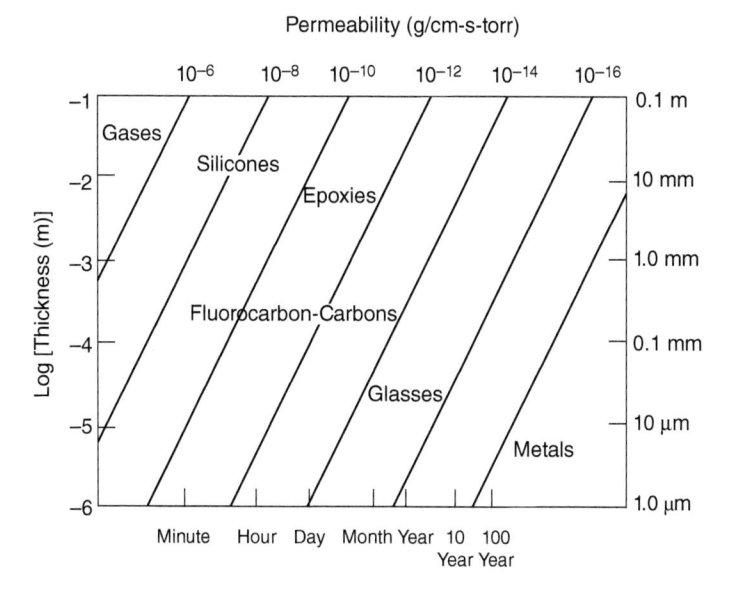

Figure 15.3 Approximate time scale for moisture to penetrate inside various materials (permeability of materials to H_2O molecules). Source: Traeger [37]. © 1976 IEEE.

to soften. At higher temperatures, they start to degrade and decompose [35, 39]. Elastomeric polymers have a useful temperature range of up to around 260 °C. They feature high peel strength but low overall strength, high flexibility, and varying chemical resistance [35, 39]. Hybrid polymers can realize a balanced combination of properties, which the other types of polymers possess. Some high-performance hybrid polymers can operate at temperatures of up to 760 °C for short times (e.g. polybenzimidazoles (PBIs)) [40], and some can have relatively low permeability for moisture (e.g. liquid crystal polymers (LCPs)), making them suitable for demanding applications [35, 39].

15.2.3 Polymers Used in Wafer Bonding

When selecting a suitable polymer adhesive for a wafer bonding application, several aspects need to be considered. For example, the polymer must be compatible with the involved wafer materials (including the materials and devices deposited on the wafers) and with the processing steps to which the bonded wafer stack will be subjected after the bonding. This includes that the polymer has to have sufficient mechanical strength, chemical resistance, and thermal stability to survive the semiconductor and MEMS processing steps involving different chemicals, gas atmospheres, and temperatures. Especially for applications where permanent bonds are formed and the polymer remains as a functional material during the lifetime of the device, the chemical stability and aging properties of the polymer are critical. In contrast, for applications where the polymer is used for temporary wafer bonding, as an intermediate step in the wafer manufacturing (e.g. for thin wafer handling), the long-term stability of the polymer is not necessarily important, but instead it is important that the polymer can be removed without leaving any polymer residues on the wafer surfaces. As another example, if polymer wafer bonding is used for lab-on-chip and biomedical applications, the polymer used for making the devices should typically be chemically inert and biocompatible.

A large number of polymers have been investigated for polymer wafer bonding applications as shown in Table 15.2. These polymers feature different reflow and hardening principles, such as cross-linking and UV curing (thermosetting) and softening by heating (thermoplastic), as discussed previously in this chapter. Many of the listed polymers are commercially available from suppliers such as Brewer Science of the United States (e.g. WaferBOND), Gersteltec Engineering Solutions of Switzerland (e.g. SU8), Dow Chemical Company of the United States (e.g. BCB (Cyclotene)), 3M of the United States (e.g. LC-Series UV-Curable Adhesives), and many more suppliers.

Importantly, polymers that outgas, produce by-products, or involve evaporation of solvents during the bonding typically are not suitable for standard wafer bonding applications [6, 44]. For instance, during the curing (imidization) process of many polyimide coatings, water vapor is produced as by-product of the imidization [6, 44]. This is problematic because typical semiconductor or glass wafers are not permeable (or porous) to gases and liquids, so the produced volatile substances would be trapped in between the wafers, causing voids and severe bond reliability issues. Such polymers can be used for wafer bonding only in exceptional cases in which either at least one of the two wafers to be bonded is permeable to gases [46, 47], such that the evaporation of volatile substances (e.g. solvents) is completed before joining the wafers to be bonded, or that the bonding areas incorporate ventilation channels, which allow the escape of the produced volatile substances [58].

Thermosetting and thermoplastic polymers that cure or reflow at elevated temperatures (e.g. up to 250 °C, depending on the polymer type) are typically suitable for bonding wafers consisting of the same material or of materials with similar coefficients of thermal expansions (CTEs). For wafers consisting of materials with large differences in their CTEs and that are bonded at high temperatures, the resulting bonded wafer stack can contain thermo-mechanical strain and strongly bend after cooling down to room temperature. This is because the wafer with the higher CTE material is expanded more than the wafer with the lower CTE material at the elevated temperature and also shrinks more after cooling. This induces stresses in the bonded wafer stack and even can cause cracking of the wafers during wafer bonding. Wafer bonding at, or near, room temperature can avoid these type of thermal stress issues, which requires the use of room temperature curable polymers such as two component or UV-curable epoxies [6].

In particular in wafer bonding with thermosetting polymers, the level of cross-linking of the polymer prior to bonding is critical for the reflow capability of the polymer and to achieve good wafer bonding results. To understand and optimize the polymer preparation for such as wafer bonding process, it can be instructive to look at the temperature and time-dependent viscosity (reflow capability) and cross-linking properties of the thermosetting polymer. As an example, Figure 15.4 shows the properties of the B-stage thermosetting polymer benzocyclobutene

Table 15.2 Polymers that have been proposed for adhesive wafer bonding (adapted from [34, 36]).

Polymer adhesives	Features	References
Epoxies	• Thermosetting polymers • Thermal curing or two-component curing • Strong and chemically stable	[6, 34, 36, 41, 42]
UV epoxies (e.g. SU8)	• Thermosetting polymers • UV curing (one of the substrates has to be transparent to UV light) or curing initiated by UV light • Strong and chemically stable • Suitable for localized bonding with patterned polymer layer	[6, 21, 34, 36]
Nano-imprint resists	• Thermosetting polymer versions (optional UV curable) and thermoplastic polymer versions available • Optimized for good reflow around micro- and nanoscale surface structures, thus typically well-adapted for wafer bonding	[6, 9, 10, 34, 36, 43]
Positive photoresists	• Typically thermoplastic polymers • Typically weak wafer bonds • Void formation can occur at the bond interface for some resists • Suitable for localized bonding with patterned polymer layer	[6, 34, 36, 43]
Negative photoresists	• Typically thermosetting polymers • Thermal curing and/or UV curing • Typically weak wafer bonds • Suitable for localized bonding with patterned polymer layer	[6, 34, 36, 43]
Benzocyclobutene (BCB, Cyclotene)	• Thermosetting polymer • Thermal curing or UV curing (photosensitive BCB) • Defect-free, very strong, and chemically and thermally stable wafer bonds • Suitable for localized bonding with patterned polymer layer	[2, 3, 6, 8, 12, 14, 15, 27–29, 31–34, 36, 44]
Polymethyl methacrylate (PMMA)	• Thermoplastic polymer	[6, 45–47]
Polydimethylsiloxane (PDMS)	• Elastomeric polymer • Thermal curing • Typically plasma-activated bonding • Biocompatible	[6, 48]
Fluoropolymers (e.g. Teflon, Flare)	• Thermosetting polymer versions and thermoplastic polymer versions available • Chemically very stable bond • Suitable for localized bonding with patterned polymer layer	[6, 49]

(*continued*)

Table 15.2 (Continued)

Polymer adhesives	Features	References
Polyimides	• Thermosetting polymer versions and thermoplastic polymer versions available • Importantly, for many polyimides, water is generated as by-product in the imidization process, which can cause voids at the bond interface • Excellent temperature stability • Suitable for localized bonding with patterned polymer layer	[6, 11, 44]
Polyetheretherketone (PEEK)	• Thermoplastic polymer • Good temperature stability	[6, 50]
Thermosetting copolyesters (e.g. ATSP)	• Thermosetting polymers	[6, 51]
Thermoplastic copolymers (e.g. PVDC)	• Thermoplastic polymers • Good gas barrier	[6, 52]
Parylene	• Thermoplastic polymer • Suitable for localized bonding with patterned polymer layer	[6, 53, 54]
Liquid crystal polymers (LCPs)	• Thermoplastic polymers • Good moisture barrier	[6, 55]
Waxes	• Thermoplastic materials • Low thermal stability • Typically used in temporary wafer bonding	[6, 56, 57]

Source: Adapted from Ramm et al. [34], Niklaus [36].

(BCB), which is a cross-linking polymer that has a temperature-dependent viscosity during the cross-linking (curing) process. This allows for some degree of reflow of the polymer for a short time during the curing process until cross-linking of the polymer chains is complete, after which the polymer cannot be reflowed. In the example of BCB, the minimum viscosity of 1000 poise is reached at around 170–190 °C. However, once the cross-linking level of the polymer is changed, the viscosity dependence on temperature is also permanently changed, which is in contrast to thermoplastic polymers, which feature constant viscosity dependence on temperature even after repeated thermal cycles. Thermosetting polymers such as BCB are typically delivered as liquid polymer precursors, which have a preset initial cross-linking level such as 20%–50% (approximately 35% for dry-etch BCB). Further increasing the percentage of cross-linking of the BCB precursor as a function of curing temperature and time is shown in Figure 15.4b. For example, an increase of the cross-linking level from the initial 35% to approximately 43% is achieved by a pre-curing time of 30 min at a curing temperature of 190 °C.

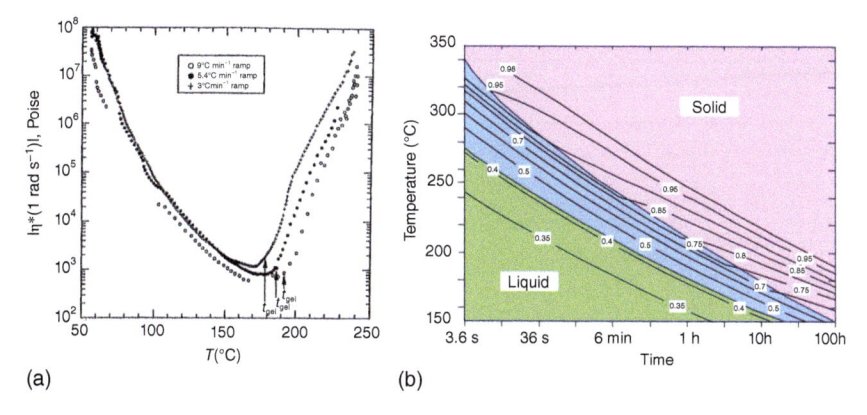

Figure 15.4 (a) Viscosity of a thermosetting polymer (BCB) in dependence of the temperature over three different temperature ramping speeds during the curing process and (b) percentage of cross-linking of the thermosetting polymer (BCB) as a function of curing times and temperatures. Source: Niklaus et al. [3]. © 2006 IOP Publishing.

15.3 Polymer Wafer Bonding Technology

To realize high-quality wafer bonding results and superior repeatability, the wafer bonding process parameters, including bonding pressure, bonding time, environmental gas pressure, and temperature ramping profiles, must be precisely controlled for the used polymer bonding material. The influences of the different bonding parameters on the resulting bond quality, e.g. defects formed at the bond interface, are qualitatively presented in this section. We also present a process scheme for a typical polymer wafer bonding process.

15.3.1 Process Parameters in Polymer Wafer Bonding

The bonding of wafers with an intermediate polymer layer can be performed using standard commercial wafer bonding equipment or hot presses. The major components of a standard wafer bonder are an atmosphere-controllable bonding chamber, two bond chucks for applying forces and heat to the wafer stack placed in between the bond chucks, and wafer fixture for handling and transferring the wafers in and out of the bonding chamber. The bond chucks are typically flat plates made of hard materials, e.g. a metal or silicon carbide. To realize more even distribution of the bonding pressure on the wafer stack and avoid pressure concentration points that, for example, can be caused by particles on the bond chucks, soft sheets (e.g. graphite or silicone) may be placed between the bond chucks and the wafer stack. Details of the basic steps of a wafer bonding process using an intermediate thermoplastic or thermosetting polymer are listed in Table 15.3.

There are different methods for coating the polymer on the wafer surface(s). Typical polymer coating thicknesses in the range of 0.1–100 μm are used for wafer

Table 15.3 Basic process steps for polymer wafer bonding [6, 34].

No.	Process step	Purpose of the process step
1.	Cleaning and drying of the wafers	Removing particles, contaminations and moisture from the wafer surfaces by e.g. ultra-sonication or spin rinsing.
2.	Optionally: Coating an adhesion promoter on the wafers	Enhancing the adhesion between the wafer surfaces and the polymer.
3.	Coating the polymer on the surface of one or both wafers. Optionally: Patterning of the polymer	Appling the polymer on the wafer surfaces. This may be done by spin-coating, screen-printing, spray-coating, etc. For optional polymer patterning, see Section 15.3.2.
4.	Soft-baking or partially cross-linking of the polymer	Removing solvents and volatile substances from the coated polymer. Thermosetting adhesives should not be fully cross-linked to remain deformable and bondable (see Figure 15.4).
5.	Placing the wafers in the bond chamber Optionally establishing a low-pressure atmosphere and joining the wafers inside the bond chamber	The low-pressure atmosphere is to prevent voids and gases from being trapped at the bond interface. Alternatively, the low-pressure atmosphere can be established after joining of wafers, as long as trapped gasses can be pumped away from the bond interface before the bond is initiated.
6.	Applying bonding pressure to the wafer stack with the bond chucks	Forcing the wafer and polymer surfaces into intimate contact over the entire wafers. For thermosetting polymers, the bond pressure should be applied before the polymer is cross-linked. For thermoplastic polymers, the bond pressure may be applied before or after the polymer reflow temperature is reached.
7.	Heating of the wafer stack through the top and bottom bond chucks	Initiating the softening, reflow or cross-linking of the polymer. If a room-temperature-curable polymer is selected, cross-linking may happen at or near room temperature.
8.	Cooling of the bond chucks near room temperature, releasing the bond pressure, and purging the chamber	Finalizing the bonding process. The sequence of cooling, bond force release, and chamber purge is usually interchangeable. However, for thermoplastic polymers, the bond pressure should not be fully released before the cooling to ensure polymer solidification before bond force release.

Source: Niklaus et al. [6], Ramm et al. [34].

bonding in microelectronics and MEMS [6, 58]. Spin-coating of a liquid polymer precursor is the prevalent method to realize a highly uniform and controlled thickness and smooth surfaces of the coated polymer. During coating and bonding, the polymer can compensate to some extent the topographic variations of the wafer surfaces [6]. A more uniform wafer surface lowers the required reflow capability of the polymer for realizing good contact between the surfaces to be bonded, and hence this is potentially beneficial for improved bond quality control.

Alternative methods for depositing the polymer coating include electro-deposition, stamping, screen-printing, brushing, and dispensing of liquid polymer precursors [6]. However, with these approaches, the resulting uniformity of the polymer coating on flat wafer surfaces is typically inferior to the spin-coating approach. In addition, chemical vapor deposition (CVD) and atomic layer deposition (ALD) processes can also be used for coating polymers on the wafer surfaces [6], but they required dedicated processing equipment. Another indirect way to coat polymers is by lamination of thin polymer films or sheets on the wafer surface [6].

The quality of the final polymer wafer bond is influenced by different factors. These factors include the polymer properties (e.g. reflow capabilities and viscoelastic behavior during bonding), the wafer surface topography, the size and number of particles at the wafer surfaces, the polymer adhesive layer thickness, the wafer bonding process parameters (bonding pressure, time, temperature, and temperature ramping profile), and the wafer stiffness. The importance and effects of these influencing parameters are summarized in Table 15.4.

High quality and void-free bonds are more achievable with polymers that obtain a low viscosity during the bonding process due to easier reflow and redistribution of these types of polymers at the bond interface. On the other hand, a high reflow capability of the polymer can result in a nonuniform thickness distribution of the polymer layer at the bond interface since such polymers tend to flow from local areas of high bonding pressure to local areas of low bonding pressure [3, 6]. If a highly uniform polymer layer should be achieved at the interface and the topographic variations of the wafer surfaces are small, polymers with low reflow capabilities can be used. For thermoplastic polymers, such uniform intermediate bonding layers can be realized using bonding temperatures, which are below the melting point of the polymer. For thermosetting polymers, this can be achieved by using partially cross-linked polymers, which already have limited reflow capabilities [3, 6]. Figure 15.5 shows an example where the same thermosetting polymer (BCB) was used in two wafer bonds; however different cross-linking degrees of the polymer before the bonding resulted in distinct bonding results. The images show top views of the bond interfaces after the top wafers were sacrificially removed by etching. It can be seen that, although both bonds are void free, the polymer that was cross-linked to a lesser degree before bonding (having a lower viscosity during bonding) resulted in a bond interface with significant thickness variations of the polymer layer after bonding, as indicated by the color fringes (Figure 15.5a). On the contrary, the polymer that had a higher level of cross-linking prior to bonding resulted in a more uniform polymer layer at the bond interface (Figure 15.5b).

The capability of polymers to reflow at the bond interface and adapt to the surface topographies of the two wafers can be described by simulation models that have been developed for nano-imprinting processes, which describes a very similar situation as in wafer bonding [6, 43, 59, 60]. Time-dependent reflow behavior of nano-imprint resists around structures, and cavities on the two contacting substrates (molds) to replicate them are shown in Figure 15.6, which is equivalent to the situation at the interface of two polymer bonded wafers. Thus, these types of models can also be used to qualitatively predict void formation due to the surface topographies in polymer

Table 15.4 Influence of various polymer wafer bonding parameters on the resulting bonds [6, 34].

Parameter	Relevance for and effects on the resulting bond	Significance
Polymer properties	Compatibility of the polymer, the involved solvents, and the bonding temperatures with the wafer materials.	Very strong
	Sufficient wetting of the wafer surface material(s) by the polymer.	
	Sufficient adhesion between the wafer surface material(s) and the polymer.	
	No outgassing or creation of volatile by-products from the polymer cross-linking process during the bonding process in order to avoid voids from being trapped at the bond interface.	
	If at least one of the wafers is permeable to gasses or if ventilation channels are incorporated at the bond interface, outgassing polymers may be tolerable.	
	The intermediate polymer layer must reflow or plastically deform to adapt to the wafer surface topography during bond. This may be achieved by heating of a thermoplastic polymer or by cross-linking of a thermosetting polymer precursor.	
	Polymers, if in a very low-viscous phase, tend to flow from areas of high bond pressure towards areas of lower bond pressure. This can result in substantial thickness variations of the polymer layer at the bond interface. Such an effect can be counteracted by controlling/optimizing the viscosity of the polymer during the heating and bonding process.	
Number and size of particles at the bond interface	The presence of particles at the wafer surfaces and bond interface can cause circular unbonded areas around the particles. Small particles with sizes on the order of the thickness of the polymer layer may to some extent be embedded in the polymer and not necessarily result in unbonded areas.	Strong
Wafer surface topography	If the wafer surface topographies are high compared to the polymer layer thickness, unbonded areas may result.	Strong
	Planarizing polymer deposition processes such as spin-coating can have a planarizing effect on wafer surface topographies and hence can reduce the tendency for unbonded areas.	
	Polymers with strong reflow capabilities (i.e. obtaining a very low-viscous phase) during the bonding process can compensate to some extent for wafer surface topographies.	

(*continued*)

15.3 Polymer Wafer Bonding Technology

Table 15.4 (Continued)

Parameter	Relevance for and effects on the resulting bond	Significance
Polymer layer thickness	Thick polymer layers compared to wafer surface topographies can reduce the tendency for void formation or unbonded areas due to stronger planarizing and reflow effects. Very thin polymer layers compensate to a lesser extent for surface topographies, nonuniformities and particles at the bond interface.	Strong
	Thicker polymer layers can to some extent alleviate the stresses induced by thermal expansion mismatch between bonded wafers, due to the increased viscoelastic deformation capabilities of a thicker polymer layer.	
Bonding pressure (force applied with the bond chucks divided by the bond area)	The bonding pressure causes the polymer layer and the wafers to deform, thereby bringing the surfaces in close contact by compensating the nonuniformities of the wafer surface topographies and wafer thickness. The bonding pressure typically has a significant impact on reducing void formation and can be adjusted in a relatively flexible way, i.e. increasing the bonding pressure typically results in reduced number of voids and defects at the bond interface.	Very strong
	An excessive bonding pressure may result in wafer cracking due to the high stresses induced on the wafers.	
	Soft sheet(s) (e.g. graphite or silicone foam sheets) can be placed between the wafer stack and the bond chuck(s) to more uniformly distribute the bonding pressure on the wafers, thus reducing the risk of stress concentration points and resulting wafer cracking.	
Temperature ramping profile and bonding temperature	The bonding temperature and temperature ramping profile must be tailored to the selected polymer.	Strong / medium
	For thermosetting polymers, the bonding and curing temperatures in combination with the curing times must be high enough for the polymer to achieve sufficient cross-linking (compare to Figure 15.4b). The low-viscosity state of the polymer typically occurs at a temperature lower than the cross-linking temperature. Thus, a wafer holding temperature and force may be introduced at the polymer reflow temperature to allow sufficient reflow of the polymer before it is fully cross-linked (compare to Figure 15.4a).	
	For thermoplastic polymers, the bonding temperature must result in a sufficiently low viscosity of the polymer to ensure reflow at the bond interface. The reflow capabilities (viscosity) of thermoplastic polymers can be tuned to some extent by the bonding temperature.	
	For bonding of two different wafers with materials that have very different coefficients of thermal expansion, low bonding temperatures and/or very slow heating and cooling cycles can reduce the thermally induced stresses and the risk for wafer cracking.	

(continued)

Table 15.4 (Continued)

Parameter	Relevance for and effects on the resulting bond	Significance
Bonding time	The bonding time in combination with the bonding temperature has an influence on the reflow and redistribution behavior of the polymer at the bond interface. Too short bonding times might result in nonuniform heating of the wafer stack and/or result in void formation at the bond interface.	Strong / medium
	For thermosetting polymers, the extent of cross-linking of the polymer can be controlled by selecting a suitable combination of bonding time and bonding temperature (see Figure 15.4b).	
Gas pressure in the bond chamber before wafer bonding is initiated	Gas pressures of below 100 mbar in the bond chamber before joining the wafer surfaces are typically sufficient to avoid trapping of gas pockets at the bond interface.	Medium
	Pumping out trapped gas after joining the wafer surfaces but before the wafers are bonded is also possible. In this case, the wafers may be joined before a low-pressure atmosphere is established in the bond chamber.	
Wafer stiffness	Thinner wafers and wafers made of materials with a lower Young's modulus are easier to elastically deform as a result of a given bonding pressure and thus can better deform and adapt to nonuniformities of the wafer thicknesses and surface topographies. This can result in reduced void formation at the bond interface.	Medium

Source: Niklaus et al. [6], Ramm et al. [34].

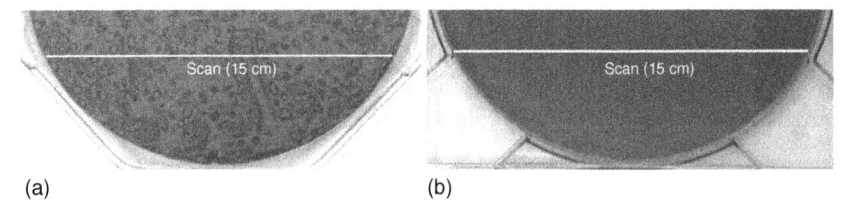

(a) (b)

Figure 15.5 Bonded wafer pairs in which the top wafers have been sacrificially removed to expose the intermediate polymer at the bond interface. (a) Resulting bond interface using less cross-linked thermal setting polymer prior to bonding. (b) Resulting bond interface using more cross-linked thermal setting polymer prior to bonding. The color fringes in (a) result from thickness variations of the polymer layer at the bond interface, which are caused by polymer redistribution during bonding. Source: Niklaus et al. [3]. © 2006 IOP Publishing.

wafer bonding. The related simulations and experiments showed that the resulting filling of a structured surface with the imprint resist is influenced by the dimensions and aspect ratios of the cavities and structures at the surface. Both in polymer bonding and in nano-imprint lithography, the polymer impacted by the protruding features of the surface to be replicated must be transported to the nearby trenches and cavities. When the polymer does not reflow sufficiently and the trenches and

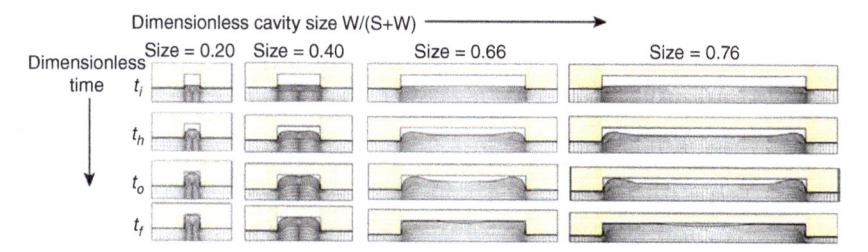

Figure 15.6 Example of a simulation of the step-by-step polymer flow into surface cavities with different aspect ratios during nano-imprinting, which also is applicable to polymer wafer bonding. S and W represent cavity spacing and cavity half width, respectively. Source: Rowland et al. [60]. © 2005 IOP Publishing.

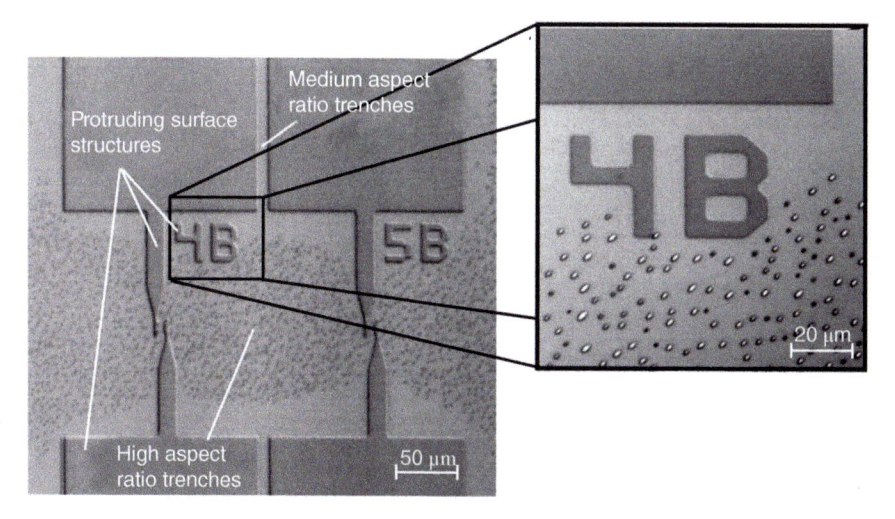

Figure 15.7 Local micro-void formation when bonding a wafer with 50 nm protruding silicon structures at the wafer surface to an unpatterned silicon wafer using a thermosetting polymer layer with a thickness of 300 nm. The top wafer has been sacrificially removed to visualize the polymer layer at the bond interface. Source: Niklaus et al. [43]. © 2009 Elsevier.

cavities are not fully filled, voids are formed. The simulated polymer filling behaviors that are dependent on the dimensions and aspect ratios of the trenches and cavities have found agreement in polymer bonding experiments, as shown in Figure 15.7 [43, 59, 60]. Micro-voids tend to form in wide, high-aspect ratio trenches rather than narrow trenches. Although this type of voids usually do not compromise the strength of the overall polymer wafer bond, they are undesirable. These simulation models also confirm the influence of some of the parameters listed in Table 15.4 on the bond quality, such as increased bonding pressure, longer bonding time (or slow temperature ramping) for the polymer reflow, thicker coated polymer layer, and higher reflow capabilities of the polymer, which enhances the filling and molding of the polymer around the surface topographies, hence reducing the risk of void formation at the bond interface.

Apart from the aforementioned wafer bonding schemes using thermosetting or thermoplastic polymers in combination with heating, polymer bonding with UV-curable thermosetting polymers is also a frequently used method. The main advantage of UV-curable polymers is that the polymer cross-linking can be completed near room temperature by UV light illumination. Therefore, such a bonding scheme is beneficial for joining wafers consisting of materials with different CTEs. However in these schemes, at least one of the wafers has to be transparent to UV light to initiate the cross-linking. In this context it should be noted that conventional wafer bonding equipment typically does not incorporate UV exposure capabilities.

15.3.2 Localized Polymer Wafer Bonding

In localized or selective polymer wafer bonding, only predefined parts of the wafer surfaces are bonded instead of bonding the wafers with a continuous (unpatterned) polymer layer. This can be realized by coating polymers only on areas of the wafer surfaces to be bonded [6, 27, 36], by using structured wafer surfaces that only allow protruding wafer areas to be bonded [29, 34, 36], by special treatments on the unwanted surface areas to locally disable the bondability of the wafer surface [61], or by incorporating localized heating at the bond interface to create local bonding [6, 62]. Four conceptual examples of common schemes for realizing localized bonding with patterned polymer layers or structured wafer surfaces are shown in Figure 15.8.

The patterning of the polymer with precisely defined structures can, for example, be performed using photolithographic techniques [6, 27]. These schemes can be typically classified into two major approaches, namely, the patterning of the polymer by either etching of the polymer using a lithographically defined mask or by using a photosensitive polymer (Figure 15.8a) [27]. Another approach is based on the patterning of the wafer surfaces to be bonded using, for example, wafer etching or material deposition on the wafer surface. The polymer can then be coated on the patterned surfaces and/or the opposing bonding surfaces, for example, by spray-coating or transfer stamping to enable localized bonding (Figure 15.8c,d) [29].

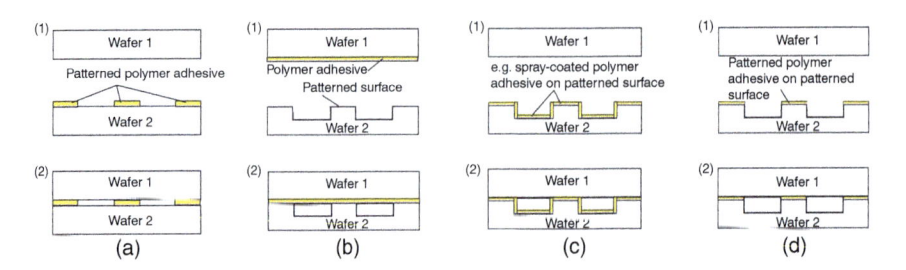

Figure 15.8 Examples for localized polymer wafer bonding with (a) a patterned polymer layer, (b) a patterned wafer surface, (c) a spray-coated polymer on a patterned wafer surface, and (d) a patterned polymer layer on a patterned wafer surface. Source: Ramm et al. [34]. © 2011, John Wiley & Sons.

15.3 Polymer Wafer Bonding Technology

In general, the bond quality of localized polymer wafer bond may be affected by the bonding parameters similar to that in wafer bonding with self-contained (unpatterned) polymer layers as discussed in Section 15.3.1. However, it should be noted that in localized wafer bonding, the equivalent bonding pressure resulted from the bonding force introduced to the wafer stack by the bond chock, divided by the patterned bonding areas (not the total wafer area). In localized wafer bonding, the polymer has to have suitable reflow capabilities to both retain the pre-patterned shape of the polymer layer but still adapt to the wafer surfaces on a nanoscopic level to form void-free and stable bonds. If the polymer has too little reflow ability, incomplete bonds due to insufficient deformation of the polymer may occur. On the other hand, if the polymer retains a too low viscosity (too high reflow capability) during the bonding process, the patterned polymer layer can lose its defined pattern, which results in loss of the patterned structures and uncontrolled thickness of the polymer layer at the bond interface [27]. For polymers that are patterned using lithographic processes (see Figure 15.8a), the useful process window for wafer bonding typically is narrower and requires more process optimization as compared to wafer bonding processes with unpatterned polymer layers [27]. An example of a localized polymer wafer bond with a pre-patterned polymer layer is shown in Figure 15.9a. Here a glass wafer was locally bonded to a Si wafer with a lithographically patterned dry-etch BCB later. The bond interface can be observed through the glass wafer, and the darker areas in Figure 15.9a indicate the bonded areas, whereas the lighter areas indicate the unbonded areas. In comparison, localized polymer wafer bonding with a broader process window can be obtained with an approach in which a conformal polymer coating is deposited on an already structured wafer surface as illustrated in the cross section (Figure 15.9b) [29].

Methods for patterning a polymer layer on a wafer surface, other than photolithographic techniques and spray coating, include screen printing of a liquid polymer, stamping of a polymer, and local dispensing of a liquid polymer.

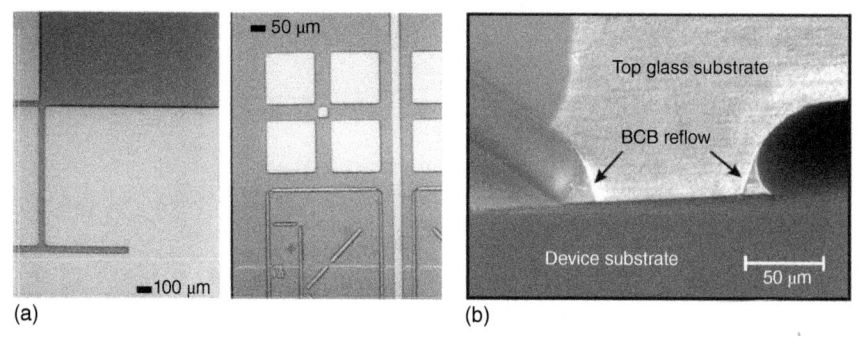

Figure 15.9 (a) Top view of a glass wafer that is selectively bonded to a silicon wafer using a lithographically patterned and etched thermosetting polymer adhesive as indicated in Figure 15.8a. The darker areas are bonded areas and the lighter areas are unbonded. Source: Oberhammer et al. [27], © 2003 Elsevier. (b) Cross section of localized bond using a spray-coated BCB layer on protruding surface structures (according to Figure 15.8c). Source: Bleiker et al. [29]. CC BY 4.0.

In addition, pre-patterned polymer sheets can be laminated on the wafers [63]. The pre-patterning of the polymer sheets may be done, for example, by punching or cutting with a laser or water jet. Localized polymer bonding can also be combined with other bonding principles, such as direct metal-to-metal bonding or solder bonding to form a hybrid wafer bond [2, 34, 64]. In these approaches, the minimum required bonding temperature for one bond type (e.g. the direct metal bond) should not exceed the maximum temperature for the polymer bond, which is formed in parallel at the bond interface.

15.4 Precise Wafer-to-Wafer Alignment in Polymer Wafer Bonding

For many wafer bonding applications, the alignment accuracy between the bonded wafers is critical. Typical wafer-to-wafer alignment requirements can range from sub-micrometer accuracy to a few tens of micrometers. To realize precise alignment of two wafers, several solutions have been implemented in commercial wafer bonding equipment [6], including optical microscopy for transparent wafers, backside alignment using digital cameras, infrared (IR) transmission microscopy, through-wafer holes with optical microscopy, inter-substrate microscopy, and Smartview© approach. It should be noted that in polymer wafer bonding, it can be challenging to retain the precise pre-bond wafer-to-wafer alignment accuracy during the wafer bonding process because the polymer typically takes on a low-viscos state during the wafer bonding. This can cause a relative shift of the joined wafers due to inevitable shear forces when the bond chuck applied the bond pressure to the wafer stack. This effect frequently causes increased post-bond misalignment between the wafers in polymer wafer bonding. One approach to address this challenge is to use a polymer that maintains relatively high viscosity during the bonding process so that the reflow and redistribution of the polymers is limited [3]. This can, for example, be realized by using a thermosetting polymer that is partially cross-linked prior to bonding [3] or by using thermoplastic polymer in combination with bonding temperatures that are below the melting point of the polymer. The limitation of these approaches is that complete reflow of the polymer during wafer bonding is avoided, which also reduces the capability of the bonding process to compensate for wafer surface topographies at the bond interface. Another approach to address this challenge is to incorporate structures at the wafer surfaces that do not reflow and that are not covered by the polymer on the wafer surfaces, as shown in Figure 15.10a. These structures introduce friction between the two wafer surfaces and thus can avoid the relative shifting of the wafers during the bonding when the intermediate polymer is in a low-viscosity phase [6, 34, 36, 65]. Materials such as metals, dielectrics, or fully cross-linked thermosetting polymers can be used for these frictional structures. A third approach employs corresponding interlocking structures at the two wafer surfaces to be bonded, which prevent the two wafers from relative shifting with respect to each other, as shown in Figure 15.10b. The interlocking structures can be formed by etching them into the wafer surfaces or by

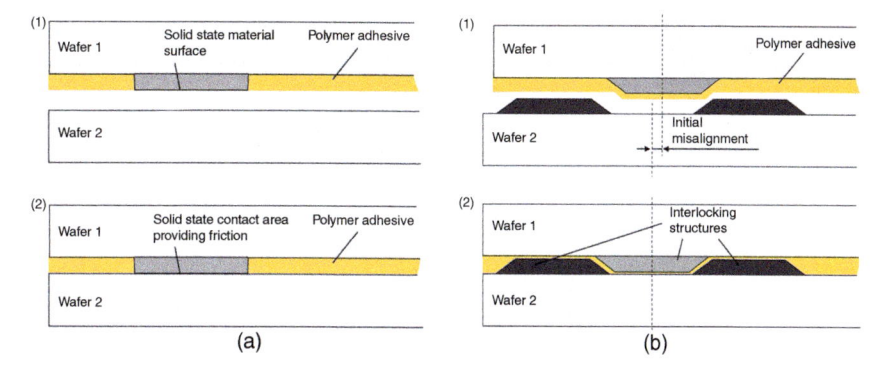

Figure 15.10 Two approaches to improve wafer-to-wafer alignment accuracy in polymer adhesive wafer bonding using (a) frictional non-reflowable structures and (b) interlocking alignment structures on the wafer surfaces. Source: Ramm et al. [34]. © 2011, John Wiley & Sons.

coating and patterning of metals, dielectric materials, or cross-linked thermosetting polymers. The self-alignment effect of such interlocking structures can help achieve sub-µm wafer-to-wafer alignment accuracy [66]. It should be noted that alignment errors can also be introduced when bonding wafers consisting of materials with different CTEs. After heating such wafer pairs to elevated temperatures during the bonding process, the thermal mismatch induced by the differential expansion of the two wafers will result in a relative shift between the two wafers, which can easily be of the order of several tens of micrometers. Therefore, if accurate alignment is desired, the wafers to be bonded should have matched CTEs or the bonding must take place at, or near, room temperature.

15.5 Practical Examples of Polymer Wafer Bonding Processes

In this section we present examples of suitable wafer bonding processes with two different types of polymers, which are commonly used for wafer bonding (BCB and a nano-imprint resist). These bonding processes can serve as possible starting points for the development of more advanced polymer wafer bonding processes. For specific applications, the bonding parameters can be adapted or changed. To counteract void formation at the bond interface, the bonding pressure can typically be increased as long as no wafer breakage is induced. In the development of new polymer wafer bonding processes, it is beneficial to use glass wafers in the bonding experiments because this approach allows easy optical inspection through the glass wafer of bond defects such as voids at the bond interface.

One polymer that has been extensively used in polymer wafer bonding is dry-etch BCB (Dow Chemical Company Inc., United States), which is a B-stage thermosetting polymer that is compatible with semiconductor manufacturing environments. Bonding with BCB typically results in defect-free, strong, and

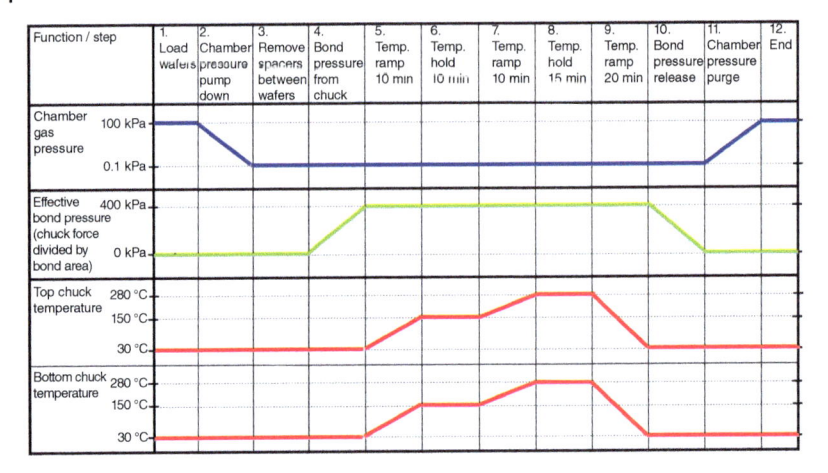

Figure 15.11 Example of bonding process details for polymer wafer bonding with thermosetting polymers. Source: Ramm et al. [34]. © 2011, John Wiley & Sons.

permanent wafer bonds, resulting from the excellent chemical resistance of BCB to various solvents and acids and its thermal stability to temperatures of up to 300 °C. Polymer wafer bonding with intermediate BCB layers that are from below 1 μm and up to >10 μm have been demonstrated [6, 7, 44]. Wafer bonding with BCB as intermediate polymer features good tolerance to topographies and particles present at the wafer surfaces, thanks to the excellent reflow capabilities of BCB (see Figure 15.4a). Dry-etch BCB typically is supplied in form of a liquid precursor, which can be spin-coated on a wafer surface (optionally in combination with an adhesion promoter). The spin-coated BCB layer can then be soft-baked (to evaporate the solvents from the polymer) and optionally partially cross-linked (cured) to tailor the reflow capabilities of the coating (see Figure 15.4b). Wafers with soft-baked BCB coatings can be stored for weeks in a particle-free environment before wafer bonding, without compromising the resulting bond quality. An example of the wafer preparation procedure and a wafer bonding process using dry-etch BCB are shown in Table 15.5 and Figure 15.11, respectively.

The nano-imprinting resist mr-I 9000 series (micro resist technology GmbH, Germany) is another B-stage thermosetting polymer, which is suitable for use in adhesive wafer bonding and is compatible with semiconductor manufacturing environments. This polymer is suitable for temporary wafer bonding applications in which the bond can be easily de-bonded or the polymer can be sacrificially etched in a post-bonding step, for example, using oxygen plasma etching. Wafer bonding with the mr-I 9000 series polymer has been demonstrated for heterogeneous integration of MEMS and nano-electromechanical system (NEMS) with ICs [6, 7, 9, 10]. The demonstrated layer thicknesses of the intermediate polymer in wafer bonding are ranging from 0.3 to 5 μm [7, 22, 58]. The resulting bonds can tolerate temperatures of approximately 100 °C and up to 250 °C under certain conditions. Wafer bonding with the mr-I 9000 series polymer typically results in void-free and uniform bonds. Similar to BCB, wafers with soft-baked mr-I 9000 polymer coatings can be stored

Table 15.5 Example of a preparation sequence for wafer bonding with thermosetting polymers (e.g. dry-etch BCB or mr-I 9000) as intermediate layer.

No.	Process step	Comments
1.	Wafer cleaning in an ultrasound bath with de-ionized water	Alternative cleaning procedures to remove particles from the wafer surfaces can be used.
2.	Wafer drying in a rinse and dry equipment	Alternative procedures, e.g. drying at elevated temperatures in an oven, can be used.
3.	Spin- or spray-coating polymer on one or both of the wafers to be bonded	The resulting polymer layer thickness depends on the viscosity of the polymer precursor and the coating parameters.
4.	Soft baking of the polymer coating(s) by placing the wafer(s) on a hotplate at 110 °C for 2 min	Evaporating solvents (without cross-linking the thermosetting polymer, e.g. 110 °C for 2 min is suitable for both BCB and mr-I 9000). Other temperatures and times may also be applicable.
5.	Option 1 Placing two wafers on the bond fixture (with or without spacers separating the bond surfaces) and transferring the fixture to the bond chamber	Option 1 The excellent reflow ability of BCB during bonding can readily compensate for topographies or particles at the wafer surfaces. However, BCB redistributes easily during bonding from areas with higher pressure to areas with lower pressure, which could lead to nonuniform layer thicknesses after bonding. Mr-I 9000 typically has a more balanced reflow capability.
	Option 2 Pre-curing of the BCB coating(s) by placing the wafer(s) on a hotplate in a non-oxidizing atmosphere (e.g. nitrogen) for 30 min at 190 °C to partially cross-link the BCB, followed by placing the two wafers on the bond fixture (with or without spacers separating the bond surfaces) and transferring the fixture to the bond chamber	Option 2 The partially cross-linked BCB layer reflows only marginally during bonding. Thus, very well-defined post-bond BCB layer thicknesses at the bond interface can be achieved. However, the BCB layer then cannot compensate for large topographies or particles at the wafer surfaces. Different combinations of pre-curing temperatures and times can be chosen to provide different reflow viscosities of the BCB during bonding. Mr-I 9000 does not need to be partially cross-linked as it already has a well-balanced reflow capability.

for weeks in a particle-free environment before the wafer bonding process, without deteriorating the resulting bond quality. Mr-I 9000 is supplied as a liquid precursor, and coatings can be prepared using the process sequences shown in Table 15.5. A bonding process using the mr-I 9000 series polymer can follow the same steps as outlined in Figure 15.11, with the exception of adapting the temperature ramping and holding time to the following parameters: Step 5 – temperature ramping to 110 °C in 5 min, Step 6 – temperature hold for 10 min at 110 °C, Step 7 – temperature ramping to 200 °C in 20 min, Step 8 – temperature hold for 30 min at 200 °C, and Step 9 – cooling down to 30 °C in 10 min.

15.6 Summary and Conclusions

Polymer wafer bonding is a generic and CMOS-compatible wafer bonding technology, which enables possible solutions to challenges in the manufacturing, integration, and functional enhancement of micro- and nano-systems. Except for CMOS compatibility, polymer wafer bonding has many other advantages, including insensitivity to surface topographies and small particles, low bonding temperatures, and ability to bond wafers made from practically any material. Polymer wafer bonding typically does not require special surface treatments, such as intensive cleaning or surface planarization, which are demanded by many other wafer bonding methods. All of these beneficial features make polymer wafer bonding a very simple, robust, and low-cost manufacturing process.

Polymers that are suitable for wafer bonding applications and that have been widely employed include B-stage thermosetting polymers, such as BCB, SU8, nano-imprint resists, and some negative photoresists, and most thermoplastic polymers, such as polymethyl methacrylate (PMMA). They can be used for polymer wafer bonding either as self-contained polymer layers or as patterned layers. Many corresponding polymer wafer bonding process schemes and bonding parameters can be readily found in literature.

Polymer wafer bonding has shown utility in many applications such as temporary bonding for thin wafer handling, transfer of 2D materials, manufacturing of 3D-ICs, photovoltaic cells and heterogeneous integration of ICs with MEMS components such as IR detector arrays, digital arrayed micro-mirrors, and micro-tips for data storage devices. Furthermore, polymer wafer bonding has been used for manufacturing and functional packaging of advanced microsystem devices, such as RF-MEMS, bio-MEMS, µTAS, microoptoelectromechanical systems (MOEMS), and photonic laser systems.

References

1 Lu, J.-Q. (2009). 3-D hyperintegration and packaging technologies for micro-nano systems. *Proc. IEEE* 97 (1): 18–30.
2 McMahon, J.J., Niklaus, F., Kumar, R.J. et al. (2005). CMP compatibility of partially cured benzocyclobutene (BCB) for a via-first 3D IC process. Proceedings MRS 2005, Vol. 863, W4.4, San Francisco, USA.
3 Niklaus, F., Kumar, R.J., McMahon, J.J. et al. (2006). Adhesive wafer bonding using partially cured benzocyclobutene (BCB) for three-dimensional integration. *J. Electrochem. Soc.* 153 (4): G291–G295.

4 Dragoi, V., Filbert, A., Zhu, S., and Mittendorfer, G. (2010). CMOS wafer bonding for back-side illuminated image sensors fabrication. Proceedings IEEE 2010 11th International Conference on Electronic Packaging Technology & High Density Packaging, Xi'an, China, 27–30.

5 Pain, B., Sun, C., Vo, P. et al. 2007. Wafer-level thinned monolithic CMOS imagers in a bulk-CMOS technology. Proc. International Image Sensor Workshop Proceedings, 158–161.

6 Niklaus, F., Stemme, G., Lu, J.-Q., and Gutmann, R.J. (2006). Adhesive wafer bonding. *J. Appl. Phys., Appl. Phys. Rev. - Focused Rev.* 99 (1): 031101.1–031101.28.

7 Lapisa, M., Stemme, G., and Niklaus, F. (2011). Wafer-level heterogeneous integration for MOEMS, MEMS and NEMS. *IEEE J. Sel. Top. Quantum Electron.* 17 (3): 629–644.

8 Forsberg, F., Lapadatu, A., Kittilsland, G. et al. (2014). CMOS-integrated Si/SiGe quantum-well infrared microbolometer focal plane arrays manufactured with very large-scale heterogeneous 3-D integration. *IEEE J. Sel. Top. Quantum Electron.* 21 (4): 30–40.

9 Niklaus, F., Kälvesten, E., and Stemme, G. (2001, 2001). Wafer-level membrane transfer bonding of polycrystalline silicon bolometers for use in infrared focal plane arrays. *J. Micromech. Microeng.* 11: 509–513.

10 Niklaus, F., Haasl, S., and Stemme, G. (2003). Arrays of monocrystalline silicon micromirrors fabricated using CMOS compatible transfer bonding. *IEEE J. Microelectromech. Syst.* 12 (4): 465–469.

11 Despont, M., Drechsler, U., Yu, R. et al. (2004). Wafer-scale microdevice transfer/interconnect: its application in an AFM-based data-storage system. *J. Microelectromech. Syst.* 13 (6): 895–901.

12 Makihata, M., Tanaka, S., Muroyama, M. et al. (2012). Integration and packaging technology of MEMS-on-CMOS capacitive tactile sensor for robot application using thick BCB isolation layer and backside-grooved electrical connection. *Sens. Actuators, A* 188: 103–110.

13 Fischer, A.C., Forsberg, F., Lapisa, M. et al. (2015). Integrating MEMS and ICs. *Microsyst. Nanoeng.* 1: 15005.

14 Christiaens, I., Van Thourhout, D., and Baets, R. (2004). Low-power thermo-optic tuning of vertically coupled microring resonators. *Electron. Lett.* 40 (9): 560–561.

15 Roelkens, G., Brouckaert, J., Van Thourhout, D., and Baets, R. (2006). Adhesive bonding of InP/ InGaAsP dies to processed silicon-on-insulator wafers using DVS-bis-benzocyclobutene. *J. Electrochem. Soc.* 153 (12): G1015–G1019.

16 Takato, H. and Shimokawa, R. (2001). Thin-film silicon solar cells using and adhesive bonding technique. *IEEE Trans. Electron Devices* 48 (9): 2090–2094.

17 Matsuo, S., Tateno, K., Nakahara, T., and Kurokawa, T. (1997). Use of polyimide bonding for hybrid integration of a vertical cavity surface emitting laser on a silicon substrate. *Electron. Lett.* 33 (13): 1148–1149.

18 Lamb, J., Kim, B., and Pargfrieder, S. (2008). Temporary bonding/debonding for ultrathin substrates. *Solid State Technol.* 51: 60–65.

19 Shuangwu, M.H., Pang, D.L.W., Nathapong, S., and Marimuthu, P. (2008). Temporary bonding of wafer carrier for 3D-wafer level packaging. Proc. Electronic Packaging Technology Conference 2008, Singapore, 405–411.

20 Badihi, A. (1999). Shellcase ultrathin chip size package. Proc. Advanced Packaging Materials: Processes, Properties and Interfaces 1999, Braselton, USA, 236–240.

21 Zoberbier, M., Hansen, S., Hennemeyer, M. et al. (2009). Wafer level cameras – novel fabrication and packaging technologies. Proc. International Image Sensor Workshop 2009, Bergen, Norway.

22 Sterner, M., Chicherin, D., Raisenen, A.V. et al. (2009). RF MEMS high-impedance tuneable metamaterials for millimeter-wave beam steering. Proc. of IEEE 22nd International Conference on Micro Electro Mechanical Systems 2009, Sorrento, Italy, 896–899.

23 Saharil, F., Wright, R.V., Rantakari, P. et al. (2010). Low-temperature CMOS-compatible 3D-integration of monocrystalline-silicon based PZT RF MEMS switch actuators on RF substrates. Proc. MEMS 2010, Hongkong, China, 47–50.

24 Somjit, N., Stemme, G., and Oberhammer, J. (2009). Deep-reactive ion-etched wafer-scale-transferred monocrystalline silicon dielectric block for ultra-broadband millimeter-wave phase shifters. *J. Microelectromech. Syst.* 19 (1): 120–128.

25 Kazlas, P.T., Johnson, K.M., and McKnight, D.J. (1998). Miniature liquid-crystal-on-silicon display assembly. *Opt. Lett.* 23 (12): 972–974.

26 Goetz, M. and Jones, C. (2002). Chip scale packaging techniques for RF SAW devices. IEEE Proc. Electronics Manufacturing Technology Symposium 2002, San Jose, USA, 63–66.

27 Oberhammer, J., Niklaus, F., and Stemme, G. (2003). Selective wafer level adhesive bonding with benzocyclobutene for fabrication of cavities. *Sens. Actuators, A* 105 (3): 297–304.

28 Oberhammer, J., Niklaus, F., and Stemme, G. (2004). Sealing of adhesive bonded devices on wafer-level. *Sens. Actuators, A* 110 (1-3): 407–412.

29 Bleiker, S.J., Visser Taklo, M.M., Lietaer, N. et al. (2016). Cost-efficient wafer-level capping for MEMS and imaging sensors by adhesive wafer bonding. *Micromachines* 7 (10): 192.

30 Tsao, C.W. and DeVoe, D.L. (2009). Bonding of thermoplastic polymer microfluidics. *Microfluid. Nanofluid.* 6 (1): 1–16.

31 Quellmalz, A., Wang, X., Wagner, S. et al. (2019). Wafer-scale transfer of graphene by adhesive wafer bonding. Proc. of 32th IEEE International Conference on Micro Electro Mechanical Systems (MEMS) 2019, Seoul, Korea, 257–259.

32 Quellmalz, A., Wang, X., Wagner, S. et al. (2020). Large-scale integration of 2D material heterostructures by adhesive bonding. Proc. of 33rd IEEE International Conference on Micro Electro Mechanical Systems (MEMS) 2020, Vancouver, Canada.

33 A. Quellmalz, X. Wang, S. Wagner, S. Sawallich, B. Uzlu, Z. Wang, M. Prechtl, O. Hartwig, S. Luo, G.S. Duesberg, M. Lemme, K.B. Gylfason, N. Roxhed, G. Stemme, F. Niklaus, "Large-area integration of two-dimensional materials and their heterostructures by wafer bonding", in review, 2020.

34 Ramm, P., Lu, J.J.-Q., and Taklo, M.M.V. (eds.) (2011). *Handbook of wafer bonding*. John Wiley & Sons.

35 Casco Nobel (1992). *Industrial Adhesives Handbook*. Fredensborg, Denmark: Casco Nobel.

36 Niklaus, F. (2002). Adhesive wafer bonding for microelectronic and microelectromechanical systems. Ph.D. Thesis. KTH-Royal Institute of Technology, Stockholm.

37 Traeger, R.K. (1977). Nonhermeticity of polymeric lid sealants. *IEEE Trans. Parts Hybrids Packag.* (2): 49.

38 Yacobi, B.G., Martin, S., Davis, K. et al. (2002). Adhesive bonding in microelectronics and photonics. *J. Appl. Phys.* 91: 6227–6262.

39 Alvino, W.M. (1995). *Plastics for Electronics: Materials, Properties, and Design*. New York, USA: McGraw-Hill Inc.

40 Vogel, H. and Marvel, C.S. (1961). Polybenzimidazoles, new thermally stable polymers. *J. Polym. Sci.* 50 (154): 511–539.

41 Ruano, J.M., Aguirregabiria, M., Tijero, M. et al. (2004). Monolithic integration of microfluidic channels and optical wave guides using a photodefinable epoxy. Proc. MEMS 2004, Maastricht, Netherlands, pp. 121–124.

42 den Besten, C., van Hal, R.E.G., Munoz, J., and Bergveld, P. (1992). Polymer bonding of micro-machined silicon structures. Proc. MEMS 1992, Travemünde, Germany, 104–109.

43 Niklaus, F., Decharat, A., Forsberg, F. et al. (2009). Wafer bonding with nano-imprint resists as sacrificial adhesive for fabrication of silicon-on-integrated-circuit (SOIC) wafers in 3D integration of MEMS and ICs. *Sens. Actuators, A* 154: 180–186.

44 Niklaus, F., Enoksson, P., Kälvesten, E., and Stemme, G. (2001). Low temperature full wafer adhesive bonding. *J. Micromech. Microeng.* 11 (2): 100–107.

45 Bilenberg, B., Nielsen, T., Clausen, B., and Kristensen, A. (2004). PMMA to SU-8 bonding for polymer based lab-on-a-chip systems with integrated optics. *J. Micromech. Microeng.* 14 (6): 814–818.

46 Lin, C.H., Fu, L.M., Tsai, C.H. et al. (2005). Low azeotropic solvent sealing of PMMA microfluidic devices. In: Proc. IEEE The 13th International Conference on Solid-State Sensors, *Actuators and Microsystems, 2005. Digest of Technical Papers. TRANSDUCERS'05*, vol. 1, 944–947. IEEE.

47 Bhattacharya, S., Datta, A., Berg, J.M., and Gangopadhyay, S. (2005). Studies on surface wettability of poly(dimethyl) siloxane (PDMS) and glass under oxygen-plasma treatment and correlation with bond strength. *J. Microelectromech. Syst.* 14 (3): 590–597.

48 Oh, K.W., Han, A., Bhansali, S., and Ahn, C.H. (2002). A low-temperature bonding technique using spin-on fluorocarbon polymers to assemble microsystems. *J. Micromech. Microeng.* 12: 187–191.

49 Shores, A.A. (1989). Thermoplastic films for adhesive bonding: hybrid microcircuit substrates. Proc. Electronic Components Conference 1989, Houston, USA, 891–895.

50 Selby, J.C., Shannon, M.A., Xu, K., and Economy, J. (2001). Sub-micrometer solid-state adhesive bonding with aromatic thermosetting copolyesters for the assembly of polyimide membranes in silicon-based devices. *J. Micromech. Microeng.* 11: 672–685.

51 Su, Y.C. and Lin, L. (2001). Localized plastic bonding for micro assembly, packaging and liquid encapsulation. Proc. of MEMS 2001, 14th IEEE International Conference on Micro Electro Mechanical Systems 2001, Interlaken, Switzerland, 50–53.

52 Noh, H., Kyoung-sik, M., Cannon, A. et al. (2004). Wafer bonding using microwave heating of parylene intermediate layers. *J. Micromech. Microeng.* 14 (4): 652–631.

53 Kim, H. and Najafi, K. (2005). Characterization of low-temperature wafer bonding using thin-film parylene. *J. Microelectromech. Syst.* 14 (6): 1347–1355.

54 Wang, X., Lu L.-H., and Liu, C. (2001). Micromachining techniques for liquid crystal polymer. IEEE Proc. MEMS 2001, Interlaken, Switzerland, 126–130.

55 Nguyen, H., Patterson, P., Toshiyoshi, H., and Wu, M.C. (2000). A substrate-independent wafer transfer technique for surface-micromachined devices. Proc. MEMS 2000, Miyazaki, Japan, 628–632.

56 Dragoi, V., Glinsner, T., Mittendorfer, G. et al. (2002). Reversible wafer bonding for reliable compound semiconductor processing. Proc. IEEE Int. Semicond. Conf. 2: 331–334.

57 Glasgow, I.K., Beebe, D.J., and White, V.E. (1999). Design rules for polyimide solvent bonding. *J. Sens. Mater.* 11: 269–278.

58 Bleiker, S.J., Dubois, V., Schröder, S. et al. (2017). Adhesive wafer bonding with ultra-thin intermediate polymer layers. *Sens. Actuators, A* 260: 16–23.

59 Schift, H. (2008). Nanoimprint lithography: an old story in modern times? A review. *J. Vac. Sci. Technol., B* 26 (2): 458–480.

60 Rowland, H.D., Sun, A.C., Schunk, P.R., and King, W.P. (2005). Impact of polymer film thickness and cavity size on polymer flow during embossing: toward process design rules for nanoimprint lithography. *J. Micromech. Microeng.* 15: 2414–2425.

61 Carlborg, C.F., Haraldsson, K.T., Cornaglia, M. et al. (2010). Large scale integrated 3D microfluidic networks through high yield fabrication of vertical vias in PDMS. Proc. MEMS 2010, Hong Kong, China, 240–243.

62 Qiu, X., Zhu, J., Oiler, J. et al. (2009). Localized Parylene-C bonding with reactive multilayer foils. *J. Phys. D: Appl. Phys.* 42 (18): 185411.

63 Abgrall, P., Lattes, C., Conédéra, V. et al. (2005). A novel fabrication method of flexible and monolithic 3D microfluidic structures using lamination of SU-8 films. *J. Micromech. Microeng.* 16 (1): 113.

64 McMahon, J.J., Lu, J.Q., and Gutmann, R.J. (2005). Wafer bonding of damascene-patterned metal/adhesive redistribution layers for via-first three-dimensional (3D) interconnect. Proc. IEEE Electronic Components and Technology, 2005. ECTC'05, 331–336.

65 Niklaus, F., Enoksson, P., Kälvesten, E., and Stemme, G. (2003). A method to maintain wafer alignment precision during adhesive wafer bonding. *Sens. Actuators, A* 107 (3): 273–278.

66 Lee, S.H., Niklaus, F., Kumar, R.J. et al. (2006). Fine keyed alignment and bonding for wafer-level 3D ICs. Proc. MRS 2006, San Francisco, USA.

16

Soldering by Local Heating

Yu-Ting Cheng[1] and Liwei Lin[2]

[1] National Chiao Tung University, Institute of Electronics, 1001 Ta-Hsueh Rd., Hsinchu 300, Taiwan
[2] University of California Berkeley, Department of Mechanical Engineering, 5135 Etcheverry Hall, Berkeley, CA 94720-1740, USA

16.1 Soldering in MEMS Packaging

Soldering is a process that enables the joint formation between two substrates via a metal filler, i.e. solder. The process has been a packaging procedure to provide electrical interconnection and mechanical protection for electronic devices. In general, a low thermal budget process is desirable to prevent dopant redistribution and interface degradation of as-fabricated electronic devices such as transistors, such that the subsequent processing temperature should be lower than the soldering temperature to avoid weakening the solder joints [1]. Micro-electro mechanical system (MEMS) devices may have free-standing multilayered structures accompanied with thermal stress issues to affect device's performances. Therefore, the temperature consideration is critical for the soldering process, including MEMS products.

Varieties of solder materials such as Au–Sn, Sn–Pb, Ag–Sn alloys, etc. have been investigated and utilized for corresponding packaging applications. In addition to the material selection, the implementation methodology for the soldering process is also important to address potential thermal issues. In general, low temperature and short processing time are desirable for the assembly and packaging technologies to minimize the thermal budget and achieve high throughputs. Localized heating in the soldering process instead of heating the whole substrate is, therefore, an advantageous approach [2–6]. Specifically, the local solder joint process can be used for both the electrical I/O connection and assembly of packaging microcaps together with MEMS devices for physical protections, and a large bonding area is typically required with good mechanical support and hermeticity. Furthermore, the local heating process is an effective way to mitigate the stress effect induced by the thermal mismatch at the interfaces of various substrates. In this section, we focus on several local soldering techniques.

3D and Circuit Integration of MEMS, First Edition. Edited by Masayoshi Esashi.
© 2021 WILEY-VCH GmbH. Published 2021 by WILEY-VCH GmbH.

16.2 Laser Soldering

The local soldering joints for MEMS packaging can be achieved using radiative, resistive, inductive, and fricative heating schemes by confining the applied energy within the bonding interfaces. For the laser soldering process, the local heating scheme can be realized via a light-transparent substrate and an energy-absorbing material such as Pyrex and solder metal, respectively. The laser irradiation power employed in the soldering regions is absorbed to form a local hot spot for the bonding process. Its transient responses can be analyzed by solving the governing Eq. (16.1):

$$\rho C \frac{\partial T}{\partial t} = k \left(\frac{\partial^2 T}{\partial r^2} + \frac{1}{r} \frac{\partial T}{\partial r} + \frac{\partial^2 T}{\partial z^2} + I(r, t, z) \right) \tag{16.1}$$

where ρ, C, t, and k are the mass density, specific heat, time, and thermal conductivity of the bonding substrate (such as silicon or silicon dioxide), respectively; r and z are the coordinate along the radius and vertical direction, and $I(r, t, z)$ represents the laser power intensity distribution.

To assemble two substrates using the laser soldering scheme, the process parameters require some optimizations such as laser pulse duration, repetition, and power. For instance, Figure 16.1 shows the simulated isothermal profiles for the transient response via a nanosecond laser heating/bonding process with adiabatic boundaries and fixed temperature constraints such as room temperature at the borders under the same pulse duration in both substrates [2]. Under a 355 nm laser irradiation with a pulse duration of 4–6 ns, power of 22 mJ, and beam size of 1 mm with the indium film as the intermediate solder layer for the glass to silicon bonding, the local temperature can be heated up to 2500 °C and reduced to 760 °C and then 43 °C within 1 μs and 1 ms, respectively, as a result of the rapid thermal heat dissipation along the silicon substrate to environment.

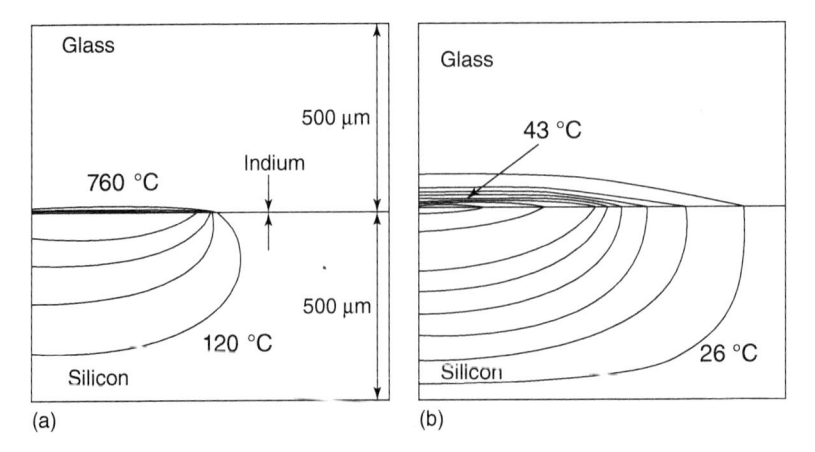

Figure 16.1 Simulated isothermal profile results for the transient response for the nanosecond laser heating and bonding process: (a) after 1 μs and (b) after 1 ms. Source: Luo and Liwei [2]. © 2002 Elsevier.

The laser soldering temperature and region can be selectively defined according to the heat absorption and conduction properties of the two substrates. For example, silicon to silicon is a major assembly system in the 3D stacked multifunctional microsystem. A continuous wave (CW) mode CO_2 laser with a wavelength of 1060 nm and a focal spot size of around 0.5 mm in diameter has been utilized for the local eutectic lead–tin soldering process to accomplish a hermetic sealing at the wafer level. Figure 16.2 shows the corresponding cross-sectional view of the sealed microcavity and corresponding soldering interface, which can exhibit a helium leakage rate lower than 2×10^{-9} atm cc s^{-1} [7]. The CO_2 laser can also assist the local soldering to hermetically seal a silicon lid with the ceramic quad flatpack (CQFP) using the $Au_{80}Sn_{20}$ alloy as shown in Figure 16.3 [7] with the MEMS chips inside the CQFPs under various atmosphere such as air, nitrogen, helium, and vacuum.

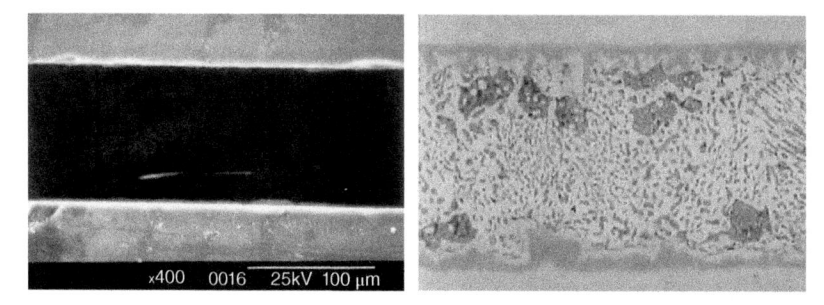

Figure 16.2 Sealed cavity microscopic cross-sectional views. Source: Tao et al. [7]. © 2004 Elsevier.

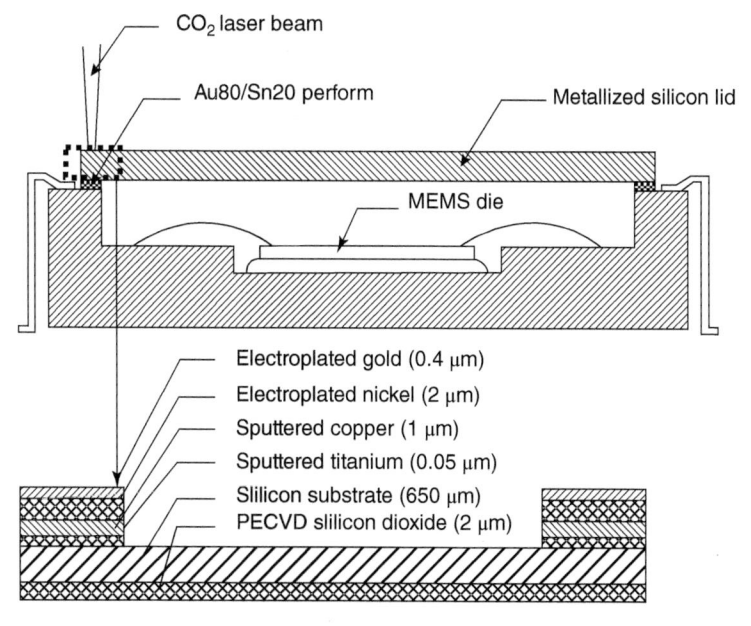

Figure 16.3 Schematic diagram of CO_2 laser-assisted metallized silicon lid encapsulation of MEMS devices in the CQFP. Source: Tao et al. [8]. © 2003, IEEE.

364 | *16 Soldering by Local Heating*

Flip chip solder bonding has facilitated high density and low parasitic reactance interconnects for high speed data communication applications. In recent developments, various optical MEMS components, such as mirror plate, membrane filter, lens, etc. have been demonstrated to boost the system performances.

Meanwhile, the solder reflow process has been proposed as a procedure to construct 3D microstructures based on the energy minimization principle. The surface tension force of the solder after the reflow process can act as a hinge to lift a planar structure to form a 3D system as schematically shown in Figure 16.4 [9]. The lifting angle is a function of solder, solder alloy volume and position, and soldered plate size. A high speed fluxless solder jet process with a tool combining precise placement of solder material and laser heating functions as shown in Figure 16.5 has been successfully developed for the local laser-assisted solder reflow

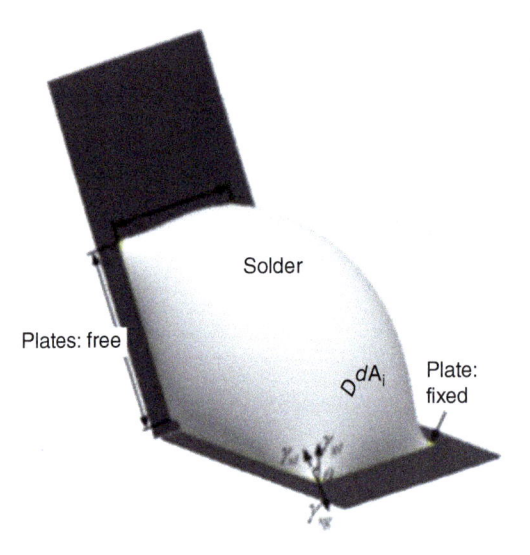

Figure 16.4 Schematic diagram for the local solder assembly process: the final equilibrium rotating angle of the solder can be designed based on the energy minimization for the solder shapes with different initial angles to achieve the lowest global energy state. Source: Yang et al. [9].

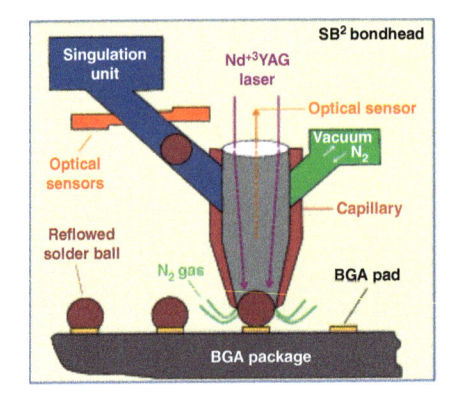

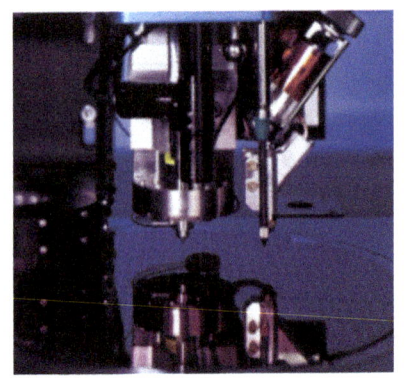

Figure 16.5 Schematic diagram of (a) the solder jet bumping process and (b) the corresponding actual machine jetting solder balls on a wafer. Source: Oppert et al. [10].

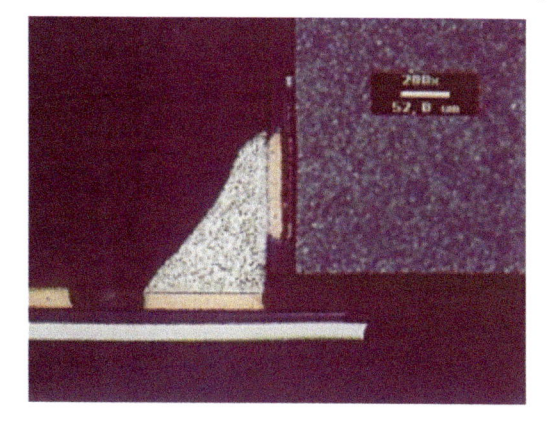

Figure 16.6 3D MEMS interconnects. Source: Oppert et al. [10].

process for optical MEMS and optoelectronic components [10]. Figure 16.6 shows the newly developed process for the realization of 3D interconnects using the scheme.

16.3 Resistive Heating and Soldering

Soldering by the local heating scheme can also be accomplished using a built-in resistive microheater. The time constant of the resistive microheater is in the range from several hundred μs to ms depending on the thermal conductivity, total thermal capacitance of the microheater, corresponding area, and effective thermal conductivity of the surrounding medium, such as glass, silicon substrate, etc. Such a soldering process is similar to a conventional process for the solder reflowing joint under a steady-state condition. Figure 16.7 shows the steady-state isotherms for the cross-sectional diagram to demonstrate that the high temperature region is confined in a small area surrounding the microheater and the silicon substrate maintains at room temperature as long as its bottom is constrained at room temperature during

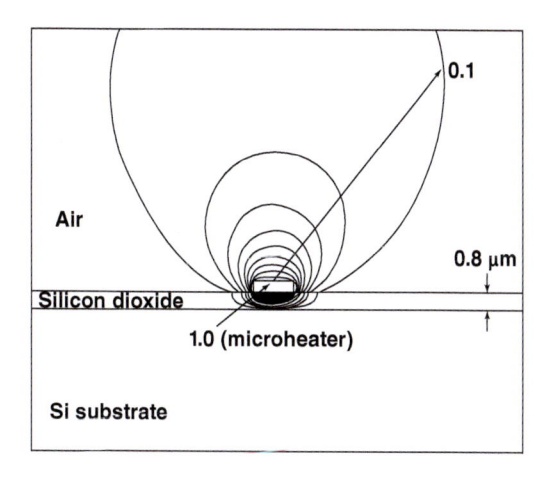

Figure 16.7 The cross-sectional view of isotherms around a micro line-shaped resistor. Source: Lin [11]. © 2000 IEEE.

the process [11]. The average temperature of the microheater can be estimated as

$$T(x) = T_r - (T_r - T_\infty)\frac{\cosh\left[\sqrt{\varepsilon}\left(x - \frac{L}{2}\right)\right]}{\cosh\left(\sqrt{\varepsilon}\frac{L}{2}\right)} \qquad (16.2)$$

where $T(x)$ is the temperature along the microheater with a total length of L and ε and T_r are functions of the structure dimensions, thermal properties, input current, and heat conduction shape factor.

Figure 16.8 shows the glass-to-silicon substrate assembly by the local indium/Au soldering process after forcefully breaking the bond [12]. A portion of the indium solder originally on the top surface of a polysilicon microheater has been stripped off and attached to the glass substrate. This demonstrates that a strong soldering joint has been formed between the glass and silicon substrate. An input current of 20 mA for a 5 µm wide microheater can be heated at 300 °C for the reflow of the indium solder. Solder reflowing characteristic as shown in Figure 16.9 also overcomes the requirement of planarization for bonding as a result of the bottom

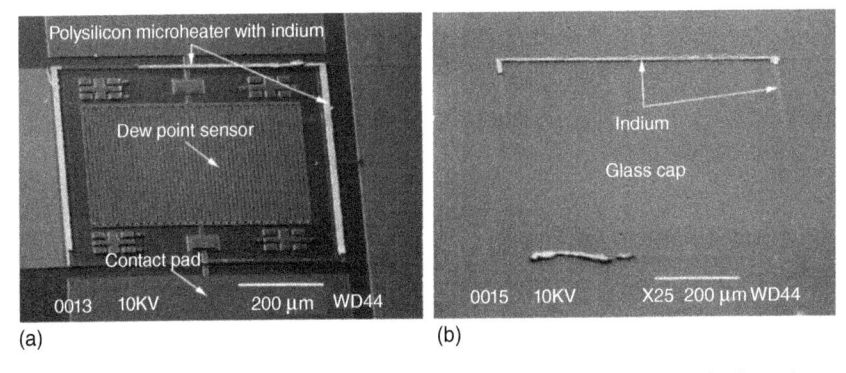

Figure 16.8 A SEM micrograph of silicon substrate (a) after breaking the indium glass bond; parts of indium layer are attached to the glass cap (b). A dew point sensor is designed and fabricated to characterize the hermeticity of the package as shown in (a). Source: Cheng et al. [12]. © 1999 IEEE.

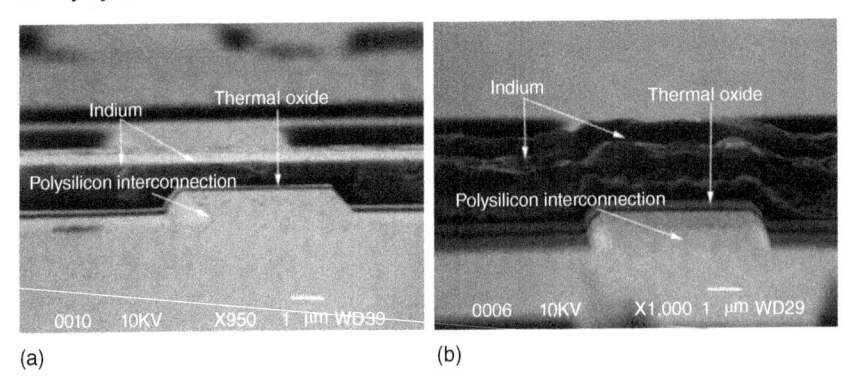

Figure 16.9 (a) A close-up SEM photograph of the silicon substrate before the localized bonding experiment. The step-up surface of the indium solder is caused by the underneath polysilicon interconnection line; and (b) after the localized soldering bonding experiment. Very good step coverage is achieved. Source: Cheng et al. [12]. © 1999 IEEE.

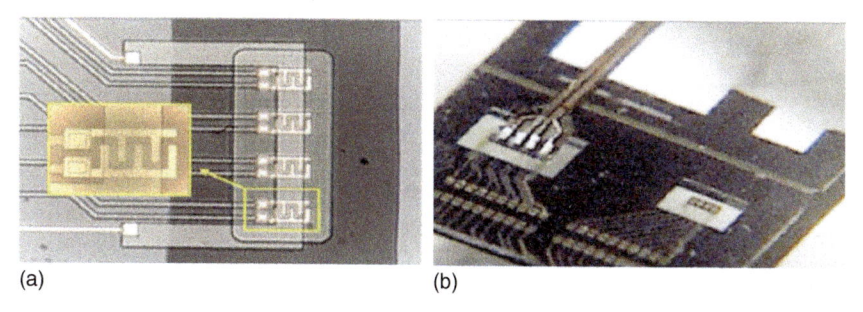

Figure 16.10 (a) A photograph of microconnection pads with the inset showing a close-up view of an interdigitated electrode pair; and (b) a photograph of cable leads after the solder bonding process to a pad island test structure which is suspended from a dielectric window. Source: Lemmerhirt and Wise [13].

1 µm thick polysilicon interconnection line. After the localized heating process, excellent flat step coverage has been achieved. Similar scheme has been applied and combined with automatic alignment functions with proper circuitry for the assembly of off-board sensors and a multichannel polyimide cable for microsystem integrations [13]. Via the capacitance detection between sensor I/O pads and interconnect leads, the interconnectors can be well aligned and soldered with the sensor to form mechanical/electrical connections using a polysilicon microheater underneath the sensor pads as shown in Figure 16.10. While the heater is heated again to melt the solder, the cable leads can be freely withdrawn. Such a scheme can facilitate low-cost but high density interconnects between flexible cables and delicate MEMS for packaging assembly and test applications.

Silicon optical bench (SiOB) assembly can be also achieved using the localized resistive heating and soldering method. Figure 16.11 shows electrical

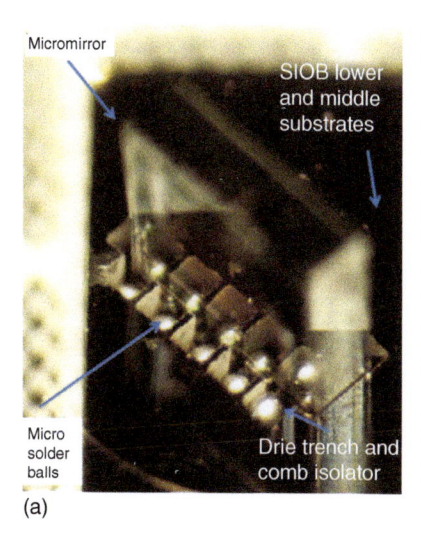

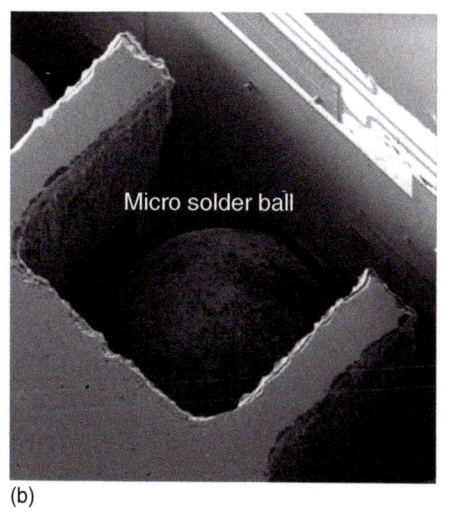

Figure 16.11 Optical images (a); and enlarged SEM image (b) of the SiOB design: a dummy micromirror is soldered on the lower substrate of SiOB using the micro soldering technology. Source: Xu et al. [14].

interconnections with the high precision alignment, which are built for a two-axis gimbal-less micromirror embedded in a silicon carrier with a nearly vertical sidewall [14]. An integrated platinum heater with a maximum output temperature of 250 °C on the carrier can be locally heated for micro Sn/Ag solder balls housed by the comb-typed insulator to realize the assembly. Owing to the compact characteristics of the SiOB assembly, the micromachined mirrors can be further implemented for endoscopic optical coherence tomography (EOCT) applications.

16.4 Inductive Heating and Soldering

One major drawback in the resistive heating approach is the microheaters, which require electrical feedthroughs for electrical current inputs. This can increase the interconnect routing and make this approach impractical for low-cost manufacturing. As a result, the inductive heating mechanism has been proposed as an alternative for the realization of localized soldering. Figure 16.12 shows the scheme of the local soldering process using inductive heating. The inductive coil driven by an AC power supply will generate a time variant magnetic field passing through the device and package substrates to be soldered. According to Lenz's law, a conductive material can generate joule heating via the eddy current in a close loop for the bonding purpose. The eddy current would be generated in conductive materials, and the magnitude relies on the electrical conductivity of the material. For microelectronics packaging application, heat generation from semiconductive silicon substrate and MEMS devices should also be considered in the packaging process design.

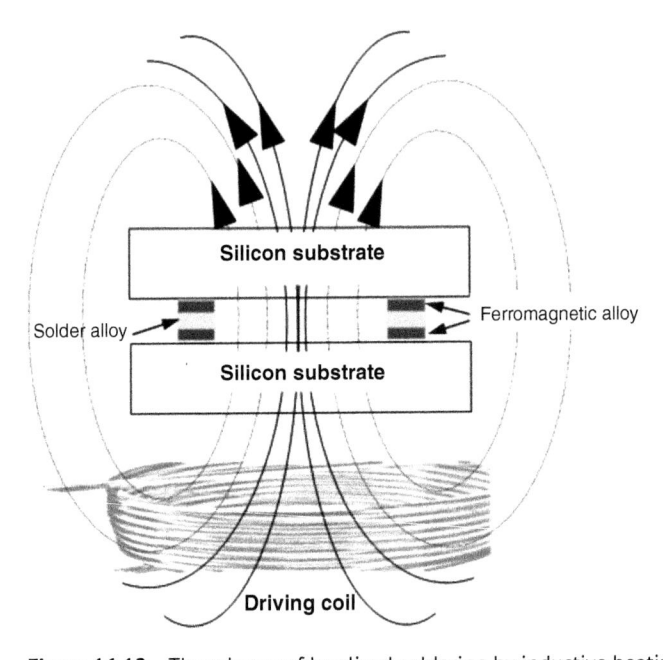

Figure 16.12 The scheme of localized soldering by inductive heating.

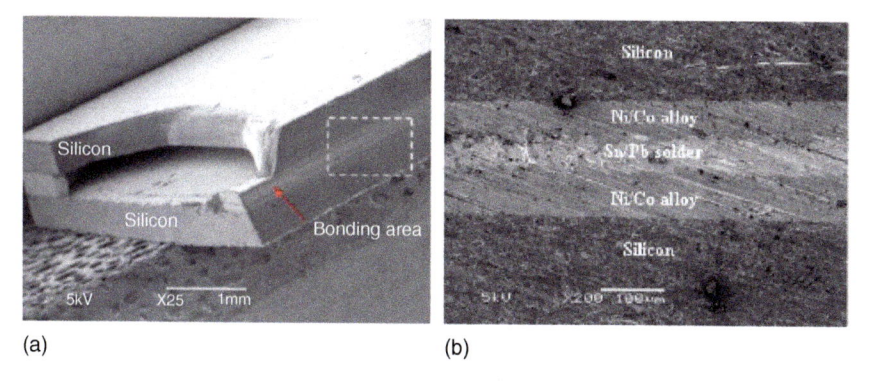

Figure 16.13 The SEM photographs of (a) the silicon substrates after bonding; and (b) the close-up of the bonded specimen in the dashed box region. Source: Yang et al. [15].

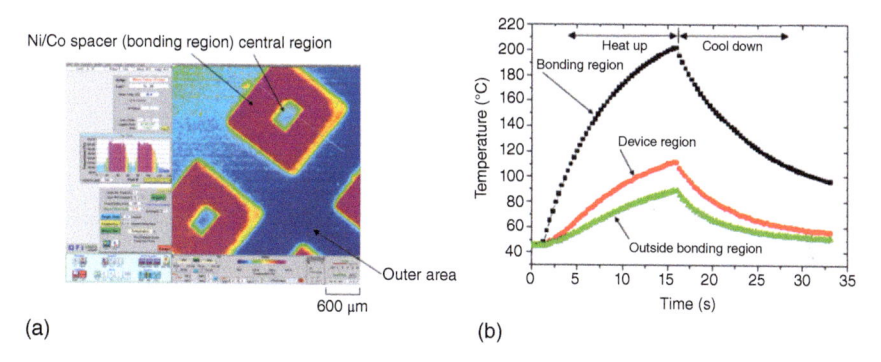

Figure 16.14 The IR measurements of the test sample temperature distribution during induction heating: (a) spatial distribution; and (b) temporal distribution. Source: Yang et al. [15].

Figure 16.13 shows two silicon substrates bonded by the inductive heating and soldering scheme [15]. The enlarged scanning electron microscopy (SEM) view on the bonding interface as indicated by the dashed box in Figure 16.13a shows three distinct layers, including two Ni/Co spacer layers and one Sn/Pb solder layer, between the silicon substrates. The bonding strength of the joint is higher than 18 MPa. In the experiment, ferromagnetic metals Ni/Co alloy and Pb/Sn alloy are utilized as the heating element and solder material, respectively. The solder joint can be achieved by maximizing the heating efficiency as the result of the introduction of hysteresis loss [16, 17]. Furthermore, Figure 16.14 shows the inductive heating measurement results where the temperature distribution during the process are characterized in three sectors: central region, bonding region, and outer area. The results show that the soldering region exhibits the highest temperature, i.e. > 200 °C suitable for solder reflow and join, the center region can be kept as low as 110 °C, and the maximum outside area temperature is less than 90 °C, which is higher than the ambient temperature. Unlike the prior resistive heating, the higher substrate temperature can be attributed to substrate heating

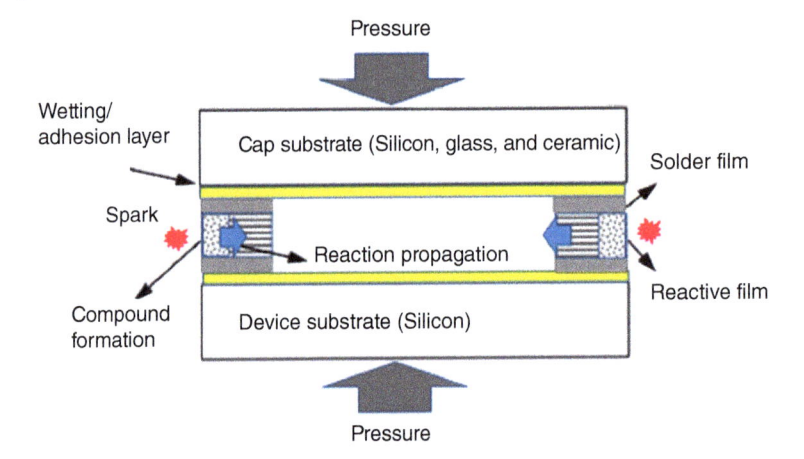

Figure 16.15 Scheme of microencapsulation formed by reactive heating and soldering, where the reactive film is triggered by sparks to start a self-propagating exothermic reaction, and the applied soldering material absorbs the released heat to accomplish the bonding process for a sealed microcavity. Source: Lee et al. [21].

due to its finite electrical conductivity and lack of substrate cooling to constrain the boundary temperature.

16.5 Other Localized Soldering Processes

16.5.1 Self-propagative Reaction Heating

Instead of the electrical heating for the local solder reflow and bonding, the thermal energy can be provided by triggering the exothermic reactions such as the multilayer Ni/Al foil or sputtered nanoscale alternatively multilayered Ni/Al or Al/Pd films, etc. [5, 18–20]. The layered materials can be activated by spark or laser impulse to generate heat to facilitate atomic diffusion and compound formation. The exothermic reactions can be divided into three categories, which are small, medium, and high by the corresponding formation of heat energy of smaller than 40 kJ mol^{-1}-atom, in between 40 and 80 kJ mol^{-1}-atom and above 80 kJ mol^{-1}-atom, respectively [5, 20]. Among these reactions, the heat generation rate by the interatomic diffusion must be faster than those removed by the thermal diffusion process to allow the reaction to self-propagate and act as a localized heating source for packaging applications as shown in Figure 16.15 [21].

Figure 16.16 shows the analytic transient temperature distribution along the thickness direction (z-axis) of the corresponding bonding interface, i.e. 5×5×0.04 mm AlNi foil as a heat source sandwiched by two SnAgCu soldering layers (25 µm), where a thin insulated-gate bipolar transistor (IGBT) chip with the size of 5×5×0.115 mm is bonded to the directbonded copper (DBC) substrate, i.e. 6×6×0.38 mm Al$_2$O$_3$ layer sandwiched by 6×6×0.3 mm Cu layers [22]. In this case, the nanofoil can reach the highest temperature at 0.11 ms and the solder

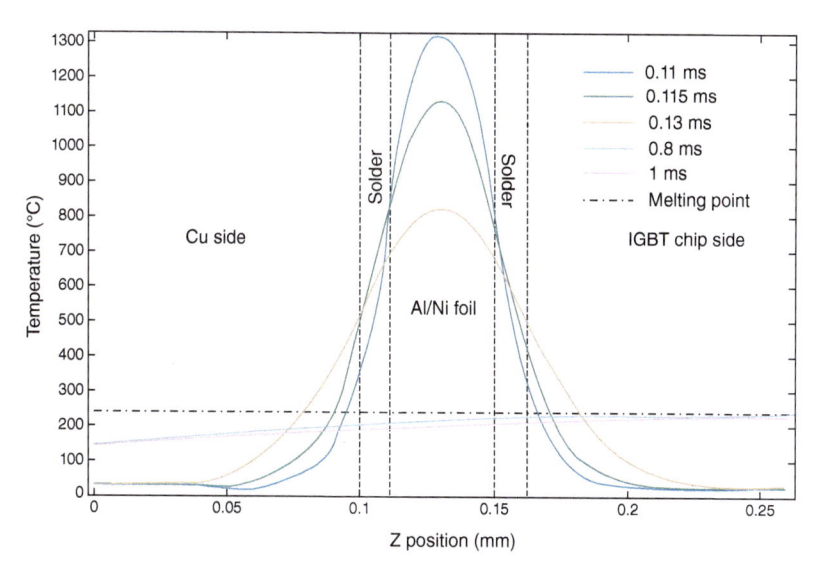

Figure 16.16 Simulated temperature distribution across the thickness direction (Z position) at different time. Source: Xiang et al. [22].

Figure 16.17 SEM micrograph of two silicon wafers that were bonded using two pieces of free-standing AuSn solder and one Ni/Al reactive foil. There was solder filling in a crack in the reacted foil. Source: Qiu et al. [18].

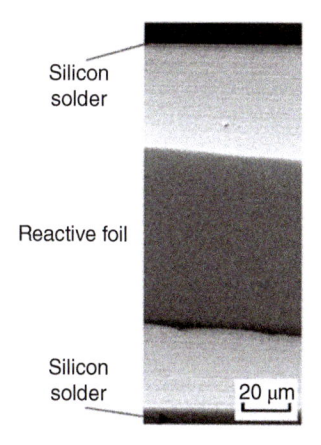

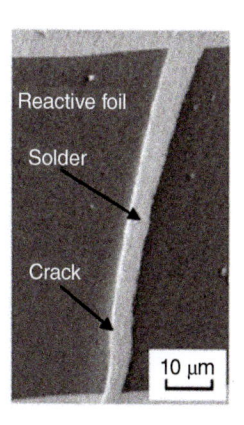

layer is melted after 0.8 ms. The heating rate of the solder is narrow and the hot zone is well confined due to large heat capacity and thermal conductivity of both IGBT and Cu substrates in comparison with that of the bonding layer. Figure 16.17 shows the cross-sectional view on a typical Si-to-Si joint using the Ni/Al-based reactive foil with two pieces of free standing AuSn solder films [18]. Solder reflow resulting from the reactive heating can also perfuse a crack in the reacted foil for good bonding. The MEMS devices susceptible to time variant magnetic field can also be encapsulated using this method.

16.5.2 Ultrasonic Frictional Heating

Through the atomic-scale friction process between two objects, mechanical energy will be transformed into heat in the contact area to result in the increase in

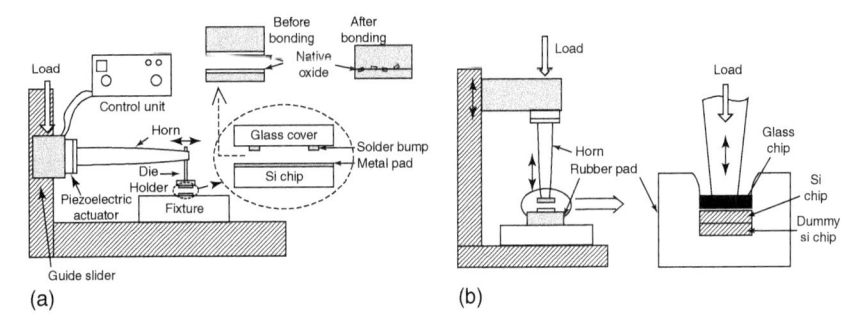

Figure 16.18 Schematic diagrams of ultrasonic bonding set up: (a) transversal load and (b) longitudinal load. Source: Kim et al. [6]. © 2018 Elsevier.

temperature of the sliding bodies. The higher the sliding velocity is, the higher the surface and subsurface temperatures will be generated in the contact interface for the possible soldering process [23–25]. Thus, the ultrasonic soldering technique can produce superior characteristics including rapid localized bonding, low temperature processing, etc. Figure 16.18 shows the schematic diagrams of one ultrasonic bonding equipment [6]. Ultrasonic force can be applied in either the transversal or longitudinal way to weld the thermoplastics by viscoelastic heating. The longitudinal ultrasonic soldering scheme has advantages over the transversal one in terms of better alignment and coplanarity between the chip and substrate for the flip chip electrical bump assembly. Negligible heat may dissipate through the thermal conduction process during the very short solder bonding process. The ultrasonic soldering temperature can be estimated as [23]

$$\Delta \theta = \frac{QT_b}{C_p \rho V_s} \tag{16.3}$$

where T_b represents the bonding time, C_p the specific heat, V_s the solder volume, and ρ the mass density of the solder. Q is the viscoelastic heat, which is correlated to the solder loss modulus, mechanical strain, and ultrasonic frequency [23–25]. The viscoelastic heating by the ultrasonication could be insufficient to melt the solder under room temperature, and preheating to an elevated temperature below the melting point can be implemented to melt the solder. The solder volume can also influence the bonding temperature. As the solder volume decreases under the same loading conditions, larger strain could result in higher temperature to help the high-density electronic package.

Figure 16.19 shows a uniform Cu/Ni/ Sn–3Ag–0.5Cu/Ni/Cu joint, which is formed by ultrasonic soldering with a load pressure of 17 MPa and processing time of three seconds to exhibit a bonding strength as high as 350 gf/bump in an area of 310 × 930 μm [24]. Figure 16.20 shows a hermetic seal accomplished by the transversal ultrasonic Al—Al bonding between silicon and glass substrates [6]. No liquid leakage is identified inside the square shape bonding ring. For polymer MEMS application, the transversal ultrasonic bonding can effectively avoid ring-shaped deformation inside the polymer substrate due to the energy absorption from the vertical bonding setup. Figure 16.21 shows a plastic bonding example. A 125 μm

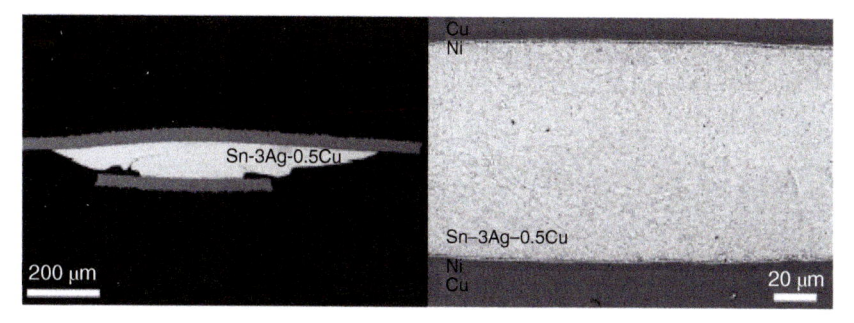

Figure 16.19 Micrograph of ultrasonic solder bumps bonded with 17 MPa pressure for three seconds. Source: Lee and Yoo [24].

Figure 16.20 Hermetically sealed cavity between silicon and glass substrate by using the ultrasonic Al–Al bonding. Source: Kim et al. [6]. © 2009 IEEE.

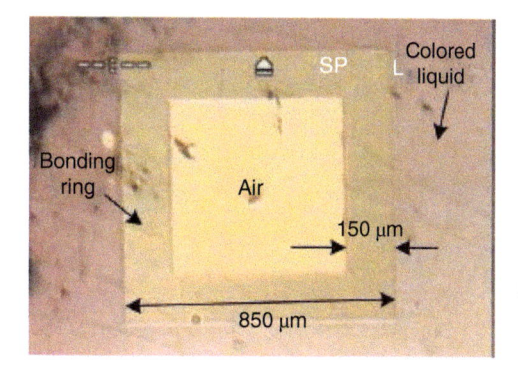

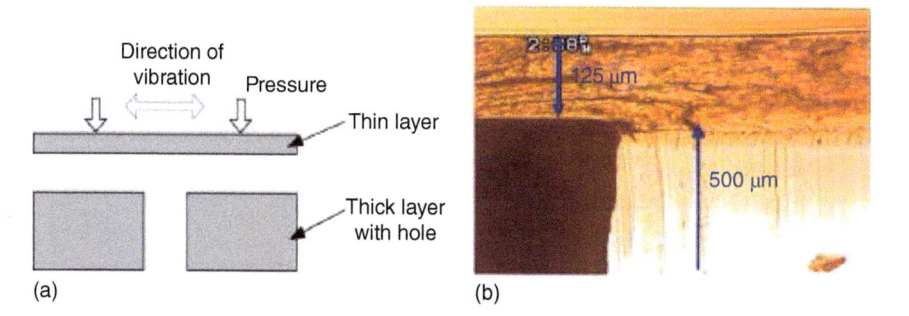

Figure 16.21 A polymer MEMS fabrication example using the transversal ultrasonic plastic bonding: (a) schematic diagram of a two-plastic substrate assembly; and (b) a close view of an optical micrograph on the bonding interface indicating excellent bonding quality without channel narrowing or clogging. Source: Kim et al. [6]. © 2009 IEEE.

thick cellulose acetate substrate is bonded to another one with 500 μm in thickness having a drilled hole of 1 mm in diameter using the transversal ultrasonic bonding technique [6]. Good sealing under leakage tests using water as the working fluid has been accomplished. Note that local heating characteristic of the bonding technique also facilitates the fabrication of the polymer-based microfluidic devices without channel narrowing or clogging.

References

1 Liu, X., Xu, S., Lu, G.Q., and Dillard, D.A. (2001). Stacked solder bumping technology for improved solder joint reliability. *Microelectron. Reliab.* 41: 1979–1992.
2 Luo, C. and Liwei, L. (2002). The application of nanosecond-pulsed laser welding technology in MEMS packaging with a shadow mask. *Sens. Actuat. A* 97: 398–404.
3 Yang, H.-A., Wu, M., and Fang, W. (2005). Localized induction heating solder bonding for wafer level MEMS packaging. *J. Micromech. Microeng.* 15: 394–399.
4 Cheng, Y.T., Hsu, W.T., Najafi, K. et al. (2002). Vacuum packaging technology using localized aluminum/silicon-to glass bonding. *IEEE/ASME J. Microelectromech. Syst.* 11: 556–565.
5 Qiu, X. and Wang, J. (2008). Bonding silicon wafers with reactive multilayer foils. *Sens. Actuat. A* 141: 476–481.
6 Kim, J., Jeong, B., Chiao, M., and Lin, L. (2009). Ultrasonic bonding for MEMS sealing and packaging. *IEEE Trans. Adv. Packag.* 32: 461–467.
7 Tao, Y., Malshe, A.P., and Brown, W.D. (2004). Selective bonding and encapsulation for wafer-level vacuum packaging of MEMS and related micro systems. *Microelectron. Reliab.* 44: 251–258.
8 Tao, Y., Malshe, A.P., Brown, W.D. et al. (2003). Laser-assisted sealing and testing for ceramic packaging of MEMS devices. *IEEE Trans. Adv. Packag.* 26: 283–288.
9 Yang, L., Liu, W., Wang, C., and Tian, Y. (2011). Self-assembly of three-dimensional microstructures in MEMS via fluxless laser reflow soldering. IEEE 12th International Conference on Electronic Packaging Technology and High Density Packaging, (4 pages).
10 Oppert, T., Teutsch, T., Azdasht, G., and Zakel, E. (2012). Micro ball bumping packaging for wafer level & 3-d solder sphere transfer and solder jetting. IEEE/CPMT 35th International Electronics Manufacturing Technology Conference, (6 pages).
11 Lin, L. (2000). MEMS post-packaging by localized heating and bonding. *IEEE Trans. Adv. Packag.* 23: 608–616.
12 Cheng, Y.T., Lin, L., and Najafi, K. (1999). Localized bonding with PSG or indium solder as intermediate layer. IEEE 12TH International Conference on Micro Electro Mechanical Systems, 285–289.
13 Lemmerhirt, D.F. and Wise, K.D. (2006). Chip-scale integration of data-gathering microsystems. Proceedings of the IEEE, 94, 1138–1159.
14 Xu, Y., Wang, M.F., Premachandran, C.S. et al. (2009). Platinum microheater integrated silicon optical bench assembly for endoscopic optical coherence tomography. *J. Micromech. Microeng.* 20: 015008.
15 s, H.A., Wu, M., and Fang, W. (2004). Localized induction heating solder bonding for wafer level MEMS packaging. *J. Micromech. Microeng.* 15: 394.
16 Hagemaier, D.J. (1990). *Fundamentals of Eddy Current Testing.* Columbus, OH: The American Society for Nondestructive Testing, Inc.
17 Stansel, N.R. (1949). *Induction Heating.* New York, NY: McGraw-Hill.

18 Qiu, X., Zhu, J., Oiler J., and Yu, H. (2009). Reactive multilayer foils for MEMS wafer level packaging. IEEE 59TH Electronic Components and Technology Conference, 1311–1316.

19 Braeuer, J., Besser, J., Wiemer, M., and Gessner, T. (2012). A novel technique for MEMS packaging: reactive bonding with integrated material systems. *Sens. Actuat. A* 188: 212–219.

20 Braeuer, J. and Gessner, T. (2014). A hermetic and room-temperature wafer bonding technique based on integrated reactive multilayer systems. *J. Micromech. Microeng.* 24: 115002–115010.

21 Lee, Y.C., Cheng, Y.T., and Ramadoss, R. (2018). *MEMS Packaging*. World Scientific Publishing Company Pte. Limited.

22 Xiang, Y., Zhou, Z., Mo, L. et al. (2017). Simulation of the temperature field for bonding IGBT chip and DBC substrate using Al/Ni self-propagating foil. IEEE 18th International Conference on Electronic Packaging Technology (ICEPT), 1016–1020.

23 Kim, J.H., Lee, J., and Yoo, C.D. (2005). Soldering method using longitudinal ultrasonic. *IEEE Trans. Comp. Packag. Technol.* 28: 493–498.

24 Lee, J. and Yoo, C.D. (2008). Thermosonic soldering of cross-aligned strip solder bumps for easy alignment and low-temperature bonding. *J. Micromech. Microeng.* 18: 125002.

25 Xiao, Y., Wang, Q., Wang, L. et al. (2018). Ultrasonic soldering of cu alloy using Ni-foam/Sn composite interlayer. *Ultrason. Sonochem.* 45: 223–230.

17

Packaging, Sealing, and Interconnection

Masayoshi Esashi

Tohoku University, Micro System Integration Center (μSIC), 519-1176 Aramaki-Aza-Aoba, Aoba-ku, 980-0845, Sendai, Japan

17.1 Wafer Level Packaging

Micro electro mechanical systems (MEMS) have moving parts and bare MEMS chips can't be molded directly with polymers, and this requires can packages or ceramic packages. Furthermore, MEMS for sensors can't be tested electrically on a wafer, and this results in high package cost if the yield is low. If bare MEMS wafer is exposed, the yield during blade dicing process is low because of its particle contamination. The packaging process prior to dicing to chips as shown in Figure 17.1 is called wafer level packaging (WLP) or 0–level packaging [1]. It is known that MEMS fabrication cost can be reduced to 20–30% by the WLP because this can solve the problems mentioned earlier. The example of the WLP in Figure 17.1 uses anodic bonding for sealing, but another wafer bonding methods can be also used for WLP. After MEMS are fabricated on a Si wafer, the Si wafer is bonded to a glass wafer, which has electrical feedthroughs for interconnections. The bonded wafer is diced to each chip and finally wires are connected. Packaging for MEMS requires sealing for encapsulation and electrical interconnection, and these will be explained in this chapter.

The WLP that uses the electrical feedthrough interconnections in the glass was applied to integrated capacitive pressure sensors [2]. The fabrication process and the photograph of the packaged chip are shown in Figure 17.2. The first process step is a patterning and etching of a n-Si wafer (1 in the figure). Complementary metal oxide semiconductor (CMOS) circuits and p^+ layers are formed on the Si wafer (2 in the figure). Holes are made in a glass wafer using sand blasting or electrochemical discharge drilling [3] (1 in the figure). The glass wafer is metalized with Ti/Pt and patterned for capacitor electrodes and light shields on the circuits (2 in the figure). The Pt on Ti was used for metallization on the glass because conventional Al metallization causes rough surface called hillocks during the thermal process of the anodic bonding. The hillocks have to be avoided for capacitors with a narrow gap. The Si wafer is anodically bonded to the glass wafer, and glass holes are metalized with Cr/Cu/Au (3 in the figure). Diaphragms are made by etching the Si wafer from the

3D and Circuit Integration of MEMS, First Edition. Edited by Masayoshi Esashi.
© 2021 WILEY-VCH GmbH. Published 2021 by WILEY-VCH GmbH.

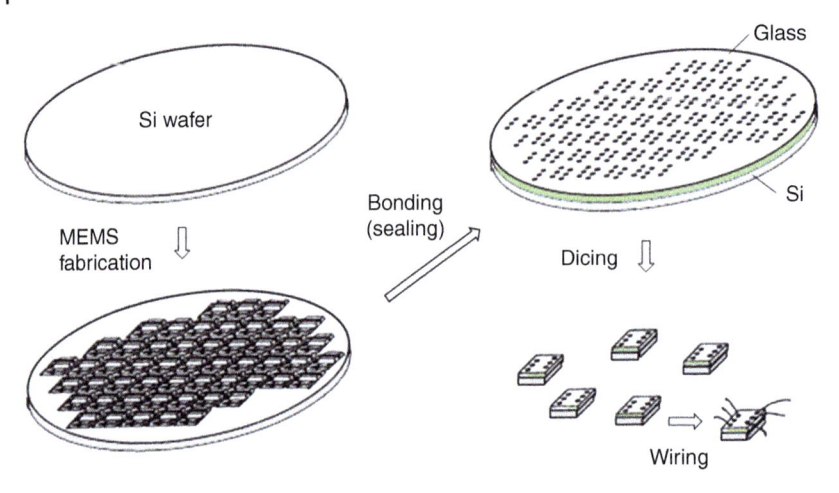

Figure 17.1 Wafer level packaging (WLP).

backside (4 in the figure). Finally the bonded glass–Si wafer is diced to each chip, and lead wires are soldered to the metalized holes in the glass (5 in the figure).

17.2 Sealing

17.2.1 Reaction Sealing

Sealing methods to make encapsulated cavities have been developed. Figure 17.3a shows a reaction sealing by thermal oxidation [4]. Narrow channels with 40 nm gap are used to etch out the sacrificial layer. These channels can be filled with thermal oxide in which Si surface is risen up by 60% of the SiO_2 thickness as shown in Figure 17.3b. A 200×200 µm laser-recrystallized piezoresistive micro-diaphragm pressure sensor of which cavity under the diaphragm is sealed by the thermal oxidation was developed. Figure 17.4 shows the fabrication process of the sensor. Sacrificial layer and etch channel made of SiO_2 are formed by thermal oxidation and patterning (1 in the figure). Si_3N_4, poly Si, and SiO_2 are deposited by chemical vapor deposition (CVD) (2 in the figure). The poly Si is recrystallized using continuous wave (CW) Ar laser (3 in the figure). P (phosphorous) ion and B (boron) ion are implanted and thermally activated (4 and 5 in the figure). After second B implantation (6 in the figure), periphery of the deposited layers is etched (7 in the figure). The sacrificial oxide is removed by etching in HF (hydrofluoric acid) through the narrow channel (8 in the figure). The narrow channel is sealed by reaction sealing using Si oxidation and by additional Si_3N_4 deposition (9 in the figure). Finally contact holes are made by etching, and Al is deposited and patterned for pads (10 in the figure).

The reaction sealing was applied to the fabrication of an integrated capacitive pressure sensor shown in Figure 17.5 [5]. Two capacitors for sensor and for reference are fabricated, and the differential capacitance of these two capacitors is measured using

1. Etching of n–Si wafer

1′. Making through holes in glass

2. Making CMOS circuit

2′. Metalization and patterning

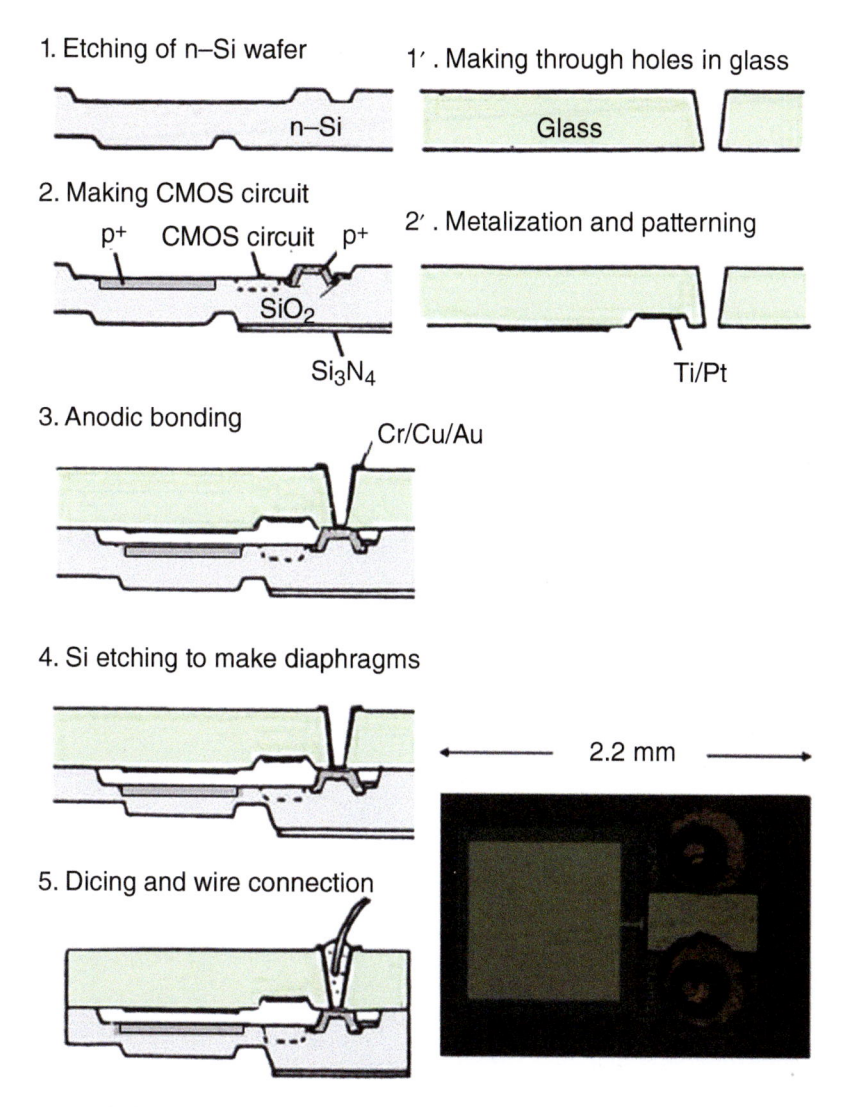

3. Anodic bonding

4. Si etching to make diaphragms

5. Dicing and wire connection

Figure 17.2 Integrated capacitive pressure sensor.

a monolithically integrated switched capacitor circuit. The fabrication process is as follows. After making a recess for capacitor gap on the bottom n-Si wafer (1 in the figure), B is diffused to make p^+ layer (2 in the figure). Oxidized n-Si wafer is bonded to the bottom Si wafer by direct (fusion) bonding and it is oxidized. Small steps on the bottom Si made during the p^+ diffusion process are filled by reaction sealing, and vacuum cavities are formed because oxygen gas inside the cavity is consumed by oxidation (3 in the figure). CMOS circuit is fabricated (4 in the figure). After patterning and SiO_2 etching (5 in the figure), Si diaphragms and isolation grooves are made by anisotropic etching (6 in the figure). Contact pads are made on the bottom wafer and wires are bonded (7 in the figure).

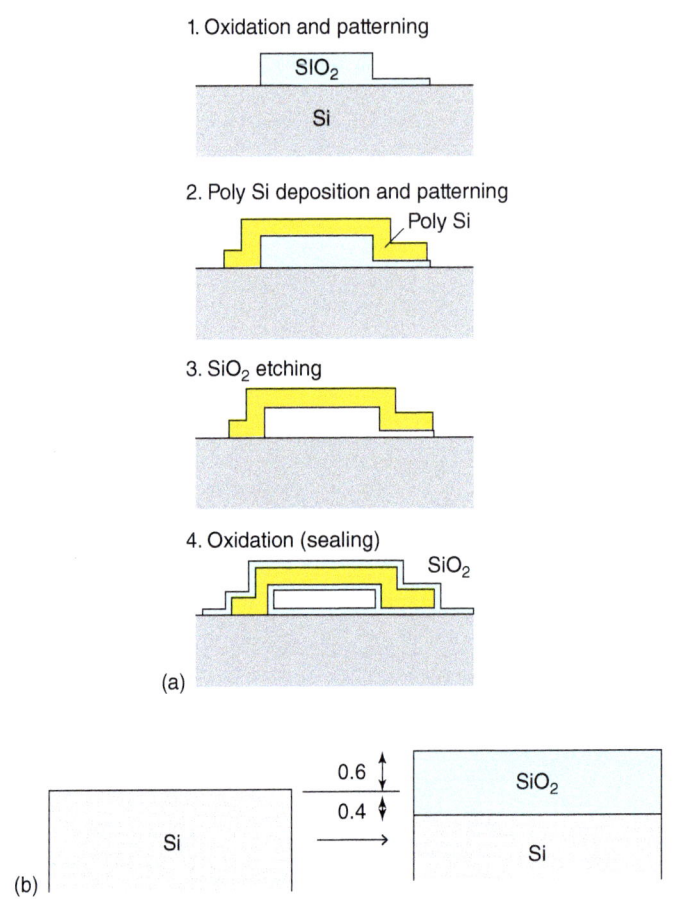

Figure 17.3 Reaction sealing by thermal oxidation. (a) Process for reaction sealing and (b) sealing principle by Si oxidation.

17.2.2 Deposition Sealing (Shell Packaging)

Sealing method by depositing some material on a porous fluid-permeable membrane is called shell packaging. Cavity structures can be fabricated as shown in Figure 17.6 [7]. Phosphosilicate glass (PSG) is deposited on a Si wafer (1 in the figure). Si_3N_4 is deposited and patterned (2 in the figure). This is covered with porous poly Si (3 in the figure). PSG under the Si_3N_4 window is etched out in HF solution because the porous poly Si is fluid permeable (4 in the figure). Finally the device is sealed by depositing Si_3N_4 (5 in the figure). Photograph of the porous poly Si surface is shown in Figure 17.7. It can be deposited at 600 °C and 73 Pa using SiH_4 (silane) as a source gas [6].

Sealing with thick Si layer can be made by the method called plug-up process [8]. Deep narrow trenches are made on the surface Si layer of a silicon on insulator (SOI) wafer by deep reactive ion etching (DRIE). The porous poly Si is deposited on its surface, and SiO_2 layer in the SOI wafer is etched through the porous poly Si in HF.

17.2 Sealing

Figure 17.4 Steps

1. Oxidation and patterning — SiO$_2$ (1.25 μm), SiO$_2$ (40 nm), n-Si
2. Si$_3$N$_4$, poly, and Si, SiO$_2$ CVD — SiO$_2$ (600 nm), Poly Si (2 μm), Si$_3$N$_4$ (400 nm)
3. Recrystallization of poly Si using CW Ar laser
4. P ion implantation and, thermal activation
5. B ion implantation and, thermal activation — Photoresist, p+-Si
6. B ion implantation — Photoresist, p-Si
7. Etching of periphery
8. Oxide removal in HF
9. Oxidation and Si$_3$N$_4$ deposition (sealing) — Vacuum cavity, SiO$_2$ (40 nm), Si$_3$N$_4$ (70 nm)
10. Contact hole etching, Al deposition, and patterning for pad and dicing — Contact hole, Al (600 nm)

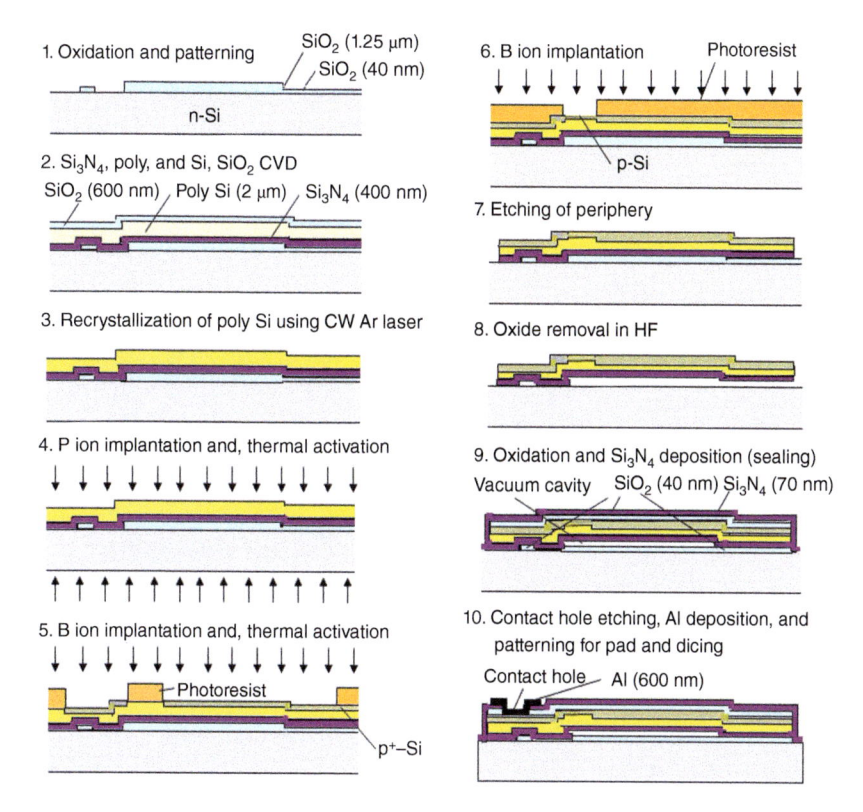

Figure 17.4 Fabrication process of laser-recrystallized piezoresistive micro-diaphragm pressure sensor by reaction sealing. Source: Based on Guckel and Burns [4].

Poly Si is deposited to fill the trench for the sealing. Sealed cavity can be fabricated by this process.

The porous poly Si can be also used for surface micromachining to make sealed vacuum structures [9]. The fabrication process is shown in Figure 17.8. After fabricating microstructures on a Si wafer (1 in the figure), PSG is deposited and patterned for a sacrificial layer (2 in the figure). The porous poly Si is deposited and patterned (3 in the figure). The PSG sacrificial layer is etched out through the porous poly Si in HF (4 in the figure). Metal is deposited by evaporation in vacuum and patterned for sealing (5 in the figure).

The MEMS structure made by surface micromachining can be sealed by plugging holes used for the etching of the sacrificial layer. Figure 17.9 is an example of vacuum encapsulation using ion beam sputter deposition of amorphous SiC [10].

Such sealing can be made by using CVD as well [11]. In the case of the CVD, H$_2$ gas remains in the cavity, but the residual H$_2$ gas can be diffused out by annealing in N$_2$ environment as will be explained in Section 18.5 (Chapter 18).

Encapsulation method based on thermal decomposition of sacrificial polymer and diffused out through a polymer overcoat was developed [12]. Figure 17.10 shows the fabrication process. Trench and insulator are formed on a Si wafer (1 in the figure).

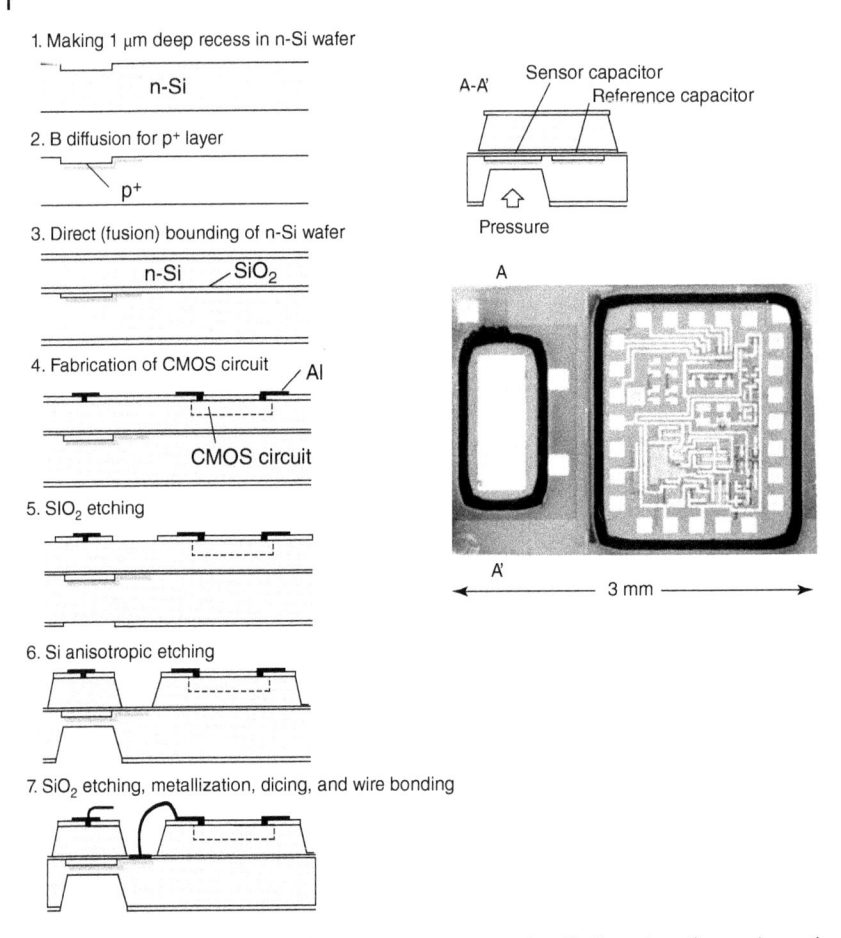

Figure 17.5 Integrated capacitive pressure sensor using Si direct bonding and reaction sealing.

Sacrificial polymer Unity (Promerus Ltd.) is coated and patterned (2 in the figure). The polymer can be patterned by exposing to UV light. The exposed part is thermally decomposed to gas. BCB (benzocyclobutene) polymer is used for overcoat on the sacrificial polymer (3 in the figure). The sacrificial polymer is thermally decomposed and the gas generated is diffused out through the BCB polymer overcoat (4 in the figure). It is metalized by depositing a metal on this structure (5 in the figure). Sealed cavity can be made by these process steps.

Two sealing methods by sealant for polymer-bonded structure are shown in Figure 17.11 [13]. Polymers are non-hermetic but the sealant such as Si_3N_4 by plasma-enhanced CVD (PE-CVD) is hermetic. The BCB was used for the polymer bonding. Si wafer with SiO_2 layer was bonded to the other Si wafer with 2.5 μm thick BCB. The PE-CVD Si_3N_4 that is 0.5 μm in thickness was deposited at 300 °C. The method (a) in the figure has a polymer exposed at the concave corner; on the other hand the method (b) has a polymer at the convex corner. The gas leakage can be reduced using the method (b) as shown in the He leak rate in the figure. This is

1. PSG deposition

2. Si$_3$N$_4$ deposition and patterning

3. Porous poly Si deposition

4. PSG etching through porous poly Si in HF

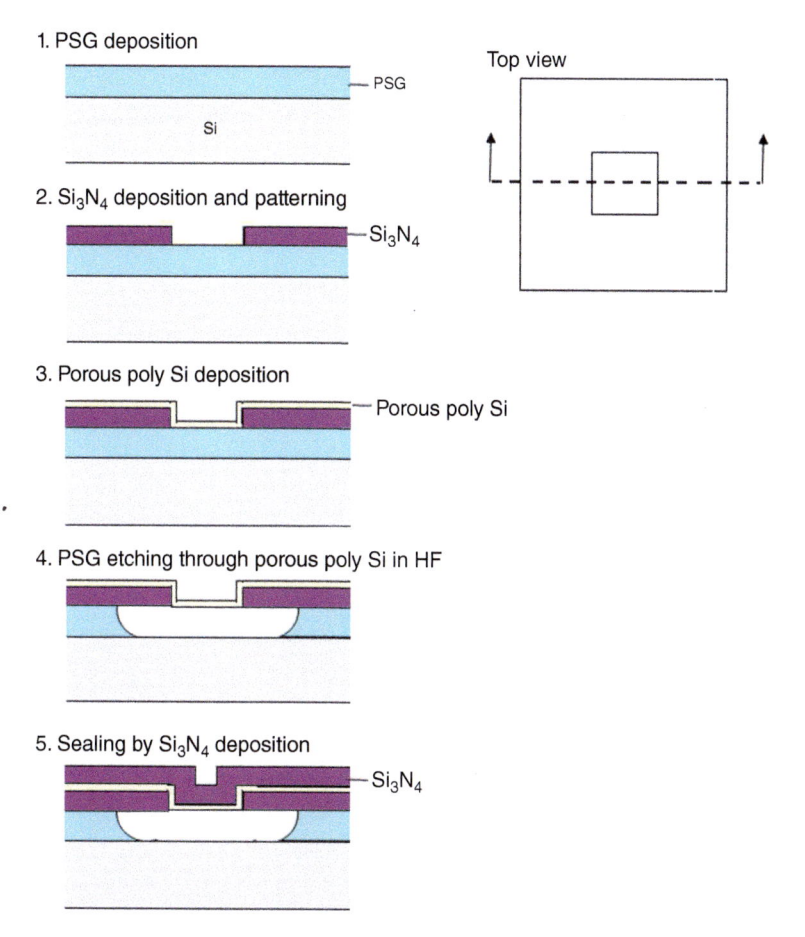

5. Sealing by Si$_3$N$_4$ deposition

Figure 17.6 Shell packaging by etching through permeable porous poly Si and Si$_3$N$_4$ deposition. Source: Modified from Lebouitz et al. [6].

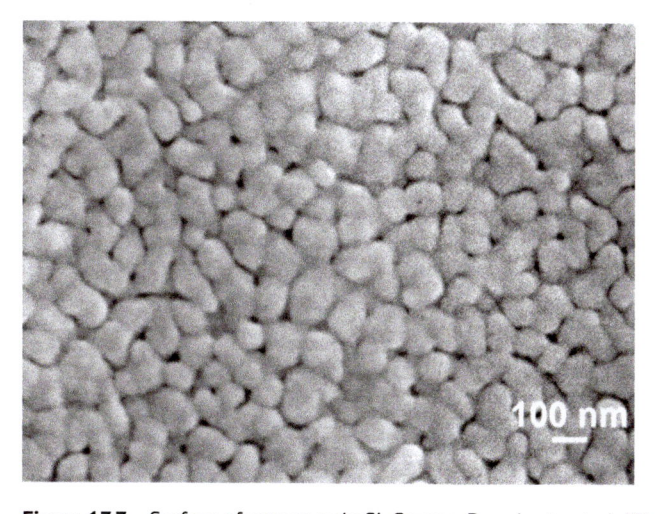

Figure 17.7 Surface of porous poly Si. Source: Dougherty et al. [7].

17 Packaging, Sealing, and Interconnection

1. Fabrication of microstructures

2. Sacrificial layer (PSG) deposition and patterning

3. Porous poly Si CVD and patterning

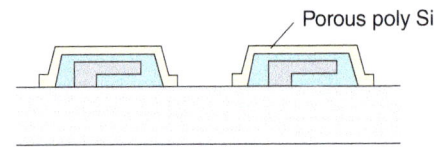

4. Sacrificial layer etching

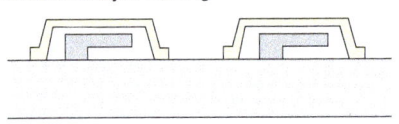

5. Sealing material deposition and patterning

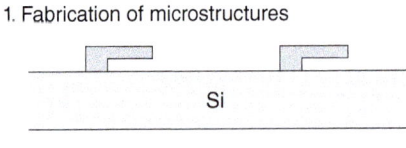

Figure 17.8 Surface micromachining using porous poly Si and deposition sealing. Source: Based on Fan et al. [9].

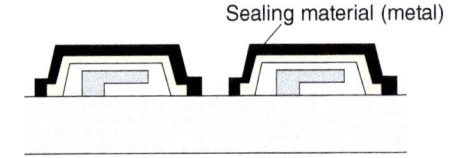

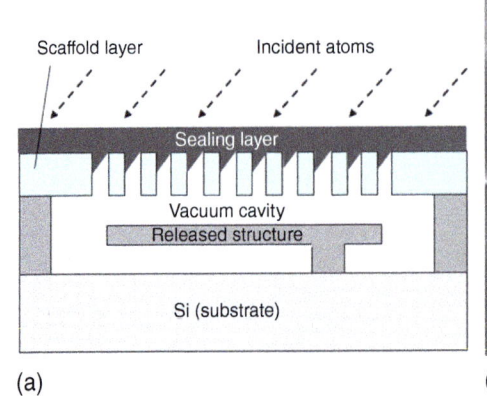

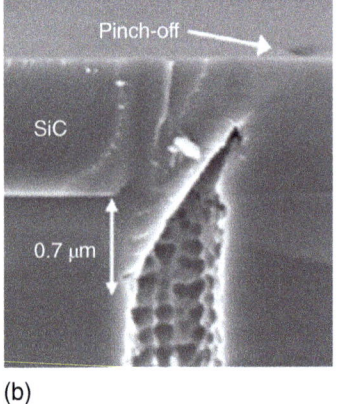

(a) (b)

Figure 17.9 Vacuum encapsulation using ion beam sputter deposition of amorphous SiC. (a) Principle of sealing and (b) photograph of the sealed part. Source: Jones et al. [10]. © 2007, IEEE.

Figure 17.10 Encapsulation based on thermal decomposition of sacrificial polymer and diffused out through polymer overcoat. Source: Modified from Monajemi et al. [12].

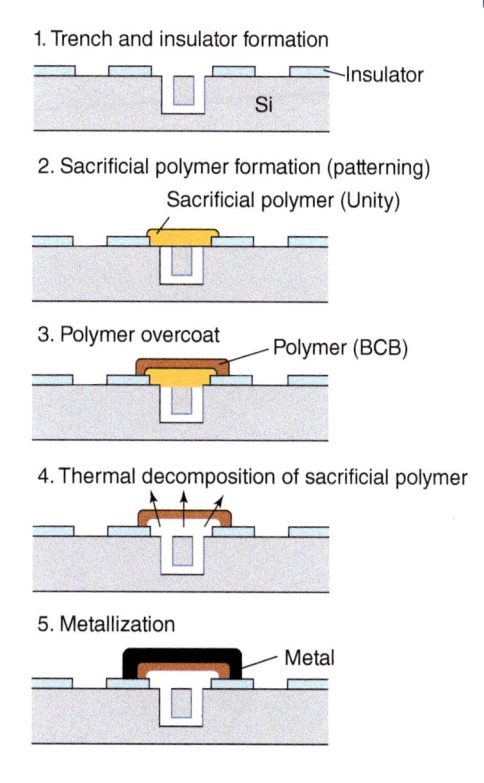

1. Trench and insulator formation
2. Sacrificial polymer formation (patterning) — Sacrificial polymer (Unity)
3. Polymer overcoat — Polymer (BCB)
4. Thermal decomposition of sacrificial polymer
5. Metallization — Metal

because the thickness of the sealant can be thick at the convex corner (method (b)) but thin at the concave corner (method (a)).

17.2.3 Metal Compression Sealing

A metal compression sealing can be used for MEMS packaging. Figure 17.12 shows the metal compression sealing for liquid using Au gasket [14]. Au gaskets are fabricated on Si and glass by electroplating using photoresist as a mask (1 in the figure). Liquid is dispensed into the reservoir and the sealing is performed by embossing to deform the Au gasket (2 in the figure). The wafer is diced to chips and the chip diced is reinforced by underfilling of epoxy (3 in the figure). Liquid can be encapsulated without thermal treatment by this method. Another sealing using side edges of Au rings was developed as well [15].

Holes can be plugged by Au stud bump, which is normally used for wire bonding [16]. The fabrication process is shown in Figure 17.13 and is as follows. Cavity is fabricated by anodic bonding of etched Si and glass (1 and 2 in the figure). Holes are made in the Si by DRIE (3 in the figure). After metalizing the surface with TiW and Au (4 in the figure), the cavity is filled with liquid through the holes (5 in the figure). Holes are plugged by the Au stud bump (6 in the figure).

Au particle and NPG (nano-porous Au) can be used for sealing and interconnection. Figure 17.14 shows the sealing method by using the Au particle [17]. The process flow is shown in Figure 17.14a. After forming rim structures (1 in the figure),

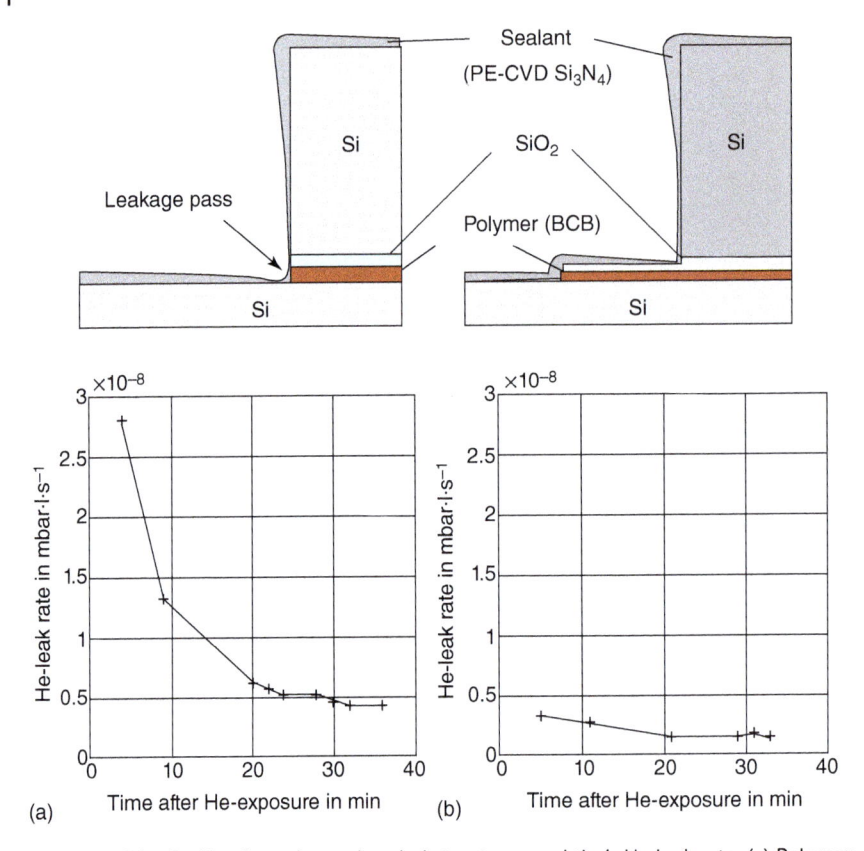

Figure 17.11 Sealing for polymer-bonded structures and their He leak rate. (a) Polymer sealing at concave corner and (b) polymer sealing at convex corner. Source: Oberhammer et al. [13]. © 2004, Elsevier.

1. Formation of electroplated Au rings for squeez sealing on lid Si wafer and liquid reservoir on glass wafer

2. Liquid dispensing into the reservoir and embossing Physically deformed squeeze ring

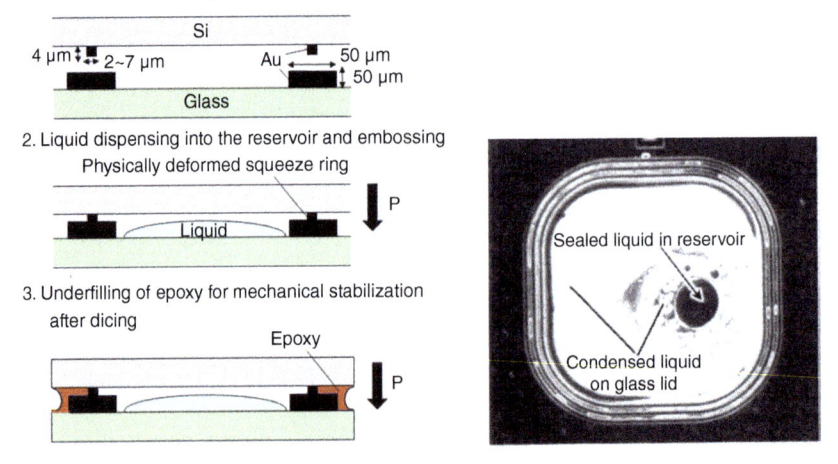

3. Underfilling of epoxy for mechanical stabilization after dicing

Figure 17.12 Wafer level liquid compression seal using Au gasket. Source: Lapisa et al. [14].

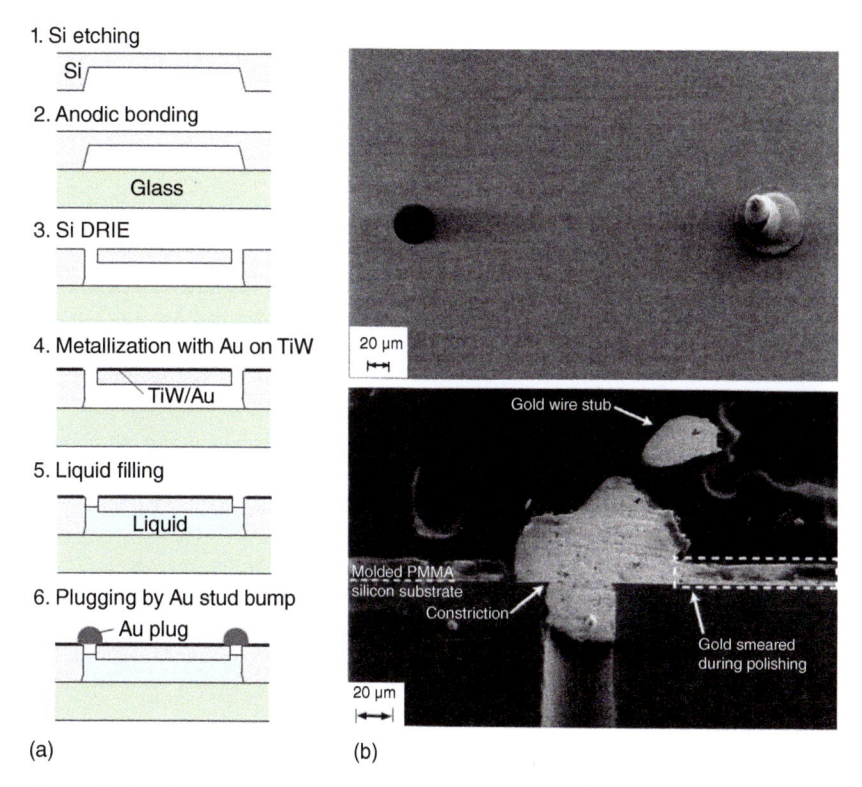

Figure 17.13 Plugging by stud bump for liquid encapsulation. (a) Process flow and (b) photographs of plug (top view and cross-sectional view). Source: Antelius et al. [16].

3–5 μm thick Au paste that have Au nanoparticle in it is printed. The mean diameter of the Au particle is from 0.2 to 0.4 μm. It is sintered at 200 °C in Ar with 4% H_2 (2 in the figure). Lid wafer is bonded by thermocompression at 200 °C for 30 min under pressure of 200 MPa.

The NPG is used for the purpose of electrical interconnection. Figure 17.15 shows the process flow for making the electrical interconnection during anodic bonding of low temperature co-fired ceramics (LTCC) and Si explained in Figure 11.19 (Chapter 19) [18]. This process takes advantage of the deformability of the NPG. The LTCC with Au feedthrough and Cr/Au surface layer is prepared (1 in the figure). After photoresist patterning on both sides (2 in the figure), Au–Sn is electroplated using the photoresist as a mold (3 in the figure). The Sn in the Au–Sn can be dissolved in 60% HNO_3 to make the NPG (4 in the figure). This process is called dealloying, and the photomicrograph of the NPG surface is shown in Figure 17.15 as well. The NPG is deformed during the anodic bonding and this can make electrical interconnections (5 in the figure).

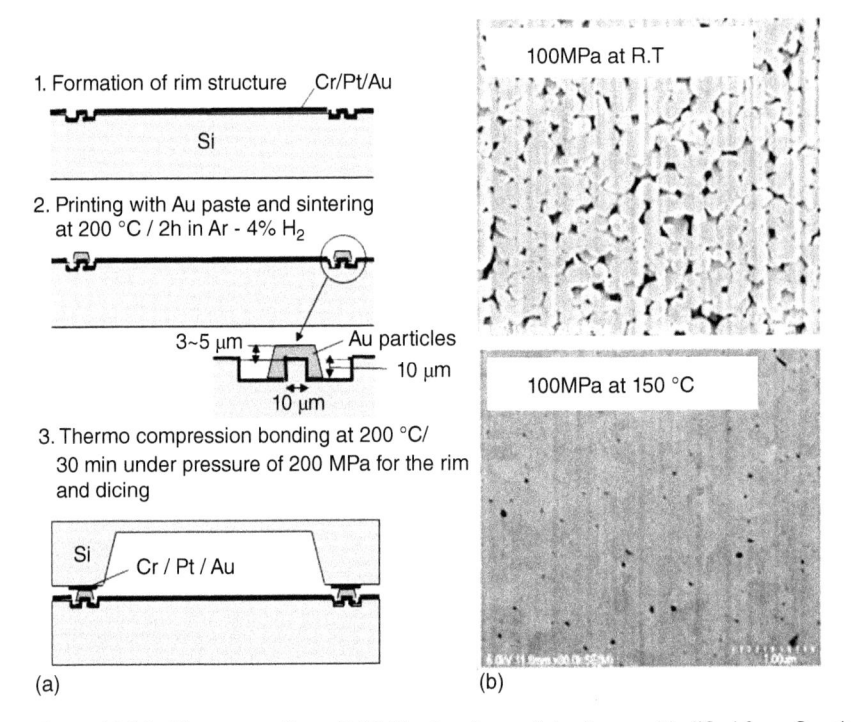

Figure 17.14 Vacuum sealing of MEMS using Au particle. Source: Modified from Ogashiwa et al. [17]. (a) Process flow and (b) photographs of the Au particle surfaces.

17.3 Interconnection

Interconnections play important roles in packaged MEMS. Vertical and lateral feedthrough interconnections are shown in Figure 17.16a,b. The former vertical interconnection enables stacking of chips and reduces the chip size. Through substrate vias (TSVs) that will be discussed in Chapter 20 are used for vertical interconnection. Through glass via (TGV) and through Si via (TSiV) used for the vertical interconnection are explained in Section 17.3.1. The lateral feedthrough structures are made between the substrate and the lid as described in Section 17.3.2. Interconnection methods by electroplating especially for nonplanar structures are described in Section 17.3.3.

17.3.1 Vertical Feedthrough Interconnection

17.3.1.1 Through Glass via (TGV) Interconnection

Implantable oximetry using infrared LEDs and photodiode was developed in Stanford University [19]. This device is packaged by using glass–Si anodic bonding and TGV interconnections. The fabrication process of the TGV is shown in Figure 17.17 [20]. SiO_2 and Al patterns are formed on a Si wafer (1 in the figure). After glass–Si anodic bonding (2 in the figure), the TGV is made by laser ablation using CO_2 laser

17.3 Interconnection | 389

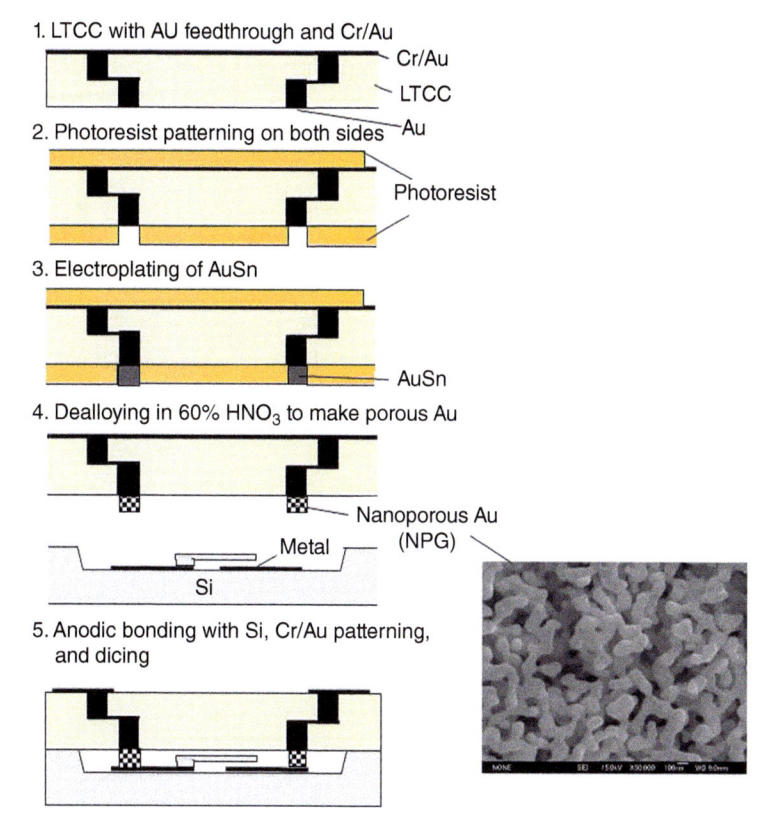

1. LTCC with AU feedthrough and Cr/Au
 — Cr/Au
 — LTCC

2. Photoresist patterning on both sides — Au
 — Photoresist

3. Electroplating of AuSn
 — AuSn

4. Dealloying in 60% HNO_3 to make porous Au
 — Nanoporous Au (NPG)
 Metal
 Si

5. Anodic bonding with Si, Cr/Au patterning, and dicing

Figure 17.15 Electrical interconnection with NPG (nano-porous Au). Source: Modified from Wang et al. [18].

Figure 17.16 Vertical feedthrough interconnection and lateral feedthrough interconnection from packaged MEMS. (a) Vertical feedthrough interconnection and (b) lateral feedthrough interconnection.

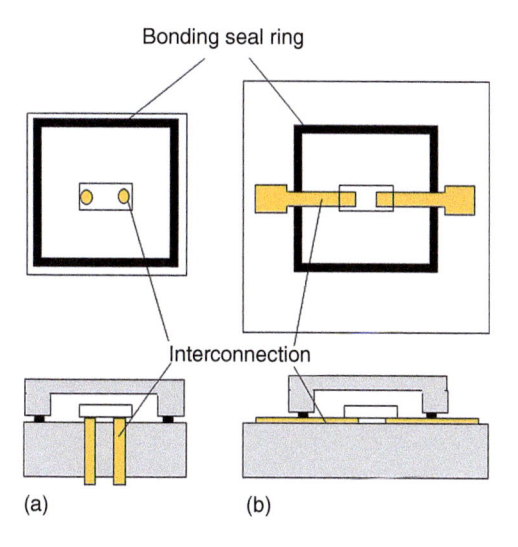

Bonding seal ring

Interconnection

(a)   (b)

17 Packaging, Sealing, and Interconnection

1. Patterning of SiO_2 and Al on Si

2. Anodic bonding of glass

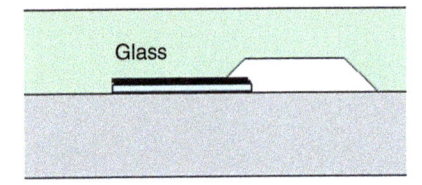

3. Laser abrasion of glass

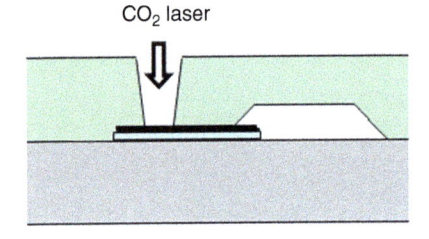

4. Al deposition using stencil mask and dicing

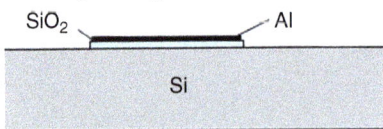

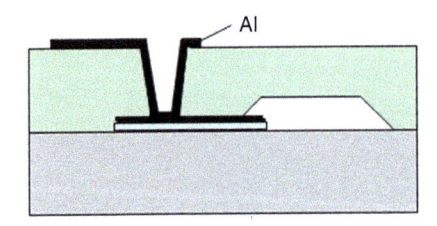

Figure 17.17 Through glass via (TGV) interconnection used for implantable oximetry. Source: Based on Bowman et al. [20].

(3 in the figure). The glass ablation stops at the bottom of the glass because the laser light is reflected on the Al surface. The glass holes are metalized by Al deposition.

MEMS switch shown in Figure 17.18 was fabricated using the WLP [21]. The surface of the electrical contact can be kept clean owing to the WLP. A bimorph thermal actuator using Al and SiO_2, which have different coefficient of thermal expansion (CTE), was adopted for the switch [22]. The switch was used for the front end of large scale integration (LSI) testers [21]. Glass that has TGV interconnections was used because these interconnections can minimize the stray capacitance for the purpose of high-frequency operation.

TGV interconnection with high aspect ratio via was developed [23]. Figure 17.19 shows its fabrication process. Metal patterns are formed on the glass surface (1 in the figure). Holes are made by laser ablation using femtosecond laser (2 in the figure). Cu is electroplated using a metal seed layer on a Si wafer under the glass (3 in the figure).

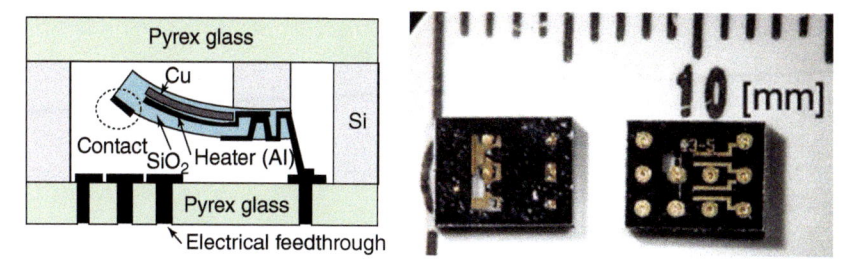

Figure 17.18 MEMS switch by wafer level packaging using TGV interconnections. Source: Modified from Nakamura et al. [21].

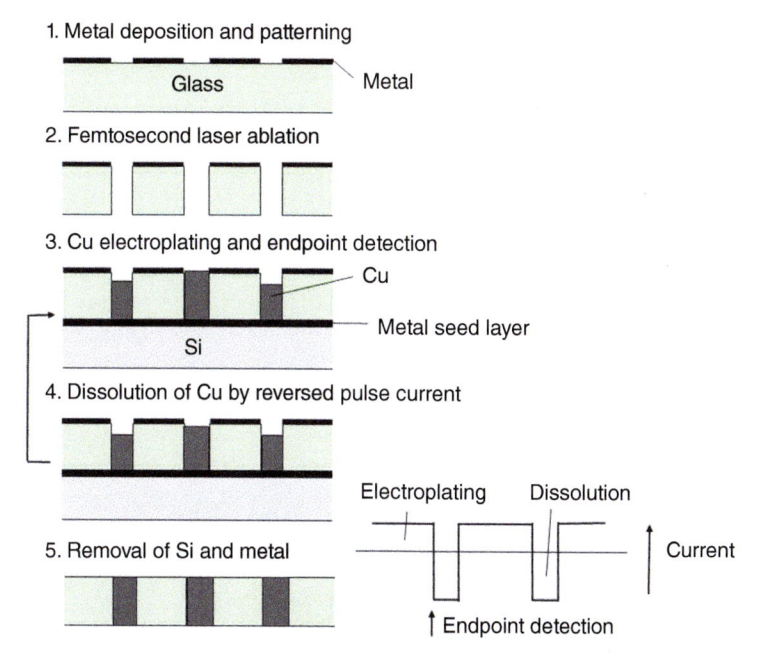

Figure 17.19 Fabrication process of the TGV interconnection by femtosecond laser. Source: Abe et al. [23]. © 2003, Elsevier.

Pulse method using electroplating current and reversed dissolution current is used to fill the holes with Cu. When the electroplated Cu makes contact with the metal on the surface, this so-called endpoint is electrically detected. The current direction is reversed for dissolving the Cu near the surface (4 in the figure). The electroplating, the endpoint detecting, and the dissolving are repeated [24]. This so-called pulse electroplating can make a planer surface without final surface polishing, and hence we can make metal patterns for interconnection in advance at the process step 1. The TGV interconnection is obtained finally by removing the Si wafer and the metal seed layer (5 in the figure).

The femtosecond laser can be used to make glass holes as shown in the process in step 2 in Figure 17.19. Glass holes with high aspect ratio such as 50 μm in diameter and 300 μm in depth can be made in about 1 sec. Figure 17.20a shows two

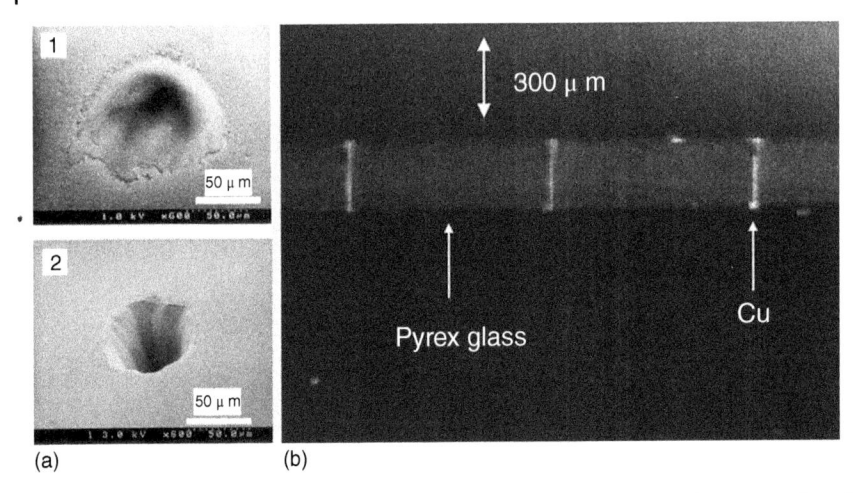

Figure 17.20 Photographs of glass via made by femtosecond laser and cross section of the TGV interconnection [23]. (a) Glass via made in air (1) and vacuum (2) and (b) cross-sectional photograph of the TGV interconnection.

photographs of the glass vias made by the femtosecond laser. Deposition of glass is observed around the glass via when the via is made in atmosphere (photograph 1). On the other hand no deposition is observed when the via is made in vacuum (photograph 2). The reason is that the glass particles ablated are not reflected in vacuum. The cross-sectional photograph of the TGV interconnection is shown in Figure 17.20b.

The metal used for the TGV interconnection should have close CTE with the glass especially in case the diameter of the TGV is not small. Otherwise a gap between the metal and the glass is made during the thermal process for the anodic bonding and causes air leakage through the gap. The LTCC with Au feedthrough explained in Figure 11.19 (Chapter 11) was developed to solve this problem [25].

Holes in a glass can be made by batch process using DRIE. However it is difficult to make deep holes in the glass wafer because of the slow etching rate ($\sim$0.5 μm min^{-1}) by DRIE and etching product deposition on the side wall of the hole [26]. DRIE of the glass wafers has been applied to the WLP using temporary bonding with polyimide, which withstand the temperature for the anodic bonding. The fabrication process is shown in Figure 17.21 [27]. A Pyrex glass wafer is etched by DRIE using Ni mask to make shallow holes ($\sim$50 μm in depth) (1 in the figure). Holes are filled with electroplated Cu (or Ni) (2 and 3 in the figure). After polishing the surface of the glass wafer, it is bonded to a handle glass wafer with the polyimide (4 in the figure). The handle wafer has grooves in it and Ge is deposited on the surface. The Cu (or Ni) is exposed by grinding and polishing the Pyrex glass wafer from the backside (5 in the figure). The Pyrex glass wafer is anodically bonded to the MEMS Si wafer (6 in the figure). The polyimide can stand the bonding temperature (400 °C). Finally the handling wafer is detached by dissolving the Ge in boiling H_2O_2, and then the polyimide is etched out (7 in the figure). The H_2O_2 penetrates in the grooves, and the handle wafer can be separated from the Pyrex glass wafer as shown in Figure 17.22.

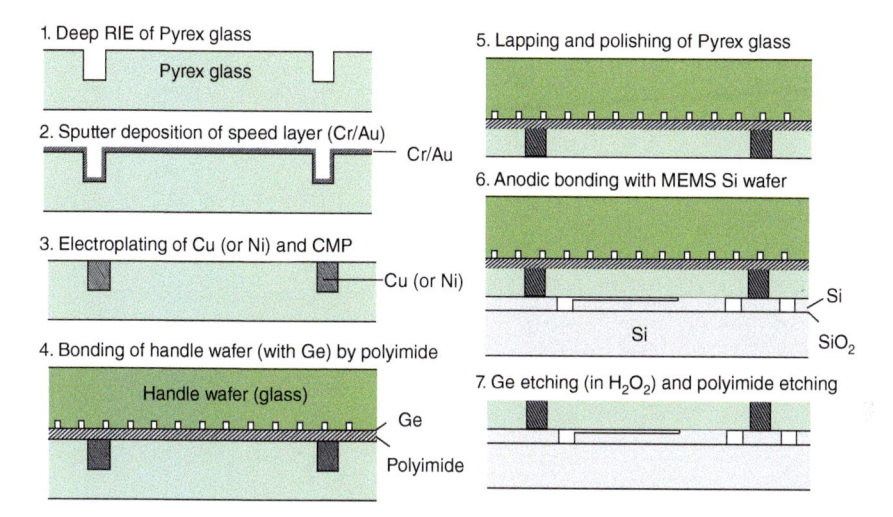

Figure 17.21 Wafer level packaging using glass RIE and temporary bonding. Source: Modified from Li et al. [27].

Figure 17.22 Detachment of the MEMS wafer from the handle glass wafer in boiling H_2O_2 ((a) → (b) → (c)) [27].

17.3.1.2 Through Si via (TSiV) Interconnection

Electrical interconnection through the Si wafer is called TSiVs. The TSiV was made by using temperature gradient zone melting [28, 29]. The fabrication process is shown in Figure 17.23. After oxidation of n-Si wafer and SiO_2 patterning (1 in the figure), Si is isotropically etched (2 in the figure). Al is deposited (3 in the figure) and the SiO_2 is etched out (4 in the figure). The backside is heated and the front side is kept at reduced temperature, which causes a temperature gradient in the thickness direction. The Al melts into the Si and moves to the backside, leaving Al-doped p^+-Si behind (5 in the figure). The TSiV that is electrically isolated from the n-Si by p–n junction is formed by thinning the Si wafer from the backside.

The TSiV can be formed by filling insulated through holes in Si with poly Si. Figure 17.24 shows the fabrication process of this poly Si TSiV. Through holes are made by DRIE (1 in the figure), and the surface is insulated with SiO_2 by oxidation (2 in the figure). Holes are filled with poly Si by CVD (3 in the figure), and the surface poly Si is etched out (4 in the figure). The poly Si can be doped to reduce series resistance. Figure 17.25 shows the cross-sectional photograph of the TSiV using poly

17 Packaging, Sealing, and Interconnection

1. Si oxidation and patterning

— SiO$_2$

n-Si

2. Si isotropic etching

3. Al deposition

— Al

4. SiO$_2$ etching

5. Temperature gradient zone melting

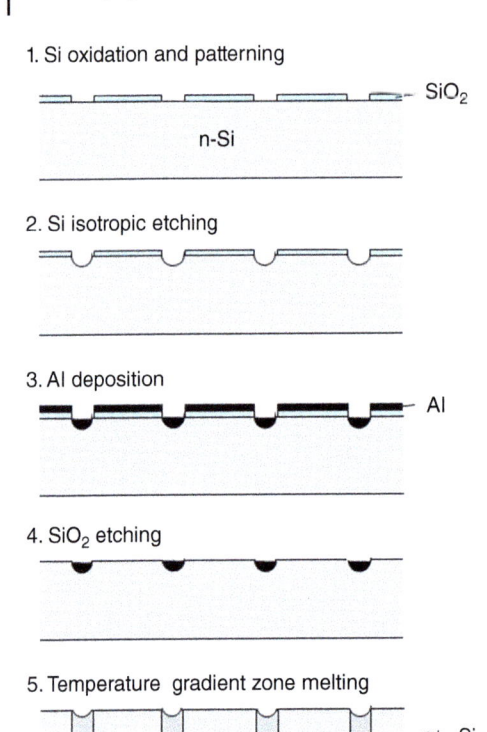

— p$^+$–Si

Figure 17.23 Through Si via (TSiV) made by temperature gradient zone melting. Sources: Based on Cline and Anthony [28]; Anthony and Cline [29].

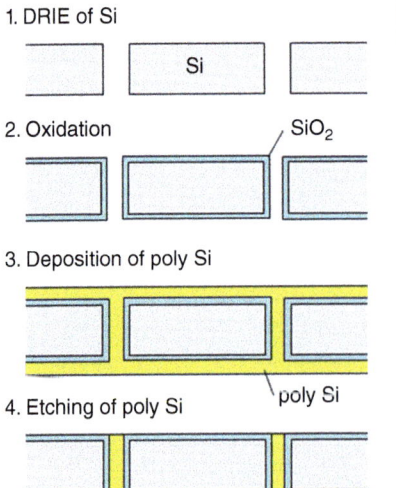

1. DRIE of Si

Si

2. Oxidation

SiO$_2$

3. Deposition of poly Si

4. Etching of poly Si

poly Si

Figure 17.24 Fabrication process of poly Si TSiV.

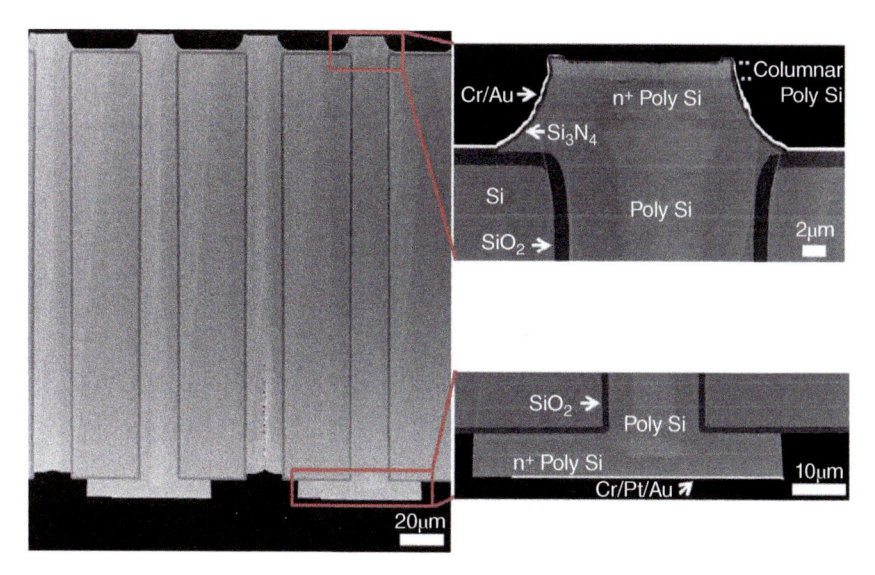

Figure 17.25 TSiV by poly Si CVD [30].

Si and n⁺-doped poly Si as the conductor and SiO$_2$ as the insulator [30]. This was developed for active matrix electron emitter array needed for massive parallel electron beam write (MPEBW) system, which is described in Section 9.4 (Chapter 9). Electron emitter called nc-Si (nano-crystalline Si) emitter is fabricated at the top of the TSiV, which is connected to the control CMOS LSI at the bottom using bumps (Figures 9.12 and 9.13 in Chapter 9).

The control LSI has also the TSiV in the MPEBW system as shown in Figure 17.16. This TSiV uses metal (Cu) as the conductor and the fabrication process is shown in Figure 17.26 [31]. The LSI has n well below surface multi-metal layers and metal vias for interconnection (1 in the figure). The Si is etched around the via from the backside using DRIE (2 in the figure), and the etched hole is filled with polymer (BCB) for insulation (3 in the figure). The Si in the polymer is etched out by DRIE (4 in the figure), and the hole is filled with Cu by electroplating (5 in the figure). Two metal interconnection layers are formed on the backside using Cu and the polymer (BCB) (6 in the figure).

17.3.2 Lateral Feedthrough Interconnection

Sealing by deposition is used to make a reference pressure chamber for a capacitive pressure sensor [32]. A glass having a capacitance measurement electrode is anodically bonded to a Si substrate, which has a diaphragm. The channel for the lateral feedthrough from the electrode to bonding pad is sealed with glass frit (low-melting-point glass).

Another example of the lateral feedthrough for the interconnection is a thermal mass-flow sensor shown in Figure 17.27 [33]. Micro-heater for the sensor is located in the gas channel, and the lateral feedthrough to the pad is sealed by depositing PECVD SiN or epoxy in the hole between the sensor and the pad.

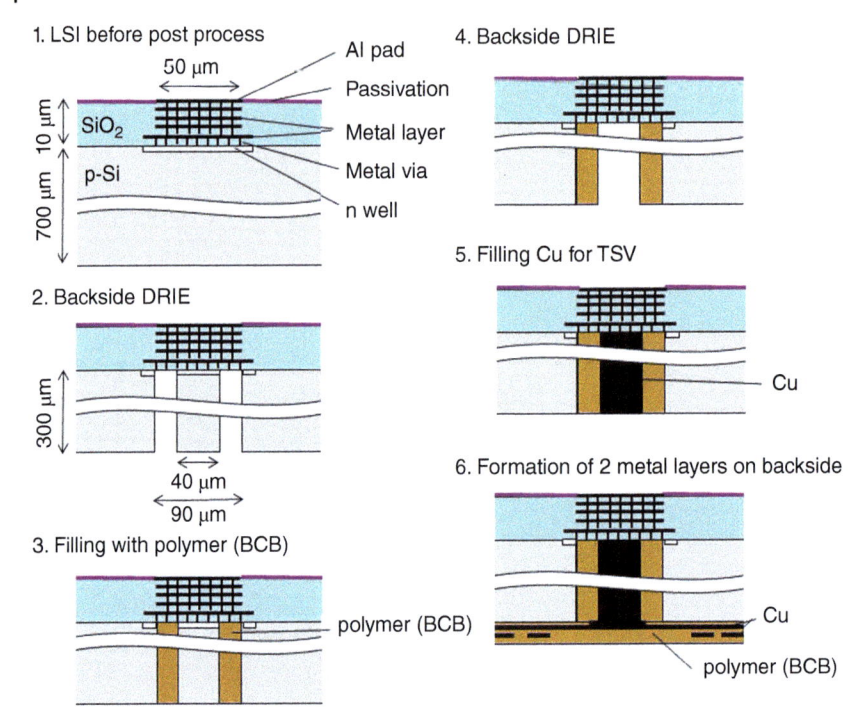

Figure 17.26 TSiV made below LSI by electroplating. Source: Based on Miyaguchi et al. [31].

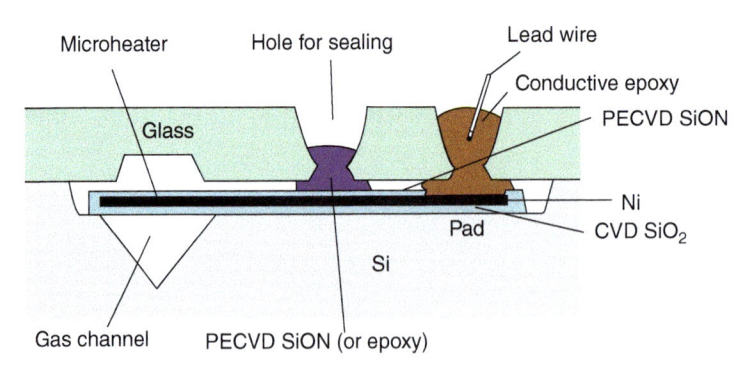

Figure 17.27 Thermal mass-flow sensor using deposition sealing. Source: Esashi et al. [33].

Lateral interconnections from hermetically sealed cavity made by anodic bonding have been developed. Figure 17.28 is a piezoresistive absolute pressure sensor [34]. A glass having Al for pad and the interconnection is anodically bonded to a Si having a thin diaphragm. The electrical interconnections from piezoresistors on the diaphragm is made by the contact of p^+-Si to Al on the glass. However the p–n junction which contacts the glass causes leakage current. Piezoresistive Si accelerometer was developed using similar lateral interconnections [35].

Figure 17.28 Absolute pressure sensor using lateral feedthrough. Source: Based on Nunn and Angell et al. [34]).

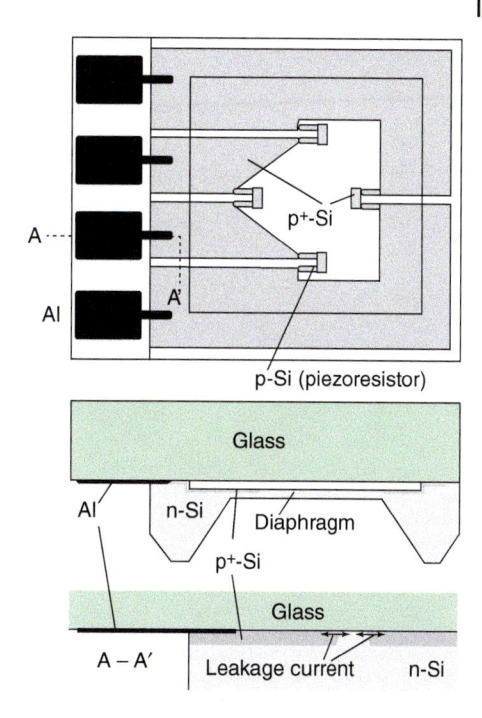

The lateral metal interconnections at anodically bonded glass–Si interface have been used. Figure 17.29a shows an encapsulated resonant structure [36]. Glass wafers that have metal patterns are anodically bonded on both sides of a Si wafer having resonators. The metal pattern is used as the lateral feedthrough. The air leakage depends on the thickness of the metal as shown in Figure 17.29b. The lateral feedthrough with thin metal (20 nm Au on 2 nm Ti in thickness and 20 µm in width) is leakage free, and it is known that metal thicker than 50 nm causes the air leakage at the step [37].

Hermetically sealed electrical feedthrough can be made at the anodically bonded glass–Si interface as shown in Figure 17.30 [37]. The leakage pass at the metal step can be sealed at the tip of the triangle pattern because the metal becomes thin by being pressed during the anodic bonding.

Figure 17.31 shows an another hermetic sealing method of the lateral feedthrough [38]. Metal pattern passes the edge of the trench in a glass as shown in the figure. The metal at the concave corner is compressed during glass to glass bonding process, and it can seal the leakage pass.

The lateral feedthrough interconnection can be made with metal in SiO_2 layer as shown in Figure 17.32 [39]. The fabrication process is as follows. Si is oxidized, and the surface oxide is partly etched to make a groove using photoresist as a mask (1 in the figure). Cr and Al are evaporated on it (2 in the figure). Metal remains in the groove by removing the photoresist, which is so-called lift-off process (3 in the figure). Spin-on-glass (SOG) is coated on it and sintered to make the surface flat (4 in the figure). Si is deposited by sputtering on it (5 in the figure), and it is

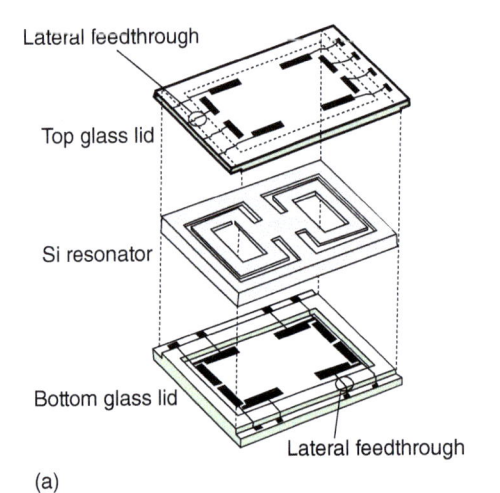

Figure 17.29 Encapsulated resonant structure with lateral feedthrough. (a) Packaged resonator and (b) lateral feedthrough without and with air leakage. Source: Modified from Corman et al. [36].

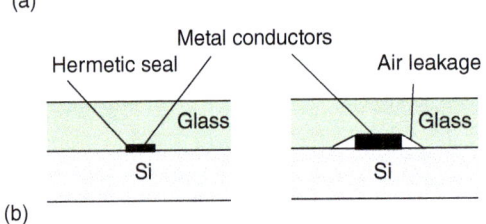

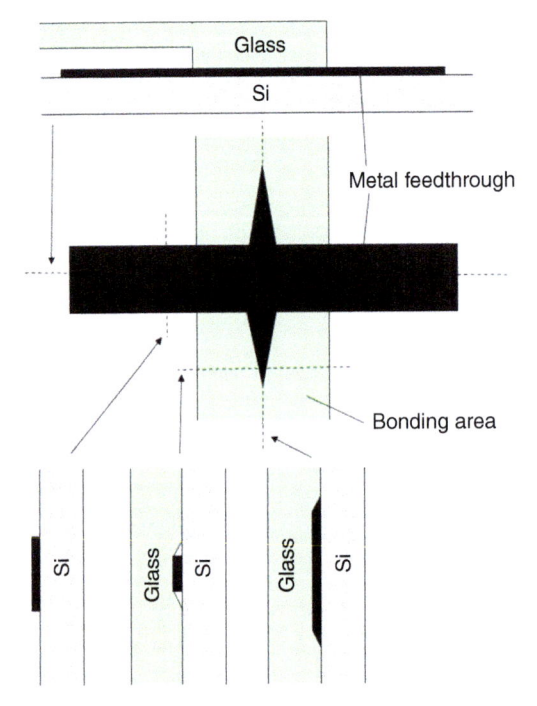

Figure 17.30 Hermetically sealed electrical feedthrough conductor. Source: Based on Petersen [37].

Figure 17.31 Sealing by metal edge at concave corner. Source: Based on Hiltmann et al. [38].

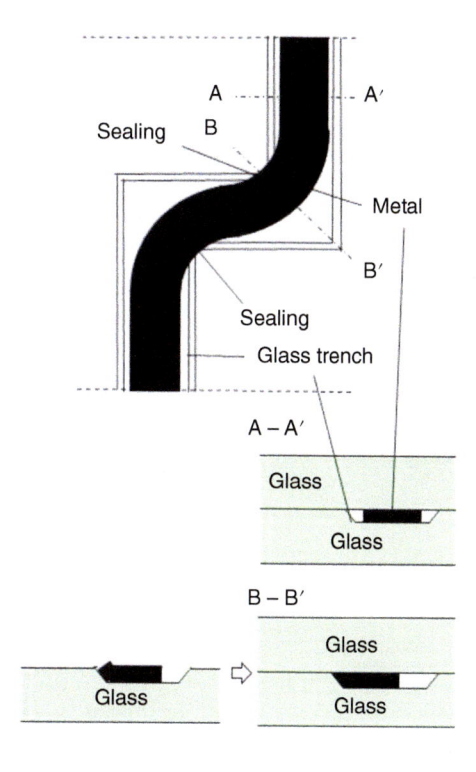

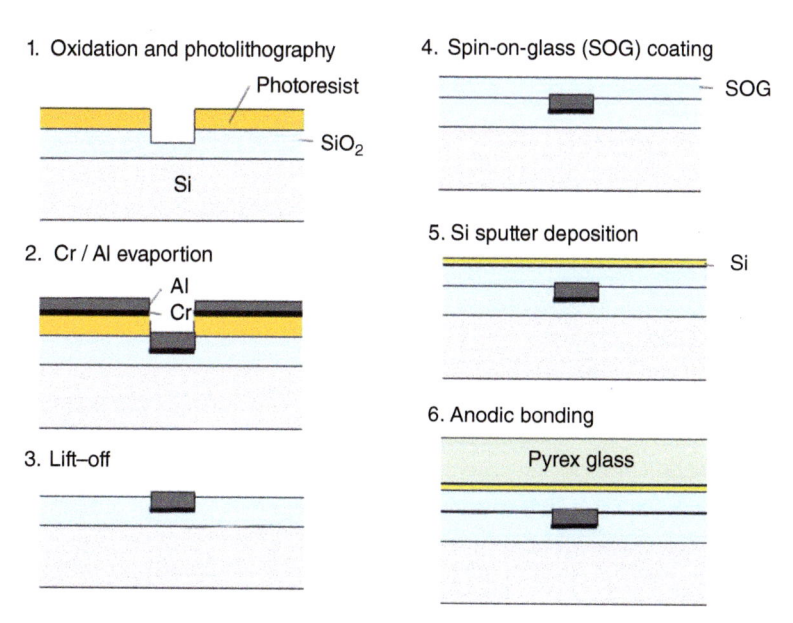

Figure 17.32 Lateral feedthrough interconnection with metal in SiO_2. Source: Modified from Esashi et al. [39].

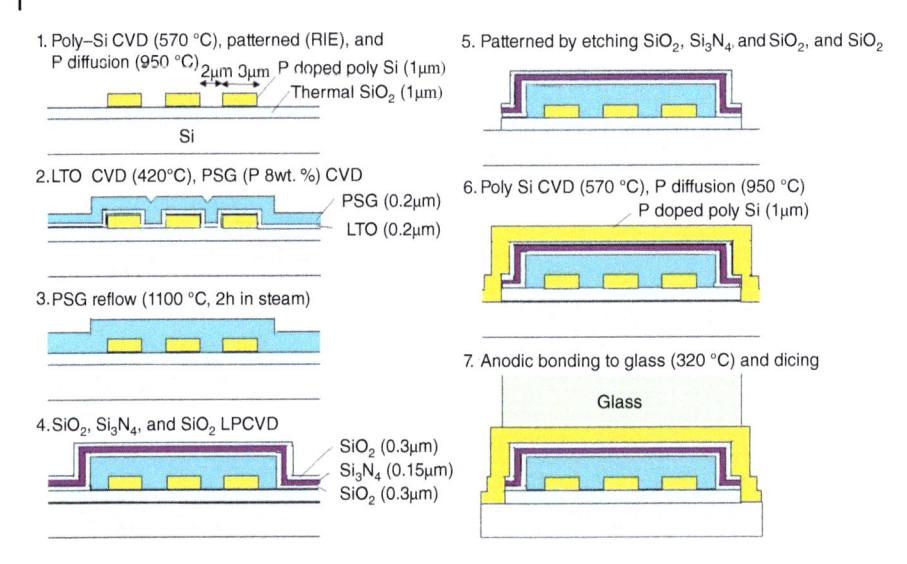

Figure 17.33 Process flow of lateral feedthrough interconnection with poly Si in phosphosilicate glass (PSG). Source: Based on Ziaie et al. [40].

anodically bonded to a Pyrex glass (6 in the figure). The process should be done after fabricating a MEMS structure on the wafer, which is difficult in many cases.

Figure 17.33 is a lateral feedthrough interconnection with poly Si in PSG [40]. This was developed for implantable telemetry capsule. The fabrication process is as follows. P-doped poly Si pattern is formed on thermal SiO_2 (1 in the figure). It is covered with low-temperature oxide (LTO) and PSG (2 in the figure) and annealed at 1100 °C for 2 h in steam to make the surface smooth by reflow (3 in the figure). SiO_2, Si_3N_4, and SiO_2 are deposited by low-pressure chemical vapor deposition (LPCVD) (4 in the figure), and these layers and the thermal SiO_2 are patterned (5 in the figure). The surface is covered with P-doped poly Si (6 in the figure), and glass capsule is anodically bonded on the poly Si for sealing (6 in the figure).

The lateral feedthrough interconnection can be made using buried diffusion layer as show in Figure 17.34. The buried diffusion layer is fabricated by making p-Si diffusion layer in a n-Si wafer (1 in the figure) and n^+-Si diffusion layer, which covers the p-Si diffusion layer (2 in the figure). Contact holes are made in the SiO_2 and then Ti and Pt are deposited and patterned (3 in the figure). After making MEMS structure such as poly Si cantilever by surface micromachining (4 in the figure), a glass cover is anodically bonded (5 in the figure). This feedthrough has some series resistance because it uses diffusion layer for the interconnection.

Metal interconnections on a chip can be covered with melted glass for the hermetically sealed packaging. The process of integrated accelerometer made by this process is shown in Figure 17.35 [41]. MEMS are made by surface micromachining on a Si wafer having integrated circuit (IC) chips. A lid wafer that has glass frit (solder glass and low-melting point glass) pattern for sealing is prepared (1 in the figure).

Figure 17.34 Lateral feedthrough interconnection with buried diffusion layer.

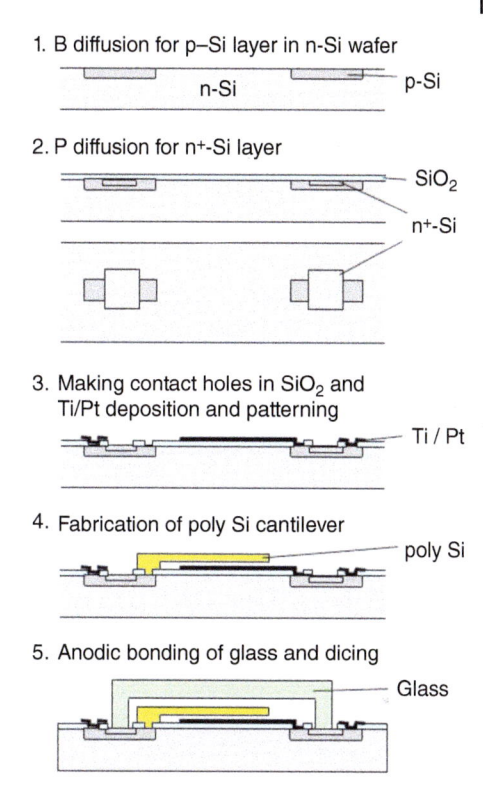

1. B diffusion for p–Si layer in n-Si wafer

   n-Si   p-Si

2. P diffusion for n+-Si layer

   SiO$_2$
   n+-Si

3. Making contact holes in SiO$_2$ and Ti/Pt deposition and patterning

   Ti / Pt

4. Fabrication of poly Si cantilever

   poly Si

5. Anodic bonding of glass and dicing

   Glass

Both wafers are thermally bonded with the melted glass frit (2 in the figure). The lid wafer is sawed to expose the bonding pads, and the bonded wafer is diced to each chips (3 in the figure). Finally wire bonding and plastic molding are performed (4 in the figure).

17.3.3 Interconnection by Electroplating

Electroplating is used for non-planer (3D) batch interconnection. Figure 17.36 is a 3D multi-electrode array in which interconnections are made by Ni electroplating [42]. The electrodes to detect neuron signals are electrically connected to the circuit for signal processing.

The interconnection by the electroplating was applied to a tactile display [43]. It uses two shape memory alloy (SMA) actuators for moving pins up and down and magnetic latch for holding the state as shown in Figure 17.37a. The SMA is joule heated by current for the actuation. SMA actuators and pins are assembled on a printed circuit board. The 3D batch interconnections are made by Ni electroplating as shown in Figure 17.37b. The photograph of the tactile display is shown in Figure 17.37c.

Another example is a batch assembly of an active catheter by the electroplating shown in Figure 17.38 [44]. The SMA coil is used as an actuator combined with a

402 | *17 Packaging, Sealing, and Interconnection*

1. MEMS wafer and lid wafer

2. Thermal bonding with glass frit

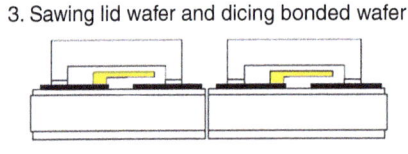

3. Sawing lid wafer and dicing bonded wafer

4. Bonding wires and plastic molding

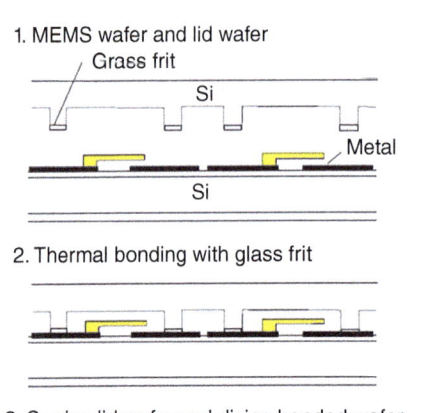

Figure 17.35 Wafer level packaging bonded to lid wafer with glass frit (solder glass). Source: Based on Judy [41].

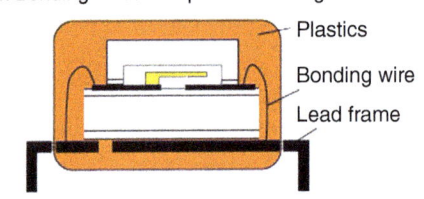

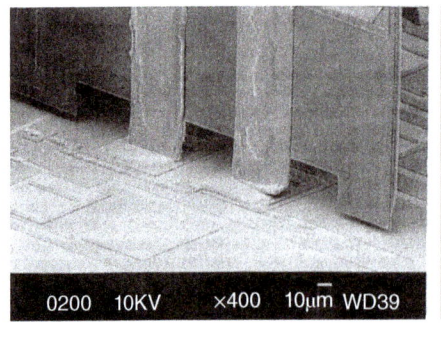

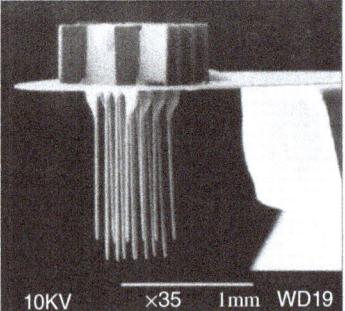

Figure 17.36 Interconnection of 3D multi-electrode array by electroplating. Source: Hoogerwerf and Wise [42] with permission.

liner coil for bias spring. The catheter can not only bend but also tilt and extend depending on the moving directions of the SMA actuators. The fabrication process is as follows. The SMA coil is coated with acrylic resin, and it is assembled with the liner coil. The polymer film on the surface of the SMA actuator is selectively removed by laser ((a) in the figure). Ni is electroplated on the exposed part of the SMA coil for the mechanical assembly and the electrical interconnection ((b) in the figure). The photograph of the assembled active catheter is shown in Figure 17.38c.

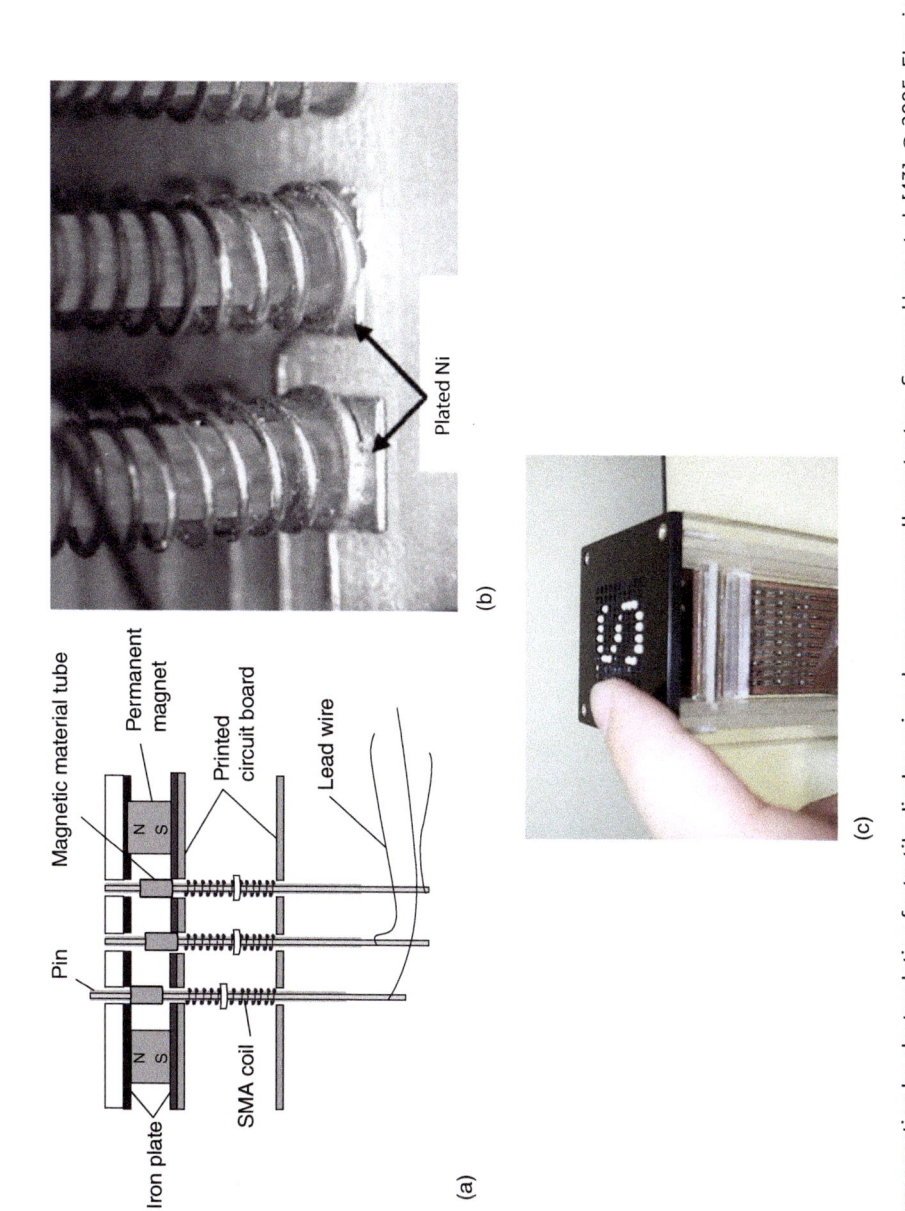

Figure 17.37 Interconnection by electroplating for tactile display using shape memory alloy actuators. Source: Haga et al. [43]. © 2005, Elsevier. (a) Principle of tactile display, (b) interconnection by Ni electroplating, and (c) tactile display.

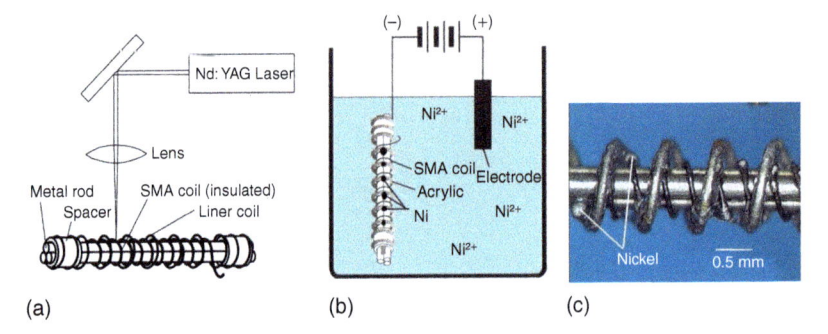

Figure 17.38 Batch assembly of active catheter by electroplating. (a) Removal of surface insulator, (b) Ni electroplating, and (c) assembly by electroplated Ni. Source: Haga et al. [44]. © 2000, IEEE.

References

1 Esashi, M. (2008). Wafer level packaging of MEMS. *J. Micromech. Microeng.* 18 (7): 073001. (13pp).
2 Matsumoto, Y., Shoji, S., and Esashi, M. (1990). A miniature integrated capacitive pressure sensor. 22nd Conf. on Solid State Devices and Materials, Sendai, Japan (22–24 August 1990), 701.
3 Shoji, S. and Esashi, M. (1990). Photoetching and electrochemical discharge drilling of Pyrex glass. Tech. Digest of the 9th Sensor Symp. Tokyo, Japan (30–31 May 1990), 25–28.
4 Guckel, H. and Burns, D.W. (1985). Laser-recrystallized piezoresistive micro-diaphragm sensor. Digest of Technical Papers, The 3rd Int. Conf. on Solid State Sensors and Actuators (Transducers'85), Stockholm, Sweden (25–29 June 1985), 182–185.
5 Shoji, S., Nisase, T., Esashi, M., and Matsuo, T. (1987). Fabrication of an implantable capacitive type pressure sensor. Digest of Technical Papers, The 4th Int. Conf. on Solid State Sensors and Actuators (Transducers'87), Tokyo, Japan (2–5 June 1987), 305–308.
6 Lebouitz, K.S., Howe, R.T., and Pisano, A.P. (1995). Permeable polysilicon etch-access windows for microshell fabrication. The 8th Int. Conf. on Solid State Sensors and Actuators, and Eurosensors IX (Transducers'95 · Eurosensors IX), Stockholm, Sweden (25–29 June 1995), 224–227.
7 Dougherty, G.M., Sands, T.D., and Pisano, A.P. (2003). Microfabrication using one-step LPCVD porous polysilicon films. *J. Microelectromech. Syst.* 12 (4): 418–421.
8 Kiihamäki, J., Dekker, J., Pekko, P. et al. (2004). "Plug-up" – a new concept for fabricating SOI MEMS devices. *Microsyst. Technol.* 10 (5): 346–350.
9 Fan, R.H., Fan, L., Wu, M.C., and Kim, C.J. (2004). Porous polysilicon shell formed by electrochemical etching for on-chip vacuum encapsulation. Tech. Digest solid-State Sensor, Actuator and Microsystems Workshop, Hilton Head Island, USA (6–10 June 2004), 332–335.

10 Jones, D.G., Azevedo, R.G., Chan, M.W. et al. (2007). Low temperature ion beam sputter deposition of amorphous silicon carbide for wafer-level vacuum sealing. 20th IEEE Intl. Micro Electro Mechanical Systems Conf. (MEMS 2007), Kobe, Japan (21–25 January 2007), 275–278.

11 Liu, C. and Tai, Y.-C. (1999). Sealing of micromachined cavities using chemical vapor deposition methods: characterization and optimization. *J. Microelectromech. Syst.* 8 (2): 135–145.

12 Monajemi, P., Joseph, P.J., Kohl, P.A., and Ayazi, F. (2006). Wafer-level MEMS packaging via thermally released metal-organic membranes. *J. Micromech. Microeng.* 16 (4): 742–750.

13 Oberhammer, J., Niklaus, F., and Stemme, G. (2004). Sealing of adhesive bonded devices on wafer level. *Sens. Actuators A* 110: 407–412.

14 Lapisa, M.A., Niklaus, F., and Stemme, G.N. (2009). Room-temperature wafer-level hermetic sealing for liquid reservoirs by gold ring embossing. The 15th International Conf. on Solid-State Sensors, Actuators and Microsystems (Transducers 2009), Denver, USA (21–25 June 2009), 833–836.

15 Decharat, A., Yu, J., Boers, M. et al. (2009). Room-temperature sealing of microcavities by cold metal welding. *J. Microelectromech. Syst.* 18 (6): 1318–1325.

16 Antelius, M., Fischer, A.C., Niklaus, F. et al. (2012). Hermetic integration of liquids using high-speed stud bump bonding for cavity sealing at the wafer level. *J. Micromech. Microeng.* 22 (4): 045021. (6pp).

17 Ogashiwa, T., Totsu, K., Nishizawa, M. et al. (2017). Wafer-to-wafer transportable gold particle plug for spot vacuum sealing of MEMS. The 19th International Conf. on Solid-State Sensors, Actuators and Microsystems (Transducers'17), Kaohsiung, Taiwan (18–22 June 2017), 1304–1307.

18 Wang, W.-S., Lin, Y.-C., Gessner, T., and Esashi, M. (2015). Fabrication of nanoporous gold and the application for substrate bonding at low temperature. *Jpn. J. Appl. Phys.* 54 (3): 030215(7).

19 Schmitt, J.M., Mihm, F.G., and Meindle, J.D. (1986). New methods for whole blood oximetry. *Ann. Biomed. Eng.* 14: 35–52.

20 Bowman, L., Schmitt, J.M., and Meindle, J.D. (1985). Electrical contacts to implantable integrated sensors by CO_2 laser-drilled vias through glass. In: *Micromachining and Micropackaging of Transducers* (eds. C.D. Fung, P.W. Cheung, W.H. Ko and D.G. Fleming), 79–84. Amsterdam: Elsevier Science Publishers B.V.

21 Nakamura, K., Takayanagi, F., Moro, Y. et al. (2004). Development of RF MEMS switch. *Advantest Tech. Rep.* 22: 9–16. (in Japanese).

22 Liu, Y., Li, X., Abe, T. et al. (2001). A thermomechanical relay with microspring contact array. The 14t IEEE Intl. Conf. on Micro Electromechanical Systems (MEMS 2001), Interlaken, Switzerland (21–25 January 2001), 220–223.

23 Abe, T., Li, X., and Esashi, M. (2003). Endpoint detectable plating through femtosecond laser drilled glass wafers for electrical interconnections. *Sens. Actuators, A* 108: 234–238.

24 Anthony, T.R. (1981). Forming electrical interconnections through semiconductor wafers. *J. Appl. Phys.* 52 (8): 5340–5349.

25 Tanaka, S., Matsuzaki, S., Mohri, M. et al. (2011). Wafer-level hermetic packaging technology for MEMS using anodically-bondable LTCC wafer. The 24t IEEE Intl. Conf. on Micro Electromechanical Systems (MEMS 2011), Cancun, Mexico (23–27 January 2011), 376–379.

26 Li, X., Abe, T., and Esashi, M. (2000). Deep reactive ion etching of Pyrex glass. IEEE The 13th Intl. Annual Intl. Conf. on Micro Electro Mechanical Systems (MEMS 2000), Miyazaki, Japan (23-27 January 2000), 271–276.

27 Li, X., Abe, T., Liu, Y., and Esashi, M. (2002). Fabrication of high-density electrical feed-through by deep-reactive-ion etching of Pyrex glass. *J. Microelectromech. Syst.* 11 (6): 625–629.

28 Cline, H.E. and Anthony, T.R. (1976). Thermomigration of aluminum-rich liquid wires through silicon. *J. Appl. Phys.* 47 (6): 2332–2336.

29 Anthony, T.R. and Cline, H.E. (1976). Random walk of liquid droplets migrating in silicon. *J. Appl. Phys.* 47 (6): 2316–2324.

30 Ikegami, N., Koshida, N., Kojima, A. et al. (2013). Active-matrix nanocrystalline Si electron emitter array with a function of electronic aberration correction for massively parallel electron beam direct-write lithography: electron emission and pattern transfer characteristics. *J. Vac. Sci. Technol.* B31 (6): 06F703(8).

31 Miyaguchi, H., Muroyama, M., Yoshida, S. et al. (2015). An LSI for massive parallel electron beam lithography: its design and evaluation. *IEEJ Trans. Sens. Micromachines* 135 (10): 374–381. (in Japanese).

32 Jornod, A. and Rudolf, F. (1989). High-precision capacitive absolute pressure sensor. *Sens. Actuators* 17: 415–421.

33 Esashi, M., Eoh, S., Matsuo, T., and Choi, S. (1987). The fabrication of integrated mass flow controller. Digest of Technical Papers, The 4th Intl. Conf. on Solid State Sensors and Actuators (Transducers'87), Tokyo, Japan (2–5 June 1987), 830–833.

34 Nunn, T.A. and Angell, J.B. (1977). An IC absolute pressure transducer with built-in reference chamber. In: *Indwelling and Implantable Pressure Transducers* (eds. D.G. Fleming, W.H. Ko and M.R. Neuman), 133–136. CRC Press.

35 Roylance, L.M. and Angell, J.B. (1979). A batch-fabricated silicon accelerometer. *IEEE Trans. Electron Devices* ED-26 (12): 1911–1917.

36 Corman, T., Enoksson, P., and Stemme, G. (1998). Low-pressure-encapsulated resonant structures with integrated electrodes for electrostatic excitation and capacitive detection. *Sens. Actuators A* 66: 160–166.

37 Petersen, K.E. (1985). Method and apparatus for forming hermetically sealed electrical feed-through conductor. WO 85/03381.

38 Hiltmann, K.M., Schmidt, B., Sandmaier, H., and Lang, W. (1997). Development of micromachined switches with increased reliability. 1997 Intern. Conf. on Solid-State Sensors and Actuators (Transducers '97), Chicago, USA (16–19 June 1997), 1157–1160.

39 Esashi, M., Ura, N., and Matsumoto, Y. (1992). Anodic bonding for integrated capacitive sensors. IEEE Micro Electro Mechanical Systems (MEMS '92), Travemunde, Germany (4–7 February 1992), 43–48.

40 Ziaie, B., Von Ark, J.A., Dokmeci, M.R., and Najafi, K. (1996). A hermetic glass-silicon micropackage with high-density on-chip feedthroughs for sensors and actuators. *J. Microelectromech. Syst.* 5 (3): 166–179.

41 Judy, M.W. (2004). Evolution of integrated inertial MEMS technology. Solid-State Sensor, Actuator and Microsystems Workshop, Hilton Head Island, USA (6–10 June 2004), 27–32.

42 Hoogerwerf, A.C. and Wise, K.D. (1991). A three-dimensional neural recording array. 1991 Intl. Conf. on Solid-State Sensors and Actuators (Transducers '91), Berkeley, USA (24–27 June 1991), 120–123.

43 Haga, Y., Makishi, W., Iwami, K. et al. (2005). Dynamic braille display using SMA coil actuator and magnetic latch. *Sens. Actuators A* 119: 316–322.

44 Haga, Y., Esashi, M., and Maeda, S. (2000). Bending, torsional and extending active catheter assembled using electroplating. IEEE The 13th Intl. Annual Intl. Conf. on Micro Electro Mechanical Systems (MEMS 2000), Miyazaki, Japan (23–27 January 2000), 181–186.

18

Vacuum Packaging

Masayoshi Esashi

Tohoku University, Micro System Integration Center (µSIC), 519-1176 Aramaki-Aza-Aoba, Aoba-ku, 980-0845, Sendai, Japan

18.1 Problems of Vacuum Packaging

Vacuum cavity is required for micro-electro mechanical systems (MEMS) devices such as resonators, thermal infrared (IR) imagers, and absolute pressure sensors for the following reasons. The resonators lose their motional energy by viscous dumping in surrounding gas and their Q (quality factor) is reduced. The thermal IR imager has arrayed IR sensor elements, and each element has an IR absorber and a temperature sensor. Thermal loss by surrounding gas has to be minimized to transduce the IR light to temperature, and hence the sensor has to be kept in vacuum cavity. The vacuum cavity is effective as well for minimizing crosstalk by thermal coupling between pixels. The absolute pressure sensors require vacuum cavity for their reference pressure.

Problems in the vacuum packaging are schematically shown in Figure 18.1 [1]. Permeation of gas molecules through the structural material is one of the problems for hermetic packaging [2]. Gas molecules are absorbed and diffused in the materials. Polymer is a non-hermetic material and it is not applicable for the vacuum packaging. Solid inorganic materials such as glasses, ceramics, metals, and Si are hermetic; however He can diffuse through glasses [3]. The other problems in the vacuum packaging are outgassing from the packaging materials and leakage through the gap between different materials as shown in Figure 18.1. The small cavity in MEMS devises requires high hermeticity for the vacuum packaging.

18.2 Vacuum Packaging by Anodic Bonding

The vacuum packaging can be made by anodic bonding in a vacuum chamber. However oxygen gas is generated from the glass at the bonding interface. Figure 18.2 is the experimental result related to the oxygen generation [4]. Si having a thin SiON diaphragm with different cavity volume is anodically bonded to a Pyrex glass in a

3D and Circuit Integration of MEMS, First Edition. Edited by Masayoshi Esashi.
© 2021 WILEY-VCH GmbH. Published 2021 by WILEY-VCH GmbH.

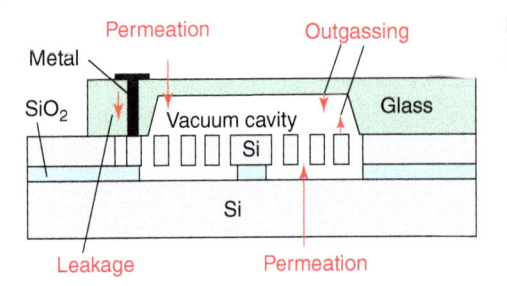

Figure 18.1 Problems in vacuum packaging.

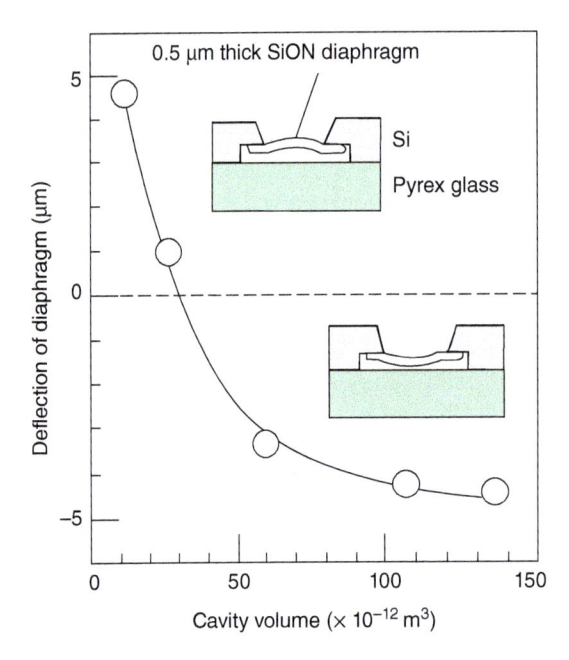

Figure 18.2 Deflection of the diaphragm in atmosphere after anodic bonding in vacuum. Source: Esashi [4].

vacuum chamber, and the deflection of the diaphragm is measured in atmosphere after the bonding. The diaphragm is deflated by the atmospheric pressure if the cavity volume is not small, which means the cavity is at reduced pressure or vacuum. On the other hand the diaphragm is inflated if the cavity volume is small. This is because the cavity is filled with oxygen generated during the anodic bonding process. We can observe displacement current by the Na^+ ion during the anodic bonding; however a small current is observed after completion of the bonding as shown in Figure 18.3. This current is caused by the following electrochemical reaction at the glass–Si interface [5].

$$2SiO^- \rightarrow 2Si + 2e^- + O_2 \uparrow$$

The SiO^- that moved from the space charge layer in the glass to the glass–Si interface generates the O_2 gas by electrochemical reaction.

Figure 18.4 shows some methods to minimize the influence of oxygen generation. The glass has a small hole. After anodic bonding the hole can be sealed with a small glass [6] or by soldering.

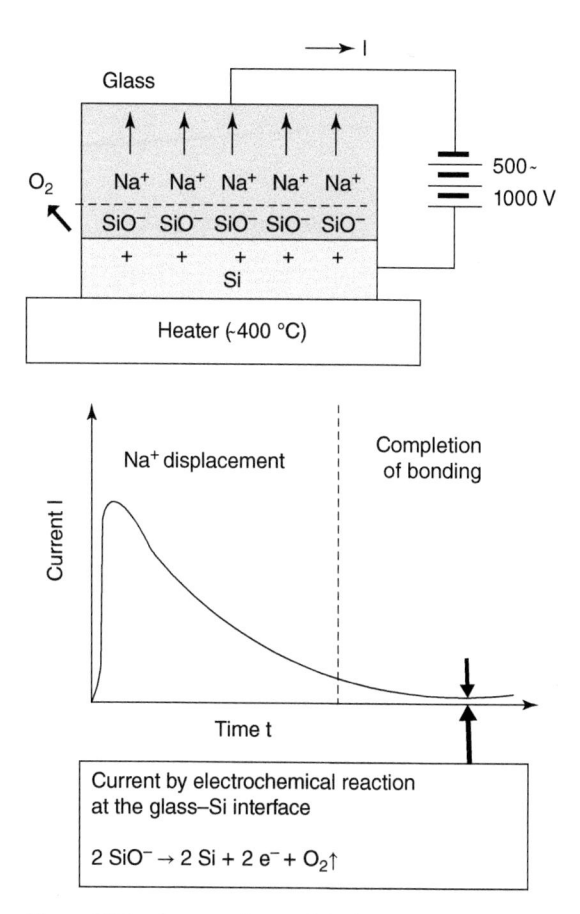

Figure 18.3 Current observed during the anodic bonding.

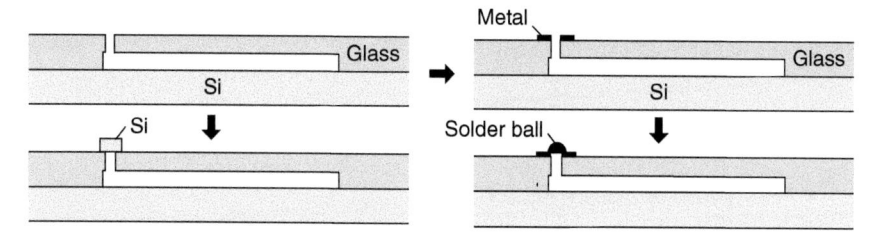

Figure 18.4 Sealing after anodic bonding.

High vacuum cavity can be obtained by using a non-evaporable getter (NEG) inside the cavity as shown in Figure 18.5 [7] [8]. During the anodic bonding in vacuum, the NEG is activated and it absorbs the oxygen gas generated. The mechanism of the NEG is shown in Figure 18.6. The NEG is composed of Ti and Zr–V–Fe alloy on a Ni-Cr ribbon, and its surface is covered with passivation layer as oxide and nitride. The surface oxide and nitride disappears during activation at

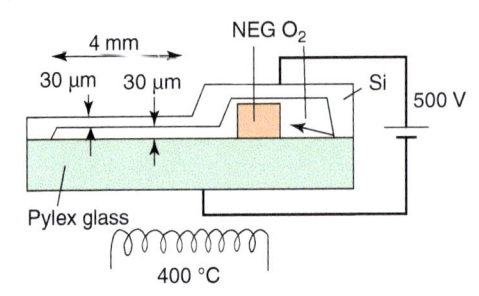

Figure 18.5 Anodically bonded vacuum package with non-evaporable getter. Source: Modified from Henmi et al. [7].

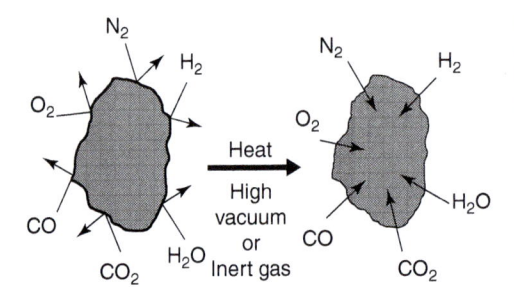

Figure 18.6 Activation of non-evaporable getter. Source: Provided by SAES Getters: Ref [8] with permission.

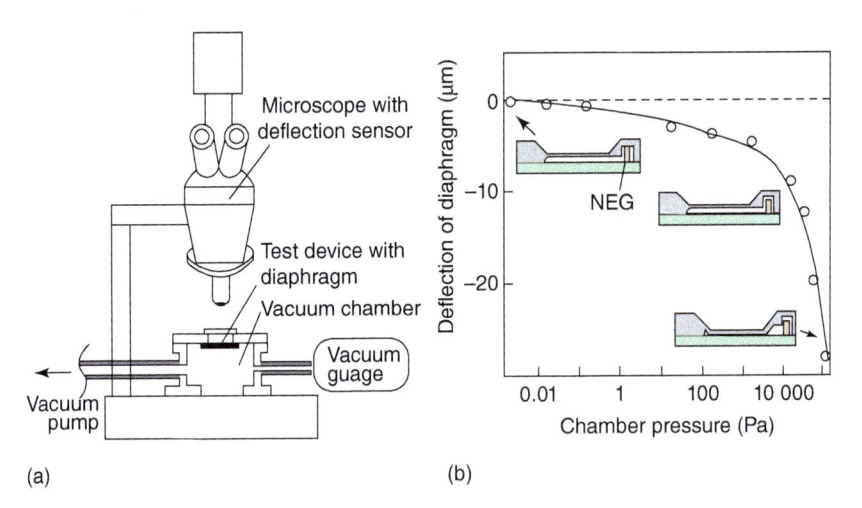

Figure 18.7 (a) Measurement system and (b) diaphragm deflection versus chamber pressure. Source: Henmi et al. [7]. © 1994, Elsevier.

400 °C in vacuum by desorption or diffusion into the bulk. Gasses can be adsorbed on the surface after the activation.

The cavity vacuum pressure can be measured by the method shown in Figure 18.7 [7]. The packaged test device which has a thin diaphragm is located in a vacuum chamber, and the deflection of the diaphragm is measured by an optical deflection sensor through a glass window. The measured deflection of the diaphragm versus the chamber pressure is shown in Figure 18.7b. The diaphragm is deflated to the cavity side at the pressure of 0.01 Pa, which means that the cavity pressure is lower than 0.01 Pa, and hence a high vacuum is achieved by using the NEG.

Figure 18.8 Si diaphragm vacuum gauge using thin diaphragm with vacuum cavity. Source: Modified from Miyashita and Kitamura [9]. (a) Chip of the Si diaphragm vacuum gauges and (b) fabrication process of the Si diaphragm vacuum gauge chip.

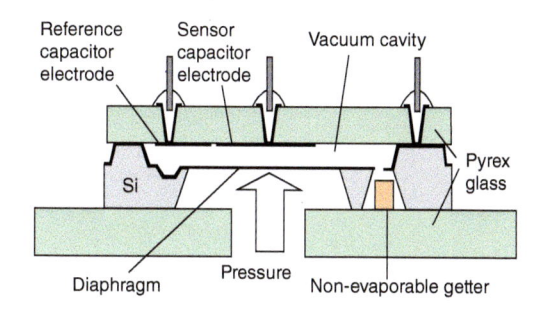

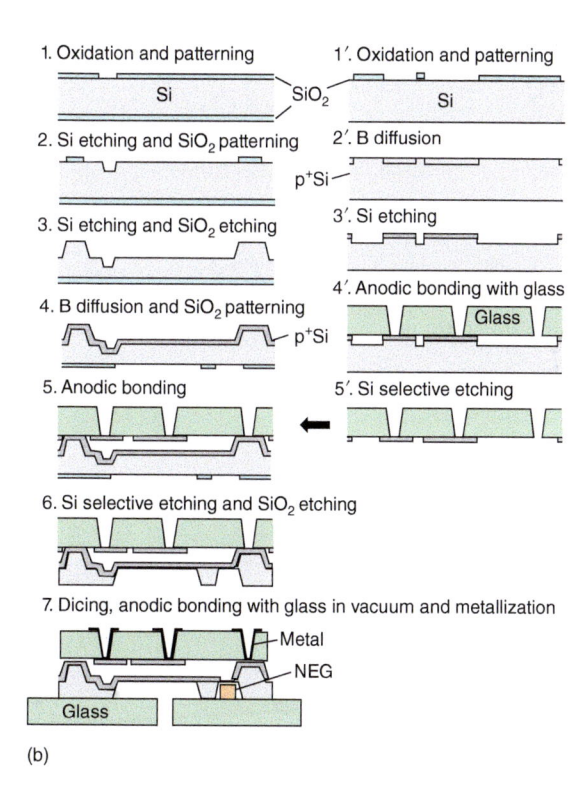

The vacuum encapsulation using the NEG was applied to Si diaphragm vacuum gauge shown in Figure 18.8a [9, 10]. This can measure low pressure from 0.1 to 133 Pa by using a thin Si diaphragm (7 μm thick, 4 mm square) and the vacuum reference cavity. Figure 18.8b shows the fabrication process of the vacuum gauge. The Si wafer for the thin diaphragm is prepared by etching and B (boron) diffusion (1–4 in the left side process in the figure). Glass wafer having through holes which have thin p$^+$ Si plates at the bottom is fabricated (1' to 5' in the right side process in the figure). These wafers are anodically bonded and the thin p$^+$ Si diaphragm is made by selective etching (5 and 6 in the left side process in the figure). The bonded glass and Si wafer is diced to a chip and a hole is made in the thin p$^+$ Si. The chip is anodically

414 | *18 Vacuum Packaging*

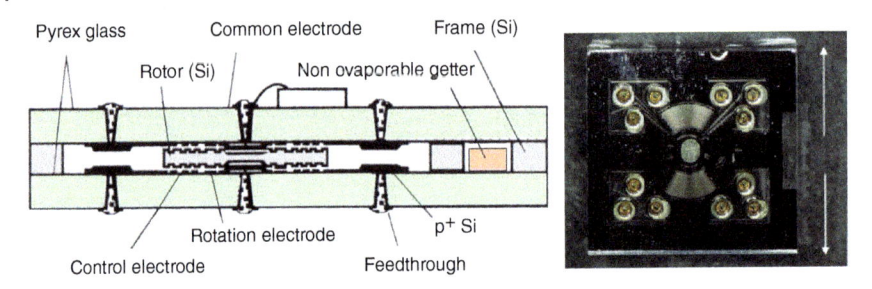

Figure 18.9 Electrostatically levitated rotational inertia measurement system using a vacuum cavity with getter. Source: Based on Fukatsu et al. [11].

bonded in vacuum to a bottom glass with NEG in the cavity, and the holes in the top glass are metalized (7 in the left side process in the figure).

The other example of vacuum packaging with the NEG is the electrostatically levitated rotational inertia measurement system shown in Figure 18.9 [11]. Si disk is electrostatically levitated and rotated inside a vacuum cavity, which has the NEG in it. The high vacuum is required to reduce a viscous dumping of the Si rotor. The Si disk rotates at 20 000 rpm by using capacitive position sensing, electrostatic actuation, and feedback control with high-speed digital signal processing. 3 axes acceleration and 2 axes rotation can be measured simultaneously using this system.

18.3 Packaging by Anodic Bonding with Controlled Cavity Pressure

Sensors such as capacitive accelerometer requires controlled optimum cavity pressure to prevent a viscous dumping at low vacuum and a resonance at high vacuum [12]. Figure 18.10 shows a force balancing Si accelerometer which has a seismic mass

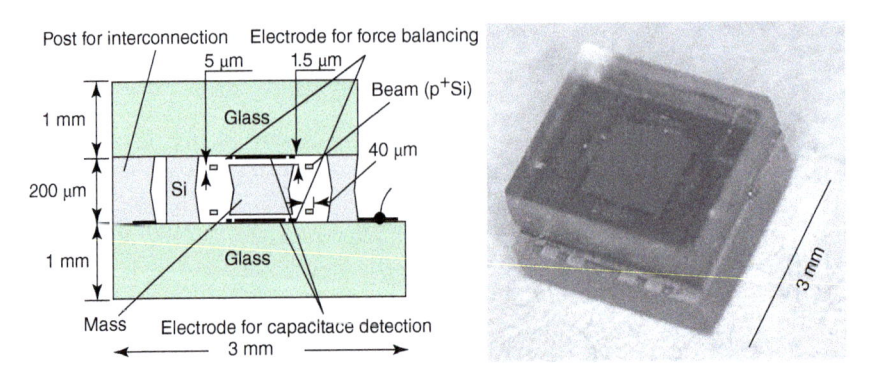

Figure 18.10 Force balancing Si accelerometer with narrow gap. Source: Modified from Minami et al. [12].

Figure 18.11 Frequency response of the Si accelerometer with different cavity pressure. Source: Modified from Lim et al. [13].

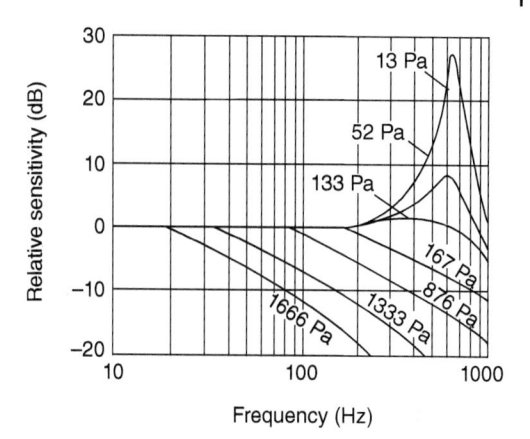

suspended with thin (5 μm thick) beams. This has a narrow 1.5 μm gap between the seismic mass and the electrode on the glass for purposes of capacitive sensing and electrostatic actuation for force balancing. Wide measurement range is achieved as far as the electrostatic force is balanced with the inertia force by the acceleration. The narrow gap is effective to increase the electrostatic force for wide acceleration range; however the cavity pressure control is required to prevent the viscous dumping at the narrow gap. Its frequency responses with different cavity pressure are shown in Figure 18.11 [13]. The vacuum pressure around 100 Pa is optimum for critical dumping as we can see in these frequency responses.

The cavity pressure can be controlled by bonding in Ar environment with NEG in the cavity as shown in Figure 18.12. The oxygen gas generated during the anodic bonding is absorbed by the NEG. On the other hand inert gas like Ar is not absorbed by the NEG. When it is bonded at 673 K (400 °C) in 13 332 Pa and in 267 Pa, the cavity pressures were around 5000 and 110 Pa, respectively. These results can be explained by the following ideal gas law.

$$PV = nRT$$

Figure 18.12 Vacuum pressure control in cavity by anodic bonding in Ar environment. Source: Modified from Lim et al. [13].

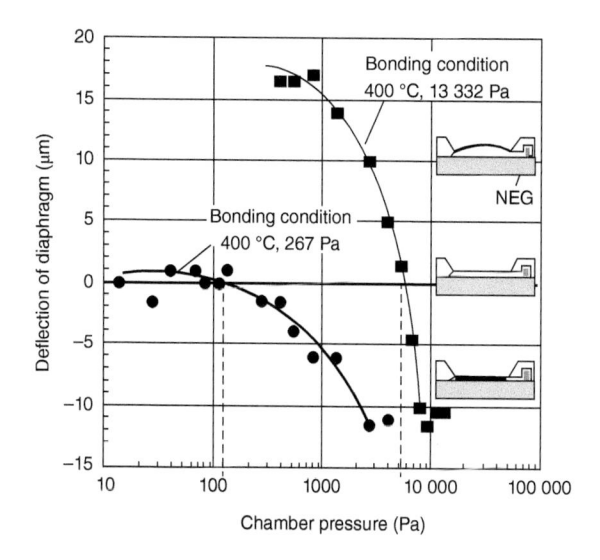

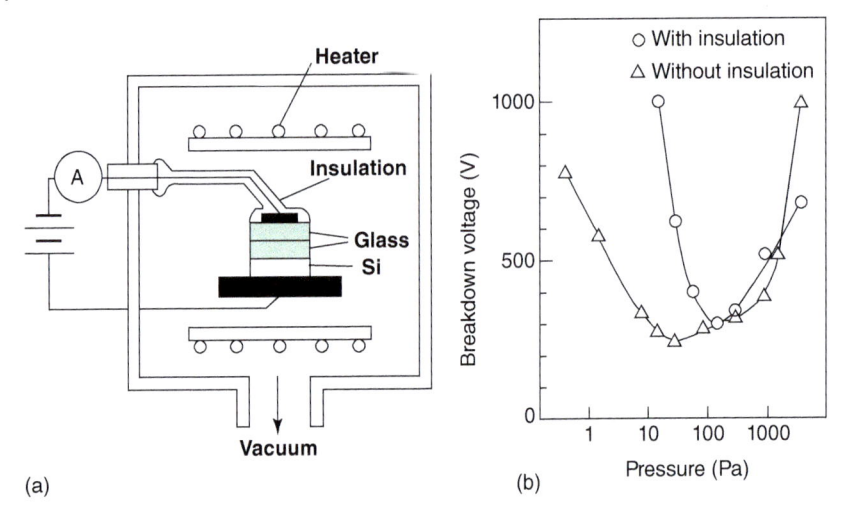

Figure 18.13 Anodic bonding for controlled vacuum pressure and breakdown voltage. (a) Bonding apparatus for controlled pressure and (b) breakdown voltage versus pressure. Source: Esashi [4].

where P is the chamber pressure, V is the chamber volume, n is the number of moles of gas, R is the gas constant, and T is the absolute temperature. When the chamber pressure was 13 332 Pa at 673 K (400 °C), the pressure calculated with this equation is 5338 Pa at 300 K (27 °C). When the chamber pressure was 267 Pa at 673 K (400 °C), the calculated pressure is 107 Pa at 300 K (27 °C). These calculated pressures (5338 and 107 Pa) at room temperature (27 °C) coincide with the obtained cavity pressures (5000 and 110 Pa). This experimental result proved that we can control the cavity pressure. However it is required to prevent an electrical breakdown during the anodic bonding in the low vacuum pressure chamber. Glass is used as the electrode to make electrical contact as shown in Figure 18.13a because metal exposed in the vacuum chamber should be covered with insulator. However maximum voltage, which can be used during the anodic bonding, is limited in some pressure range as shown in Figure 18.13b [4].

18.4 Vacuum Packaging by Metal Bonding

An example of vacuum packaging for IR imager by metal bonding is shown in Figure 18.14 [14]. IR sensor pixels composed of IR absorber and Si/SiGe quantum-well bolometer are thermally isolated and arrayed (17 μm pitch, 32 × 32). The sensor array is packaged using a Si lid, which has anti-reflective (AR) coating on both sides. These are sealed in vacuum by metal bonding using solder. Thin film NEG on the Si is used to keep the vacuum in the cavity.

Figure 18.14 Vacuum packaging of IR imager. Source: Based on Forsberg et al. [14].

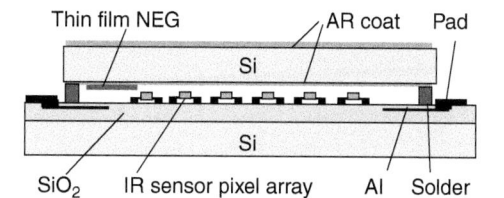

Figure 18.15 Resonating pressure sensor using a resonator in vacuum cavity in diaphragm. Source: Ikeda et al. [15]. © 1990, Elsevier. (a) Structure and (b) photograph of the resonator in vacuum cavity.

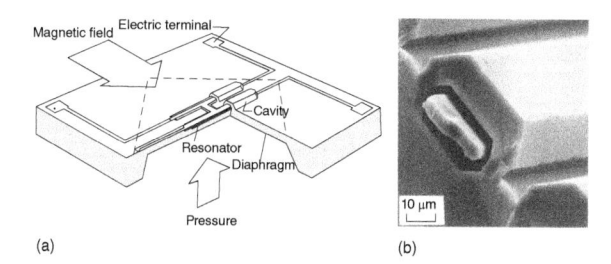

18.5 Vacuum Packaging by Deposition

Vacuum cavity can be made by sealing with deposited material. Resonating pressure sensor fabricated by Si deposition is shown in Figure 18.15 [15]. The cavity vacuum having a Si resonator is fabricated in a Si diaphragm. The resonance frequency of the resonator is modulated by a tensile stress caused by differential pressure to the diaphragm. The fabrication process is shown in Figure 18.16. The n-Si wafer is etched using SiO_2 as a mask (1 in the figure). In a low-pressure reactor p-Si, p^+-Si, p-Si, and p^+-Si are epitaxially grown successively (2 in the figure). After SiO_2 etching (3 in the figure), the p-Si is selectively etched out using $N_2H_4 \cdot H_2O$ as an etchant (4 in the figure). The selective etching is explained in Figure 18.16b. The positive voltage applied to the n-Si makes anodic oxide film on its surface, and the anodic oxide film protect the n-Si from the etching. The structure is sealed by epitaxially grown n-Si (5 in the figure). The H_2 gas remained in the cavity during the n-Si epitaxial growth is diffused out by annealing in N_2 environment (6 in the figure). Vacuum cavity is obtained using this process.

18.6 Hermeticity Testing

The hermeticity of the MEMS cavity has to be tested. The structure of Figure 18.17a can be used for self-test of the vacuum cavity [16]. The vacuum cavity is noticeable from the deflated diaphragm. Figure 18.17b is an example of a leakage test, in which stable displacement of the diaphragm confirms the hermeticity. On the other hand flat diaphragm means the cavity is filled with air because of some air leakage. The Cu feedthrough in the glass sometimes has a problem of air leakage. The Cu has higher

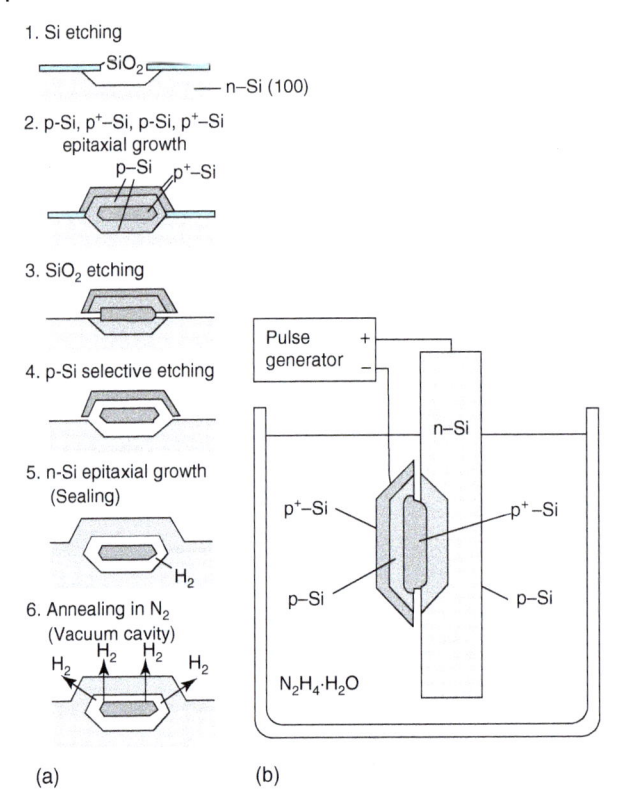

Figure 18.16 Fabrication process by selective etching and deposition vacuum seal. (a) Process flow and (b) selective etching. Source: Ikeda et al. [15]. © 1990, Elsevier.

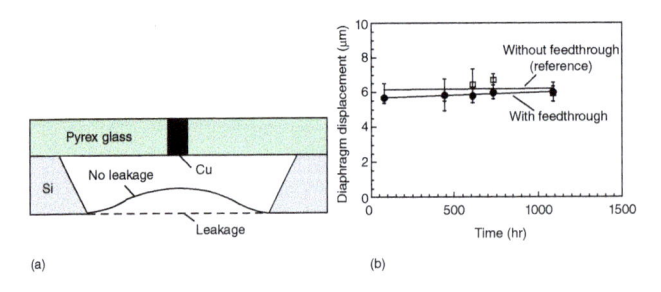

Figure 18.17 Leakage test device. (a) Device structure and (b) example of leakage test. Source: Modified from Li et al. [16].

thermal expansion than glass, and this can cause an air leakage pass between the Cu and the glass. The low-temperature co-fired ceramics (LTCC) feedthrough explained with Figure 11.19 in Chapter 11 was developed to solve this problem.

The vacuum level in the cavity can be also monitored by using micro heater in the cavity [17]. Heat dissipation of the heater due to surrounding gas molecule is measured by its resistance change, which is the principle of the Pirani vacuum gauge.

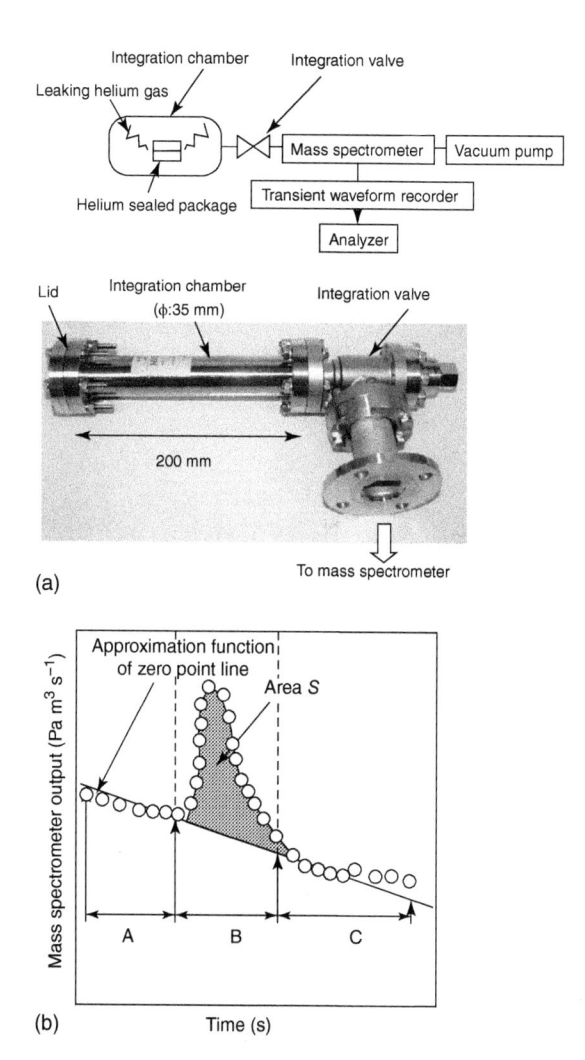

Figure 18.18 Highly sensitive He leakage tester using integration chamber. Source: Fujiyoshi et al. [20]. (a) Measuring system and (b) monitoring of accumulated He gas in the integration chamber.

Micro resonators have been used to monitor the vacuum level in the cavity as well [18, 19]. The Q (quality factor) decreases because of the dumping due to the surrounding gas molecule.

The air leakage can be directly measured by using He leakage test. Figure 18.18a is a highly sensitive He leakage tester for small cavity volume using integration chamber [20]. A sample of MEMS package kept in a chamber filled with He gas is transferred into the integration chamber. After keeping the sample in the integration chamber, the integration valve is opened, and the accumulated He gas in the chamber is measured by the mass spectrometer. An example of the monitoring of the accumulated He gas is shown in Figure 18.18b. The hermeticity can be evaluated from the area S in the graph. The leak rate resolution of 1×10^{-15} Pa m^3 s^{-1} was obtained when the accumulation time was 1200 s. This resolution is three orders of magnitude better than that of a conventional leakage testing methods.

References

1 Choa, S.-H. (2005). Reliability of MEMS packaging : vacuum maintenance and packaging induced stress. *Microsyst. Technol.* 11: 1187–1196.
2 Tummala, R.R. (2001). Chapter 15: fundamentals of sealing and encapsulation. In: *Fundamentals of Microsystems Packaging* (eds. C.P. Wong and T. Fang), 586–610. McGraw-Hill.
3 Urry, W.D. (1932). Further studies in the rare gases. 1. The permeability of various glasses to helium. *J. Am. Ceram. Soc* 54: 3887–3901.
4 Esashi, M. (1993). Complex micromechanical structures by low temperature bonding. 183rd Electrochemical Society Meeting, Honolulu, Hawaii, 1233–1234.
5 Carlson, D.E., Hang, K.W., and Stockdale, G.F. (1972). Electrode "polarization" in alkali-containing glasses. *J. Am. Ceram. Soc.* 55 (7): 337.
6 Hara, T., Kobayashi, S., and Ohwada, K. (1999). Fabrication of micromachined wafer level low-pressure package and its stability. IEEJ Technical Report, SMP-99-6, 11–15 (in Japanese).
7 Henmi, H., Shoji, S., Shoji, Y. et al. (1994). Vacuum packaging for microsensors by glass-silicon anodic bonding. *Sensor. Actuat. A* 43: 243–248.
8 Tominetti, S. and Amiotti, M. (2002). Getters for flat-panel displays. *Proc. IEEE* 90 (4): 540–558.
9 Miyashita, H. and Kitamura, Y. (2005). Micromachined capacitive diaphragm gauge. *Anelva Technical Report* 11 (4): 37–41. (in Japanese).
10 Esashi, M., Sugiyama, S., Ikeda, K. et al. (1998). Vacuum-sealed silicon micromachined pressure sensors. *Proc. IEEE* 86 (8): 1627–1639.
11 Fukatsu, K., Murakoshi, T., and Esashi, M. (1999). Electrostatically levitated micro motor for inertia measurement system. Technical Digest of the Transducers'99, 3P2.16 (7–10 June 1999) 1558–1561.
12 Minami, K., Moriuchi, T., and Esashi, M. (1997). Dumping control for packaged micro mechanical devices. *Trans. IEEJ* 117-E (2): 109–116. (in Japanese).
13 Lim, G., Baek, S., and Esashi, M. (1998). A new bulk-micromachining using deep RIE and wet etching for an accelerometer. *Trans. IEEJ* 118-E (9): 420–424.
14 Forsberg, F., Lapadatu, A., Kittilsland, G. et al. (2015). CMOS-integrated Si/SiGe Quantum-well infrared micro bolometer focal plane arrays manufactured with very large-scale heterogeneous 3-D integration. *IEEE J. Sel. Top. Quantum Electron.* 21 (4): 2700111.
15 Ikeda, K., Kuwayama, H., Kobayashi, T. et al. (1990). Three-dimensional micromachining of silicon pressure sensor integrating resonant strain gages on diaphragm. *Sens. Actuat.* A21–A23: 1007–1010.
16 Li, X., Abe, T., and Esashi, M. (2004). An integrated encapsulating technology with high-density plated-through-holes in Pyrex glass. ICEE 2004/APCOT MNT 2004, Sapporo (5–6 July 2004), 634–637.

17 Mitchell, J., Lahiji, G.R., and Najafi, K. (2006). Long-term reliability, burn-in and analysis of outgassing in Au-Si eutectic wafer-level vacuum packages. Solid-State Sensors, Actuators, and Microsystems Workshop, Hilton Head Island, USA (4–8 June 2006) 376–379.

18 Chiao, M. and Lin, L. (2002). Accelerated hermeticity testing of a glass-silicon package formed by rapid thermal processing aluminum-to-silicon nitride bonding. *Sens. Actuat. A* 97–98: 405–409.

19 Candler, R.N., Matthew, M.A., Hopcroft, A. et al. (2006). Long-term and accelerated life testing of a novel single-wafer vacuum encapsulation for MEMS resonators. *J. Microelectromech. Syst.* 15 (6): 1446–1455.

20 Fujiyoshi, M., Nonomura, Y., and Senda, H. (2007). High-sensitivity leak testing method with high-resolution vacuum integration technique, Proc. of the 24th Sensor Symposium, Tokyo, Japan (16–17 October 2007) 99–102.

19

Buried Channels in Monolithic Si

Kazusuke Maenaka

University of Hyogo, Department of Electronics and Computer Science, Graduate School of Engineering, 2167 Shosha, Himeji 671–2280, Japan

19.1 Buried Channel/Cavity in LSI and MEMS

A micro-electro mechanical system (MEMS) technology has been developed with the integrated circuit (IC) technology. Accordingly, many similar technologies are used both for MEMS and IC technologies. In the following, buried channels in large-scale integration (LSI) will first be described before the buried channel in MEMS. For the IC, the parasitic capacitance and leak current through the p–n junction between the active transistor and the substrate deteriorate the operation speed and power efficiency. Some devices, where the transistors are formed on the insulator, such as sapphire [1] and silicon dioxide [2], were developed to reduce the parasitic capacitance and the leak current. These devices are called as silicon on sapphire (SOS) and silicon on insulator (SOI). These devices have low parasitic capacitance and low leakage current between the transistor and the substrate compared to conventional devices (Figure 19.1a–c). For further improvement of the device characteristics, a silicon on nothing (SON) device was developed [3]. The transistor is now placed on an empty cavity (Figure 19.1d), drastically reducing the parasitic capacitance, resulting in an increased device performance.

These structures (i.e. devices with an insulator layer or cavity/channel) are also useful for MEMS devices (Figure 19.2). The insulator layer (commonly SiO_2) is useful for etching stop layer or sacrificial layer, which realizes a membrane of the pressure sensor, movable parts for acceleration sensor, and so on. Some methods are used to make cavity/channels for MEMS devices. In this chapter, the SON technology, which is formed by surface migration, will be discussed after briefly describing the common MEMS devices with an empty cavity or a channel. The special feature of the SON includes the surface of the fabricated SON being atomically flat and being a single crystal. The buried channel can be vacuum sealed, and wafers with the SON structure can be the starting material for IC.

Commonly, sacrificial etching has become a major method of fabricating microstructures with channels by multiple deposition/bonding of a structural layer and etching the sacrificial layer over a substrate to realize a movable structure and

3D and Circuit Integration of MEMS, First Edition. Edited by Masayoshi Esashi.
© 2021 WILEY-VCH GmbH. Published 2021 by WILEY-VCH GmbH.

19 Buried Channels in Monolithic Si

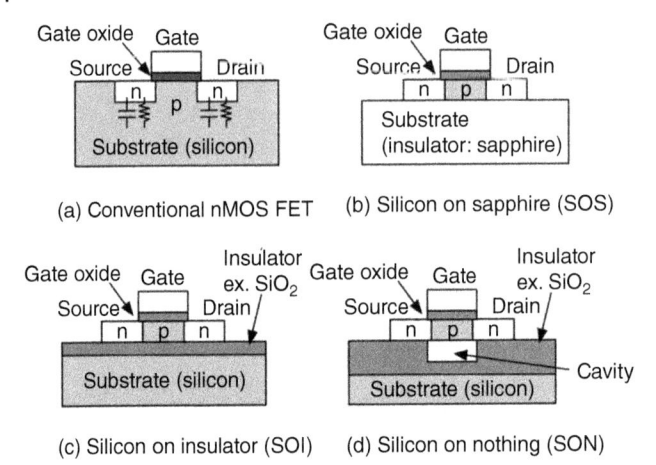

Figure 19.1 Evolution of structure of MOS transistor for low parasitic capacitance. (a) conventional, (b) SOS, (c) SOI, and (d) SON technologies.

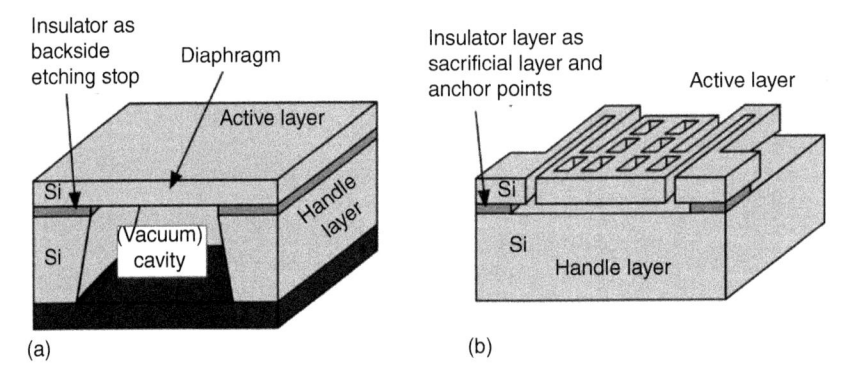

Figure 19.2 MEMS devices with cavity or channel. (a) Cavity for pressure sensor etc. using SOI and (b) movable structure of SOI by sacrificial etching.

channel [4]. Figure 19.2b shows an example of such structure. The fabrication process is basically (i) preparation of SOI as a starting wafer, (ii) patterning and etching the active layer and forming both the floating and anchor regions, and (iii) releasing of the structure by laterally etching the sacrificial layer using selective wet or gas etchant. The etching needs to be timed such that the anchor region is not etched away.

The release etching must be carefully controlled to successfully implement the fabrication process. If etching is continued for too long a period, the anchor region will also be etched out, resulting in device failure. To avoid this failure, the process may be extended to a two-step lithography with layer deposition. Figure 19.3 shows the process flow, in which the sacrificial layer is patterned for an anchor opening followed by a conformal deposition of the structural layer. The device structure is then patterned and etched, and sacrificial etching is finally performed.

Figure 19.3 Process follow of alternate device of Figure 19.2b. (a) Patterning for anchors, (b) active layer deposition, (c) patterning of active layer and (d) release etching.

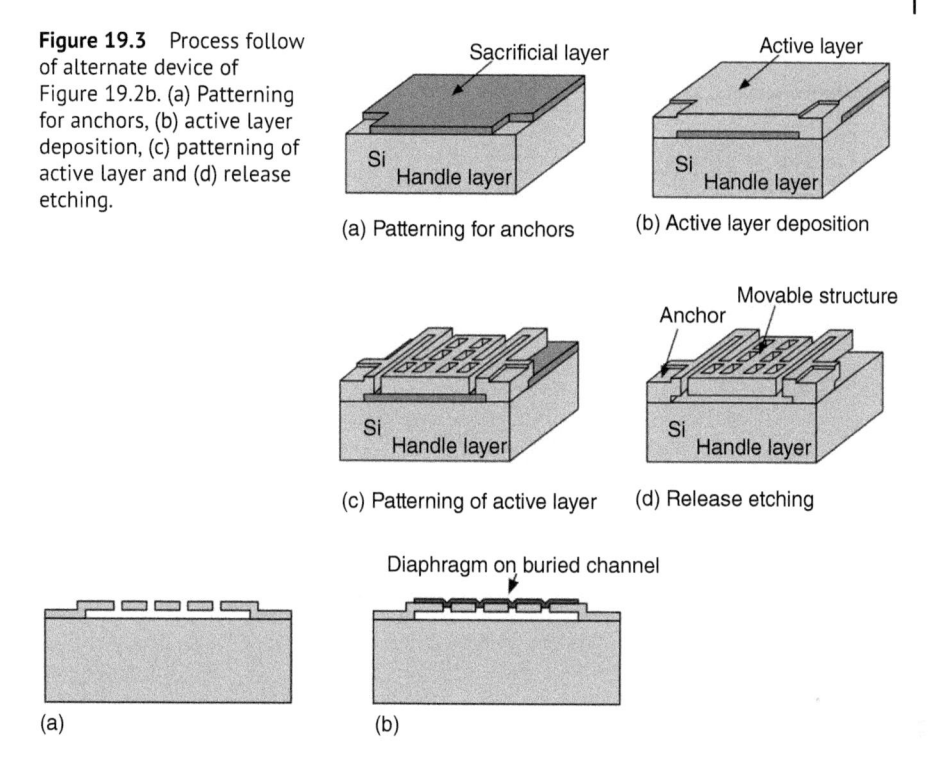

(a) Patterning for anchors
(b) Active layer deposition
(c) Patterning of active layer
(d) Release etching

Figure 19.4 Hermetic sealing by additional deposition. (a) Fabricated by basic process and (b) additional deposition makes hermetic seal.

Many types of practical MEMS devices, especially sensors, can be designed based on the aforementioned fabrication process. Almost all current MEMS inertial sensors (i.e. acceleration sensors and gyroscopes) are basically being manufactured by the process shown in Figure 19.3. Moreover, the additional process to this basic process can extend a new function of the structure (e.g. a hermetic seal or buried channel/cavity), as shown in Figure 19.4 [5]. This structure can be used for a pressure-sensitive diaphragm or sealing of a movable device.

19.2 Monolithic SON Technology and Related Technologies

The common MEMS fabrication processes were described in the previous section. The channels or floating structures were fabricated by complex processes (i.e. deposition of the layers, bonding, and sacrificial etching) and generally composed of multiple materials. On the contrary, another technique can be used to realize buried channels or a floating structure, in which a complex process is not required, and the resultant structure is composed of a single material (single crystalline silicon). The method is performed using the surface migration of silicon atoms and often called as the SON or empty space in silicon (ESS) technology.

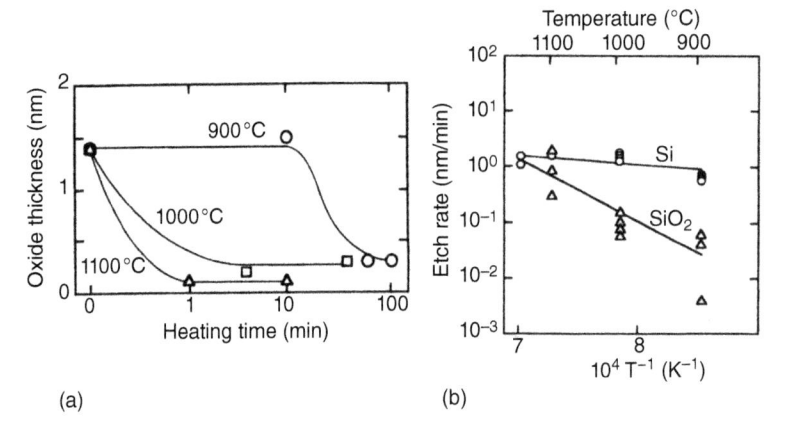

Figure 19.5 (a) Thickness of native oxide after heating and (b) etching rate of Si and SiO_2 for hydrogen annealing under atmospheric pressure. Source: Habuka et al. [6]. © 1995, IOP Publishing.

The effect of thermal treatment to the silicon surface will be described prior to discussing the SON. Thermal treatment is one of the important processes for LSI. A Si wafer with an atomically flat surface bears great concern as regards to the current LSI technology because the rough surface of Si degenerates the device characteristics. Thus, the effort of obtaining a flat surface of Si was positively performed. Hydrogen annealing is one of the processes used to make a flat surface, and it offers surface etching [6] and migration of Si atoms [7–10].

The chemical etching reactions for hydrogen–silicon and hydrogen–silicon dioxide during annealing are expressed as follows:

$$Si + 2H_2 \rightarrow SiH_4$$
$$Si + H_2 \rightarrow SiH_2$$
$$SiO_2 + H_2 \rightarrow SiO + H_2O$$

where the right-hand equations are the volatile products to be etched out. The Si surface is commonly covered by native oxide, and the hydrogen annealing process etches both SiO_2 and Si as shown in the equations earlier. However, the remaining SiO_2 on the Si surface acts as an etching mask for Si, resulting in some pits on the Si surface. Thus, the etching rate of SiO_2 should be high enough for the pit-less surface. The etching rates and their ratio of Si and SiO_2 are a function of the temperature; hence, the annealing temperature is important. Figure 19.5 shows the thickness of the native oxide after heating in hydrogen ambient with atmospheric pressure and etching rates of SiO_2 and Si as a function of the annealing temperature [6]. Reference [6] indicated that some pits arise at the annealing temperature under 1000 °C because of the low etching rate of SiO_2 compared to that of Si. In high-temperature annealing, the pits disappear, and surface flattening is performed. The silicon etching phenomena are from hydrogen; hence, the etching rate will decrease under a low-hydrogen partial pressure condition.

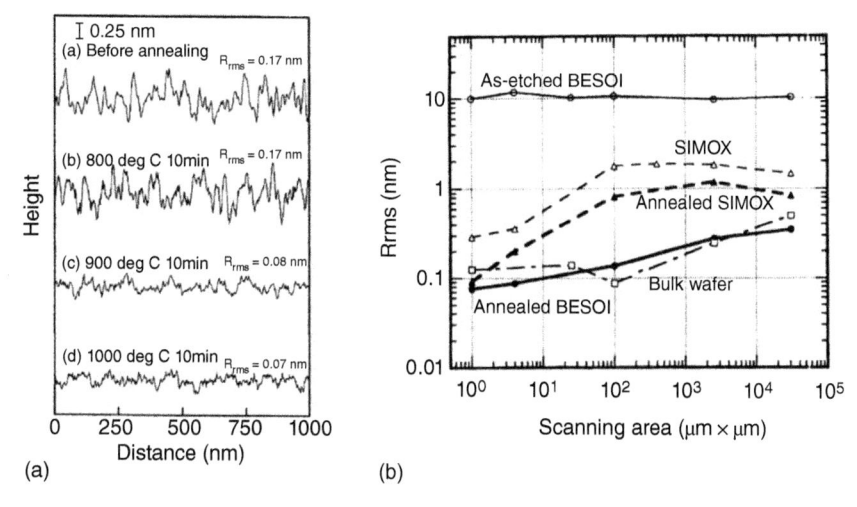

Figure 19.6 Example of the flattening effect by annealing. (a) Si(100) surface flattening by hydrogen annealing (3% H_2 in He for low hydrogen partial pressure) in the atmospheric pressure measured by the AFM. (b) Root mean square of the surface roughness for several wafers measured by the AFM. The samples include bond and etch-back SOI (BESOI), separation by implantation of oxygen (SIMOX), and bulk wafer. The AFM scanning area affects the results. Source: Sato and Yonehara [7], with the permission of AIP Publishing.

In the hydrogen annealing process, not only etching but also a surface migration or diffusion of Si atoms also progresses if the surface atoms are in an active state (without surface oxide). The migration is a phenomenon, wherein the atoms tend to minimize their energy state, and the surface will be flattened. The main mechanism of surface flattening is surface migration rather than etching in hydrogen annealing in vacuum or low-hydrogen partial pressure [7–10]. Figure 19.6 shows an example of the Si surface flattening, where the atomic-level flattening is observed. Atomic migration can occur when the surface atom is in an active state; hence, annealing in high vacuum with high-temperature pre-cleaning (deoxidation) also shows the same phenomenon [10].

In the case of ultrathin (<30 nm) crystalline silicon on silicon dioxide, the silicon can be agglomerated and forms nanoscale islands or wires (Figure 19.7) [11–13]. The samples in Figure 19.7 were annealed in ultra-high vacuum. The final structure and crystallinity depend on the Si layer thickness, crystal orientation, and annealing ambient and temperature.

The aforementioned flatting technology was also used to smooth the side wall of a trench capacitor for high-density DRAMs. Sato et al. [14, 15] used hydrogen annealing for this purpose. As described before, the annealing gave rise to surface migration, and the side wall roughness became smooth. Moreover, when the annealing was in a high temperature or low pressure for sufficient time, the deep trench was reshaped to several pieces of voids (Figure 19.8) [15] because the Si atoms diffused, and the trench tended to minimize its surface area to minimize the surface energy. The final shape can be controlled by the diameter and depth of the initial trench and the thermal treatment condition. This phenomenon was applied to make an ESS or

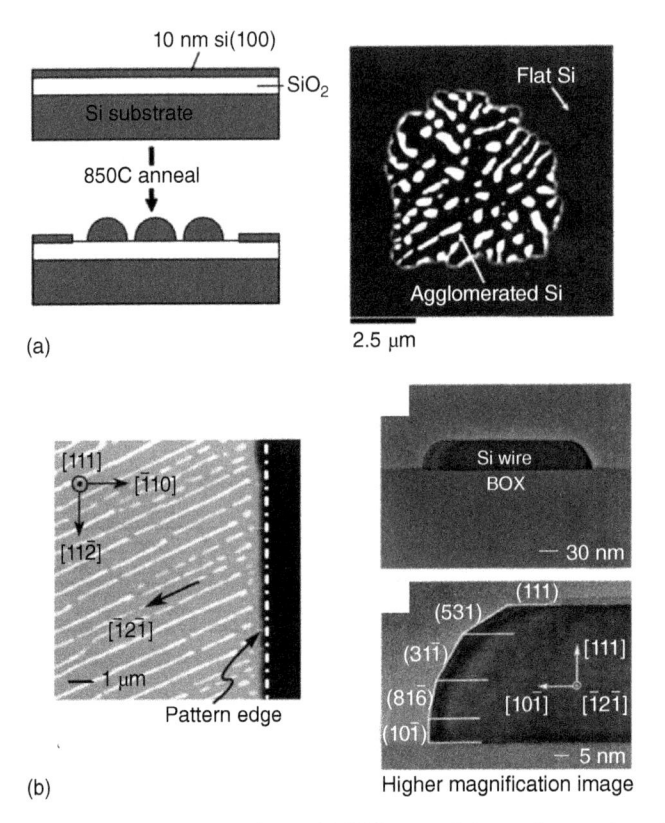

Figure 19.7 (a) Illustration and AFM image of the agglomeration of the 10 nm-thick (100) SOI and (b) top and side views of the Si nano wire array fabricated from (111) SOI at the annealing temperature of 950 °C. Source: (a) Reproduced from Danielson et al. [11], with the permission of AIP Publishing, (b) Reproduced from Burhanudin et al. [12] with the permission of AIP Publishing.

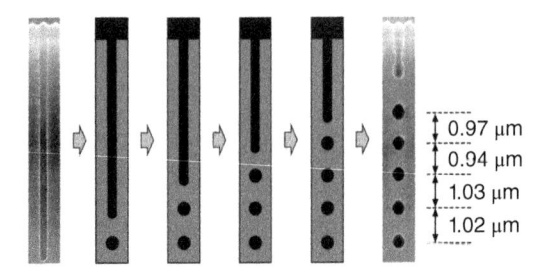

Figure 19.8 Trench transformation by hydrogen annealing under high-temperature and low-pressure conditions. Source: Modified from Sato et al. [15]. The scale is added to original figure.

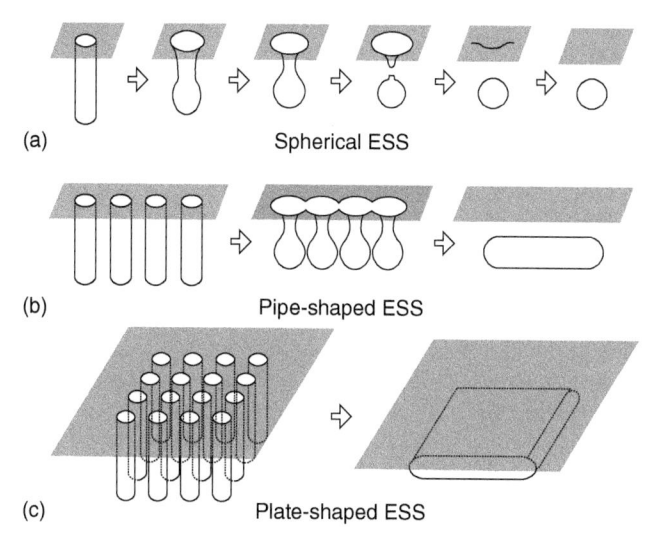

Figure 19.9 Formation of the buried channels by trench annealing. Channel shapes can be (a) spherical, (b) pipe, and (c) plate. Source: Sato et al. [17]. Copyright © 2004 The Japan Society of Applied Physics.

Figure 19.10 Illustration of the void formation for (a) low, (b) medium, and (c) high aspect ratios of the initial trench.

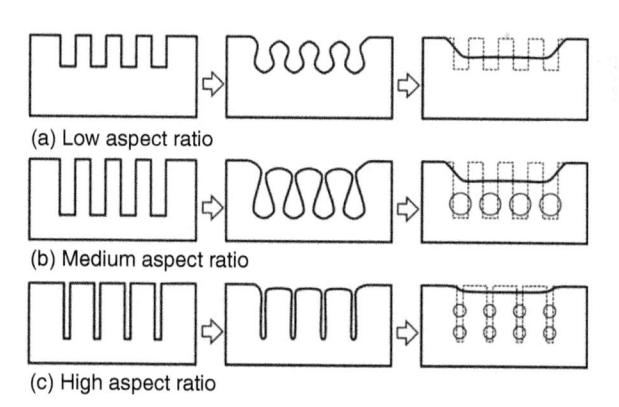

SON [16–18]. As shown in Figure 19.9, for the proper dimension and arrangement of the initial trenches, a single trench can become a single spherical cavity, trenches arranged in a line become a lying pipe cavity, and trenches arranged in a lattice become a plate-shaped cavity after the proper thermal treatment. The initial shape, especially the aspect ratio (depth to diameter ratio of the trench), is very important in forming the cavity. For the low-aspect ratio, the trenches vanish, and only the flat surface with some subsidence appears. The medium-aspect ratio trenches make a single cavity, while the high-aspect ratio trenches become more than one cavity. Figure 19.10 illustrates the formation of cavities for different aspect ratios. Some experimental results from the early work are quoted for a quantitative discussion [17]. In this experiment, an isolated trench was tested with the trench diameter and depth as the parameters. The diameter and depth (distance between the surface of the silicon substrate and the center of the spherical cavity) of the final spherical cavity were measured. The annealing conditions were the temperature of 1100 °C and 10 Torr hydrogen ambient for 10 minutes. Figure 19.11 shows the results, where the aspect ratio of the trench was used as a parameter. The cavity did not appear at the low-aspect ratio of less than 3, whereas multiple cavities arose for the aspect ratio larger than 9.5. A single cavity can be obtained between these ratios. Figure 19.11a depicts the measured diameters of the resultant cavities for different initial diameters of the trench as a function of the aspect ratio. For the same initial diameter of the trench, the resultant diameter of the cavity increased as the aspect ratio increased. These results are on the single line on the graph when the cavity diameter was normalized by the initial trench diameter (Figure 19.11b). This result indicates that the cavity diameter that is approximately two times larger than the initial diameter of the trench can be obtained. Figure 19.11c,d shows the cavity depth. In (d), the depth was normalized by the initial trench diameter. The dashed line in (d) indicates the cavity depth equal to the initial trench depth multiplied by 0.6. The measured data agreed well with this dashed line. Thus, the cavity will be placed at 0.6 of the initial depth, regardless of the aspect ratio.

Pipe- and plate-shaped cavities, rather than spherical-shaped ones, can be formed by arranging the trenches close to each other. The cavities merged together, and the pipe- or plate-shaped cavity can be obtained according to the trench arrangement when the distance between the neighbor trenches was less than the spherical cavity diameter. Figure 19.12 shows an example of the formation of the plate-shaped cavity with time, in which the smooth silicon layer on the empty space, SON, was obtained at the final stage. The effects of the annealing conditions were experimentally and theoretically investigated. As for the hydrogen pressure in annealing, Kuribayashi et al. showed the dependence of the migration speed on the hydrogen pressure by observing the deformation of the trench corner [19, 20]. Figure 19.13 shows the experimental results. The curvature of the migrated trench corner was measured under several conditions, where the large curvature meant less migration of the Si atoms. As shown in (b), the curvature decreased in proportion to $-1/4$ power of time, implying that the rounding was caused by the surface self-diffusion [21]. In other words, a too large deformation is difficult to obtain under a limited time. Panel (c) shows the dependence of the curvature on the hydrogen pressure, where the annealing time was kept to three minutes. A high hydrogen pressure clearly reduced the Si diffusion, especially for the low annealing temperature. Kuribayashi et al. considered that adsorbed hydrogen suppresses the surface diffusion of Si adatoms even at high temperatures. The diffusion constants for several annealing conditions were estimated from the aforementioned experiments (Table 19.1) [19]. In conclusion, low H_2 pressure or vacuum during annealing is preferable in forming SON. Inversely, if the aim of annealing is not shape deformation but surface flattening, a high H_2 pressure is preferable [22]. Numerical simulations on the shape transformation were also presented to help understand the mechanisms and predict the final structure [23–26]. Figure 19.14 shows an example of the simulation results, where the migration phenomena are well represented.

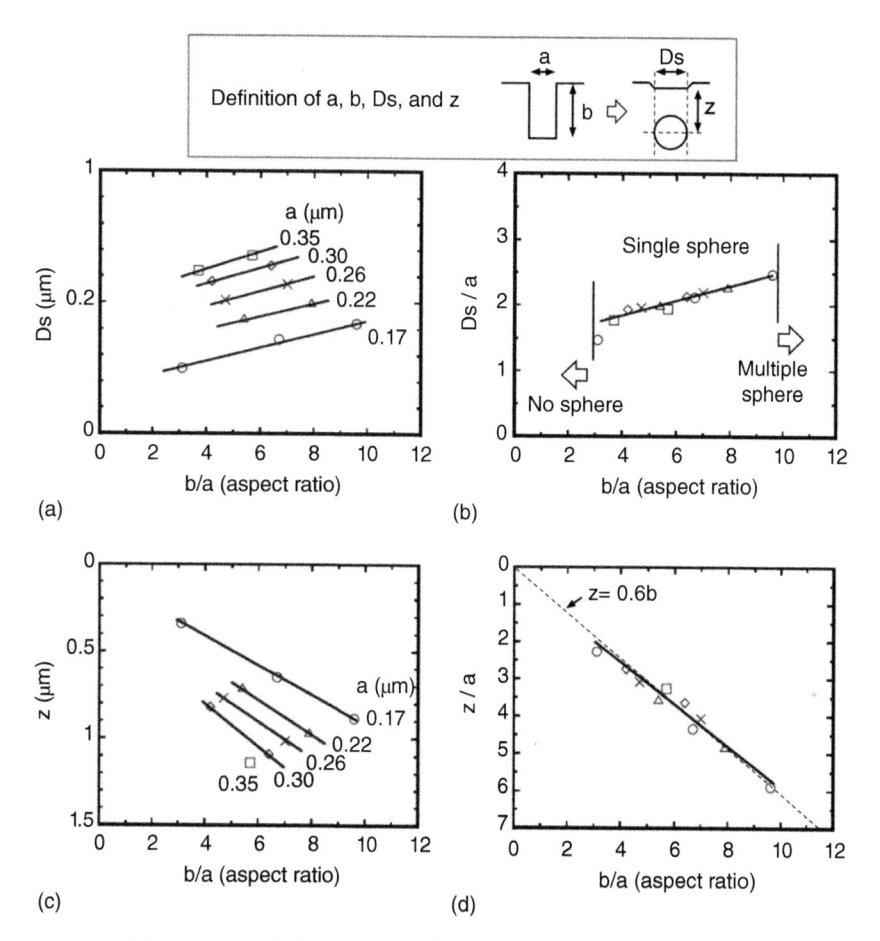

Figure 19.11 Relationship between the initial trench and the diameter of the spherical void and depth of the void. Some modifications are done from the figures in [17]. Source: Sato et al. [17]. Copyright © 2004 The Japan Society of Applied Physics.

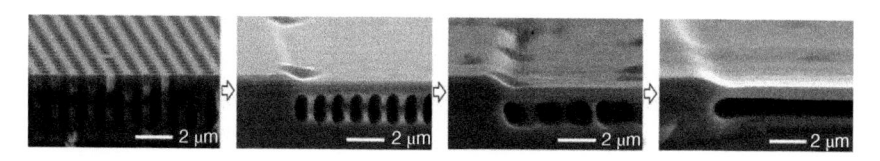

Figure 19.12 Formation of the plate-shaped cavity.

The specific features of this technique are as follows:

(1) The fabricated thin silicon layer, or diaphragm, on the empty space is stress-free, and a single crystal inherited the original crystal orientation [24, 25] because the migration phenomenon is driven by surface energy minimization. A single crystal silicon has excellent electrical and mechanical properties for MEMS devices

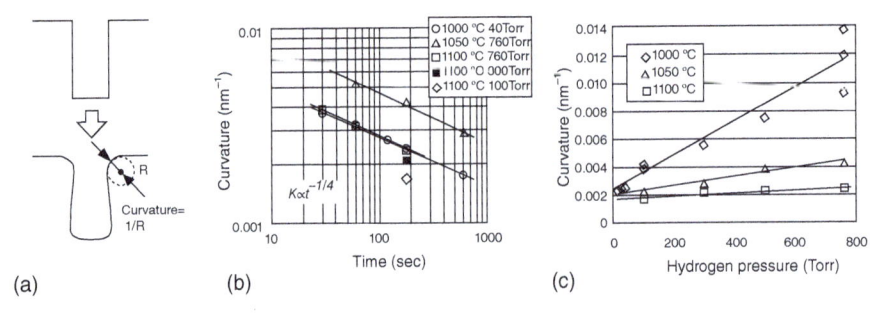

Figure 19.13 Dependence of migration on the annealing condition. (a) Migration shape, (b) time and temperature dependence, and (c) hydrogen effect. Source: Reproduced from Kuribayashi and Shimizu [19] with the permission of the American Vacuum Society.

Table 19.1 Diffusion constant of the Si atoms for various annealing conditions.

Temperature	Diffusion coefficient (nm² s⁻¹)				
	H_2 760 Torr	H_2 100 Torr	H_2 40 Torr	Ar 760 Torr	UHV
1000 °C	3.4×10^3	2.0×10^5	1.9×10^6	2.0×10^6	2.8×10^8
1100 °C	1.9×10^6	6.5×10^6	–	–	5.7×10^8

Source: Data from Kuribayashi and Shimizu [19].

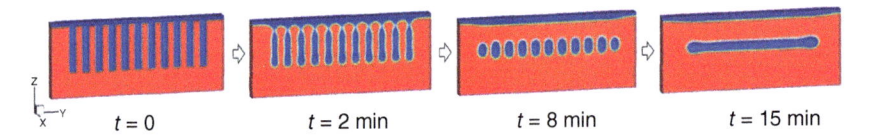

Figure 19.14 Simulation results of the migration phenomena. Source: Song et al. [23].

and can be used as a starting material of the IC so as to monolithically incorporate stress-sensitive gauges and/or LSI with the diaphragm.

(2) The interface between the thin silicon layer on the empty space and the substrate is smoothly connected together and keeps a single crystalline silicon, resulting in mechanically strong structure and no generation of thermal stress. It is quite different to other devices fabricated by the conventional method (Figures 19.2–19.4).

(3) Both open and closed (sealed) cavities can be realized according to the arrangement and size of the initial trenches. For the closed cavity, the empty space will be vacuum (hydrogen is easily diffused toward the outside of the silicon) because annealing was performed in low-pressure hydrogen ambient or high vacuum. This is useful for the applications involving absolute pressure sensors [26] and so on.

The limitations of this technique are the difficulty to make a deep cavity and a large closed cavity. For the deep cavity, the depth of the initial trenches must be large with a large diameter. However, trenches with a very large diameter of over 5 μm

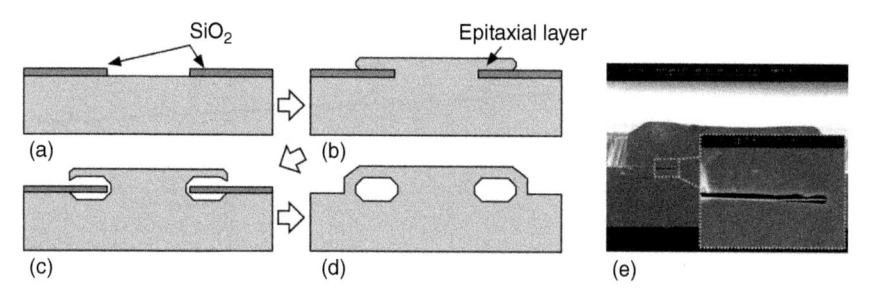

Figure 19.15 (a–d) Formation of the buried channels during selective epitaxy. (e) Selective epitaxial lateral overgrowth followed by H_2 annealing at high temperature forming a 0.7 μm-high and 25 μm-wide buried channel. Source: Sagazan et al. [27]. © 2005 Springer Nature.

cannot merge under practical annealing time or conditions because of the diffusion coefficient limit. This limits the maximum depth of the cavity to several μm. For the closed cavity with a large area, the fabricated thin diaphragm receives atmospheric pressure when the sample is exposed to the air ambient because the cavity was in vacuum. This makes the diaphragm deformation result in the contact between the diaphragm and the substrate if the diaphragm has a large area. Thus, the maximum dimension of the diaphragm is limited to less than several tens to hundreds of μm. Some modifications to the basic fabrication process of the SON are applied to avoid these limitations.

A cavity under the epitaxial layer can be realized by combining a migration phenomenon and the epitaxial growth of the silicon. Commonly for the silicon epitaxial growth, tetrachlorosilane and hydrogen gases are supplied in a high-temperature reactor as a precursor and a buffer/reducing agent gas, respectively. The hydrogen will act not only as a buffer gas but also as an etching gas in the reactor, resulting in the clean surface. Figure 19.15a–d [27] shows the steps of the buried channel formation under the epitaxial layer, for an example. First, the silicon surface was oxidized and patterned for selective epitaxy (a). Second, the sample was placed in an epitaxial reactor, and the selective single crystalline silicon growth started from the area where the oxide was stripped. The silicon growth is an isotropic phenomenon, and a lateral silicon epitaxy on the oxide occurs (b). The process was then switched to higher-temperature annealing. The migration at the interface between silicon and oxide arose during this step (c). The oxide etching process by the high-temperature annealing in hydrogen and the selective epitaxial growth were combined, and the oxide finally disappeared. An additional epitaxial growth closed the cavity, resulting in the buried channels. The depth of the buried channel can be increased by the additional epitaxial growth. The merits of using the epitaxial growth were that a high-quality single crystal layer can easily be obtained, and it has fine controllability of the thickness and impurity concentration of the layer.

Not only trenches but also porous silicon can be used as an initial structure for the Si migration. The porous silicon is a network of air holes with an interconnected silicon matrix. The porous silicon can be obtained by anodizing the silicon

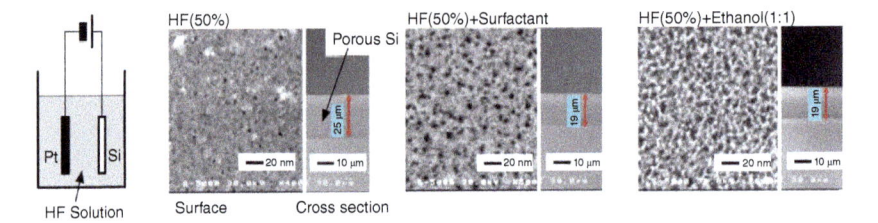

Figure 19.16 Examples of the porous silicon. In this experiment, the anodized current and the time are 50 mA cm^{-2} and 5 minutes, respectively, for the p-type (100) Si with 0.02 Ωcm resistivity.

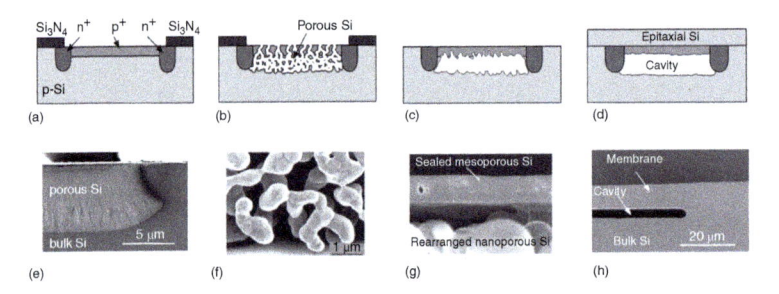

Figure 19.17 Fabrication steps for the buried channel using porous silicon. Panels (a–d) show the process flow, while (e–h) depict the cross section for the individual process steps. Source: Armbruster et al. [29]. © 2003 IEEE.

in the hydrofluoric acid solution [28]. The porosity and size of pores can be controlled, depending on impurity concentration of the starting material, mixture of solution, and anodizing current (Figure 19.16). The atoms inside the remaining pore kept their original crystal orientation when the starting material was a single crystalline silicon. Thus, the epitaxial growth on the porous silicon can be available. The technique introduced herein combined the migration of porous silicon and epitaxial growth of the silicon [29]. Figure 19.17a–d shows the process flow. First, using the standard IC process, the shallow p$^+$ and deep n$^+$ layers were formed in the p-Si substrate. The n$^+$ layer acted as a stopping layer for the following anodization. After the Si$_3$N$_4$ layer used for the protection layer of anodization was deposited and patterned, the anodization was performed at the low anodizing current. In this step, the p$^+$ layer was anodized and became a mesoporous layer with a porosity of approximately 45%. The p-substrate was then anodized using a higher current, making a buried porous silicon layer with a porosity of approximately 70% under the mesoporous layer (Figure 19.17b). After cleaning and drying, the sample was annealed in the epitaxial reactor with hydrogen ambient, resulting in sealing of the surface pores and merging of the buried pores (c). Finally, the top layer was deposited by the epitaxial growth, resulting in a diaphragm structure with vacuum cavity. This process was introduced by Robert Bosch GmbH and called as advanced porous silicon membrane (APSM).

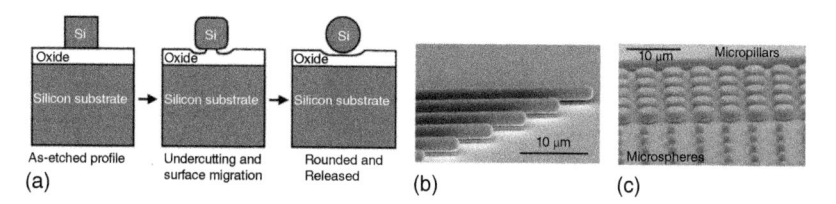

Figure 19.18 Deformation of the silicon beam on the oxide: (a) principle, (b) released beams with a circular cross section, and (c) microspheres. Source: Lee and Wu [30]. © 2006 IEEE.

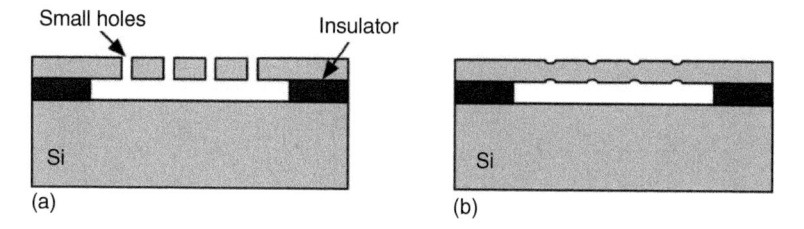

Figure 19.19 Cavity formation of SOI by annealing. (a) SOI with small holes (after sacrificial etching) and (b) After annealing.

Not only the cavity but also other unique structures can be obtained using the migration phenomenon of the Si atoms. For example, a thin and narrow silicon pattern on the oxide will be rounded and released by annealing (Figure 19.18) [30], resulting in beams with a circular cross section (b) and microspheres (c). Hydrogen annealing induced not only surface diffusion but also chemical etching at the interface of silicon and silicon dioxide [31]. Thus, the final structure will be released from the substrate. The beams can be anchored to the substrate using a large-area silicon feature, which was still unreleased. The annealing time for the microspheres was controlled to prevent the complete release of spheres. In the common MEMS technology, the microstructures with round profiles are difficult to fabricate, and this technique may be useful in forming such structures.

For SOI, a closed cavity can be realized by annealing. As shown in Figure 19.19, after sacrificial etching of the insulator layer through small holes, the hydrogen annealing can close the holes, resulting in closed cavity [32]. In contrast to standard SONs that have a unified diaphragm and substrate, this structure has the feature of electrical isolation between the diaphragm and the substrate. This makes capacitor between them and could be useful for capacitive-type sensors.

19.3 Applications of SON

The SON is especially useful for applications of pressure sensors because the basic structure involving a diaphragm with a vacuum cavity is directly applicable to the pressure-sensing mechanism. Practical devices have been presented in the literature and in the market. Some example devices will be introduced herein.

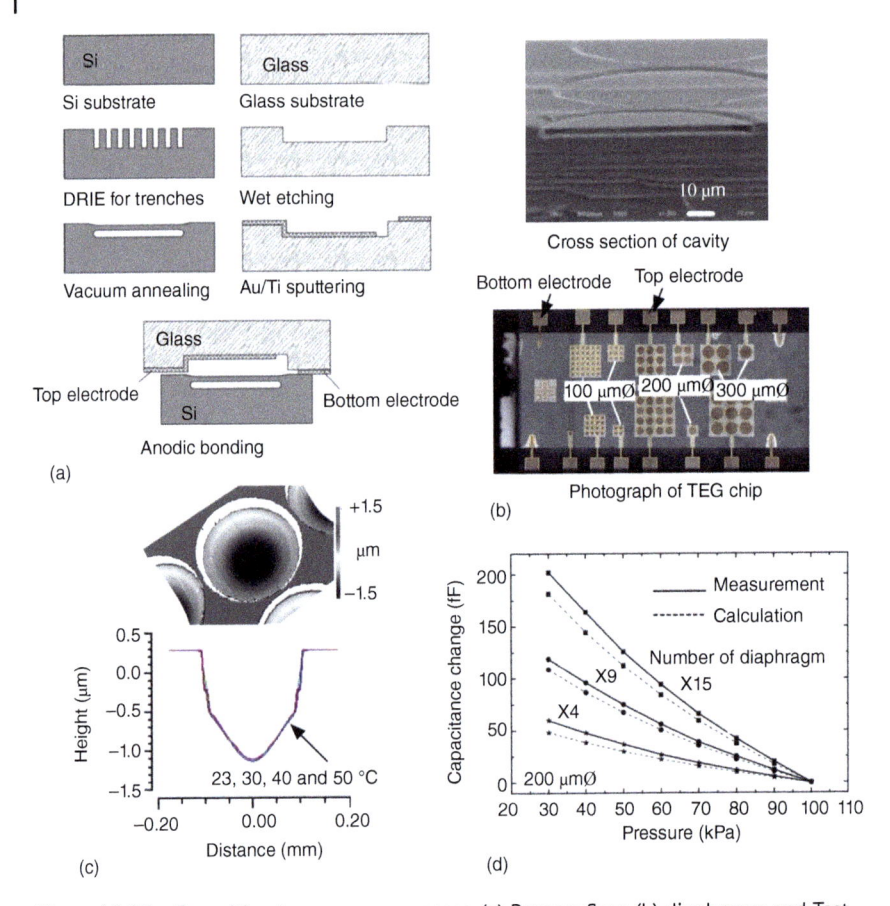

Figure 19.20 Capacitive-type pressure sensor. (a) Process flow, (b) diaphragm and Test Element Group (TEG) device (viewing from glass side), (c) surface profiles of diaphragm for several temperatures, and (d) sensor output. Source: Hao et al. [33] © 2014 IEEE.

Hao et al. proposed a capacitive-type absolute pressure sensor by directly using the basic SON structure explained in Figures 19.9–19.12. The detection of capacitance between the electrode and the diaphragm was realized by anodically bonding the glass substrate with an electrode to the SON (Figure 19.20a) [33]. The diaphragm was deformed, and the distance between the electrode and the diaphragm was determined according to the ambient pressure. This result defines the capacitance and the output signal of this sensor. The diameter of the diaphragm was limited to 300 μm to avoid contact between the diaphragm and the bottom of the cavity by atmospheric pressure. Additionally, multiple diaphragms were connected in parallel to increase the capacitance change. The diaphragm and cavity thicknesses were 2.5 and 2.4 μm, respectively. The gap between the electrode and the diaphragm was 2 μm. Figure 19.20c shows the surface profiles of the diaphragm at several temperatures, which are caved in by the atmospheric pressure, while (d) shows the sensing output.

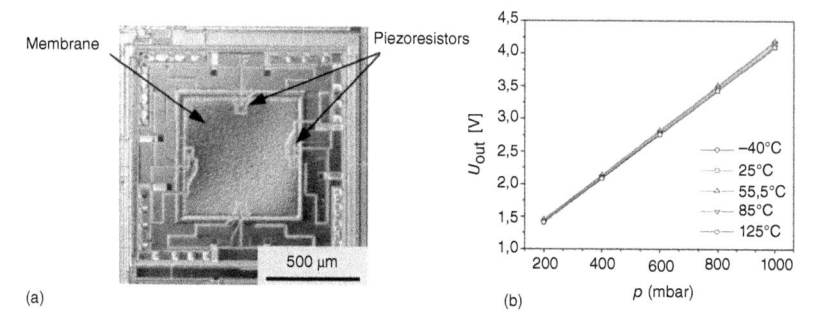

Figure 19.21 (a) Pressure sensor chip fabricated by the APSM technology from Bosch and (b) its characteristics. Source: (a) Armbruster et al. [29] adapted with permission, (b) Armbruster et al. [29]. © 2003, IEEE.

Armbruster et al. from Bosch used a migration of the porous silicon (APSM process), as previously shown in Figure 19.17, to fabricate their pressure sensors [29]. Figure 19.21a shows the top view of the sensor chip, where the inner rectangular area is the SON structure and forms the pressure-sensitive diaphragm with vacuum chamber. Four piezoresistors, whose resistances change according to the applied stress, were peripherally placed on the diaphragm. The device was tested under a 1.5×10^5 pressure cycle between 200 mBar and 2 Bar at a temperature of 150 °C. The test ensured that the cavity was successfully hermetically sealed for the long-time operation. Panel (b) shows the output of the device operating under a temperature range from −40 to 125 °C. In the commercial device, an interface LSI, which controls the sensor device and converts the analog sensing signal to a digital output, is combined with the sensor device in a small package. There are some other reports using the APSM technology for pressure sensors [34].

STMicroelectronics also utilizes the SON technology for pressure sensors [35]. They call the technology as VENSENS. Figure 19.22 schematically shows the internal structure of LPS22HB as an example. In a small plastic package measuring $2 \times 2 \times 0.76$ mm^3, the sensor and LSI dies are vertically stacked and wired together (a). The silicon sensor die consisted of a cap layer with small air-intake holes, a sensing element, and a substrate (b). One side of the cap layer was directly exposed to the outer environment. The diaphragm was fabricated by the migration of the silicon trenches. The piezoresistors on the diaphragm detected the deformation of the diaphragm because of ambient pressure. In this device, a penetrated groove surrounds the diaphragm region, resulting in the floating or isolated sensing region with one point of fixed area. This avoids the additional and unexpected stress to the sensing region from inside and outside of the sensor die, resulting in a high stability of the sensor characteristics.

The SON technology is applicable not only for pressure sensors but also for other MEMS devices. A filament or heater was fabricated using a thin monocrystalline layer as a material. The merits of the SON filament include the low thermal conductivity caused by the thin material, leading to a low power operation with a high operation speed and the monocrystalline silicon showing stable characteristics as a

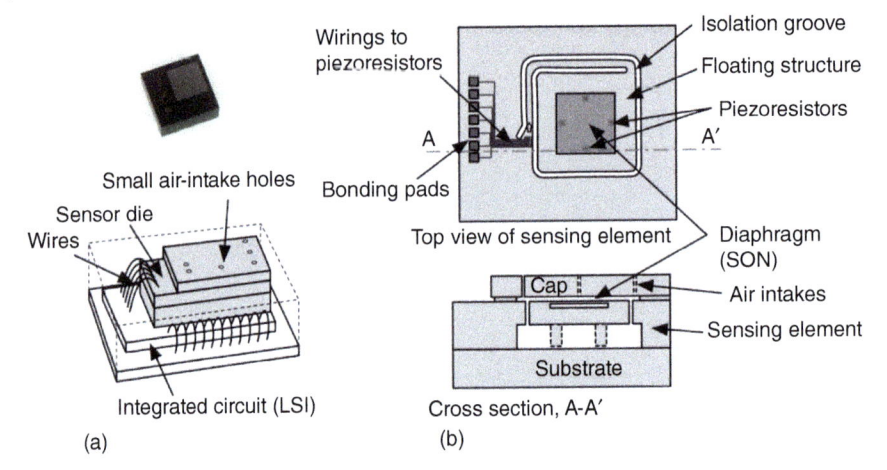

Figure 19.22 Example of a commercialized pressure sensor. (a) Outside view and inside image of the sensor and (b) conceptual structure of sensor die.

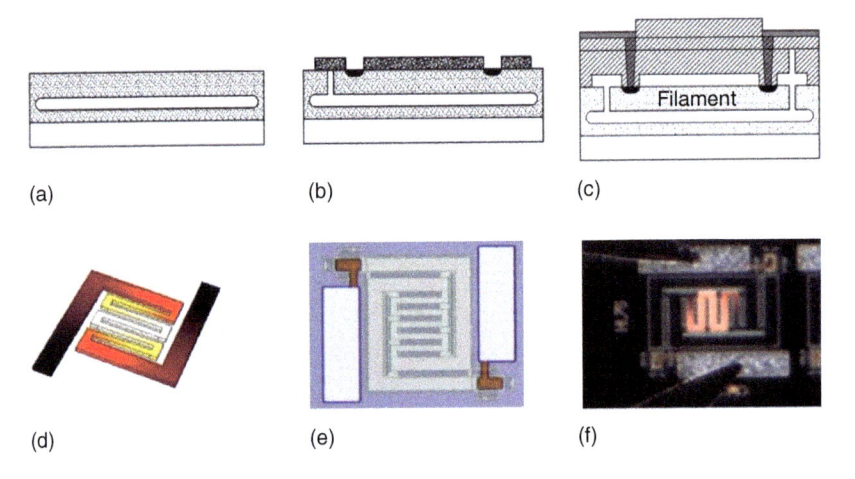

Figure 19.23 IR emitter using SON: (a) SON as a starting material; (b) sacrificial layer deposition; (c) final device via the lid layer formation, deep etching for the filament shape definition, and sacrificial etching; (d) filament structure; (e) top view of the device; and (f) operation as an IR emitter. Source: Komenko et al. [37]. CC BY 4.0.

heater. The filament was designed for a Pirani gauge [36] and an IR emitter [37]. In these cases, in contrast to the one with a vacuum channel, the channel can be open, and the SON area has no limitation. For an example of the IR emitter [37], the simplified process flow, filament shape, and operation are shown in Figure 19.23. The final device has a lid structure that supports only the ends of the filament to minimize the thermal conductivity.

The thin layer of the SON is a high-quality single crystalline silicon. Thus, the SON technology is one method used to obtain a low-cost thin Si film. For the solar cells, a low-cost thin Si film is heavily required, and the SON thin film is considered to be

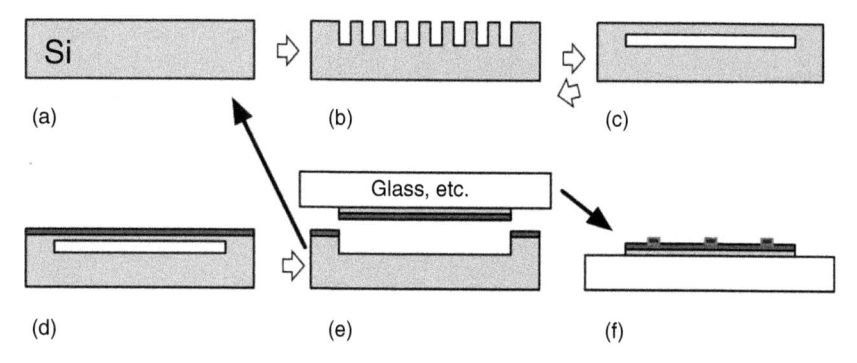

Figure 19.24 Process flow for the SON solar cell: (a) starting material, (b) trench etching, (c) SON formation, (d) processing of the rear-side of the cell, (e) bonding and detachment, and (f) processing of the topside of the cell. Source: Based on Depauw et al. [38].

used as a substrate material for one example. Figure 19.24 shows this concept [38]. The thin layer of SON is bonded to a handle substrate, such as a glass wafer, and detached from the SON substrate. The remaining SON substrate without the thin layer can be recycled for the next process. This is also one of the applications of SON.

The concept and the applications of silicon monolithic buried cavity or channel in the MEMS technology utilizing hydrogen annealing were mainly discussed in this chapter. The SON technology can be realized using simple process steps: only one lithography with etching and hydrogen annealing. The monocrystalline silicon structures, such as silicon beam and diaphragms or membranes, can be realized by using this technology. This technology would be one of the promising technologies for future MEMS devices.

References

1 Burgener, M.L. and Reedy, R.E. (1995). Minimum charge FET fabricated on an ultrathin silicon on sapphire wafer. US Patent 5,416,043A.
2 Kuo, J.B. and Lin, S.-C. (2001). *Low-Voltage SOI CMOS VLSI Devices and Circuits*. Wiley.
3 Kilchytska, V., Chung, T.M., Olbrechts, B. et al. (2007). Electrical characterization of true silicon-on-nothing MOSFETs fabricated by Si layer transfer over a pre-etched cavity. *Solid-State Electron.* 51 (9): 1238–1244.
4 Bustillo, J.M., Howe, R.T., and Muller, R.S. (1998). Surface micromachining for microelectromechanical systems. *Proc. IEEE.* 86 (8): 1552–1574.
5 Liu, C. and Tai, Y.C. (1999). Sealing of micromachined cavities using chemical vapor deposition methods: characterization and optimization. *IEEE J. MEMS* 8 (2): 135–145.
6 Habuka, H., Tsunoda, H., Mayusumi, M. et al. (1995). Roughness of silicon surface heated in hydrogen ambient. *J. Electrochem. Soc.* 142 (9): 3092–3097.
7 Sato, N. and Yonehara, T. (1994). Hydrogen annealed silicon-on-insulator. *App. Phys. Lett.* 65: 1924–1926.

8 Kumagai, Y., Namba, K., Komeda, T., and Nishioka, Y. (1998). Formation of periodic step and terrace structure on Si(100) surface during annealing in hydrogen diluted with inert gas. *J. Vac. Sci. Technol. A* 16 (3): 1775–1778.

9 Komeda, T. and Kumagai, Y. (1998). Si(001) surface variation with annealing in ambient H_2. *Phys. Rev. B* 58 (3): 1385–1391.

10 Keeffe, M.E., Umbach, C.C., and Blakely, J.M. (1994). Surface self-diffusion on Si from the evolution of periodic atomic step arrays. *J. Phys. Chem. Solids* 55 (10): 965–973.

11 Danielson, D.T., Sparacin, D.K., Michel, J., and Kimerling, L.C. (2006). Surface-energy-driven dewetting theory of silicon-on-insulator agglomeration. *J. Appl. Phys.* 100: 083507.

12 Burhanudin, Z.A., Nuryadi, R., Ishikawa, Y., and Tabe, M. (2005). Thermally-induced formation of Si wire array on an ultrathin (111) silicon-on-insulator substrate. *Appl. Phys. Lett.* 87: 121905.

13 Yang, B., Zhang, P., Savage, D.E. et al. (2005). Self-organization of semiconductor nanocrystals by selective surface faceting. *Phys. Rev. B* 72: 235413.

14 Sato, T., Mizushima, I., Kito, M. et al. (1998). Trench transformation technology using hydrogen annealing for realizing highly reliable device structure with thin dielectric films, Symposium on VLSI tech. Dig. Tech. Papers: 206–207.

15 Sato, T., Mitsutake, K., Mizushima, I., and Tsunashima, Y. (2000). Micro-structure transformation of silicon: a newly developed transformation technology for patterning silicon surfaces using the surface migration of silicon atoms by hydrogen annealing. *Jpn. J. Appl. Phys.* 39 (Part 1, 9A): 5033–5038.

16 Mizushima, I., Sato, T., Taniguchi, S., and Tsunashima, Y. (2000). Empty-space-in-silicon technique for fabricating a silicon-on-nothing structure. *Appl. Phys. Lett.* 77 (20): 3290–3292.

17 Sato, T., Mizushima, I., Taniguchi, S. et al. (2004). Fabrication of silicon-on-nothing structure by substrate engineering using the empty-space-in-silicon formation technique. *Jpn. J. App. Phys.* 43 (1): 12–18.

18 Sato, T., Matsuo, M., Mizushima, I. et al. (2007). Method of making empty space in silicon, US Patent 7,235,456 B2.

19 Kuribayashi, H. and Shimizu, R. (2004). Hydrogen pressure dependence of trench corner rounding during hydrogen annealing. *J. Vac. Sci. Technol. A* 22 (4): 1406–1409.

20 Kuribayashi, H., Hiruta, R., Shimizu, R. et al. (2004). Investigation of shape transformation of silicon trenches during hydrogen annealing. *Jpn. J. Appl. Phys.* 43 (4A): 468–470.

21 Mullins, W.W. (1957). Theory of thermal grooving. *J. Appl. Phys.* 28 (3): 333–339.

22 Hiruta, R., Kuribayashi, H., Shimizu, R. et al. (2006). Flattening of micro-structured Si surfaces by hydrogen annealing. *Appl. Surf. Sci.* 252: 5279–5283.

23 Song, J., Zhang, L., and Kim, D. (2016). Design of silicon-on-nothing structure based on multi-physics analysis. *Multiscale Multiphys. Mech.* 1 (3): 225–231.

24 Sudoh, K., Iwasaki, H., Hiura, R. et al. (2009). Void shape evolution and formation of silicon-on-nothing structures during hydrogen annealing of hole arrays on Si(001). *J. Appl. Phys.* 105: 083536.

25 Hiruta, R., Kuribayashi, H., Shimazu, S. et al. (2004). Evolution of surface morphology of Si-trench sidewalls during hydrogen annealing. *Appl. Surf. Sci.* 237: 63–67.

26 Su, J., Zhang, X., Zhou, G. et al. (2018). A review: crystalline silicon membranes over sealed cavities for pressure sensors by using silicon migration technology. *J. Semicond.* 39 (7): 071005.

27 Sagazan, O.D., Denoual, M., Guil, P. et al. (2005). Horizontal buried channels in monocrystalline silicon. Proceedings of the International Conference on MEMS, 661–664. (Sagazan, O.D, Denoual, M., Guil, P., et al. (2006). Horizontal buried channels in monocrystalline silicon. Microsyst. Technol. 12 (10–11) 959–963).

28 Lehmann, V. (2002). *Electrochemistry of Silicon*. Weinheim: Wiley-VCH.

29 Armbruster, S., Schafer, F., Lammel, G. et al. (2003). A novel micromachining process for the fabrication of monocrystalline Si-membranes using porous silicon. International Conference on Transducers, 246–249

30 Lee, M.-C.M. and Wu, M.C. (2006). Thermal annealing in hydrogen for 3-D profile transformation on silicon-on-insulator and sidewall roughness reduction. *J. Microelectromech. Syst.* 15 (2): 338–343.

31 Liu, S.T., Chan, L., and Borland, J.O. (1987). Reaction kinetics of SiO_2/Si(100) interface in H_2 ambient in a reduced pressure epitaxial reactor. Proceedings of the 10th International Conference on Chemical Vapor Deposition, Pennington, NJ, 428–434.

32 Ebschke, S., Poloczek, R.R., Kallis, K.T., and Fiedler, H.L. (2013). Creating a Monocrystalline membrane via etching and sealing of nanoholes considering its sealing behavior. *J. Nano Res.* 25: 49–54.

33 Hao, X.C., Tanaka, S., Masuda, A. et al. (2014). The application of silicon on nothing structure for developing a novel capacitive absolute pressure sensor. *IEEE Sens. J.* 14 (3): 808–815.

34 Knese, K., Armbruster, S., Weber, H. et al. (2009). Novel technology for capacitive pressure sensors with monocrystalline silicon membranes. 22th IEEE International Conference on Micro Electro Mechanical Systems, 697–700.

35 Villa, F.F., Barlocchi, G., Corona, P., Vigna, B., and Baldo, L. (2004). Halbleiterdrucksensor und Verfahren zur Herstellung, Patent, EP1577656B1.

36 Kravchenko, A., Komenko, V., and Fischer, W.-J. (2018). Silicon-on-nothing micro-Pirani gauge for interior-pressure measurement. *Proceedings* 2 (13): 1079. https://doi.org/10.3390/ proceedings2131079.

37 Komenko, V., Kravchenko, A., and Fischer, W.-J. (2018). Silicon-on-nothing IR-emitter for gas sensing applications. *Proceedings* 2 (13): 1080. https://doi.org/10.3390/proceedings2131080.

38 Depauw, V., Gordon, I., Beaucarne, G. et al. (2009). Proof of concept of an epitaxy-free layer-transfer process for silicon solar cells based on the reorganisation of macropores upon annealing. *Mat. Sci. Eng. B* 159–160: 286–290.

20

Through-substrate Vias

Zhyao Wang

Tsinghua University, Institute of Microelectronics, Beijing 100084, China

A through-substrate via (TSV) is an electrical interconnect, which passes vertically through a substrate to conduct the electrical signals from one side of the substrate to the other. As an enabling technology, TSVs allow micro electro mechanical system (MEMS) devices and systems to achieve new integration schemes and wafer-level vacuum packaging (WLVP) with high performance, multifunctions, low costs, high reliability, and small package sizes. These functions have significantly boosted the technological advances for the successful story of MEMS commercialization in the past 10 years [1].

The appearance of TSVs in MEMS areas can be dated back to the early 1990s when Tohoku University, MIT, Technical University of Denmark, etc. used wet etching and metal filling to fabricate slant cavities in glass or silicon wafers for sensor backside contacts [2–6]. A representative at the time was a capacitive pressure sensor reported in 1992 [2]. It used a metal-filled, tapered TSV fabricated in a glass wafer to conduct the electrical signals of a plate capacitor in a vacuum cavity, which was formed by bonding the glass wafer and a silicon integrated circuit (IC) wafer. This pressure sensor showed the pioneer concepts of TSVs for 3D MEMS integration and WLVP. Due to the large footprints of wet etched TSVs at the early stage, TSVs at that time were unsuitable for applications that need high TSV density.

The emergence of Bosch deep reactive ion etching (DRIE) technology in the mid 1990s made etching of vertical and deep vias in silicon wafers into reality, enabling high-aspect-ratio TSVs filled with tungsten (W) or polysilicon conductors to be achieved [7–11]. Using DRIE technology, vertical and high-aspect-ratio TSVs with polysilicon or single crystalline silicon conductors were developed for micromachined ultrasonic transducer arrays, microcantilevers, and microengines by the end of the last century [12–15]. Then MEMS manufacturers successively commercialized inertial sensors [1, 16] and film bulk acoustic resonators (FBARs) [17], which used air-gap silicon TSVs or hollow metal TSVs for WLVP. Since 2000 when copper (Cu) electroplating was developed for filling high-aspect-ratio TSVs [18–20], 3D integrated MEMS sensors and complementary metal oxide semiconductor (CMOS) were reported, and integration scheme of MEMS and CMOS shifted

3D and Circuit Integration of MEMS, First Edition. Edited by Masayoshi Esashi.
© 2021 WILEY-VCH GmbH. Published 2021 by WILEY-VCH GmbH.

from conventional single-chip integration, which is quite technically challenging, and multi-chip system in package, which is of large sizes, to the route of 3D stacking integration.

20.1 Configurations of TSVs

Typically a TSV consists of a through-hole etched in a substrate, a conductor in the through-hole, and an insulator isolating the conductor from the substrate. The conductors are normally made of low-resistivity polysilicon or single crystalline silicon or metals such as copper (Cu), tungsten (W), aluminum (Al), and nickel (Ni). The insulator can be SiO_2, polymer, or even an air-gap. The substrate can be a silicon wafer or a glass wafer. Normally the substrates are thinned to a reasonable thickness to facilitate TSV fabrication and to reduce the thickness of the bonded wafers.

Although the configuration of a typical TSV is quite simple, the TSVs in MEMS areas have large varieties in resistance, diameters, and configuration, making the TSVs in MEMS show great diversities in configurations, materials, and fabrication methods. According to the configurations, TSVs can be classified into solid TSVs, hollow TSVs, and air-gap TSVs, as shown in Figure 20.1, and according to the conductor materials, TSVs can be classified into polysilicon TSVs, silicon TSVs, and metal TSVs. Each TSV has different materials and fabrication methods and has different properties and applicable areas.

20.1.1 Solid TSVs

A solid TSV means that the through-holes etched in the substrate are completely filled with insulators and conductors, as shown in Figure 20.1a. The conductors in a solid TSV can be polysilicon, silicon, Cu, W, Ni, etc., and the insulators are normally SiO_2. The fabrication methods for solid TSVs vary with the conductor materials. For Cu, the TSVs are normally fabricated using deep hole etching, deposition of a SiO_2 or a polymer insulator, and filling the holes with metals. For polysilicon or W, the conductors are filled using low-pressure chemical vapor deposition (LPCVD) or plasma enhanced chemical vapor deposition (PECVD), respectively. For silicon, the TSVs are fabricated using deep etching of an annular trench and filling the trench with a SiO_2 insulator.

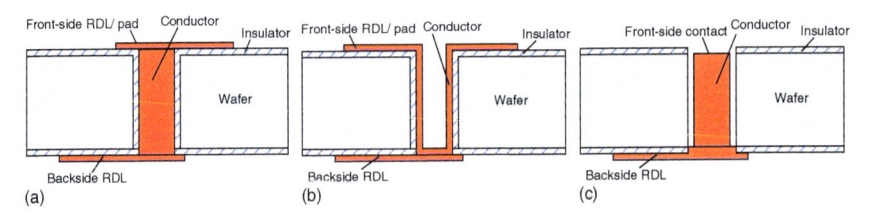

Figure 20.1 TSV Configurations. (a) Solid TSV, (b) hollow TSV with open via, and (c) air-gap TSV.

Solid TSVs can be in either a MEMS substrate or a CMOS substrate for connecting the devices or circuits on the front side of the substrate to the backside. The advantages of the solid Cu TSVs include low resistance, high density, small diameters, and good mechanical stability. The main drawbacks of solid Cu TSVs are their complex fabrication processes, high costs, and large thermal stresses in the substrate caused by the difference in the coefficients of thermal expansion between silicon and copper. Due to the high costs and complex fabrication processes, Cu TSVs are only used in the cases where necessary, e.g. in CMOS substrates for integration of MEMS and CMOS or high-density MEMS arrays, which need TSVs with low resistance, high density, or small diameters.

20.1.2 Hollow TSVs

Figure 20.1b shows a hollow TSV, which has a conductor film deposited on the sidewalls of a through-hole. The insulator is normally SiO_2, and the conductor may be polysilicon [21] or metals such as Cu, Al, and W [22–26]. The fabrication of hollow TSVs is more cost-effective because they need only metal deposition to fabricate the thin metal films instead of time-consuming electroplating. In addition, the hollow cavity in the TSV provides a free space for the thermal expansion of the metal films, such that the stresses in the substrate caused by the thermal expansion of the metal film is much lower than that of solid TSVs [22, 23], which is preferred by MEMS devices that are sensitive to substrate stresses. However, for the same TSV diameters, hollow TSVs have an inferior capability in carrying electrical currents compared with solid TSVs due to the relatively small cross-sectional areas. Once hollow TSVs are fabricated, further lithography on the surface can be implemented by using dry film photoresist [26].

20.1.3 Air-gap TSVs

Air-gap TSVs use an annular trench (air-gap) instead of SiO_2 or polymers as the insulator, as shown in Figure 20.1c. An air-gap TSV using silicon conductors can be fabricated with only one step deep etching of an annular trench in a highly conductive silicon substrate, with the trench as the insulator and the silicon post as the conductor. This leads to a much easy and simple fabrication method, at a cost of relatively large resistance compared with metal TSVs. The air-gaps also provide a free space for thermal expansion of the conductors, but the lack of mechanical support to the conductor sidewalls may deteriorate the mechanical robustness to vibration and shock.

20.2 TSV Applications in MEMS

As an enabling technology, TSVs have found wide applications in MEMS areas. Basically, TSVs function as interconnects between the two sides of a substrate; however,

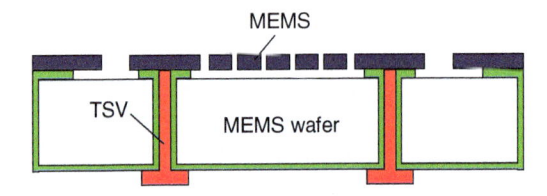

Figure 20.2 Signal routing to wafer backside using TSVs.

the TSVs in MEMS area have more functions, which can be classified as the following according to the TSV functions, namely, backside routing of electrical signals, CMOS-MEMS 3D integration, 2.5D integration using TSV-embedded interposers, and TSV-based WLVP.

20.2.1 Signal Conduction to the Wafer Backside

TSVs can be used to conduct the electrical signal of a MEMS device from one side of the substrate to the other by embedding TSVs in the substrate, as shown in Figure 20.2. Signal routing to backside do not require bonding wires on the front side, avoiding the damage of the bonding wires when the devices are operated in harsh conditions such as high temperatures or corrosive environment. The ultrasonic agitation used in welding process for Au wire metallization causes abrasion to take place and microscopic holes to develop. Under high temperatures, the Au migrate and form Au–Si eutectic, which causes the leads to fail. In addition, the long-term reliability of the devices, which are operated in aggressive chemicals may degrade due to the corrosion of bonding wires in direct contact to the chemicals.

Using TSVs to replace bonding wires may also improve the performance of MEMS devices. Conventional bonding wires on the front side may affect the measurand by causing obstruction to flow or acoustic fields due to their geometrical sizes and interferences to magnetic and electrical fields due to the metallic properties. Therefore, replacing bonding wires with TSVs can avoid the influences of the wires on the measurement results.

Since the output signals of many MEMS sensors are extremely low, short TSVs allow much closer distance between MEMS and readout integrated circuit (ROIC) to reduce parasitics, noise, and interference. For example, TSVs can significantly reduce the large parasitic capacitance of long horizontal interconnects and bonding wires, which is seriously detrimental to capacitive sensors.

20.2.2 CMOS-MEMS 3D Integration

TSV technology allows integration of MEMS and CMOS chips, which was previously quite difficult due to the vulnerability of the fragile and suspended MEMS structures and due to the incompatibility in processes and/or materials between MEMS and CMOS. With wafer stacking and TSVs, vertical (3D) integration of multiple MEMS and CMOS chips can be achieved without the compatibility issues, as shown in Figure 20.3. 3D integration of MEMS and CMOS is able to achieve integrated microsystems with high performance. For example, high-density and short

Figure 20.3 MEMS-CMOS 3D integration using TSVs.

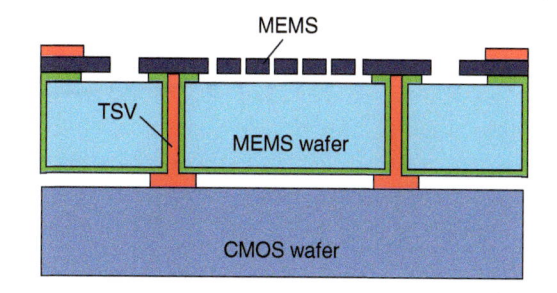

TSV arrays allow each element in a MEMS array with thousands or even millions elements to achieve element-level signal processing, such that in-pixel signal processing could be implemented and the noise, parasitics, and delay caused by long wires can be avoided. If the noisy digital circuits are further separated from the sensitive analog circuits on different layers, noise coupling via the conductive substrate noise reduction can be eliminated for better performance.

3D integration of MEMS and CMOS also enables compact sizes, high filling factors, and high reliability. For example, MEMS arrays with performance sensitive to the areas occupied by the arrays, such as ultrasonic sensor arrays, infrared focal planes, and active pixel arrays, benefit greatly from 3D integration to achieve high filling factors by moving the CMOS circuits, which are originally placed on the same surface of the MEMS array, to the backside. For the MEMS arrays that the elements are suspended from the substrate, TSVs can not only function as the electrical interconnects but also act as the mechanical supports to suspend the MEMS array.

20.2.3 MEMS and CMOS 2.5D Integration

Although 3D integration of MEMS and CMOS has attractive advantages, the fabrication of 3D integration is quite complex, and for some applications 2.5D interposer integration is highly preferred. 2.5D integration refers to using a TSV-embedded interposer as a substrate to integrate MEMS and CMOS, which are side-by-side placed on the interposer, as shown in Figure 20.4. The 2D metal wires on the interposer surfaces connect all the chips placed on the interposer in a manner of 2D placement, and TSVs embedded in the interposer connect the packaging substrate. 2.5D integration provides a way for MEMS to revolute from a single mechanical device to a microsystem by integrating various chips including multi-MEMS/sensors, analog and digital chips, memories, microprocessors, optoelectronic devices, RF, as well as many others.

Figure 20.4 Multi-chip 2.5D integration using TSV interposers.

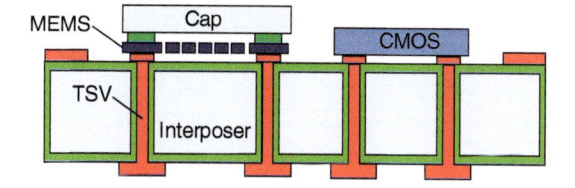

Integration of multiple sensors is highly required for both multi-parameter sensing and sensor fusion, which may achieve performance and functions that cannot be reached before. For example, inertial measurement units (IMUs) need three accelerometers, three gyroscopes, three magnetometers, and an altimeter (pressure sensors). In sensor fusion, multi-sensor integration is needed to compensate the inherent drawbacks of each sensor using the merits of others. Using multi-sensors to obtain redundant data and employing data fusion, the final output is more accurate than that of any individual sensors. For example, by taking the signals of all the sensors in an IMU and processing the signals using data fusion algorithms such as extended Kalman Filter, the deficiencies of single sensors can be avoided and the output is more accurate than single ones [27].

Integration of multiple MEMS/sensors with various IC chips such as logic chips, RF chips, memories, and microprocessors allows autonomous microsystems to be achieved with the ability to implement data processing and communication. The powerful hardware obtained by integrating advanced ICs also allow to run complex software so that flexible functions, on-chip data processing, smart decisions, and artificial intelligence can be achieved.

However, due to the complexity and cost issues of 3D integration of multiple layers, it is uneconomical, for most cases, to integrate multilayers of MEMS and CMOS chips, which are fabricated with different technologies. Literature reports and industrial products have indicated that vertical 3D configuration is cost-effective in integrating one layer of MEMS with one layer of ICs. This layer of ICs could be a 3D stack of other two IC chips, which can be 3D integrated with acceptable complexity and cost. Such a three-layered vertical integration has been successfully demonstrated for image sensors and active pixel sensors.

2.5D integration has emerged as a powerful tool for integrating multiple MEMS and other chips. As a compromise between system performance and implementation expense, 2.5 integration yields significant gains in reducing process complexity and cost, at expenses in chip size and bandwidth. Fortunately, 2.5D integration still satisfies the requirements of most MEMS applications on bandwidth and speed. Further, the chips on an interposer could be a 3D integrated chip stack of two or three chips, which can be vertically integrated. From this point of view, 2.5D integration is not a transition stage to 3D integration; it is instead parallel to 3D integration in multi-chip integration in the future.

20.2.4 Wafer-level Vacuum Packaging

The last but not least function of TSVs is to conduct the electrical signals of MEMS devices from a vacuum cavity in WLVP. Hermetic (vacuum) packaging is indispensable to most MEMS devices to protect the fragile structures and provide proper operation conditions. For example, a large variety of resonant or thermal devices, such as accelerometers, gyroscopes, switches, resonators, infrared focal plane arrays (FPAs), and micromirrors, needs vacuum environments from tens to 10^{-2} Pa [28]. WLVP is currently a dominating technology to hermetically package MEMS due to the high throughputs and low costs.

Employing TSVs in WLVP allows to conduct the electrical signals of MEMS vertically out of the vacuum cavities, as shown in Figure 20.5a. TSVs avoid the confliction

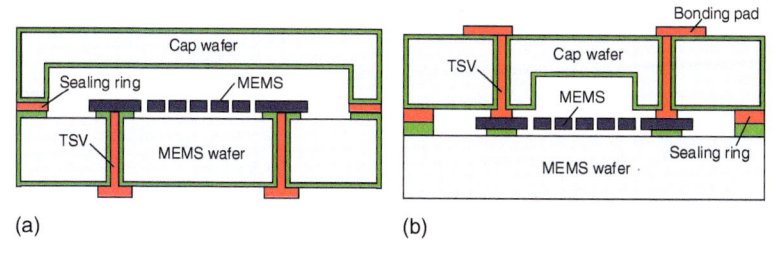

Figure 20.5 Wafer-level vacuum packaging using TSVs. (a) TSVs in MEMS wafer and (b) TSVs in cap wafer.

between the planar metal wires and the sealing rings, facilitating wire routing and sealing material selection. For example, it allows to use narrow metal sealing rings for hermetic bonding, which have good sealability and reliability while occupying small chip areas. In addition, incorporating TSVs allows to place the bonding pads that are originally on the chip surface onto the cap wafer or the backside of the device wafer, as shown in Figure 20.5b, reducing both the chip and the package sizes as well as the costs. Although TSV fabrication increases the processing costs, the cost reduction due to chip size shrinkage as well as the resulting compactness of the board systems could be more significant. Therefore, TSV-based WLVP is one of the most important technical advances, which have greatly promoted the applications of MEMS in mobile and consumer electronics in the last 10 years.

WLVP uses wafer bonding to bond a MEMS wafer and a cap wafer to form a hermetic cavity for MEMS devices. As a simple MEMS device needs only two TSVs, TSVs with various diameters, configurations, and conductors may be applicable. Several considerations for TSV-involved WLVP include the material choice of the cap wafer, the location of the TSVs, the sealing ring materials, and the metal bonding method. The cap wafer can be a silicon wafer, a glass wafer, or even a CMOS wafer. If the TSVs do not require small diameters and high density, glass wafers using laser ablation for TSV fabrication and anodic bonding for vacuum sealing could be a choice from the cost point of view. For optical applications that need the cap wafers to be transparent, glass wafers with through glass vias (TGVs) are the best or the only choice.

The TSVs can be placed either in the substrate or in the cap wafer. Because the cap wafer has no devices, placing the TSVs in the cap wafer decouples TSV fabrication from MEMS fabrication but requires simultaneous bonding of the sealing rings and the metal bumps. Placing the TSVs in the MEMS wafer only needs sealing ring bonding but needs to coordinate the MEMS and TSV processes. To achieve simultaneous vacuum packaging and CMOS integration, one option is to use CMOS as the cap wafer. Although such an integration achieves small chip sizes, fabrication TSVs in CMOS wafers is costly.

An interposer with TSVs can be used to bond with a cap wafer to form a vacuum cavity. Once WLVP is finished, the bonded package can be 3D integrated with CMOS by using the areas beneath the devices for solder connection. TSV-after-bonding has an advantage of simple processes and low cost by using hollow TSVs or silicon pillar TSVs [29]. If air-isolated silicon TSVs are used, TSV-after-bonding is the only choice. Because only one step of DRIE is used for fabrication silicon pillar TSVs, the TSV process has no influence on MEMS [1].

20.2.5 Other Applications

Besides the aforementioned mainstream applications, TSVs can also serve as a functional component of a MEMS device rather than only as an electrical conductor. The materials, structures, and the mechanical, thermal, and physical properties of various TSVs can be exploited for sensing or actuating. For example, TSVs have been used as thermal conductors to measure the temperature of an embedded layer by transferring its heat to temperature sensors [30]. Using TSVs as a functional component of MEMS may reduce the device complexity, improve the performance, simplify the fabrication processes, or realize the functions, which were impossible previously.

20.3 Considerations for TSV in MEMS

Although TSVs facilitate MEMS and CMOS integration, process compatibility is still a concern in MEMS applications. Because most MEMS devices are fragile and suspended, they are subject to damage by TSV processes. It is highly preferred to separate TSVs and MEMS on independent wafers. However, once TSVs and MEMS have to be fabricated on the same wafer, TSVs should be fabricated before MEMS. If this is also impossible, the MEMS should not be released to suspending before TSV processes are finished.

Incorporating TSVs into MEMS makes the mechanical and thermal issues of MEMS, which have already been complex, even more complex because TSVs may cause serious stress or warpage of substrates due to the mismatch in the coefficient of thermal expansions (CTEs) between the metal conductors and the substrates. As many MEMS devices are sensitive to stress and strain, it is important to control the stresses caused by TSVs. To avoid the influences of TSVs on MEMS, deep trenches or flexible structures can be employed to isolate MEMS devices from the stressed areas.

Considering that a large portion of MEMS devices has relaxed requirements on TSV diameters, densities, resistance, and aspect ratios, the mainstream TSVs in IC 3D integration, solid Cu TSVs, are not the first choice in MEMS applications because the fabrication processes of Cu TSVs are lengthy, complex, and expensive. Instead, simple TSV configurations, materials, and fabrication processes may be more appropriate. For example, highly-doped polysilicon or silicon TSVs are preferred by MEMS sensors for signal transmission because of their simple and lost-cost fabrication processes. Although the resistance of such TSVs are relatively high compared with Cu TSVs, their resistance could be negligible because many sensors have an internal resistance several orders higher.

20.4 Fundamental TSV Fabrication Technologies

The methods for TSV fabrication highly depends on the TSV configurations and the materials. A simple TSV using a silicon conductor and an air-gap insulator can be fabricated by one step deep silicon etching of an annular trench, but solid Cu TSVs need more complex processes including etching deep holes in a substrate, depositing a dielectric insulator and adhesion/barrier/copper seed layers on the hole sidewalls, Cu electroplating to fill the holes, flipping and backside thinning the substrate to expose the TSVs, and fabricating redistribution layers (RDLs) and metal bumps on both sides.

20.4.1 Deep Hole Etching

20.4.1.1 Deep Reactive Ion Etching

Deep hole etching in silicon wafers mainly uses high-density SF_6 plasma-based DRIE technology, which can be classified into time-multiplexed etching (Bosch process) [31] and steady-state etching, which can be further grouped into cryogenic [32] and non-cryogenic etching [33].

Bosch process employs periodical alternation between isotropic SF_6 etching and C_4F_8 passivation to achieve deep and anisotropic silicon etching [34], as shown in Figure 20.6. Deep and anisotropic etching is achieved essentially by multiple repeating and stacking of shallow isotropic etching vertically. The multiplexed nature causes 50–150 nm height scallops on the sidewalls of the etched structures, depending on the duration of etching cycles. The scallops may induce concentrations of stress and electric fields in the insulator and the barrier layer of TSVs [35], causing possible reliability problems such as dielectric breakdown or Cu diffusion [36]. High-frequency switching between etching and passivation cycles benefits to reducing the scallops [37], at a cost of low etching rate. State-of-the-art Bosch etchers employ advanced RF sources and ultrafast gas modulation techniques to minimize downtime between etching and passivation steps, reducing the loss in the etching rates. The Teflon-like residues deposited on via sidewalls may cause polymer-mediated defects and traps, influencing the C-V characteristics of TSVs. Dry etching using oxygen plasma or wet cleaning using dilute hydrofluoric acid (HF) and/or amine or hydrofluoroether can remove the residues.

State-of-the-art Bosch etchers can attain, not simultaneously, etching rate up to 50 μm/min, aspect ratio greater than 100 : 1, scallops lower than 5 nm, profile perpendicularity of $90 \pm 0.1°$, nonuniformity less than 5% across a 300 mm wafer, and selectivity to photoresist up to 100 : 1.

Cryogenic and non-cryogenic steady-state etchers employ a delicate balance between simultaneous etching and passivation to achieve deep etching. Cryogenic etching, at −110∼−130 °C, uses reaction of SF_6 and O_2 to produce SiO_xF_y for sidewall

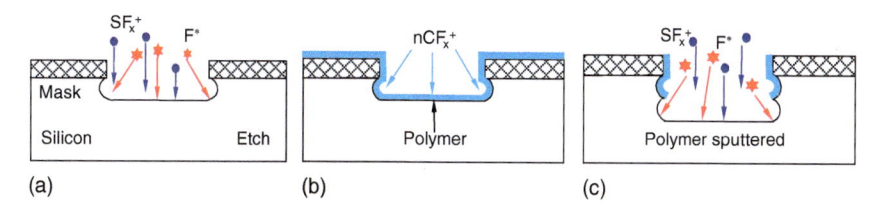

Figure 20.6 Bosch deep reactive ion etching. (a) SF_6 plasma for isotropic etching, (b) C_4F_8 plasma for passivation layer deposition, and (c) next SF_6 plasma etching to remove the passivation layer on the bottom by directional ion bombardment and to perform isotropic etching once again.

protection, whereas the SiO_xF_y on the bottom is bombarded by the directional ions simultaneously with its deposition. Cryogenic temperature improves the selectivity about 10 times by reducing the etching rates of SiO_2 and photoresist masks [32]. Cryogenic etching is able to etch vias with diameters down to 1 μm and aspect ratios up to 30 : 1, and the etching rate can reach 20 μm/min, the selectivity to photoresist is up to 150 : 1, and the nonuniformity between wafers is less than 1% [38].

Non-cryogenic steady-state etching using $HBr+SF_6+O_2$, $SF_6+C_4F_8$, or SF_6+O_2 have attracted attention in recent years [39, 40]. Using magnetic neutral loop discharge (NLD), uniform plasma density and etching rates can be obtained, and deep vias with a diameter of 30 μm and depth of 300 μm in a 300 mm wafer have a nonuniformity less than 3%, but the etching rate of silicon 1.5 μm/min is much low. Using NLD, etching quartz with a rate more than 1 μm/min and Pyrex more than 0.8 μm/min are possible [41].

One significant advantage of steady-state etching is that it is able to achieve scallop-free etching [42]. In addition, it can tailor the tapered angles of the vias by changing the flow rate of O_2 and the temperatures [43]. This is attractive because tapered vias with sidewall angles ranging from 83–85° can facilitate PVD deposition of barriers and seed layers, as well as subsequent void-free electroplating of Cu plugs [44]. In Bosch processes, controlling the sidewall angle is much more complex as more interacting process parameters are involved [45].

20.4.1.2 Laser Ablation

Laser ablation (drilling) employs photons from high-energy directional laser beams to erode the substrates. In recent years, nano-, pico-, and femto-second lasers have been widely used for creating vias on silicon, glass, and polymer substrates [46–52]. The drilling mechanism of femto-second lasers (FSLs) differs from that of nanosecond lasers (NSLs) due to the difference in pulse duration. The laser pulse duration of NSL (around 10^{-9} s) is much longer than photon-to-electron conversion time of most materials (about 10^{-12} s) and allows heat generation during each pulse to instantly melt the substrates, as shown in Figure 20.7.

In comparison, the pulse duration of FSL (10^{-12} s or less) is insufficient for photon-to-electron conversion, and FSL removes the substrate material by changing it to plasma. Because no heat transfer occurs, FSL can drill holes with uniform diameters, smooth sidewalls, less solid debris, and small thermal zones in the

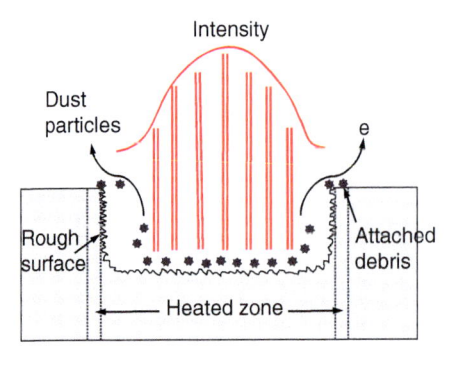

Figure 20.7 Laser ablation using nanosecond laser.

Table 20.1 Lasers with different pulse durations and wavelengths.

Laser	CO$_2$	NSL	PSL	FSL
Type	CO$_2$/N$_2$/He	Nd:YAG	Nd:YAG	Ti-sapphire
Wavelength	10.6 µm	1064 nm	532 nm	~800 nm
Pulse duration	>100 ns	190 ns	16 ps	~100 fs
Pulse frequency	20 kHz	100 kHz	50 kHz	1 kHz
Power per pulse	1 W	1 mJ	1–100 nJ	nJ
Average power	~kW	50–100 W	10–500 mW	100 mW
Via diameter	~50 µm	~50 µm	~30 µm	~20 µm
Via depth	500 µm	500 µm	500 µm	200 µm
Drill method	Percussion	Percussion	Trepanning	Trepanning

substrates. However, FSL delivers a limited amount of energy to the substrate because of the low pulse frequency and the low power per pulse, and thus they are inferior to NSLs in drilling depth. Table 20.1 compares the lasers with different pulse duration and wavelengths.

Due to the rough surfaces and the residual debris inherent to laser ablation, HF-HNO$_3$ wet treatment after laser ablation is preferentially needed to clean/polish the surfaces. The laser-drilled holes have a tapered shape, preferred by subsequent sidewall deposition and metal filling. However, localized compressive stresses are generated in the substrate around each via due to the thermal effects of laser heating. It has been found that for a 15 µm via, the residual stresses disappear in the regions of 3 µm away from the via edge.

The smallest TSV diameter obtained by laser ablation has reached 10 µm, but further shrinking to 5 µm is difficult due to the challenge in focusing laser beams. The highest aspect ratios achieved by NSLs exceed 20 : 1 [47], and this satisfies the requirements of most applications in MEMS areas. The depth, diameter, pitch, and profile angle of vias can be controlled approximately to ±5 µm, 2 µm, 2–5 µm, and 88°, respectively. Due to the thermal effects, the wafer warpage may increase after laser drilling.

Laser ablation is a serial process and the drilling rates decrease with increases in the via diameter and depth [53]. So far the drilling rate is capable of producing TSVs with a 30 µm diameter and 50 µm depth at a rate more than 2000 vias/s.

Since laser drilling does not need etching masks and is capable of drilling through different materials (including metals, dielectrics, and silicon/glass) in the same run, the total cost of ownership is much lower than that of DRIE. These traits make laser drilling attractive for TSV etching in many MEMS applications, in particular those with low TSV densities and large diameters or in glass wafers. However, the energy efficiency of laser ablation highly depends on the absorptivity of the substrates at the laser wavelength.

For wafers with 100,000 TSVs or less, the production rate and the cost of laser drilling are about 2–3 times higher and 10 times lower with respect to DRIE, respectively. For a wafer with approximately 250 000 (depending on specific TSV parameters) or more TSVs, DRIE etching surpasses laser ablation in terms of both production rate and cost [53].

20.4.2 Insulator Formation

A dielectric insulator is indispensable for TSVs to electrically isolate the conductors from the substrate. The most commonly used materials for TSV insulators are SiO_2 and polymers. For TSVs with high aspect ratios, the first concern for insulator deposition is the conformality, as well as the processing temperatures and the dielectric properties.

20.4.2.1 Silicon Dioxide Insulators

Silicon dioxide is the most commonly used insulator material because of the good dielectric properties, good conformality, and ease of deposition. Three deposition methods have been employed for SiO_2 insulators: (i) thermal oxidation at temperatures around 1000 °C, (ii) sub-atmospheric chemical vapor deposition (SACVD) using tetraethoxysilane (TEOS) and O_3 at temperatures around 400 °C, and (iii) PECVD at 200–350 °C.

If high temperature is allowed, thermal SiO_2 at temperatures around 1000 °C is preferred because of the excellent conformality and outstanding dielectric properties. Such a method is normally used for polysilicon or W TSVs. Normally a thickness of 100~200 nm is sufficient for insulation and conformation deposition.

SACVD using TEOS/O_3 deposits SiO_2 at much lower temperatures around 400 °C while still achieving adequate conformality, acceptable dielectric properties, fast deposition rates, and good smoothing ability to 100 nm scallops [54, 55]. As shown in Figure 20.8, the conformality of a SiO_2 insulator deposited in a 16 : 1 via using TEOS/O_3 SACVD reaches 43% [54]. The sidewall scallops after Bosch etching are well isolated by the SACVD SiO_2, and the surface of the SiO_2 is smooth. Typically SiO_2 insulators with thicknesses ranging from 0.2 to 0.5 μm meet most requirements for conformality and breakdown voltage while still allowing high throughput [37].

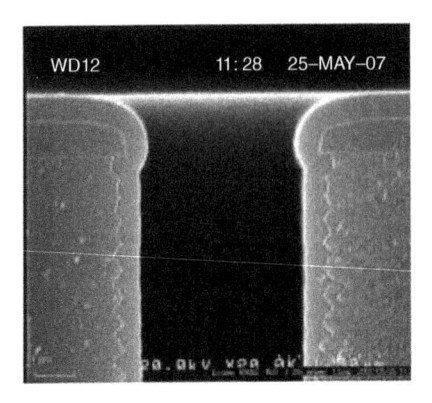

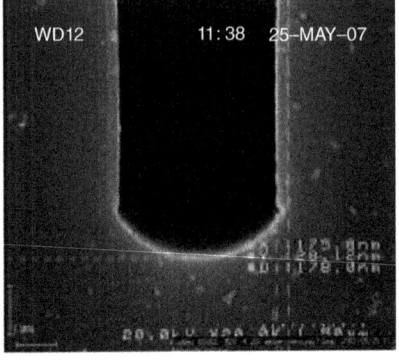

Figure 20.8 SiO_2 insulator deposited using SACVD. Source: Ramm et al. [54]. © 2008 IEEE.

SACVD suffers from large residual stresses and thus the thickness should be no more than 0.5 μm [56].

PECVD is preferred for its ease of availability and low deposition temperatures at 200–350 °C, without much deterioration in breakdown field or leakage current density. One drawback of PECVD deposition is the relatively low conformality and step coverage. For small diameters, atomic layer deposition (ALD) was used to deposit highly conformal SiO_2 insulators [57].

20.4.2.2 Polymer Insulators

Polymers have been used as TSV insulators by either completely filling an annular trench or depositing a thin film on the via sidewalls [58–60]. Complete filling is applicable to silicon TSVs by filling polymers into deep trenches, and depositing is applicable to Cu TSVs by coating polymer on the sidewalls of a hole. Polyimide and benzocyclobutene (BCB) are two common materials for insulators because of their good chemical and thermal stabilities, low out-gassing, low permeability, low dielectric constants, and ease of use. To facilitate filling or coating, the viscosities of polymers should be tailored to proper ranges before spin coating. For polymer filling, vacuum treatment after polymer dispersing is helpful to removing the trapped air bubbles in the trenches [58].

Polymer insulators, though have large thermal expansion coefficients relative to SiO_2, are better able to mitigate stresses induced by thermal expansion of Cu due to their low elastic moduli. In addition, their low dielectric constants benefits to reducing the TSV capacitance, which could be attractive to the applications that deal with small capacitive signals. It should be noted that the out-gassing properties of polymers, though may be quite low, could deteriorate the high vacuum and affect the performance of the devices.

20.4.2.3 Air-gaps

Air-gap insulators can be used for both silicon and Cu conductors. Several MEMS manufacturers and foundries provide TSV products or fabrication services using air-gap insulators and silicon conductors. Such kind of TSVs can be fabricated with only one step of DRIE etching on the substrate, significantly reducing the fabrication costs.

Annular air gaps have also been demonstrated as the insulators of Cu TSVs to mitigate thermal stress and improve high-frequency performance [61]. For Cu conductor TSVs, the air-gap can be fabricated using non-conformal SiO_2 deposition or polymer sacrificial layer techniques, which are released by either reactive ion etching (RIE) or thermal decomposition to form the air-gaps between the Cu conductor and the substrate [61].

20.4.3 Conductor Formation

Unlike in 3D IC applications where Cu is the mainstream TSV conductor material, the TSV conductor materials in MEMS applications show large diversity. Tungsten, nickel, and heavily doped polysilicon and single crystalline silicon have also found wide applications due to their simple fabrication processes and low costs.

20.4.3.1 Polysilicon

Low-resistivity (heavily doped) polysilicon is a widely used conductor for TSVs [15, 62, 63]. Polysilicon TSVs are normally fabricated by CMOS-compatible LPCVD and in situ doping at temperatures around 650 °C or higher to achieve aspect ratios filling (>10 : 1), low stress, and low resistance. The planar cross sections of polysilicon TSVs can be either of a closed-shape-like circular or rectangular rings or of an I- or U-shape trench. Such shapes facilitate filling high-aspect-ratio trenches. Due to the high deposition temperatures and the high tolerable temperatures (>1000 °C), polysilicon TSVs are normally fabricated before other devices, for which thermal oxidation is used to grow SiO_2 insulators. Chemical mechanical polishing (CMP) is normally performed to planarize the wafer by removing the polysilicon deposited on the wafer surface. After that the MEMS devices can be fabricated as usual on the wafers, which have been incorporated with polysilicon TSVs.

The CTEs of polysilicon and silicon are almost identical, so the thermal stresses induced by the mismatch in the CTEs between TSV conductors and silicon wafers are minimum [64]. Polysilicon does not need adhesion and barrier layers, significantly simplifying processing and reducing costs. Therefore, polysilicon is far superior to both Cu and W in terms of manufacturability, thermal stress, reliability, and fabrication cost. However, the resistivity of heavily doped polysilicon is normally around 1–20 $\Omega\cdot$cm, several orders greater than that of Cu [14], and thus the TSV resistance is normally about 5–50 Ω. High-temperature annealing can reduce the resistance [65], but it is still quite difficult to achieve resistances less than 1 Ω for reasonable sizes. In addition, polysilicon TSVs have large noise, in particular for p-type polysilicon [66]. Even though, polysilicon TSVs have found wide applications in piezoelectric, piezoresistive, or capacitive microsensors and MEMS devices, because the internal resistances of these devices are about 1 or 2 orders of magnitude higher than that of the polysilicon TSVs. For such applications, even Cu TSVs with resistance on the order of mΩ do not offer much superiority.

20.4.3.2 Single Crystalline Silicon

Low-resistivity single crystalline silicon (hereafter referred as silicon) has similar properties to polysilicon in TSV applications but with some added benefits. Silicon TSVs can be fabricated by one step etching of a closed trench in low-resistivity silicon wafers using DRIE, with the silicon pillar as the TSV conductor and the surrounding trench as the insulator. This even allows TSV to be achieved by taking the DRIE etching of MEMS devices, making the TSV fabrication simplest and cost lowest. The trenches can also be optionally filled with SiO_2 or polymer to enhance the stability and reliability of the silicon conductors.

Silicon TSVs perfectly match to the wafer, minimizing any reliability and stress issues caused by the mismatch in CTE. The resistivity of silicon is lower than that of polysilicon at the same doping level. In addition, the single crystalline feature enables the noise of silicon TSVs to be lower than polysilicon, which may suffer quite large 1/f noise. In the case of using air trenches as insulators, the air trenches isolate the silicon pillars and the wafer, completely eliminating the thermal and stress-related issues. These advantages make silicon TSVs widely used in MEMS areas.

20.4.3.3 Tungsten

Tungsten, widely used as W plugs in IC technologies to connect the transistors and the first metal, can be filled into high-aspect-ratio holes using chemical vapor deposition (CVD) [67–72]. Tungsten has a high melting point of 1650 °C and thus allows MEMS fabrication after W TSVs, eliminating the process compatibility issues. For TSVs with extremely high aspect ratios, ALD can be used for W deposition with reaction gases of WF_6 and SiH_4 [68]. A thin polysilicon liner is deposited before W deposition to improve the step coverage and adhesion of W, since WF_6 dissolves at the polysilicon liner that acts as a catalyst.

As-deposited W exhibits large compressive residual stresses, which increase with increasing TSV diameters and decreasing TSV pitches [72]. Annealing at 400 °C does not significantly relieve the stress. The influences of the residual stresses can be avoided by using hollow TSV configuration. Therefore W is suitable for hollow TSVs by depositing thin layers or for solid TSVs with small diameters by filling narrow trenches.

The small difference in the thermal expansion coefficients between W (4.5 ppm) and Si (2.5 ppm) mitigates the thermal stresses and reliability issues associated with TSVs. The resistivity of W (5.65 $\mu\Omega\cdot cm$) is about 3.3 times that of Cu, and thus the resistance of small diameter W TSVs, normally in the range of 0.1–1 Ω, is about one order higher in magnitude than that of Cu TSVs. Despite this, the resistance is still low enough to satisfy the need of most MEMS applications. One drawback is that W is difficult to etch using plasma and the W overburdens on the substrate after deposition must be removed using CMP, making TSV fabrication somehow complex and expensive. Although W diffusion in silicon is quite low, an adhesion/ barrier layer such as Ti/TiN is preferentially deposited to ensure reliability. Tungsten TSVs with small diameters down to 1 µm and high aspect ratios up to 50 : 1 have been demonstrated by using CVD process.

20.4.3.4 Copper

Adhesion and the barrier layers, such as Ti–TiN, Ti–TiW, and Ta–TaN, are indispensable for Cu TSVs [73, 74]. Ionized physical vapor deposition (iPVD) is the most commonly used method for depositing these layers as well as Cu seed layers in vias with aspect ratios up to 10 : 1 in batch production due to the advantages of low cost, high productivity, and good conformality [75]. ALD and metalorganic chemical vapor deposition (MOCVD) are alternative technologies for depositing thin and conformal metal layers TSVs with small diameters and high aspect ratios.

The commonly used method for Cu filling is electrochemical deposition (ECD) or electroplating. The most critical challenge in plating Cu TSVs is the high aspect ratios. Because of the faster plating rates at the via openings than the via bottom [76], large overburden on wafer surfaces and pinch-off effect and sealing plating solution in TSVs may inherently occur [77, 78]. To address this challenge, superconformal Cu

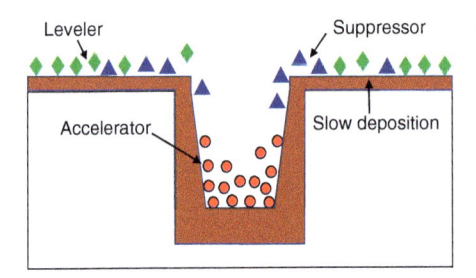

Figure 20.9 Distribution and effects of additives in Cu plating.

plating in blind-vias [79–82] and unidirectional Cu plating in through-vias [83–88] have been developed to obtain void-free Cu TSVs.

Superconformal Cu plating depends on complex chemical additives in plating solutions and proper plating current waveforms [89–93]. Accelerators, such as bis-(3-sulfopropyl) disulfide (SPS) and 3-mercapto-1-propanesulfonic acid (MPS) [94, 95], are small molecules, which diffuse faster than other additives. Suppressors, such as polyethylene glycol (PEG) and polypropylene glycol (PPG) [96, 97], are long-chain polymers with low diffusivity, which adsorb mainly around the TSV openings instead of the TSV interior, inhibiting Cu deposition at the TSV openings. Levelers, such as Janus Green B (JGB) and 1,2,3-Benzotriazole (BTA), are polymers with high viscosities, which preferentially adsorb on Cu surfaces to decrease the local accelerator action [98, 99].

Superconformal plating is the result of the interaction and competition among the additives [100], as shown in Figure 20.9. Accelerators and suppressors dominate the electrochemical reaction of Cu plating inside the vias, whereas levelers plays the leading role on the surface after the vias are completely filled [101]. At the convex edges of via openings, the suppressors surpass accelerators and levelers in adsorption competition, ensuring a low deposition rate at the opening. In the vias, accelerators diffuse much faster than suppressors, accumulate at the bottom corners, and promote the Cu plating rate there. Continuous Cu deposition shrinks the via surface and volume and further increases the accelerator concentrations at the bottom corners, leading to a faster deposition rate there. Therefore, Cu deposition in vias is superconformal and the Cu profile is of V-shape. Once the Cu plugs reach the wafer surface, levelers tend to displace the accelerators and suppressors, inhibiting the formation of Cu mushrooms.

Proper current waveforms, such as periodic pulse reverse (PPR) currents, are also critical to the conformality of Cu plating [91–93]. PPR helps conformal deposition by providing thickness-dependent etch back and providing a term with only ion transportation but without consumption. As shown in Figure 20.10, during the positive current term, the Cu deposition rate at via openings is high. During the negative current term, electrochemical reaction etches the Cu at the edge of openings with a much faster rate than other areas due to the large current concentration. Hence, pulse reverse

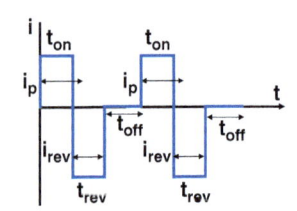

Figure 20.10 Periodic pulse reverse (PPR) currents.

Figure 20.11 Unidirectional bottom-up Cu electroplating.

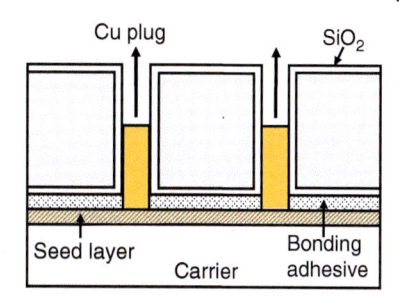

currents balances the Cu deposition rates by thickness-proportional etching back. During the silent term, electroplating that consumes Cu ions stops, while Cu ion transportation into vias continues, supplementing ions inside vias for the next plating term. With additives and PPR currents, TSVs with aspect ratios up to 20 : 1 can be filled without void formation, and the TSV diameter is also down to less than 1 μm [81].

Unidirectional bottom-up plating fills Cu from the bottom of a through-via (via etched through a substrate) [83–88]. If a wafer containing through-vias is bonded to a carrier wafer deposited with a Cu seed layer, the Cu seed layer is exposed only at the via bottom by etching away the bonding adhesive from the via opening, as shown in Figure 20.11. Thus, Cu deposits along the height of the through-via if the resistance of the sidewall barrier is high enough. This unidirectional deposition completely avoids void formation, and TSVs with aspect ratios larger than 10 : 1 can be readily achieved without the need for complex additives [85]. Etching the bonding adhesive or insulators at the via bottom should be anisotropic; otherwise lateral plating will occur at the bonding interface. To avoid Cu deposition on the sidewalls, it is possible to use thin barriers, silicon nitride for long-term-reliability undemanding applications, or glass substrates, which do not need barriers [86, 87].

20.4.3.5 Other Conductor Materials

Other materials have also been used as TSV conductors. Nickel TSVs have been developed by electroless plating for MEMS applications [102]. Ni has a conductivity similar to W, and its CTE is approximately 25% lower than that of Cu. One attractive advantage of electroless plating is that it does not need seed layers, simplifying the fabrication processes for high-density TSVs in large-scale array applications. Besides electroless plating, Ni TSVs were also realized by using external magnetic-field-assisted self-assembly of ferromagnetic Ni wires [103, 104].

Molten solder filling has been developed as a low-cost method to fill TSVs. Sn-3.0Ag-0.5Cu (SAC305) solder can be filled into vias by either with solder paste coating or a dip into a molten solder bath, and subsequently applying vacuum to the wafer backside to suck the solder into the vias [105]. A capillary-based solder pump has been developed to fill solder balls into through-holes [106]. Conductive polymers such as DuPont CB100, filled into through-holes with screen printing, have also been used as TSV conductors [107]. Their sheet resistance is measured at $0.13 \pm 0.02 \, \Omega/\square$ (ohm/square). TSV filling using gold particles [108], vacuum sucking indium [109], and electrodepositing Invar alloys [110] were also reported.

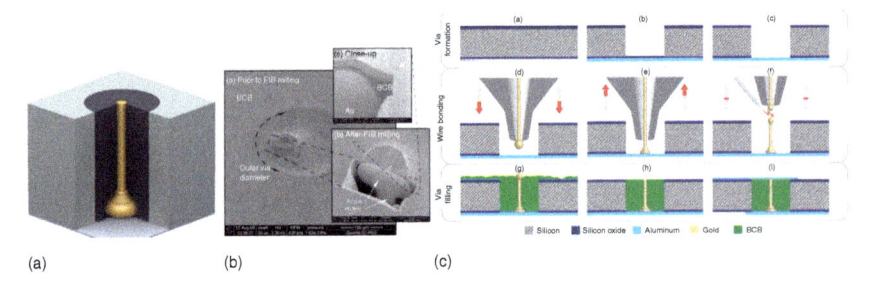

Figure 20.12 Wire bonded Au wires. (a) Configuration, (b) SEM photo, and (c) fabrication processes. Source: Fischer et al. [111]. © 2012 IEEE.

For through-vias with large diameters, it is also possible to use bonded Au wires as the conductor and polymer filled in the hole as the insulator [111]. Figure 20.12 shows a bonded Au-wire with a 25 μm diameter in 200 μm diameter holes [111]. The holes are then filled with BCB to enhance the stability of the Au wires. Such kind of TSVs have the advantages of low costs and ease of fabrication for packaging applications.

20.5 Polysilicon TSVs

Polysilicon TSVs have minimized thermal stresses, good process compatibility, and low fabrication costs. Polysilicon TSV can be fabricated readily in high-aspect-ratio trenches with dimensions down to a few micrometers. Currently several MEMS fabs provide polysilicon TSV foundry services [112–114], and polysilicon TSV technologies are becoming mature.

Almost all polysilicon TSVs are of solid configurations because of the ease of polysilicon filling and the concern on TSV resistance; however, polysilicon TSVs with hollow or air-gap configurations are also feasible.

20.5.1 Solid Polysilicon TSVs

Figure 20.13 shows a U-shaped solid polysilicon TSV. The trench is filled with P-doped polysilicon and isolated from the bulk silicon by a 1 μm thick thermal SiO_2 [114]. The TSV has a height of 300 μm, and each segment of the U-shape has a size of 7 × 70 μm. After thermal oxidation in the trench, a 1 μm undoped polysilicon is deposited by LPCVD inside the trench, followed by P-doping using $POCl_3$, resulting in a sheet resistance of 5.5 Ω/□ (ohm/square). The polysilicon deposition and P-doping were repeated until the TSVs are completely filled, followed by RIE to remove the polysilicon on the wafer surfaces. The TSV resistance is 1.2 Ω with a standard deviation of 0.1 Ω.

Solid polysilicon TSVs can be used for signal routing to wafer backside, CMOS-MEMS integration, and WLVP of MEMS. Figure 20.14 shows a 100 × 100 active nanocrystalline silicon emitter array 3D integrated with a driving IC using

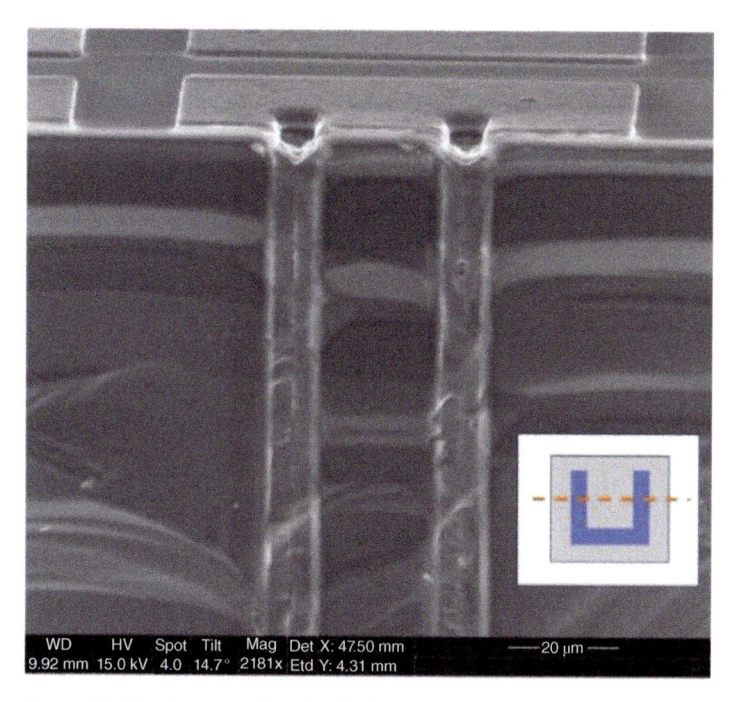

Figure 20.13 Cross section of a U-shaped polysilicon TSV. Source: Lietaer et al. [114]. © 2010 IEEE.

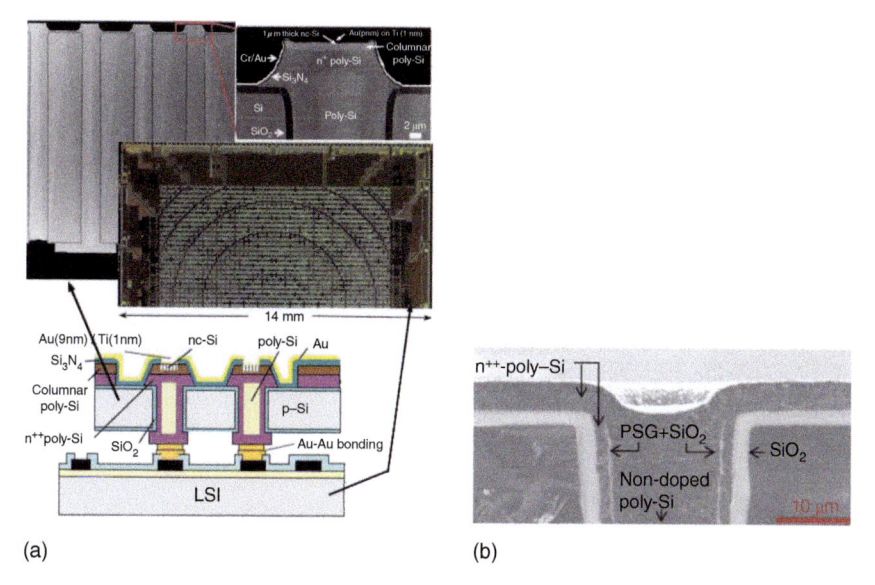

Figure 20.14 Integrated nanocrystalline Si emitter array using polysilicon TSVs. (a) Configuration, and (b) TSV cross section. Source: Ikegami et al. [115]. © 2016 IEEE.

solid polysilicon TSVs [115]. The TSV fabrication includes DRIE etching of through-holes with a diameter of 25 μm and a height of 300 μm, thermal growing of a 2 μm SiO_2 as insulators, LPCVD deposition of a 1.5 μm undoped polysilicon, deposition of PSG and SiO_2 as a doping source and barrier layer, and deposition of 16 μm undoped polysilicon to completely fill the holes. With driven-in at 1100 °C, the first polysilicon layer is n++ doped by the PSG to form the conductor. With the barrier layer, the second polysilicon in the holes is undoped, maximizing the doping concentration of the first polysilicon. The resistance of the TSV is about 150 Ω, corresponding to a resistivity of 4×10^{-3} Ω-cm. The device wafer is bonded with the IC wafer using Au–Au bonding. Such a TSV has the advantage of high-temperature endurability, small diameters, and low cost inherent to via-first features.

Figure 20.15 shows a thermoelastic actuator integrated with piezoresistive sensors using solid polysilicon TSVs for monitoring the gaseous cavity around underwater high-speed supercavitating vehicles [116]. Operation in supercavitating flow necessitates TSVs to avoid front-side bonding wires. Polysilicon TSVs with 20 μm diameter are fabricated through 450 μm thick silicon-on-insulator (SOI) wafers using SiO_2 as insulators. The sensors are then fabricated by thermal oxidation of an insulator,

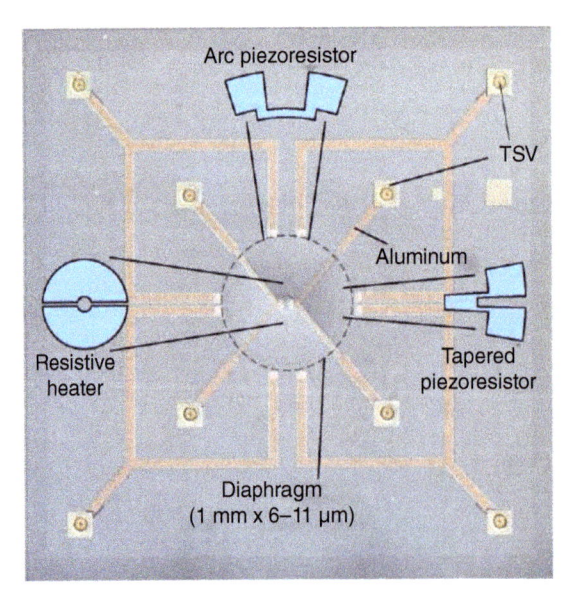

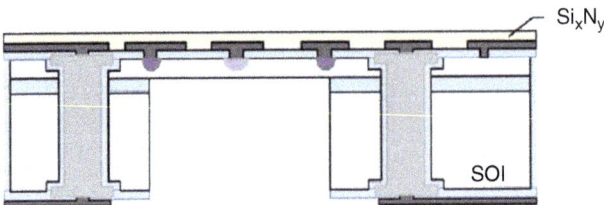

Figure 20.15 Top view and cross section of thermoelastic actuator/piezoresistive sensors with TSVs. Source: Griffin et al. [116]. © 2012 IEEE.

Figure 20.16 Polysilicon TSVs and electrodes fabricated using dual damascene-like processes. (a) Cross section photo and (b) schematic configuration. Source: Midtbø et al. [117]. © 2012 IEEE.

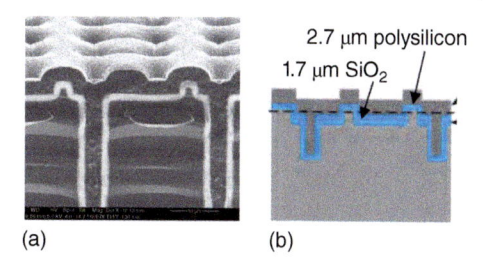

implantation of piezoresistors, and backside etching the handle layer of the SOI to form a membrane. The polysilicon TSVs can withstand 1100 °C annealing and thermal oxidation required by device fabrication, and the resistance is suitable for this application.

Figure 20.16 shows a capacitive micromachined ultrasonic transducer (CMUT) array using polysilicon TSVs [117]. Blind vias and shallow trenches are etched using two-stage DRIE etching, followed by thermal oxidation of 1.7 μm SiO_2. Heavily doped polysilicon is then deposited into the blind vias and onto the undulated top surface. CMP is then used to planarize the surface and remove the polysilicon on the top of the insulating walls, singulating the polysilicon layer into independent elements defined by the shallow trenches. By bonding with another wafer, two electrodes with a small gap is created to form the sensing capacitor. The TSVs have a 4 μm diameter, 25 μm depth, and 25 μm pitch. This idea to simultaneously form horizontal electrodes and vertical TSVs resembles the dual damascene process used in CMOS technology for Cu interconnect fabrication.

Figure 20.17 shows solid polysilicon TSVs fabricated in a cap wafer for WLVP of six-degree-of-freedom inertial sensors [118]. The sensors are fabricated on a 30 μm thick 0.01 Ω·cm device layer of an SOI wafer. A 0.7 μm thick Al is deposited on top of the device layer as the bonding pads and the sealing ring. TSVs in the cap wafer has a 2 μm thick thermal SiO_2 insulator and a heavily doped LPCVD polysilicon conductor. The polysilicon is polished after deposition and a Ge layer is deposited, followed by etching to fabricate both the sealing ring and the standoff. The remaining polysilicon acts as the bottom RDLs on the cap wafer. After the cap wafer and the MEMS wafer are bonded using the Al–Ge eutectic bonding, the cap wafer is grinded to expose the TSVs, followed by SiO_2 deposition, TSV opening, and Al RDL and pad fabrication.

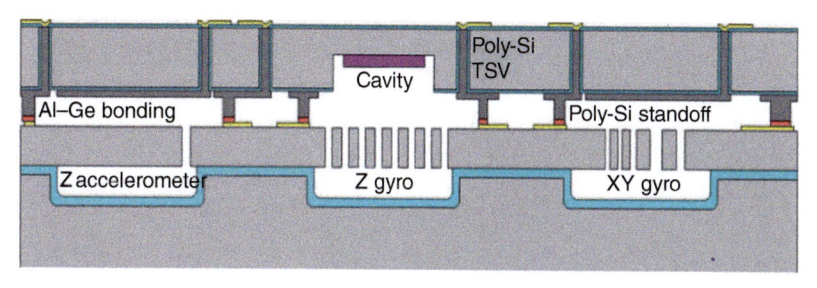

Figure 20.17 Solid polysilicon TSVs in cap wafers for WLVP. Source: Wu et al. [118]. © 2019 IEEE.

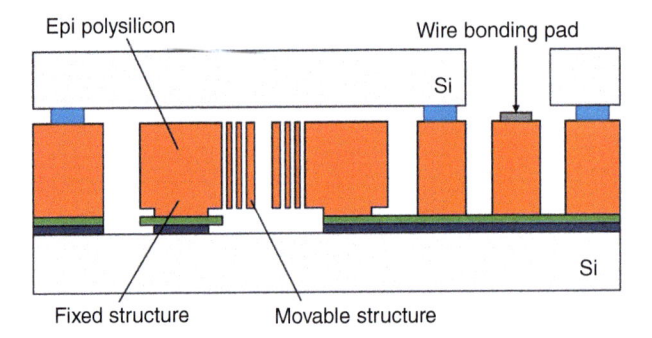

Figure 20.18 Air-gap polysilicon TSV.

20.5.2 Air-gap Polysilicon TSVs

Normally polysilicon TSVs have a solid configuration with SiO$_2$ insulators. However, it is possible to fabricate air-gap polysilicon TSVs by etching a closed trench on a polysilicon layer and using the stand-alone polysilicon posts as TSVs. Strictly speaking, such a TSV is not a through-wafer interconnect; instead, it is a through-layer interconnect but with the features of normal TSVs.

Figure 20.18 shows an air-gap polysilicon TSVs fabricated in a thick epitaxial polysilicon layer developed by STMicroelectronics. The thick polysilicon with 3 MPa low stress is deposited on a 125 nm thick polysilicon seed layer by low pressure epitaxy at 1000 °C [119]. The air-gap TSVs are formed by etching a closed trench in the polysilicon layer together with the MEMS devices [2]. The TSVs are used to connect the bonding wires and the MEMS devices, which are sealed in a vacuum cavity.

An air-gap polysilicon TSV technique, so-called Epi-Seal, has been developed for WLVP of silicon resonators. Figure 20.19 shows a schematic illustration of the structure [120]. The resonators are DRIE etched in the 20 μm thick device layer of an SOI wafer. The trenches are filled with TEOS SiO$_2$, which is etched to expose the contacts on the device layer. A 6 μm thick epi-polysilicon is deposited on top of the SiO$_2$ and etched with small venting holes, through which the SiO$_2$ in the trenches is etched using HF vapor to release the resonators. After high-temperature baking, a 20 μm thick epi-polysilicon is deposited to seal the venting holes. The TSVs are formed in the thick polysilicon by etching trenches at the contact positions down to the underlying SiO$_2$ layer. Covering the trenches with another SiO$_2$ layer enhances the stability and reliability of the air-gap polysilicon TSVs.

20.6 Silicon TSVs

Silicon TSVs use highly conductive silicon posts, which are fabricated by etching annular trenches on a silicon wafer as conductors, and the insulators can be SiO$_2$, polymer, glass filled in the trenches [121], or even the trench itself [16, 122]. As the resistance of silicon TSVs are much smaller than the internal resistance of many

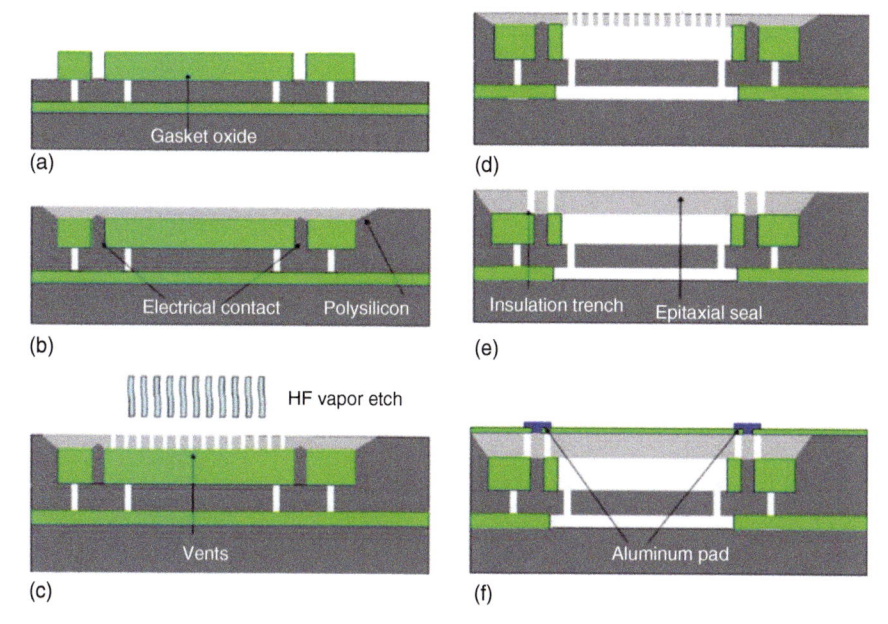

Figure 20.19 Epi-Seal process. (a) Etching the devices in an SOI wafer and TEOS SiO$_2$ deposition as a sacrificial spacer, (b) epi-polysilicon deposition and TSV filling, (c) etching releasing holes on epi-poly, (d) etching SiO$_2$ using HF vapor, (e) second epi-polysilicon deposition and TSV etching, and (f) SiO$_2$ deposition and metal contact fabrication. Source: Ayanoor-Vitikkate [120], © 2009 Elsevier.

MEMS devices, silicon TSVs are widely used in MEMS areas. Compared with metal and polysilicon TSVs, silicon TSVs have simpler processes and lower costs.

Most silicon TSVs are used for routing signals to the wafer backside or conducting the MEMS signals to the top of the cap wafer in WLVP. By using silicon TSVs, the wire bonding pads that are originally placed on the MEMS wafer can be moved to the top of the cap wafer to save chip areas and reduce the chip sizes. The silicon TSVs for signal routing to wafer backside are fabricated in MEMS wafers, and the TSVs for WLVP are fabricated in cap wafers [123]. As it is impossible to alter the conductivity of the wafer in whole thickness after wafer manufacturing, silicon TSVs with height larger than ~10 µm can only be fabricated on heavily doped wafers.

20.6.1 Solid Silicon TSVs

Figure 20.20 shows a capacitive pressure sensor using a solid silicon TSV as the back electrode of the sensing capacitor [124]. The TSV is a thick, large diameter, and highly conductive silicon column insulated from the substrate by a narrow SiO$_2$ insulator. The TSV acts as the plate electrode of the sensing capacitor, and a deformable diaphragm fabricated from the device layer of an SOI wafer serves as the opposite electrode. The TSV wafer and the SOI wafer are bonded to form the gap between the two electrodes. The thickness of the deformable membrane can be

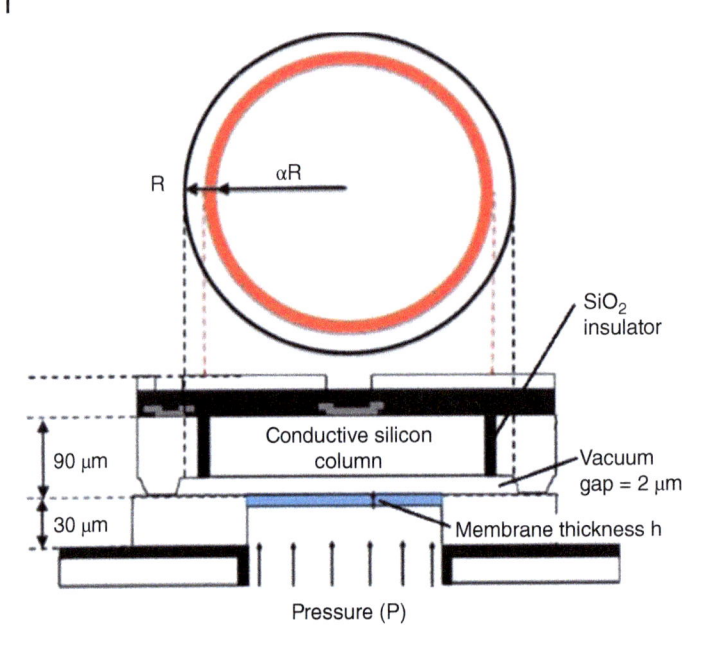

Figure 20.20 Capacitive pressure sensor with TSV electrode. Source: Merdassi et al. [124].

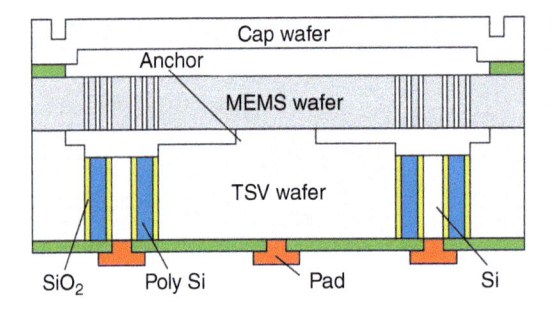

Figure 20.21 WLVP using poly-Si or Si TSVs. Source: Redraw after Marx et al. [125]. © 2017 Springer Nature.

reduced to several micrometers. Using the TSV as the electrode avoids wiring for the back plate electrode.

A 3-layered structure using silicon TSVs or polysilicon TSVs has been developed for vacuum packaging [125]. Figure 20.21 schematically shows the configuration. The TSVs in the TSV wafer are either polysilicon TSVs isolated by SiO_2 insulators or low-resistivity silicon TSVs isolated by SiO_2 filled trenches. A recess is etched on the TSV wafer surface, followed by face-to-face fusion bonding the TSV wafer with a bare MEMS wafer. After fabrication of the MEMS devices, the MEMS wafer is bonded with a cap wafer using metal sealing rings. The recesses on the TSV wafer and on the cap wafer form the vacuum cavity, providing a free space for the out-of-plane motion of the MEMS devices. The TSVs also function as a vertical capacitive sensing electrode to measure the out-of-plane motion without need of an extra layer in the cavity. This allows a better utilization of the space and larger areas for sensors. Using this method, a 6-axis IMU has been developed [126].

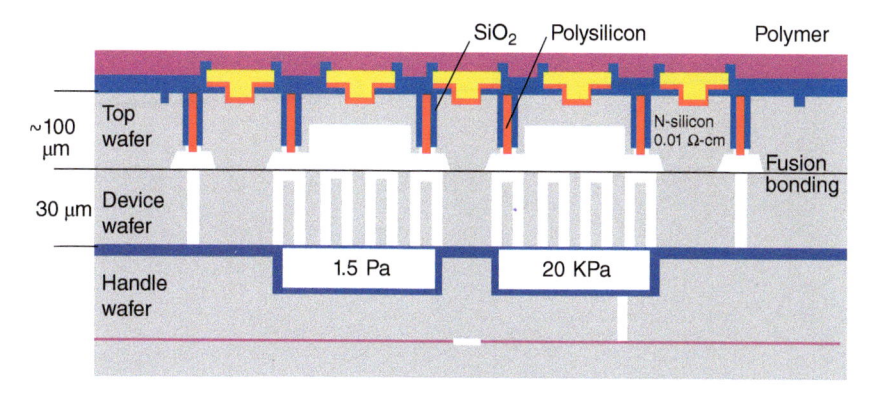

Figure 20.22 WLVP using silicon TSVs for inertial sensors. Source: Merdassi et al. [127]. © 2015 Elsevier.

Figure 20.22 shows the so-called MEMS integrated design for inertial sensors (MIDIS) technology developed by Teledyne DALSA for WLVP with an inner pressure of 1.5 Pa [127]. The vacuum cavity is formed by fusion bonding two cap wafers on the two sides of a MEMS wafer. The TSVs are formed in one cap wafer by etching holes, depositing SiO_2 liners as insulators, and filling in situ doped polysilicon as conductors. If TSVs with large diameters are acceptable, the TSVs use the silicon pillar as conductors, which are isolated from the low-resistivity wafer by SiO_2-filled trenches. The MIDIS technology can be used to fabricate vacuum packaged inertial sensors and resonators.

20.6.2 Air-gap Silicon TSVs

Air-gap silicon TSVs can be fabricated readily by etching a through-trench with a closed shape (most often annular) in a silicon wafer to form a silicon post conductor insulated from the wafer by the trench. Several MEMS manufacturers use air-gap silicon TSVs in commercialized products such as inertial sensors, and foundries provide fabrication services for air-gap TSVs [161]. Silex Microsystems was the first MEMS foundry to provide air-gap silicon TSV processes since 2005. The fabrication technique, so-called Sil-Via™, is able to fabricate TSVs with pitch less than 50 μm and depth of 600 μm. The gap formed by DRIE etching can be optionally filled with a proprietary insulating material, which gives TeraOhm level DC isolation of the via to the substrate. Sil-Via™ TSVs with a 100 μm diameter and a 430 μm thickness exhibit typical resistances of 0.5–1 Ω [122].

In 2011, STMicroelectronics released a TSV-embedded 3-axis accelerometer as a special version of LIS320DL, representing the first mass-produced inertial sensor, which incorporates TSVs [128]. The accelerometer is fabricated using SMERALDO technology [1], an updated version of the thick epi-polysilicon THELMA technology [129] with added via-last TSV processes. The TSVs are fabricated by etching a trench in a heavily doped silicon substrate, with the trench as the insulator and the remaining silicon pillar as the conductor, as shown in Figure 20.23 [1]. One end of the TSV provides metal pads for wire bonding, and the other connects the MEMS devices

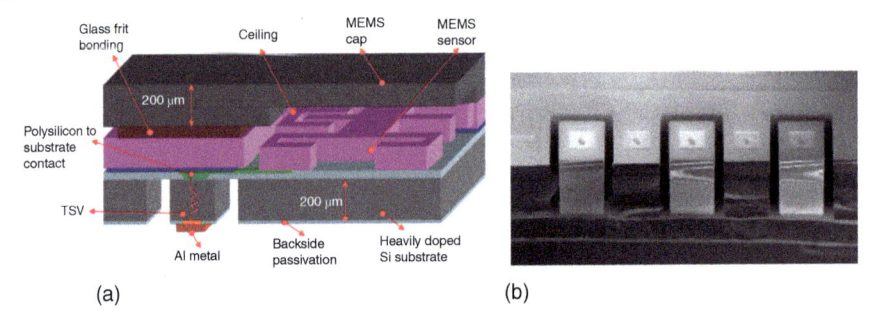

Figure 20.23 SMERALDO using silicon TSVs. (a) Schematic configuration, and (b) air-gap silicon TSVs. Source: Hirama et al. [1]. © 2015 IEEE.

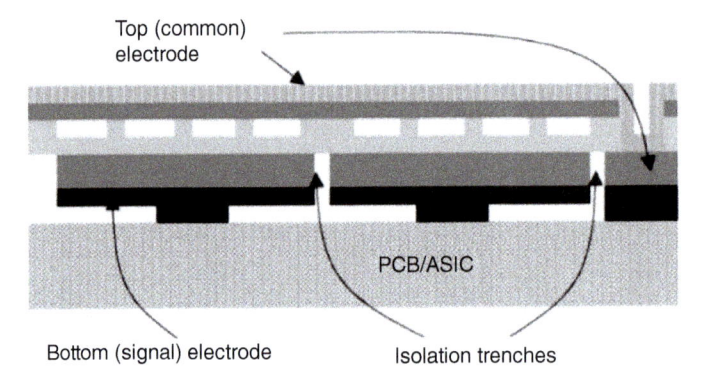

Figure 20.24 CMUT array in 3D integration using low-resistivity silicon substrate blocks as TSVs. Source: Zhuang et al. [130], © 2007 Elsevier.

through a doped polysilicon contact, which is fabricated from the front side before thick polysilicon is deposited. Such a method reduced chip size as much as 40%.

CMUT arrays benefit from TSVs in minimizing the parasitic capacitance of interconnects as well as improving the filling factors. Figure 20.24 shows a CMUT array using air-gap silicon TSVs to connect the sensor array and the ICs [130]. Air-gaps are closed trenches etched on the silicon wafer to define the silicon conductors. The thickness of the silicon conductors is 120~180 μm, the diameter is 250 μm, and the corresponding resistance is about 4.5 Ω. It is insignificant when compared to the CMUT impedance at many kilo Ohms. If a silicon wafer with low resistivity ($\sim$0.01 Ω·cm) is used, the resistance of the TSVs is theoretically below 1 Ω. Air trenches significantly simplifies the fabrication processes by avoiding the processes for insulator deposition, polysilicon filling, and CMP.

Figure 20.25 shows a CMUT array integrated with application specific integrated circuit (ASIC) using air-gap silicon TSVs [131]. The air-gap silicon TSVs are fabricated in the CMUT array wafer and bumped with eutectic Sn–Pb bump solders to bond the array and a Teflon-based organic interposer. The TSVs have a size of 85 μm and a height of 250 μm. The other side of the interposer is flip-chip bonded with an

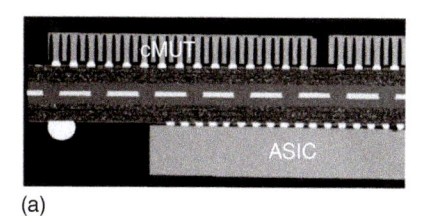

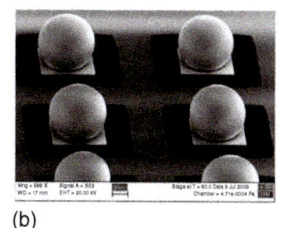

Figure 20.25 Air-gap silicon TSVs for CMUT array. (a) Cross section of the CMUT array and (b) silicon TSVs with bumps. Source: Wodnicki et al. [131]. © 2009 IEEE.

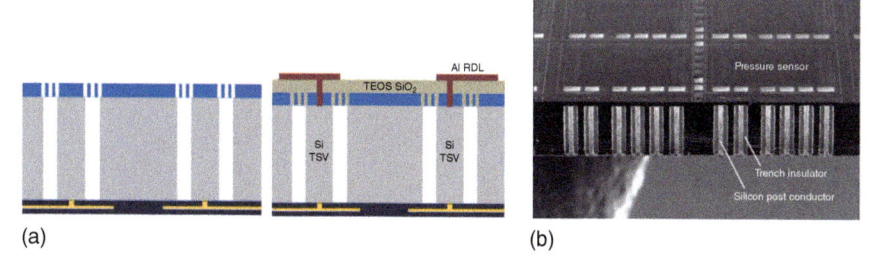

Figure 20.26 Pressure sensors using air-gap silicon TSVs. (a) TSV fabrication processes and (b) pressure sensor with TSVs. Source: Bergmann et al. [132]. © 2013 IEEE

ASIC chip with more than 4000 I/O. The silicon TSVs electrically connect the CMUT array and the interposer and mechanically support the CMUT array.

Figure 20.26 shows sealed air-gap silicon TSVs and the application in a pressure sensor [132]. A SiO_2 hard mask is first deposited and etched with many high-density small holes, of which the envelope forms a trench shape. Bosch DRIE is performed to etch the silicon wafer through the small holes. The slight undercut in silicon etching expands the holes etched in the silicon wafer and merges the small holes into a closed trench defined by the envelop of the holes. After DRIE etching, another SiO_2 layer is deposited and the small holes on the mask are sealed as a continuous layer. This sealing step enhances the fixation to the TSVs and allows deposition of RDL metal, and no stress is induced on the device substrate. This TSV is applied to a pressure sensor, which consists of a silicon diaphragm and four piezoresistors to connect the piezoresistors on the wafer front side to the backside.

20.7 Metal TSVs

Metal TSVs have more significant diversities than other TSVs in metal materials, TSV configurations, and fabrication methods. Solid metal TSVs are mainly used for CMOS integration, which can be fabricated in CMOS wafers with low-density or in MEMS wafers with high density. Hollow metal TSVs can be used for either CMOS integration or WLVP.

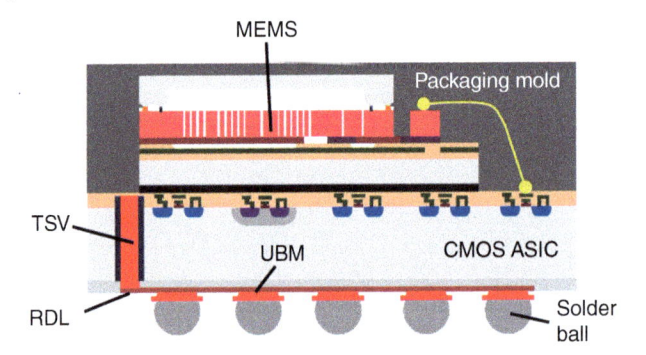

Figure 20.27 Cu TSVs in CMOS chip for MEMS-CMOS integration. Source: Redraw after [133].

20.7.1 Solid Metal TSVs

In 2014, Bosch Sensortec released BMA355, the smallest 3-axis accelerometer on the market at the time and the first mass-produced MEMS device, which incorporates TSVs fabricated in CMOS chips with via-middle processes [133]. As shown in Figure 20.27, the MEMS structures are hermetically packaged with a cap wafer and are electrically connected to the ASIC chip using wire bonding. High-aspect-ratio Cu TSVs (10 μm diameter and 100 μm height) are fabricated in the ASIC chip to connect the CMOS circuity to the backside, such that the solder balls can be fabricated on the backside to save chip areas. The chip has a size of $1.2 \times 1.5 \times 0.8$ mm, about 60% smaller than a common chip at the time.

Metal TSVs have been used as a part of 3D spiral inductors and Faraday cages [19, 134]. Figure 20.28 shows a 3D air-core inductor, which uses vertical TSVs together with planner metal wires on the front side and the backside of a wafer to constitute a multi-turn spiral inductor [134]. The inductor is formed by multiple fan-shaped metal wires on the double side of a wafer and many solid Cu TSVs, which connect the metal wires vertically. The TSVs are fabricated using unidirectional bottom-up Cu electroplating by first sealing the through-holes at the wafer surface. The TSVs

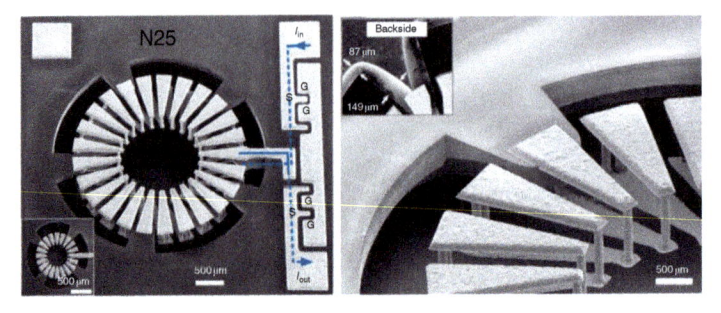

Figure 20.28 Overview and detail view of a 3D air-core MEMS inductor. Source: Le et al. [134]. © 2018 Springer Nature.

Figure 20.29 CMUT with TGVs for signal routing and anodic bonding and on-site film deposition for vacuum packaging. Source: Zhang et al. [135]. © 2017 IEEE.

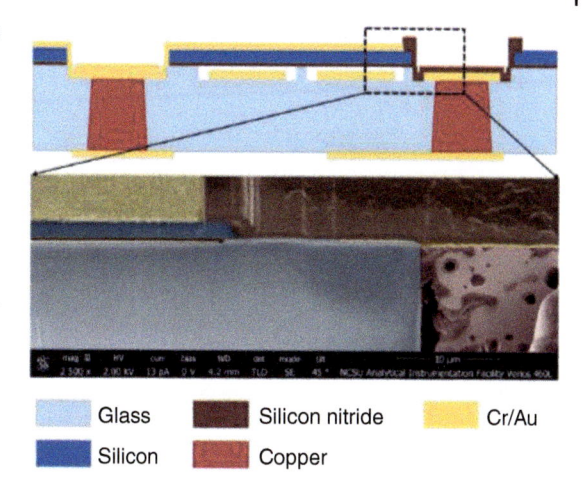

have a diameter of 35 μm and an aspect ratio about 10 : 1. By removing the silicon core surrounded by the inductor metals, the quality factor is improved 1.4 times and the operation frequency increases 2.3 times.

Figure 20.29 shows a CMUT array using TGVs for backside electrical signal conduction [135]. The lower electrodes of the CMUT elements are Au films deposited on the bottom of shallow cavities etched on a glass wafer, and the top electrodes are also Au films deposited on the top of the vibration plate, which is a thin and low-resistivity silicon transferred from an SOI wafer onto the glass wafer by anodic bonding. All the top electrodes are connected together and then connected to the wafer backside using a TGV. Another TGV connects a bottom electrode to the backside, which is further connected to other bottom electrodes through the Au wires, which pass through the channels between neighboring cavities. The TGVs are fabricated in a 0.7 mm glass wafer using laser drilling and Cu electroplating. The pitch of the TGV is 250 μm, the diameters of the two openings are 70 and 50 μm, and the parasitic resistance the shunt parasitic capacitance are 2Ω and 20fF.

With proper materials, metal TSVs can be used in high-temperature environment. For example, metal-glass frits-filled TSVs have been used to develop high-temperature pressure sensors [136]. The sensor consists of a silicon diaphragm and island-like piezoresistors, which are etched from the silicon substrate and the device layer of an SOI wafer, respectively. A glass cover wafer etched with a recess is face-to-face bonded with the sensor wafer to form the pressure cavity. TSVs are fabricated in the glass wafer by etching through-holes and filling the holes with a mixture of metal powder and glass frit. The TSVs connect the piezoresistors at the bonding interface to the backside of the glass wafer. The TSVs improve the reliability in high-temperature conditions by avoiding the problems associated with metal wires and ball bonds.

If the MEMS device is movable or adiabatic, TSVs can act as both mechanical supports to suspend the MEMS devices and as electrical conductors to connect the MEMS and CMOS. A common way to realize such a configuration uses wafer transfer technology, which essentially transfers, in wafer level, a device layer fabricated

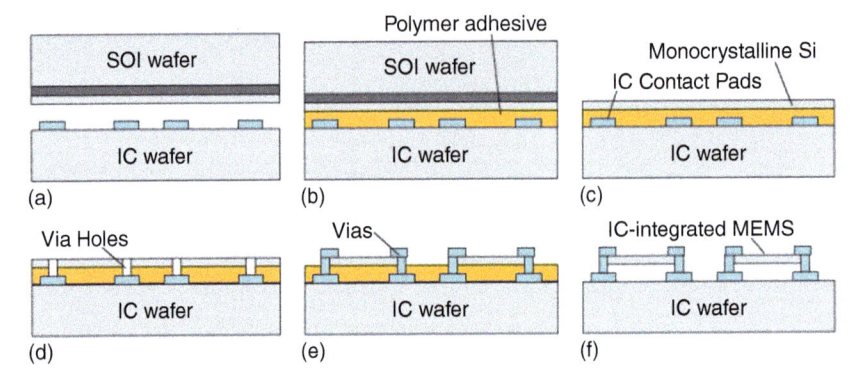

Figure 20.30 3D integration of SOI MEMS and CMOS. (a) SOI wafer and CMOS wafer, (b) adhesive bonding the two wafers, (c) removing the SOI wafer substrate, (d) etching vias, (e) filling vias with metal for form TSVs, and (f) etching the adhesive to release the MEMS device to suspending. Source: Niklaus et al. [137], © 2009 Elsevier.

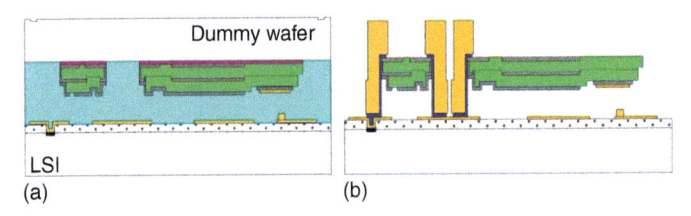

Figure 20.31 PZT switch with TSVs as mechanical supports. (a) Transfer PZT devices fabricated on a dummy wafer to a CMOS wafer and (b) Au pillar to support the PZT device to suspending. Source: Matsuo et al. [142]. © 2012 IEEE.

on one substrate to another substrate using wafer bonding and thinning techniques [61, 65]. As shown in Figure 20.30 [137], an SOI wafer is flip-bonded to a CMOS wafer using polymer adhesive, and then the SOI substrate is removed, followed by etching vias through the SOI device layer and bonding adhesive until the metal pads on the CMOS wafer. TSVs are then formed by filling metals to connect the metal pads and the SOI layer. After the MEMS array is fabricated in the SOI layer, the polymer is removed to release the MEMS. The gap between MEMS and CMOS provides a free space for MEMS movement, prevents thermal dissipation to CMOS, or reduces electromagnetic loss in CMOS substrates. Using this technology, a silicon micromirror array, a SiGe focal plane, and a silicon diode FPA have been successfully realized [138, 139].

Besides silicon layers, materials that are incompatible to CMOS can also be transferred on another wafer, such as piezoelectric devices, carbon nanotubes (CNTs), or other nanomaterials [140, 141]. Figure 20.31 shows a suspended piezoelectric MEMS switch with TSVs acting as both the electrical conductor and the mechanical support [142]. The piezoelectric switch is made of lead zirconate titanate (PZT) films deposited on a dummy wafer and then transferred onto the circuit wafer using polymer bonding, substrate removing, and Au electroplating. With the support of the Au TSVs, the PZT switch is released to suspending by removing the bonding polymer.

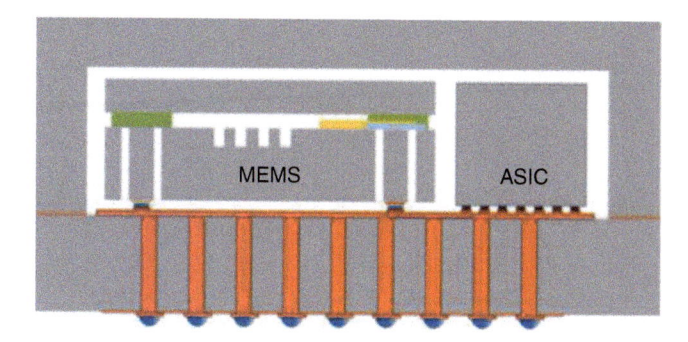

Figure 20.32 Interposer for simultaneous CMOS integration and vacuum packaging. Source: Steller et al. [143]. © 2014, IEEE.

Figure 20.32 shows a CMOS integration and vacuum packaging of MEMS using an interposer and both air-gap silicon TSVs and solid metal TSVs [143]. The MEMS wafer is first flip-bonded with a cap wafer and then is fabricated with air-gap silicon TSVs. The MEMS/cap stack and an ASIC chip are bonded on an interposer using metal bumps/solders. The Cu TSVs in the interposer conduct the electrical signals outside of the vacuum cavity, which is formed by bonding the interposer with another cap wafer. The silicon TSVs function as both the electrical conductors between the MEMS and the interposer and the mechanical supports to the MEMS/cap stack. The silicon TSVs can tilt in the air-gaps in some extent, isolating the MEMS from being affected by the stresses and thermal expansion of the interposer.

Figure 20.33 shows an ultra-miniaturized wireless sensor nodes (e-CUBES) realized by 2.5D integration of multi-MEMS and CMOS chips, with an tire pressure monitoring system example using a pressure sensor, an RF transceiver, a microcontroller, and a bulk acoustic resonator [54]. A silicon interposer bridges a microcontroller and

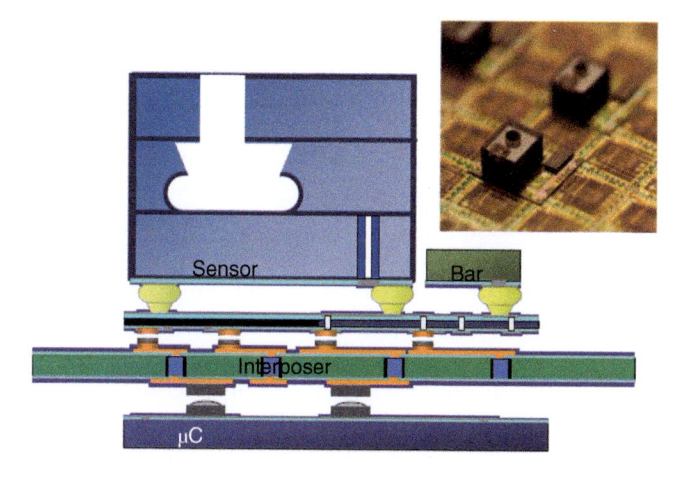

Figure 20.33 A TMPS with pressure sensor, BAR, controller and memory on a TSV-embedded interposer. Source: Ramm et al. [54]. © 2008 IEEE.

a 3D die-stack on the two sides of the interposer through TSVs. The 3D die-stack consists of a pressure sensor and a resonator, which are wafer-level bonded onto a 60 µm thick RF transceiver die. The cap of the sensor die and the transceiver die are processed with TSVs. The TSVs in transceiver are high-aspect-ratio (20 : 1) W TSVs. Top and bottom wafers are stacked using Cu/Sn solid–liquid inter-diffusion (SLID) layers or SnAg microbumps.

Normally separate wafers are needed to fabricate MEMS and CMOS. However, it is theoretically feasible for some applications to fabricate MEMS and CMOS on the opposite sides of a wafer and connect them using TSVs [144, 145]. A ZnO nanowire sensor fabricated on the front side of a wafer was integrated with signal processing circuits fabricated on the backside of the same wafer [146]. The MOSFETs are first fabricated on the wafer backside, and then Cu TSVs are fabricated in the wafer. Finally ZnO nanowires are grown on the wafer front side using vapor–liquid–solid method at 300 °C and are electrically connected to the MOSFETs on the backside with TSVs.

20.7.2 Hollow Metal TSVs

Most hollow metal TSVs are fabricated either in the cap wafer for WLVP or in the CMOS wafer to conduct the signals to the wafer surfaces to save the chip areas occupied by the wire bonding pads. They can also be fabricated in MEMS wafers for 3D integration of MEMS and CMOS.

A typical fabrication process for hollow metal TSVs is shown in Figure 20.34. After the two wafers (say MEMS and CMOS wafers or MEMS and cap wafers) are bonded, the top wafer is etched until the conductive pads on the bottom wafer, followed by TSV insulator deposition. The common insulator material is 1–2 µm thick SiO_2 insulator layer deposited by PECVD. Then the insulator on the via bottom is anisotropically and masklessly etched away using RIE to expose the conductive pads. Finally a conformal metal film is deposited and patterned to form the TSV conductor. Typical conductors are Cu, which can be deposited by electroplating in a mold

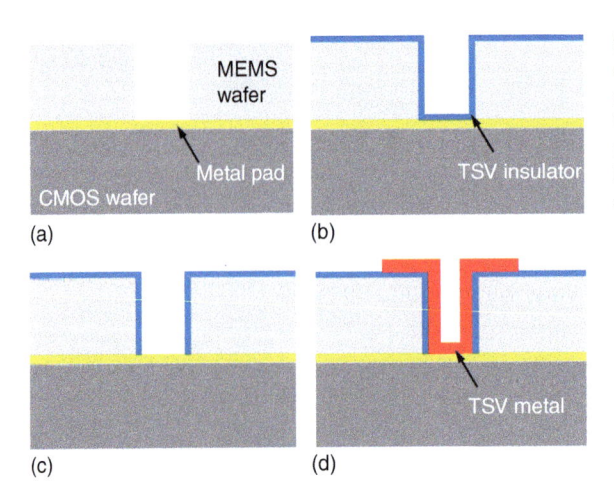

Figure 20.34 Fabrication of hollow-metal TSVs. (a) Via etching, (b) deposition of insulator film, (c) etching of the insulator at the via bottom, (d) deposition of conductor film.

Figure 20.35 Hollow Cu TSVs. (a) Annular and (b) X shape. Source: (a) Kraft et al. [24] © 2011 IEEE, (b) de Veen et al. [148]. © 2015 Elsevier.

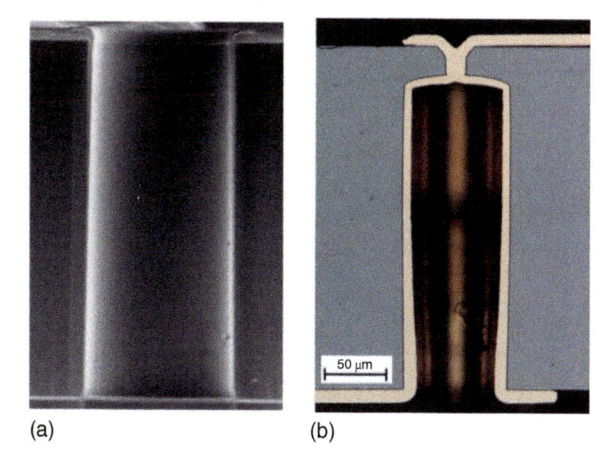

fabricated with dry-film photoresist after adhesion/barrier/seed layers are sputtered. Hollow Cu TSVs were widely used in CMOS image sensors to connect the pixels to the backside of carrier wafers [147].

Besides regular hollow TSVs shown in Figure 20.35a, X-shaped hollow TSV was also developed by leaving silicon with a certain thickness unetched at one end, as shown in Figure 20.35b [148]. A vertical and large diameter hole is etched from one side and stopped at a large portion of the wafer, and a coaxial, small diameter, and tapered hole is etched from the other side to get breakthrough. Using double side Cu plating to deposit a 10 μm thick conformal Cu film on the sidewalls creates X-shaped Cu conductor. Compared with the hollow TSVs, the remaining silicon improves the mechanical strength of the X-shaped TSVs, while maintaining the fabrication relatively simple.

Figure 20.36 shows an integrated capacitive 3-axis tactile sensor using hollow metal TSVs fabricated in a CMOS large scale integration (LSI) chip [149]. The MEMS chip consists of a diaphragm-supported boss and four seesaw electrodes, which together with four fixed electrodes fabricated on the CMOS LSI chip constitute the sensing capacitors. Tactile force deflects the boss and changes the capacitance, which is readout by the LSI chip. Hollow TSVs are fabricated in the CMOS chip for power supply, GND, and data transfer. Direct power supply to the tactile sensor through the TSVs prevents voltage drops caused by the LSI to ensure the sensor performance. The main TSV processes include DRIE etching a 110 μm wide and 400 μm depth trenches, TEOS PECVD depositing a 2 μm SiO_2 insulators, Ar-ion directional etching of the SiO_2 at trench bottom, sputtering a 400 nm Ta barrier and a 1 μm and adhesion layers, electroless plating a 500 nm Ni seed layers, and electroplating a 10 μm Cu layer.

Figure 20.37 shows an FBAR using TSV-based WLVP [17]. A CMOS wafer fabricated with circuits and etched with cavities is bonded with a MEMS wafer using Au–Au thermal compression bonding to seal the suspended FBARs. The tapered hollow vias are fabricated in the CMOS wafer probably by laser drilling and deposited with a thin metal film on the via sidewalls as conductors. The FBAR is connected to the circuits by Au–Au bump bonding. As packaging typically occupies 90% of the

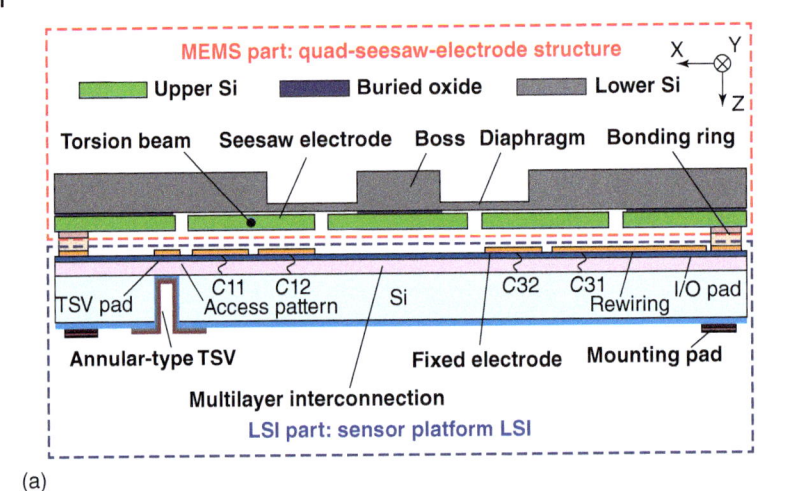

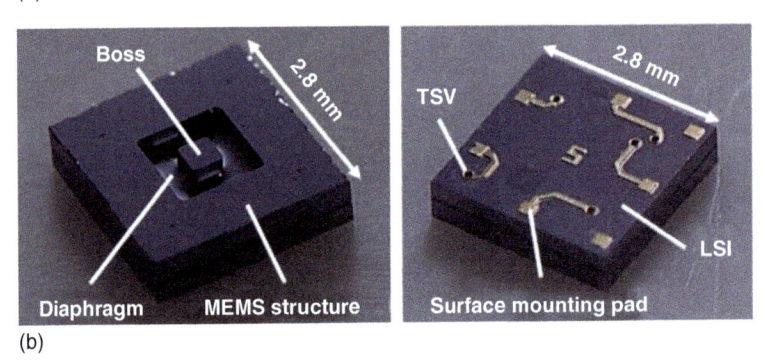

Figure 20.36 3D integration of capacitive 3-axis tactile sensor with CMOS. (a) 3D configuration and (b) front side and backside of the 3D integrated sensor. Source: Hata et al. [149], © 2018 Elsevier.

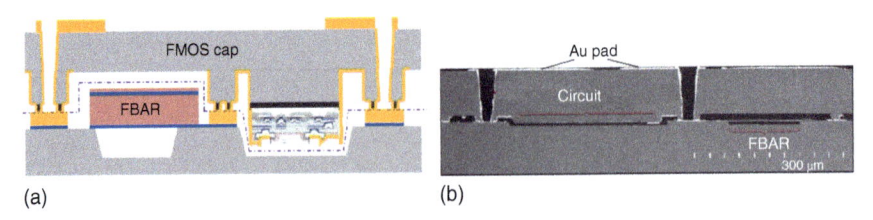

Figure 20.37 WLVP of FBAR. (a) Schematic of WLVP and (b) Cross-sectional view of the FBAR chip. Source: Small et al. [17]. © 2011 IEEE.

areas of MEMS resonator chips, placing the wire bonding pads on the cap wafer surface can reduce the chip size significantly [150].

Hollow W TSVs has been developed by mCube to integrate a 3-axis accelerometer with CMOS in batch production, as shown in Figure 20.38 [151]. The hollow W TSVs are fabricated in the MEMS chip using via-middle processes. A blank wafer used for MEMS fabrication is first fusion bonded onto the polished dielectric surfaces of a

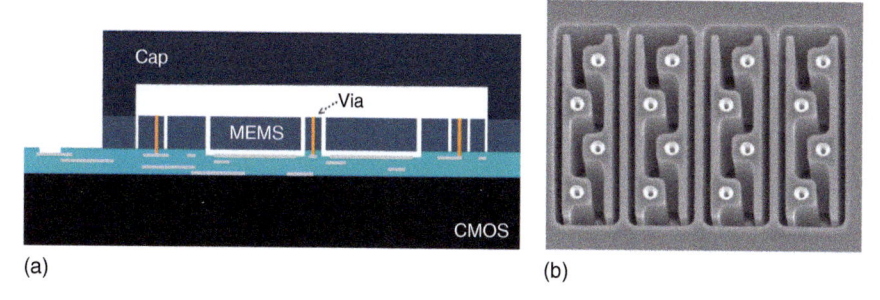

Figure 20.38 Hollow W TSVs fabricated in MEMS layer for CMOS integration. [151]. (a) Configuration and (b) SEM photo of TSVs. Source: Image courtesy of mCube, Inc. Copyright 2014. All rights reserved.

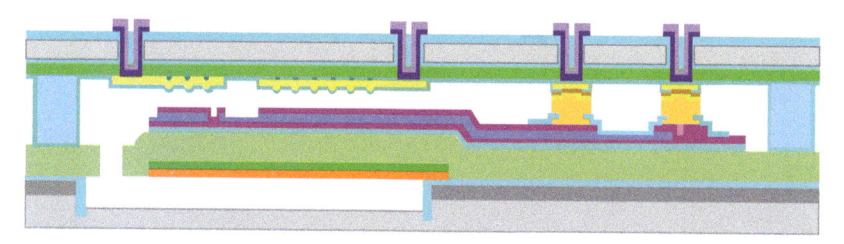

Figure 20.39 RF switches packaged with TSV-last process. Source: Ferrandoni et al. [152]. © 2010 IEEE.

CMOS wafer etched with shallow cavities in the dielectrics. After wafer thinning, hollow W TSVs are fabricated in the top wafer using deep etching and CVD. The top wafer is then etched to fabricate MEMS structures, followed by bonding a cap wafer for hermetic packaging. The hollow W TSVs, with a diameter of 3 μm and an aspect ratio of 10 : 1, connect the MEMS devices to the underlying CMOS. The final package size is $1.1 \times 1.3 \times 0.74$ mm.

The TSVs in cap wafers can be fabricated before wafer bonding; TSVs can also be fabricated after bonding, as shown in Figure 20.39 [152]. Unlike common methods which use etched cavities on the cap wafer to provide space for movable MEMS devices, this method uses 5 μm thick TEOS PECVD SiO_2 on the cap wafer as the spacer between the MEMS and the cap wafers. The cap wafer is electroplated with Au–Sn sealing rings and bonding bumps and then is flip-bonded with the MEMS wafer using eutectic bonding. After the cap wafer is thinned to 80 μm, through-holes with 80 μm diameter are etched on the cap wafer, followed by Cu plating to form hollow TSVs.

Figure 20.40 shows 3D integration of a CMOS chip and a CNT resonator. The CNT resonator is fabricated on a nano-electromechanical system (NEMS) chip and packaged in a 0.1 mTorr vacuum using WLVP [153]. As CNTs are fragile and sensitive to post processes, TSVs are fabricated at first using KOH etching of slant cavities, Pt sputtering, and LPCVD deposition of a SiN passivation layer. CNTs grown on a separate wafer are transferred onto the NEMS chip. A glass cap is bonded on top of the device chip using Au–Si eutectic bonding to form a vacuum cavity for CNTs, followed by bonding a CMOS chip on the backside of the NEMS chip using Ag conductive

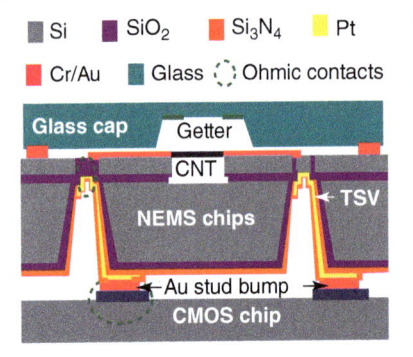

Figure 20.40 CNT resonator with TSV-incorporated vacuum packaging and CMOS integration. Source: Gueye et al. [153].

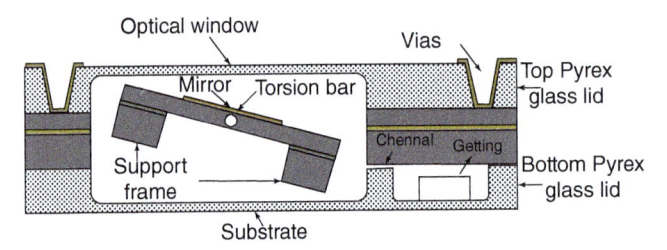

Figure 20.41 WLVP packaged 2D rotational micromirror. Source: Chu et al. [154]. © 2018 Springer Nature.

paste. Placing TSVs in the NEMS chip instead of in the cap chip facilitates electrical connection of TSVs to CNT resonators.

Figure 20.41 shows a 2D micromirror with TSV-based WLVP [154]. The micromirror is fabricated on the device layer of an SOI and sealed in a vacuum cavity formed by anodic bonding of two glass wafers with etched deep cavities which allows the micromirror to achieve a rotation angle of 25° at a driving voltage around 20 V. The hollow TSVs are fabricated on the cap glass wafer using HF solution wet etching.

Using TSVs in the glass cap wafer also facilitates double side anodic bonding. As shown in Figure 20.42 [155], once a silicon wafer is anodic bonded with a glass wafer Pyrex 1, a capacitor is formed at the silicon-Pyrex interface as a result of formation of depletion. This capacitor then prevents the bonding voltage to be applied to the silicon wafer during anodic bonding of Pyrex 2 on the other side of the silicon wafer,

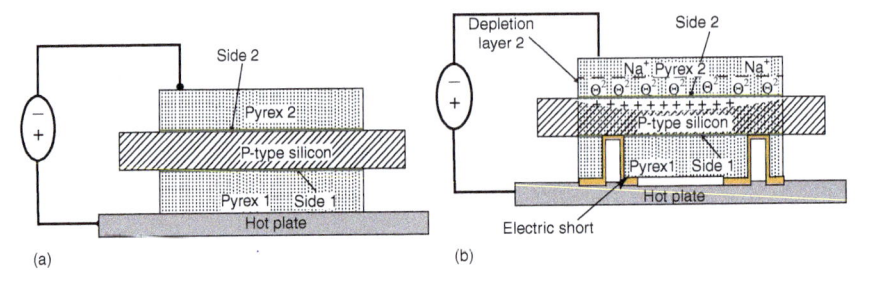

Figure 20.42 TSVs for double-side anodic bonding. (a) Two glass wafers without TSVs, (b) Two glass wafers with TSVs. Source: Chu et al. [155], © 2013 Elsevier.

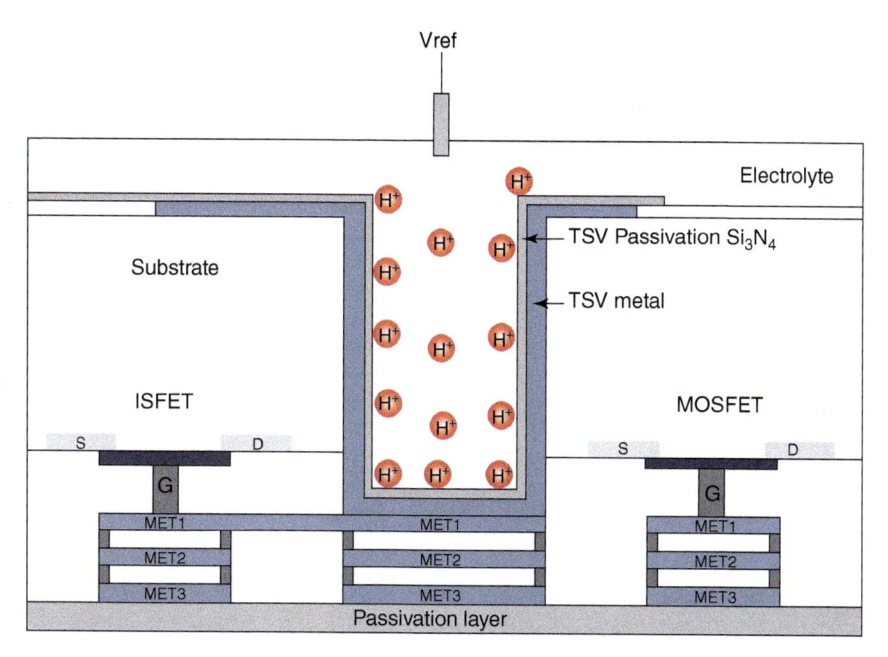

Figure 20.43 Hollow TSV for expanding the gate surface of an ISFET sensor. Source: Xiao et al. [156]. © 2017 IEEE.

so a second anodic bonding is difficult to be performed. By fabricating TSVs in the Pyrex 1, the supply voltage can be applied to the silicon wafer through the TSVs during anodic bonding of Pyrex 2.

Figure 20.43 shows an ion-sensitive field-effect transistor (ISFET) pH sensor for DNA sequencing [156]. The channel currents of the ISFET change as a result of the gate voltage variations induced by the [H$^+$], which are released by incorporation of known deoxynucleotides with shorter strands broken from target DNA at the passivation layer, which is capacitively coupled to the floating gate [156]. A hollow TSV connected to the ISFET metal 1 significantly increases the gate surface of the ISFET, i.e. the sensing layer by using the sidewall surface of the TSV, and thus significantly improves the sensitivity of the ISFET.

Most hollow metal TSVs are fabricated by depositing metals on the conductive bottoms and the sidewalls of through-holes; it is also possible to metalize the bottoms and the sidewalls of bland-vias followed by backside thinning to expose the metals, as shown in Figure 20.44 [157], just like the processes for fabricating solid metal TSVs. Although hollow metal TSVs feature a metal film conductor and a hollow space in the TSV center, the hollow space can be filled with polymer. Small interposers fabricated with hollow TSVs can also been embedded into a polymer substrate, as shown in Figure 20.45 [158], so-called chip-in-polymer. By using the TSVs fabricated in the interposers, the double sides of the polymer substrate can be electrically connected. This technique shows attractive advantages of low cost, high yield, and flexibility to integrate silicon dies fabricated with different technologies and dimensions.

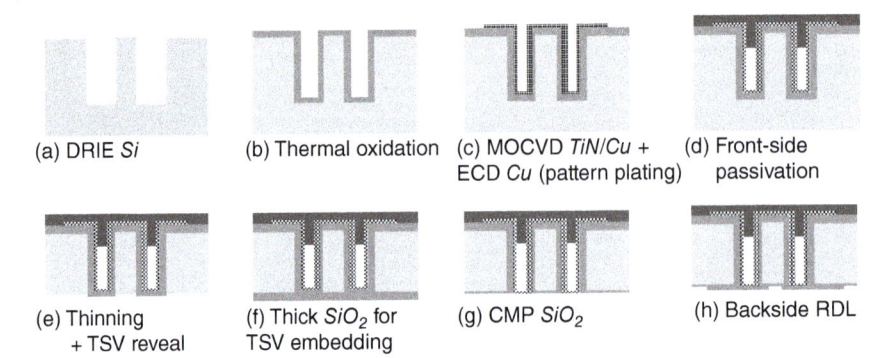

(a) DRIE Si
(b) Thermal oxidation
(c) MOCVD TiN/Cu + ECD Cu (pattern plating)
(d) Front-side passivation
(e) Thinning + TSV reveal
(f) Thick SiO_2 for TSV embedding
(g) CMP SiO_2
(h) Backside RDL

Figure 20.44 Double side processes for fabricating hollow metal TSVs. Source: Hofmann et al. [157].

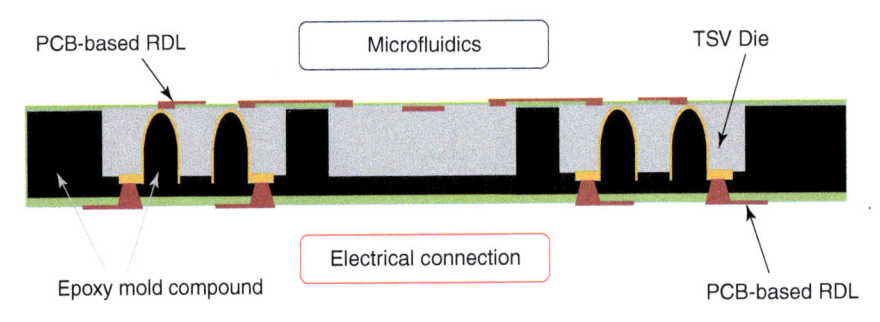

Figure 20.45 Chip-in-polymer: polymer molded TSV interposers and biosensors. Source: Silex Microsystems [158].

20.7.3 Air-gap Metal TSVs

Although air-gap metal TSVs can be fabricated by using a sacrificial polymer and Cu electroplating techniques [61], such a fabrication method is not cost-effective in MEMS applications. Instead of metal plating, TSV conductors can be formed by using wire bonding on the metalized bottom of through-holes. This method avoids TSV filling and thus save processing costs but at an expanse of wasting MEMS wafer areas because the hole diameters should be large enough for wiring bonding.

The holes can be either in the cap wafer, as shown in Figure 20.46a, or in the MEMS wafer, as shown in Figure 20.46b [159]. If the through-vias are etched before bonding, the cap wafer and the MEMS wafer can be bonded with aligning, such that the metal pads deposited on the MEMS layer can locate in the through-vias. Then metal wires are bonded on the metal pads to act as conductors. If the through-vias are etched after wafer bonding, a recess etched on the spacer layer between the substrate and the MEMS layer is needed, as shown in Figure 20.46c [159], such that the metal pads deposited on the bottom of the via is disconnected to the substrate. Figure 20.46d shows a vacuum packaging using two types of TSVs. Solid metal TSVs are fabricated in the MEMS wafer to conduct the MEMS to the metal pads, which

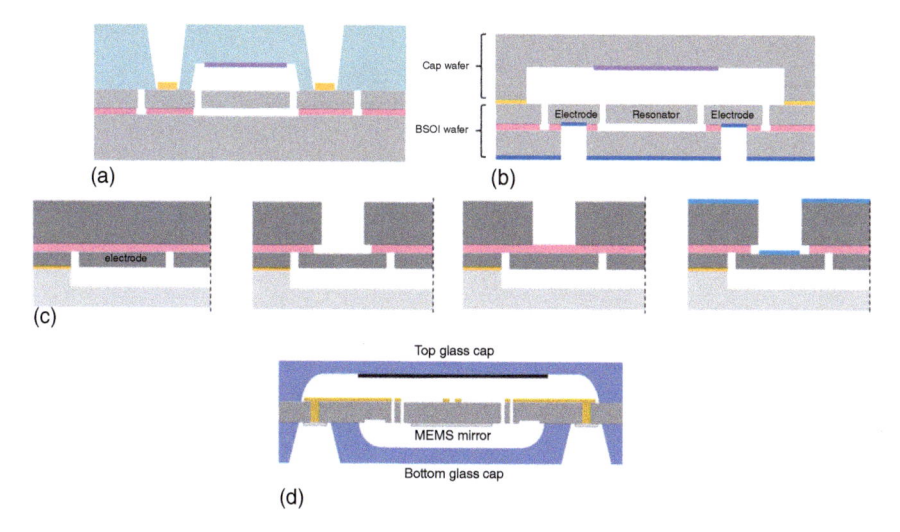

Figure 20.46 Air-gap metal TSVs using bonding wires as the conductors. (a) TSVs in the cap wafer, (b) TSVs in the MEMS wafer, (c) fabrication processes, and (d) TSVs in the MEMS wafer and in the bottom cap glass wafers. Source: (a–c) Nicolas et al. [159] © 2014 IEEE, (d) Chung et al. [160]. © 2019 Elsevier.

are then connected outside using air-gap metal wire TSVs fabricated in the bottom cap wafer.

References

1 Hirama, I. (2015). New MEMS sensor process by TSV technology for smaller packaging, IEEE International Conference on Electronics Packaging iMAPS All Asia Conference, 456–459.
2 Esashi, M. (2012). Revolution of sensors in micro-electromechanical systems. *Jpn. J. Appl. Phys.* 51: 080001.
3 Goldberg, H.D, Breuer, K.S., Schmidt, M.A.. et al. (1994). A silicon-wafer bonding technology for microfabricated shear-stress sensor with backside contacts. Technical Digest Solid-State Sensor and Actuator Workshop, 111–115.
4 Henmi, H., Shoji, S., Shoji, Y. et al. (1994). Vacuum packaging for microsensors by glass-silicon anodic bonding. *Sens. Actuat. A* 43: 243–248.
5 Linder, S, Baltes, H., Gnaedinger, F.. et al. (1994). Fabrication technology for wafer through-hole interconnections and three-dimensional stacks of chips and wafers. IEEE Micro Electro Mechanical Systems Conference, 349–354.
6 Jono, K., Minami, K., Esashi, M. et al. (1995). An electrostatic servo-type three-axis silicon accelerometer. *Meas. Sci. Technol.* 6: 11–15.
7 Ruhl, G., Fröschle, B., Ramm, P. et al. (1995). Deposition of titanium nitride/tungsten layers for application in vertically integrated circuits technology. *Appl. Surf. Sci.* 91: 382–387.

8 Ramm, P., Bollmann, D., Braun, R. et al. (1997). Three dimensional metallization for vertically integrated circuits. *Microelect. Eng.* 37/38: 39–47.

9 Kurino, H, Fukushima, T., Tanaka, T.. et al. (1997). Three-dimensional integration technology for real time micro-vision system. IEEE International Conference Innovative Systems in Silicon, 203–212.

10 Matsumoto, T., Fukushima, T., Tanaka, T. et al. (1998). New three-dimensional wafer bonding technology using the adhesive injection method. *Jpn. J. Appl. Phys.* 37: 1217–1221.

11 Soh, H., Yue, C., McCarthy, A. et al. (1999). Ultra-low resistance, through-wafer via technology and its application in three dimensional structures on silicon. *Jpn. J. Appl. Phys.* 1 (38): 2393–2396.

12 Chow, E., Soh, H.T., Lee, H.C. et al. (1999). Two-dimensional cantilever arrays with through-wafer interconnects. *Transducers*: 1886–1887.

13 Calmes, S, Cheng, C.-H., Degertekin, F.L.. et al. (1999). Highly integrated 2-D capacitive micromachined ultrasonic transducers. IEEE International Ultrasonics Symposium, 1163–1166.

14 Ok, S., Kim, C., Baldwin, D. et al. (2003). High density, high aspect ratio through-wafer interconnect vias for MEMS packaging. *IEEE Trans. Adv. Packag.* 26: 302–309.

15 Mehra, A., Zhang, X., Ayon, A. et al. (2000). Through-wafer electrical interconnect for multilevel microelectromechnical system devices. *J. Vac. Sci. Technol B.* 18: 2583–2589.

16 Rimskog, M. (2007). Through wafer via technology for MEMS and 3D integration. IEEE International Electronics Manufacturing Technology Symposium, 286–289.

17 Small, M., Ruby, R., Ortiz, S. et al. (2011). Wafer-scale packaging for FBAR-based oscillators. Joint Conference IEEE International Frequency Control European Frequency and Time Forum, 1–4.

18 Takahashi, K., Terao, H., Tomita, Y. et al. (2001). Current status of research and development for three-dimensional chip stack technology. *Jpn. J. Appl. Phys.* 40: 3032–3037.

19 Wu, J, del Alamo, J.A., Jenkins, K.A.. et al. (2000). A high aspect-ratio silicon substrate-via technology and applications. IEEE IEDM, 477–480.

20 Sasaki, K, Matsuo, M., Hayasaka, N.. et al. (2001). 128Mbit NAND flash memory by chip-on-chip technology with Cu through plug. International Conference on Electronics Packaging, 3943-3947.

21 Lietaer, N., Storås, P., Breivik, L. et al. (2006). Development of cost-effective high-density through-wafer interconnects for 3D microsystems. *J. Micromech. Microeng.* 16: S29–S34.

22 Ebefors, T, Fredlund, J., Perttu, D.. et al. (2013). The development and evaluation of RF TSV for 3D IPD applications, IEEE International 3D System Integration Conference 1–8.

23 Ko, S.C., Min, B.-G., Park, Y.-R. et al. (2013). Micromachined stress-free TSV hole for AlGaN/GaN-on-Si (111) platform-based devices. *J. Micromech. Microeng.* 23: 035011.

24 Kraft, J, Schrank, F., Teva, J.. et al. (2011). 3D sensor application with open through silicon via technology. IEEE Electronic Components and Technology Conference, 560–566.

25 Hofmann, L, Schubert, I., Gottfried, K.. et al. (2013). Investigations on partially filled HAR TSVs for MEMS applications. IEEE International Interconnect Technology Conference, 1–3.

26 Lietaer, N, Summanwar, A., Herum, S.R.. et al. (2014). Dry-film resist technology for versatile TSV fabrication for MEMS, tested on blind dummy TSVs. Symposium On Design, Test, Integeration and Packaging of MEMS/MOEMS, 1–5.

27 Zhao, H. and Wang, Z. (2012). Motion measurement using inertial sensors, ultrasonic sensors, and magnetometers with extended Kalman filter for data fusion. *IEEE Sens. J.* 12: 943–953.

28 Esashi, M. (2008). Wafer level packaging of MEMS. *J. Micromech. Microeng.* 18: 073001.

29 Hofmann, L., Fischer, T., Werner, T. et al. (2016). Study on TSV isolation liners for a Via Last approach with the use in 3D-WLP for MEMS. *Microsyst. Technol.* 22: 1665–1677.

30 Li, D., Joshi, S., Kim, J.-H. et al. (2017). End-to-end analysis of integration for thermocouple-based sensors into 3-D ICs. *IEEE Trans. Very Large Scale Integrated Syst.* 25: 2498–2511.

31 Laermer, F, Schilp, A. (1996). Method for anistropic plasma etching of substrates, US patent 5498312.

32 Tachi, S., Tsujimoto, K., and Okudaira, S. (1988). Low-temperature reactive ion etching and microwave plasma etching of silicon. *Appl. Phys. Lett.* 52: 616–618.

33 Morikawa, Y, Akazawa, M., Yamada, S.. et al. (2017). High-density via fabrication technology solution for heterogeneous integration. Pan Pacific Microelectronics Symposium, 1–6.

34 Wu, B., Kumar, A., and Pamarthy, S. (2010). High aspect ratio silicon etch: a review. *J. Appl. Phys.* 108: 051101.

35 Lin, P.R., Zhang, G.Q., van Zeijl, H.W. et al. (2015). Effects of silicon via profile on passivation and metallization in TSV interposers for 2.5D integration. *Microelect. Eng* 134: 22–26.

36 Ranganathan, N., Lee, D.Y., Liu, Y. et al. (2011). Influence of bosch etch process on electrical isolation of TSV structures. *IEEE Trans. Comp. Pack. Manuf. Tech.* 1: 1497–1507.

37 Ramaswami, S., Dukovic, J., Eaton, B. et al. (2009). Process integration considerations for 300 mm TSV manufacturing. *IEEE Trans. Dev. Mat. Reliab.* 9: 524–528.

38 Teh, W.H., Caramto, R., Arkalgud, S. et al. (2010). 300-mm production-worthy magnetically enhanced non-Bosch through-si-via etch for 3-D logic integration. *IEEE Trans. Adv. Semicond. Manuf.* 23: 293–302.

39 Gomez, S. and Belen, R.J. (2004). Etching of high aspect ratio structures in Si using SF_6/O_2 plasma. *J. Vac. Sci. Tech. A* 22: 606–615.

40 Ranganathan, N., Liao, E.B., Linn, L. et al. (2009). Integration of high aspect ratio tapered silicon via for silicon carrier fabrication. *IEEE Trans. Adv. Packag.* 32: 62–71.
41 https://www.ulvac.co.jp/products_e/equipment/products/etching-system/nld-5700
42 Morikawa, Y, Murayama, T., Sakuishi, T.. et al. (2012). A novel scallop free TSV etching method in magnetic neutral loop discharge plasma. IEEE Components and Technology Conference (ECTC), 794–795.
43 Kamto, A., Divan, R., Sumant, A.V. et al. (2010). Cryogenic inductively coupled plasma etching for fabrication of tapered through-silicon vias. *J. Vac. Sci. Tech. A* 28: 719–725.
44 Ranganathan, N., Ranganathan, N., Lee, D.Y. et al. (2008). The development of a tapered silicon micro-micromachining process for 3D microsystems packaging. *J. Micromech. Microeng.* 18: 115028.
45 Tezcan, D.S, De Munck, K., Pham, N.. et al. (2006). Development of vertical and tapered via etch for 3D through wafer interconnect technology. IEEE Electronics Packaging Technology Conference, 22–28.
46 Dubey, A. and Yadava, V. (2008). Experimental study of Nd:YAG laser beam machining-an overview. *J. Mat. Process. Tech.* 195: 15–26.
47 Tan, B. (2006). Deep micro hole drilling in a silicon substrate using multi-bursts of nanosecond UV laser pulses. *J. Micromech. Microeng.* 16: 109–122.
48 Tang, C., Young, H.T., Li, K.M. et al. (2012). Innovative through-silicon-via formation approach for wafer-level packaging applications. *J. Micromech. Microeng.* 22: 045019.
49 Le, V.N.-A. (2017). Investigation on drilling blind via of epoxy compound wafer by 532 nm Nd:YVO$_4$ laser. *J. Manuf. Proc.* 27: 214–220.
50 Grob, T, Grob, T., Hovenkamp, R.A.. et al. (2004). Comparison of via-fabrication techniques for through-wafer electrical interconnect applications. IEEE Electronic Components and Technology Conference, 1466–1470.
51 Rieske, R, Landgraf, R., Wolter, K.-J.. et al. (2009). Novel method for crystal defect analysis of laser drilled TSVs. IEEE Electronic Components and Technology Conference, 1139–1146.
52 Laakso, P., Penttilä, R., Heimala, P. et al. (2010). Effect of shot number on femtosecond laser drilling of silicon. *J. Laser Micro/Nanoeng.* 5: 273–276.
53 Rodin, A.M., Callaghan, J., and Brennan, N. (2008). High throughput low CoO industrial laser drilling tool. 4th International Conference Exhibition Device Packaging, Arizona, USA http://www.emc3d.org/documents/ library/technical/XSil_Laser_Drilling_A_Rodin.pdf
54 Ramm, P, Wolf, M.J., Klumpp, A.. et al. (2008). Through silicon via technology-processes and reliability for wafer-level 3D system integration. IEEE Electronic Components and Technology Conference, 841 846.
55 Van Olmen, J., Huyghebaert, C., Coenen, J. et al. (2011). Integration challenges of copper Through Silicon Via (TSV) metallization for 3D-stacked IC integration. *Microelect. Eng.* 88: 745–748.
56 Suu, K. (2016). High-density packaging technology solution for smart ICT. Pan Pacific Microelectronics Symposium, 1–6.

57 Li, Y.L., Van Huylenbroeck, S., Roussel, P. et al. (2016). Dielectric liner reliability in via-middle through silicon vias with 3 micron diameter. *Microelect. Eng.* 156: 37–40.

58 Chen, Q., Huang, C., Tan, Z. et al. (2013). Low capacitance through-silicon-vias (TSVs) with uniform benzocyclobutene (BCB) insulation layers. *IEEE Trans. Comp. Packag. Manuf. Technol.* 3: 724–731.

59 Civale, Y., Tezcan, D.S., Philipsen, H.G.G. et al. (2011). 3-D wafer-level packaging die stacking using spin-on-dielectric polymer liner through-silicon vias. *IEEE Trans. Comp. Packag. Manuf. Technol.* 1: 833–840.

60 Wang, W.J., Yan, Y.Y., Ding, Y.T. et al. (2015). Electrical characteristics of a novel interposer technique using ultra-low-resistivity silicon-pillars with polymer insulation as TSVs. *Microelect. Eng.* 137: 146–152.

61 Huang, C., Chen, Q., Wang, Z. et al. (2013). Air-gap through-silicon vias (TSVs). *IEEE Elect. Dev. Lett.* 34: 441–443.

62 Agarwal, A, Murthy, R.B., Lee, V.. et al. (2009). Polysilicon Interconnections (FEOL): fabrication and characterization. IEEE Electronics Packaging Technology Conference, 317–320.

63 Dixit, P., Vehmas, T., Vähänen, S. et al. (2012). Fabrication and electrical characterization of high aspect ratio poly-silicon filled through-silicon vias. *J. Micromech. Microeng.* 22: 055021.

64 Pares, G, Bresson, N., Moreau, S.. et al. (2010). Effects of stress in polysilicon via - first TSV technology. IEEE Electronics Packaging Technology Conference, 333–337.

65 Tomozeiu, N., Antohe, S., and Modreanu, M. (2000). Electrical properties of LPCVD polysilicon deposited in the vicinity of amorphous polycrystalline phase. *J. Optoelect. Adv. Mat* 2: 657–663.

66 Deen, M.J., Rumyantsev, S., Orchard-Webb, J. et al. (1998). Low frequency noise in heavily doped polysilicon thin film resistors. *J. Vac. Sci. Technol. B* 16: 1881–1884.

67 Liu, F, Yu, R.R., Young, A.M.. et al. (2008). A 300-mm wafer-level three-dimensional integration scheme using tungsten through-silicon via and hybrid Cu-adhesive bonding. IEEE Electron Devices Meeting, 1–4.

68 Kikuchi, H, Yamada, Y., Mossad Ali, A.. et al. (2007). Tungsten Through-Si Via (TSV) technology for three-dimensional LSIs. International Conference on Solid State Devices Materials, 482–483.

69 Wieland, R., Bonfert, D., Klumpp, A. et al. (2005). 3D Integration of CMOS transistors with ICV-SLID technology. *Microelect. Eng.* 82: 529–533.

70 Ramm, P., Bonfert, D., Gieser, H. et al. (2001). InterChip via technology for vertical system integration. IEEE International Interconnect Technology Conference, 160–162.

71 Koyanagi, M., Nakamura, T., Yamada, Y. et al. (2006). Three-dimensional integration technology based on wafer bonding with vertical buried interconnections. *IEEE Trans. Elect. Dev.* 53: 2799–2808.

72 Dao, T, Triyoso, H., Mora, R.. et al. (2010). Thermo-mechanical stress characterization of tungsten-fill through-silicon-via, VLSI Design Auto. Test Symposium, 7–10.

73 Shen, W.-W. and Chen, K.-N. (2017). Three-dimensional integrated circuit (3D IC) key technology: through-silicon via (TSV). *Nanoscale Res. Lett.* 12: 56.

74 Garrou, P., Bower, C., Ramm, P. et al. (2008). *Handbook of 3D Integration*. Wiley.

75 Civale, Y., Croes, K., Miyamori, Y. et al. (2013). On the thermal stability of physically-vapor-deposited diffusion barriers in 3D Through-Silicon Vias during IC processing. *Microelect. Eng.* 106: 155–159.

76 Xiao, H., He, H., Ren, X. et al. (2017). Numerical modeling and experimental verification of copper electrodeposition for through silicon via (TSV) with additives. *Microelect. Eng.* 170: 54–58.

77 Song, C., Wang, Z., Tan, Z. et al. (2012). Moving boundary simulation and experimental verification of high aspect-ratio through-silicon-vias for 3D integration. *IEEE Trans. Comp. Packag. Manuf. Technol.* 2: 23–31.

78 Beica, R, Sharbono, C., Ritzdorf, T. (2008). Through silicon via copper electrodeposition for 3D integration. IEEE Electronics Components and Technology Conference, 577–583.

79 Radisic, A., Lühn, O., Philipsen, H.G.G. et al. (2011). Copper plating for 3D interconnects. *Microelect. Eng.* 88: 701–704.

80 Hwang, G, R. Kalaiselvan (2017). Development of TSV electroplating process for via-last technology. IEEE Electronics Components and Technology Conference, 68–72.

81 Abbaspour, R., Brown, D.K., and Bakir, M.S. (2017). Fabrication and electrical characterization of sub-micron diameter through-silicon via for heterogeneous three-dimensional integrated circuits. *J. Micromech. Microeng.* 27: 025011.

82 Zhang, D., Smith, D., Kumarapuram, G. et al. (2015). Process development and optimization for 3 μm high aspect ratio via-middle through-silicon vias at wafer level. *IEEE Trans. Semicond. Manuf.* 28: 454–460.

83 Dixit, P. and Miao, J.M. (2006). Aspect-ratio-dependent copper electrodeposition technique for very high aspect-ratio through-hole plating. *J. Electrochem. Soc.* 153: G552–G559.

84 Wang, Z., Wang, L., Nguyen, N.T. et al. (2006). Silicon micromachining of high aspect ratio, high-density through-wafer electrical interconnects for 3-D multichip packaging. *IEEE Trans. Adv. Packag.* 29: 615–622.

85 Song, C., Wang, Z., Chen, Q. et al. (2008). High aspect ratio copper through-silicon-vias for 3D integration. *Microelect. Eng.* 85: 1952–1956.

86 Chang, H.H, Shih, Y.C., Hsiao, Z.C.. et al. (2009). 3D stacked chip technology using bottom-up electroplated TSVs. IEEE Electronics Components and Technology Conference, 1177–1180.

87 Eun, C.K., Luo, X., Wang, J.-C. et al. (2014). A microdischarge-based monolithic pressure sensor. *IEEE J. Microelectromech. Syst.* 23: 1300–1310.

88 Zervas, M, Temiz, Y., Leblebici, Y.. et al. (2010). Fabrication and characterization of wafer-level deep TSV arrays. IEEE Electronics Components and Technology Conference (ECTC), 1625–1630.

89 Moffat, T.P. and Josell, D. (2012). Extreme bottom-up superfilling of through-silicon-vias by damascene processing: suppressor disruption, positive feedback and turing patterns. *J. Electrochem. Soc.* 159: D208–D216.

90 Hoang, V.-H. and Kondo, K. (2017). Acceleration kinetic of copper damascene by chloride, SPS, and cuprous concentration computation in TSV filling. *J. Electrochem. Soc.* 164: D564–D572.

91 Hofmann, L., Ecke, R., Schulz, S.E. et al. (2011). Investigations regarding through silicon via filling for 3D integration by periodic pulse reverse plating with and without additives. *Microelect. Eng.* 88: 705–708.

92 Tian, Q., Cai, J., Zheng, J. et al. (2016). Copper pulse-reverse current electrodeposition to fill blind vias for 3-D TSV integration. *IEEE Trans. Comp. Packag. Manuf. Technol.* 6: 1899–1904.

93 Zhu, Q.S., Zhang, X., Liu, C.Z. et al. (2018). Effect of reverse pulse on additives adsorption and copper filling for through silicon via. *J. Electrochem. Soc.* 166: D3006–D3012.

94 Hayashi, T., Kondo, K., Saito, T. et al. (2011). High-speed through silicon via (TSV) filling using diallylamine additive. *J. Electrochem. Soc.* 158: D715–D718.

95 Willey, M.J. and West, A.C. (2007). SPS adsorption and desorption during copper electrodeposition and its impact on PEG adsorption. *J. Electrochem. Soc.* 154: D156–D162.

96 Dow, W.-P., Yen, M.-Y., Lin, W.-B. et al. (2005). Influence of molecular weight of Polyethylene Glycol on microvia filling by copper electroplating. *J. Electrochem. Soc.* 152: C769–C775.

97 Tsai, T.-H. and Huang, J.-H. (2011). Electrochemical investigations for copper electrodeposition of through-silicon via. *Microelect. Eng.* 88: 195–199.

98 Cao, Y., Taephaisitphongse, P., Chalupa, R. et al. (2001). Three-additive model of superfilling of copper. *J. Electrochem. Soc.* 148: C466–C472.

99 Tantavichet, N. and Pritzker, M. (2006). Copper electrodeposition in sulphate solutions in the presence of benzotriazole. *J. Appl. Electrochem.* 36: 49–61.

100 Beica, R, Siblerud, P., Sharbono, C.. et al. (2008). Advanced metallization for 3D integration. IEEE Electronics Packaging Technology Conference, 212–218.

101 Moffat, T.P. and Yang, L.-Y.O. (2010). Accelerator surface phase associated with superconformal Cu electrodeposition. *J. Electrochem. Soc.* 157: D228–D241.

102 Fischer, A.C., Lapisa, M., Roxhed, N. et al. (2010). Selective electroless nickel plating on oxygen-plasma-activated gold seed-layers for the fabrication of low contact resistance vias and microstructures. IEEE Micro Electro Mechanical Systems Conference, 472–475.

103 Laakso, M.J., Bleiker, S.J., Liljeholm, J. et al. (2018). Through-glass vias for glass interposers and MEMS packaging applications fabricated using magnetic assembly of microscale metal wires. *IEEE Access* 6: 44306–44317.

104 Fischer, A.C., Bleiker, S.J., Haraldsson, T.K. et al. (2012). Very high aspect ratio through-silicon vias (TSVs) fabricated using automated magnetic assembly of nickel wires. *J. Micromech. Microeng.* 22 (10): 105001.

105 Ko, Y.-K., Fujii, H.T., Sato, Y.S. et al. (2012). High-speed TSV filling with molten solder. *Microelect. Eng.* 89: 62–642.

106 Gu, J., Pike, W., and Karl, W.J. (2009). A novel capillary-effect-based solder pump structure and its potential application for through-wafer interconnection. *J. Micromech. Microeng.* 19 (7): 074005.

107 Saint-Patrice, D, Jacquet, F., Bridoux, C.. et al. (2011). Ultra low cost wafer level via filling and interconnection using conductive polymer. IEEE Electronics Components and Technology Conference (ECTC), 1711–1716.

108 Shih, K., Nimura, M., Kanehira, Y. et al. (2013). Simple through silicon interconnect via fabrication using dry filling of sub-micron Au particles for 3D MEMS. *IEEE Micro Elect. Mech. Syst. Conf.*: 299–302.

109 Alfaro, J.A., Sberna, P.M., Silvestri, C. et al. (2018). Vacuum assisted liquified metal (VALM) TSV filling method with superconductive material. *IEEE Micro Electro Mech. Syst. Conf.*: 547–550.

110 Dubin, V.M., Lisunova, M.O., Walton, B.L. et al. (2017). Invar electrodeposition for controlled expansion interconnects. *J. Electrochem. Soc.* 164: D321–D326.

111 Fischer, A.C., Roxhed, N., Stemme, G. et al. (2010). Low-cost through silicon vias (TSVs) with wire-bonded metal cores and low capacitive substrate-coupling. *IEEE Micro Elect. Mech. Syst. Conf.*: 480–483.

112 http://www.teledynedalsa.com/semi/mems/toolbox/

113 Lietaer, N., Taklo, M.V., Schjølberg-Henriksen, K. et al. (2010). 3D interconnect technologies for advanced MEMS/NEMS applications. *ECS Trans.* 25: 87–95.

114 Lietaer, N., Summanwar, A., Bakke, T. et al. (2010). TSV development for miniaturized MEMS acceleration switch, IEEE 3D Syst. *Integr. Conf.*: 1–4.

115 Ikegami, N, Yoshida, T., Kojima, A.. et al. (2016). Fabrication of through silicon via with highly phosphorus-doped polycrystalline Si plugs for driving an active-matrix nanocrystalline Si electron emitter array. 11th IEEE Annual International Conference on Nano/Micro Engineered and Molecular Systems (IEEE NEMS), 578–582.

116 Griffin, B.A., Chandrasekaran, V., and Sheplak, M. (2012). Thermoelastic ultrasonic actuator with piezoresistive sensing and integrated through-silicon vias. *IEEE J. Microelectromech. Syst.* 21: 350–358.

117 Midtbø, K., Rønnekleiv, A., Ingebrigtsen, K.A. et al. (2012). High-frequency CMUT arrays with phase-steering for in vivo ultrasound imaging. *IEEE Sens. Conf.*: 1–5.

118 Wu, G., Han, B., Cheam, D.D. et al. (2019). Development of six-degree-of-freedom inertial sensors with an 8-in advanced MEMS fabrication platform. *IEEE Trans. Industrial Elect.* 66: 3835–3842.

119 Kirsten, M., Wenk, B., Ericson, F. et al. (1995). Deposition of thick doped polysilicon films with low stress in an epitaxial reactor for surface micromachining applications. *Thin Solid Films* 259: 181–187.

120 Ayanoor-Vitikkate, V., Chen, K.-L., and Park, W.-T. (2009). Development of wafer scale encapsulation process for large displacement piezoresistive MEMS devices. *Sens. Actuat. A* 156: 275–283.

121 Kuisma, H. (2014). Glass isolated TSVs for MEMS. IEEE Electronics Systems-Integration Technology Conference, 1–5.

122 Himes, P. (2013). Vertical through-wafer insulation: enabling integration and innovation. *Solid State Technol.* 56: 13–17.

123 Jeong, Y., Serrano, D.E., and Ayazi, F. (2018). A wide-bandwidth tri-axial pendulum accelerometer with fully-differential nano-gap electrodes. *J. Micromech. Microeng.* 28: 115007.

124 Merdassi, A., Allan, C., Harvey, E.J. et al. (2017). Capacitive MEMS absolute pressure sensor using a modified commercial microfabrication process. *Microsyst. Technol.* 23: 3215–3225.

125 Marx, D.L, C. Acar, S. Akkaraju et al. (2010). Micromachined devices and fabricating the same. US Patent 8,710,599 2010.3.8.

126 Acar, C. (2016). High-performance 6-Axis MEMS inertial sensor based on through-silicon via technology. IEEE International Symposium on Inertial Sensors Systems, 62–65.

127 Merdassi, A., Kezzo, M.N., Xereas, G. et al. (2015). Wafer level vacuum encapsulated tri-axial accelerometer with low cross-axis sensitivity in a commercial MEMS process. *Sens. Actuat. A* 236: 25–37.

128 Castoldi, L. The MEMS revolution. http://www.semi.org/eu/sites/semi.org/files/docs/STM.pdf

129 Grieco, B., Ausilio, D., Banfi, F. et al. (2004). A low-g 3 axis accelerometer for emerging automotive applications. In: *Advanced Microsystems for Automotive Applications*, 211–222. Springer.

130 Zhuang, X., Ergun, A.S., Huang, Y. et al. (2007). Integration of trench-isolated through-wafer interconnects with 2D capacitive micromachined ultrasonic transducer arrays. *Sens. Actuat. A* 138: 221–229.

131 Wodnicki, R., Woychik, C.G., Byun, A.T.. et al. (2009). Multi-row linear cMUT array using cMUTs and multiplexing electronics. IEEE International Ultrasonics Symposium, 2696-2699.

132 Bergmann, Y., Reinmuth, J., Will, B.. et al. (2013). Integration of a new through silicon via concept in a microelectronic pressure sensor. 14th International Conference Thermal, Mechanical, & Multi-Physics Simulation and Experiments in Microelectronics and Microsystems, 1-5.

133 http://www1.semi.org/eu/sites/semi.org/files/images/Eric%20Mounier%20-%20Future%20of%20MEMS.%20A%20Market%20and%20Technologies%20Perspective.pdf

134 Le, H.T., Mizushima, I., Nour, Y. et al. (2018). Fabrication of 3D air-core MEMS inductors for very-high-frequency power conversions. *Microsyst. Nanoeng.* 4: 17082.

135 Zhang, X., Yamaner, F.Y., Oralkan, Ö. et al. (2017). Fabrication of vacuum-sealed capacitive micromachined ultrasonic transducers with through-glass-via interconnects using anodic bonding. *IEEE J. Microelectromech. Syst.* 26: 226–234.

136 Kurtz, A.D, Ned, A.A., Epstein, A.H.. et al. (2004). Ultra high temperature, miniature, SOI sensors for extreme environments, IMAPS International HiTEC Conference, 1-11.

137 Niklaus, F., Decharat, A., Forsberg, F. et al. (2009). Wafer bonding with nano-imprint resists as sacrificial adhesive for fabrication of silicon-on-integrated-circuit (SOIC) wafers in 3D integration of MEMS and ICs. *Sens. Actuat. A* 154: 180–186.

138 Zimmer, F., Lapisa, M., Bakke, T. et al. (2011). One-megapixel monocrystalline-silicon micromirror array on CMOS driving electronics manufactured with very large-scale heterogeneous integration. *IEEE J. Microelectromech. Syst.* 20: 564–572.

139 Xue, X., Xiong, H., Song, Z. et al. (2019). Silicon diode uncooled focal plane array with three-dimensional integrated CMOS readout circuits. *IEEE Sens. J.* 19 (2): 426–434.

140 Esashi, M. and Tanaka, S. (2016). Stacked integration of MEMS on LSI. *Micromachines* 7: 137.

141 Tanaka, S., Park, K.D., and Esashi, M. (2012). Lithium-niobate-based surface acoustic wave oscillator directly integrated with CMOS sustaining amplifier. *IEEE Trans. Ultrason. Ferroelectr. Freq. Contr.* 59: 1800–1805.

142 Matsuo, K., Moriyama, M., Esashi, M.. et al. (2012). Low-voltage PZT-actuated MEMS switch monolithically integrated with CMOS circuit. IEEE International Conference On Micro Electro Mechanical Systems, 1153-1156.

143 Steller, W., Meinecke, C., Gottfried, K.. et al. (2014). SIMEIT-Project: high precision inertial sensor integration on a modular 3D-interposer platform. IEEE Electronics Components and Technology Conference, 1218–1225.

144 Santagata, F., Farriciello, C., Fiorentino, G. et al. (2013). Fully back-end TSV process by Cu electro-less plating for 3D smart sensor systems. *J. Micromech. Microeng.* 23: 055014.

145 Chou, L.-C, Lee, S.-W., Huang, P.-T.. et al., (2014). Integrated microprobe array and CMOS MEMS by TSV technology for bio-signal recording application. IEEE Electronic Components and Technology Conference, 512–517.

146 Lam, K.-T., Chen, Y.-H., Hsueh, T.-J. et al. (2016). A 3-D ZnO-nanowire smart photo sensor prepared with through silicon via technology. *IEEE Trans. Elect. Dev.* 63: 3562–3566.

147 Charbonnier, J., Henry, D., Jacquet, F.. et al. (2008). Wafer level packaging technology development for CMOS image sensors using through silicon vias. IEEE Electronic System-Integration Technology Conference, 141–148.

148 de Veen, P.J., Bos, C., Hoogstede, D.R. et al. (2015). High-resolution x-ray computed tomography of through silicon vias for RF MEMS integrated passive device applications. *Microelect. Reliab.* 55: 1644–1648.

149 Hata, Y., Suzuki, Y., Muroyama, M. et al. (2018). Integrated 3-axis tactile sensor using quad-seesaw-electrode structure on platform LSI with through silicon vias. *Sens. Actuat. A* 273: 30–41.

150 Hsu, W.-T. (2008). Resonator miniaturization for oscillators. IEEE International Frequency Control Symposium, 392-395.

151 The advantages of integrated MEMS to enable the internet of moving things. http://www.mcubemems.com/wp-content/uploads/2014/06/mCube-Advantages-of-Integrated-MEMS-Final-0614.pdf

152 Ferrandoni, C., Grecoi, F., Lagouttei, E.. et al. (2010). Hermetic wafer-level packaging development for RF MEMS switch. IEEE Electronics System-Integration Technology Conference, 1-6.

153 Gueye, R., Lee, S.W., Akiyama, T. et al. (2013). High-temperature compatible 3D-integration processes for a vacuum-sealed CNT-based NEMS. *Proc. SPIE* 8614: 86140H.

154 Chu, H.M., Sasaki, T., and Hane, K. (2018). Wafer-level vacuum package of two-dimensional micro-scanner. *Microsyst. Technol.* 24: 2159–2168.

155 Chu, H.M., Vu, H.N., Hane, K. et al. (2013). Electric feed-through for vacuum package using double-side anodic bonding of silicon-on-insulator wafer. *J. Electrostat.* 71: 130–133.

156 Xiao, W, Miscourides, N., Georgiou, P. (2017). A novel ISFET sensor architecture using through-silicon vias for DNA sequencing. IEEE International Symposium of Circuits and Systems, 1-4.

157 Hofmann, L., Dempwolf, S., Reuter, D. et al. (2015). 3D integration approaches for MEMS and CMOS sensors based on a Cu through-silicon-via technology and wafer level bonding. *SPIE* 9517: 951709.

158 https://silexmicrosystems.com/2012/04/17/silex-microsystems-met-via-technology-enables-through-mold-via-applications/

159 Nicolas, S, Greco, F., Caplet, S., et al. (2014). High vacuum wafer level packaging for high-value MEMS applications. IEEE Electronics Components and Technology Conference, 1714–1721.

160 Chung, S.-H., Lee, S.-K., Ji, C.-H. et al. (2019). Vacuum packaged electromagnetic 2D scanning micromirror. *Sens. Actuat. A* 290: 147–155.

161 Bauer, T. (2007). High density through wafer via technology, NSTI-Nanotech, 116–119.

Index

a

absolute pressure sensor 30, 49, 54, 396–397, 409, 432, 436
accelerometer ADXL50 99, 111
active matrix electron emitter array 236, 238, 395
actuator 244
adhesive wafer bonding 352
advanced piezoresistive pressure sensor 49
advanced porous silicon membrane (APSM) 42, 49, 434
air-gap insulators 450, 455
air-gap polysilicon TSVs 464
air-gap TSVs 445, 464, 467
AlN Contour Mode Resonators (CMR) 117
amorphous silicon deposition 73
amperometry 226
amplitude-to-phase modulation (AM-PM) phase noise conversion 205
analog-to-digital conversion (ADC) 183
anchor loss 201
anisotropic wet etching 17–19, 25, 29
anodic bonding 10, 20, 392
anodic bonding (chapter 11) 259–288
adhesion 274
charge and voltage distribution 261
distortion 262–263
Fe-Ni-Co alloy 266
GaAs and glass surfaces 266
glass reflow process 274–276
glass-Si anodic bonding 259
influence to circuit 263–265
intermediate thin films 269–271
low temperature co-fired ceramics (LTCC) 269
Mallory bonding 259
variation of 271–274
atomic layer deposition (ALD) 93, 343, 455

b

back-end-of-line (BEOL) 139
ballistic electrons 231
benzocyclobutene (BCB) 224, 228, 236, 338, 382, 455
bimorph thermal actuator 390
binary-weighted capacitor array 187
biochemical sensing 173
bis-(3-sulfopropyl) disulfide (SPS) 458
blade test 282
bond-and etch-back 279
boron-doped diamond (BDD) 224

Index

Bosch process 19
BST varactor 241
bulk micromachining (chapter 2)
    15–48
  active DWP (ADWP) 30, 32
  advanced porous silicon membrane
      (APSM) 42, 49, 434
  anisotropic wet etching 17–18
  anodic bonding 20
  bond and etched back SOI (BESOI)
      22
  cavity SOI technology 27–29
  dissolved wafer process (DWP)
      29–34
  electrochemical etch-stop 18
  empty-space-in silicon (ESS) 40
  EPW (ethylenediamine
      pyrocatechol water) 17
  etching techniques of Si 16
  HF/HNO3/CH3COOH (HNA) 16
  intermediate-layer bonding 20
  KOH etching 18
  micro-holes interetch and sealing
      (MIS) 42
  open circuit potential (OCP) 18
  plasma based etching 19
  sacrificial bulk micromachining
      (SBM) 38
  single-crystal reactive etching and
      metallization (SCREAM)
      process 34
  silicon fusion bonding 20
  silicon on glass processes 29
  silicon on nothing (SON) 40
  SOI MEMS 20–27
  SOI multiuser MEMS process
      (SOIMUMPs) 23
  three-degree-of-freedom (3-DOF)
      accelerometer 25
  TMAH (tetramethylammonium
      hydroxide) 17

Venice process for sensors
    (VENSENS) 41, 437
bump bonding 233
buried channel in MEMS (chapter 19)
    423–442
  application of SON 435–439
  atomic migration 427
  chemical etching reactions 426
  empty space in silicon (ESS) 425
  hydrogen annealing 427, 428
  migration phenomena 432, 435
  monolithic SON technology
  pipe-and plate-shaped cavities
      430
  plate-shaped cavity 431
  porous silicon 434
  silicon etching phenomena 426
  silicon on nothing (SON) 423
  thermal treatment to silicon surface
      426
buried oxide (BOX) 321

C

cantilever piezoresistive sensor 183
capacitance detection circuits 7
capacitive accelerometers 63
capacitive avian influenza virus
    sensor 174
capacitive micromachined ultrasonic
    transducer (CMUT) 176, 468
capacitive pressure sensor 274
capacitive sensing 154
capacitive sensors 184
capacitive transduction 144
capacitive-type pressure sensor 436
capillary force based biosensor 175
cavity pressure 415
cavity SOI technology 27
ceramic quad flatpack (CQFP) 363
chemical mechanical planarization
    (CMP) 115, 193

chemical mechanical polishing
(CMP)   80, 275
chemical vapor deposition (CVD)
   343
chip-in-polymer   480
chip level transfer   236, 239
coefficient of thermal expansion
(CTE)   155, 271, 338
combo sensors   132
compressive stress   70
conductive polymers   459
copper   457
Coriolis force   63
creatinine detection   173
Cu deposition rates   459
cyclic voltammogram   225

d

deep-etch shallow-diffusion process
   32, 33
deep reactive ion etching (DRIE)   19,
   182, 233, 254, 320, 322, 451
deposition   380, 417
depressing thermal-mismatch stress
   49
design of experiment (DoE)   117
diaphragm bending stress   51
digital micromirror device (DMD)
   104
digital offset trimming   186
direct bonded copper (DBC) substrate
   370
direct bonding (chapter 12)
   279–288
  bond and etch back SOI (BESOI)
  high-resolusion transmission
     electron microscope (HRTEM)
     281–284
  hydrophilic wafer bonding
     279–283
  silicon-on-insulator (SOI) wafer
     279

surface activated bonding (SAB)
   283–285
wafer direct bonding   279
wafer fusion bonding   279
direct current (DC) magnetron
   sputtering   313
direct wafer bonding   20
dissolved wafer process (DWP)   29,
   30
distributed tactile sensors   228
dopamine detection   173
doped poly Si   7
dry-etching processes   25
dual damascene-like processes   463

e

elastomeric polymers   335
electrical conductor   450
electrochemical deposition (ECD)
   321, 323
electrochemical etch-stop   19
electrochemical etch-stop technique
   (ECE)   146
electrochemical reaction   410
electrode module   85
electron beam exposure system   235
electron beam lithography   231
electron cyclotron resonance (ECR)
   source   33
electroplating   401
electrostatic bonding   259
electrostatic recoil detection analysis
   263
encapsulation method   381
endoscopic optical coherence
   tomography (EOCT)   368
enhance crystallization   78
enhanced bulk micromachining
   based on MIS process
   (chapter 3)   49–60
  pressure sensor with PS^3 structure
     50

enhanced bulk micromachining based on MIS process (chapter 3) (*contd.*)
    advanced porous silicon in membrane (APSM) 49
    conventional single-unit pressure sensor 53
    cross-unit Wheatstone-bridge 50
    depressing thermal-mismatch stress 49
    diaphragm bending stress 51
    dual-unit pressure sensor 51
    dual-unit sensor 50
    empty-space-in-silicon (ESS) 49
    fabricated dual-unit PS3 sensor 50
    fabricated sensor 52
    Kovar-alloy substrate 51
    micro-holes interetch and sealing (MIS) 49
    silicon-on-nothing (SON) 49
    temperature coefficient of offset (TCO) 50
    P+G integrated sensors 52
    fabrication complexity and cost 52
    in-plane X-axis cantilever-mass accelerometer 53
    PinG configuration 53
    tire pressure monitoring systems (TPMS) 52
    vertical Z-axis cantilever-mass accelerometer 53
    surface mounting technique (SMT) 58
    TUB (thin-film under bulk) 55
epi-seal process 64, 465
epitaxial poly Si surface micromachining (chapter 4) 61–68
    application-specific integrated circuit (ASIC) 64
    bipolar and CMOS (BiCMOS) 62
    capacitance detection circuit 61
    dual thick poly Si 66
    epi-poly Si 61
    low-pressure chemical vapor deposition (LPCVD) 61
    MEMS devices 61–67
etching techniques 16
eutectic bonding 289

f

fabrication process 4, 226, 229, 274, 385, 394, 397, 417
femtosecond laser 391, 392
field assisted bonding 259
film bulk acoustic resonator (FBAR) 475
flip-chip bonding 121
flip-chip bump-bonding process 123
force feedback loop 186
four piezo-resistors 437
front-end-of-line (FEOL) 139

g

gap measurement 265
gas sensors 180
germanium content 70
germanium sacrificial layer Ge(SA) 85
glass bonding process 397
glass frit 8
gold wire bonding 104

h

hermetically sealed electrical feedthrough conductor 398
hermeticity testing 417
hermetic packaging 409
hermetic sealing 64, 397

heterogeneously integrated aluminum nitride MEMS resonators and filters (chapter 7) 113–130
  aluminum nitride (AlN) 113
  bandwidth (BW) 123
  CMOS electronics 114
  contour mode resonators (CMRs) 117
  design of experiment (DoE) 117
  encapsulation process 116–118
  filter frequency responses 119–121
  finite element analysis (FEA) 119
  flip-chip bonding 121–123
  free-free beam (FFB) 144
  hybrid integration 114
  insertion loss 123
  out of band-rejection (OBR) 123
  probability density function (PDF) 124
  thin-film encapsulation (TFE) 114
  process flow 115–116
  redistribution layer (RDL) 114
  self-healing filters 123
  statistical element selection (SES) 123–12
  thin film bulk acoustic resonator (TFBAR) 113
  thin film encapsulation (TFE) 114
  3D hybrid integrated chip stack 124–126
  universal bump metallization (UBM) 118
  repeatable physical vapor deposition techniques 113
heterojunction bipolar junction transistors 70
high aspect ratio 19, 33, 34, 40, 147, 322, 347, 390

high density plasma (HDP) oxide 85
high-resolution transmission electron microscope (HRTEM) 281
hollow TSVs 445
hybrid polymers 335
hydrogen annealing 428, 435
hydrogen dilution 72
hydrophilic bonding 285
hydrophilic wafer bonding 279, 281

i

implantable telemetry capsule 400
inductive coupled plasma source 19
inductive heating 368
inductive sensors 188
inductively coupled plasma (ICP) 19, 322
infrared (IR) sensors 156
insulated-gate bipolar transistor (IGBT) 370
integrated capacitive pressure sensor 378
integrated capacitive sensors 264
integrated reactive material systems (iRMS) 309
Interconnection methods 10
interdiffusion coefficient 298
interdigital fringing electrodes 178
interdigital sensing capacitors 172
interdigital sensing electrodes 174
intermediate-layer bonding 20
intermediate thin films 269
isotropic etching method 35
isotropic phenomenon 433

j

Janus Green B (JGB) 458

k

Kovar-alloy substrate 51

l

laser ablation   452, 388, 390
laser–recrystallized piezoresistive micro-diaphragm pressure sensor   381
laser soldering   362
liquid encapsulation   387
lithography/etching method   22
local solder assembly process   364
low-melting-point glass   8
low-pressure chemical vapor deposition (LPCVD)   71, 93
low temperature SiO2 (LTO)   93

m

magnetic flux   155
Mallory bonding   259
manifold absolute pressure (MAP)   183
Marangoni effect dryer   86
mass flow controller (MFC)   75
mass flow meters (MFM)   75
Maszara method   282
MEMS using CMOS wafer (chapter 8)   131–220
  accelerometer   148–149
  analog-to-digital conversion (ADC)   183
  anchor loss   201
  biochemical sensors   173–175
  buffer oxide etch (BHF/BOE)   143
  bulk micromachining technology   132
  capacitive micromachined ultrasonic transducers (CMUT)   176
  coefficient of thermal expansion (CTE)   155
  creep   192
  CTE mismatch   192
  design rule check (DRC)   143
  dielectric charging   203–204
  double-ended tuning fork (DETF)   145
  dry film resist (DFR)   179
  force feedback loop   186
  gas sensors   180–181
  gas and humidity sensors   164
  generic fabrication platform   136
  Hepatitis B virus (HBV)   175
  implementation and monolithic integration   138
  infrared (IR) sensors   156–158
  initial deformation   192
  interface loss   201
  KOH & TMAH (wet anisotropic)   146–147
  long-time stability   197
  manifold absolute pressure (MAP)   183
  material and interface loss   201–203
  material loss   199
  material properties   201
  metal oxide   164–170
  metal sacrificial   140–142
  microelectronics and micro-mechanical components   135
  microfluidic structures   178–180
  multi project wafer (MPW)   132
  multi-sensor integration   180–183
  nonlinearity   204–205
  1P6M (one poly Si layer and six metal layers) process   135
  oxide sacrificial   142–143
  phase locked loop (PLL)   191
  phase noise   204
  physical sensors   181–183
  polymers   170–172
  planar fabrication technology   131
  poly dimethylsiloxane (PDMS)   178
  poly etherurethane (PEUT)   181

post CMOS processes 136
pressure and acoustic sensors 175
pressure sensor 149–150
quality factor 199–203
readout circuit integration
   183–191
residual stresses of thin films
   192–199
resonators 150–152, 158–160, 165
RIE & DRIE 147
sensors 165
SF6 and XeF2 (dry isotropic)
   145–146
single-axis and tri-axis
   accelerometer 162
SOI micromachining MPW
   processes 134
surface micromachining processes
   132
tactile sensors 154–156
TC_E (temperature coefficient of
   elastic modulus) 144
TC_f (temperature coefficient of
   frequency) 144
temperature stability 200
thermal deformation 195
thermal expansion coefficient
   mismatch 195
thermoelastic damping (TED) 201
TiN-composite (TiN-C) 143–145
tire pressure monitoring system
   (TPMS) 137
thick epi-poly silicon layer 133
2P4M (two poly Si layers and four
   metal layers) process 135
vertical-parallel-plate (VPP) 171,
   187
volatile organic compounds (VOC)
   181
metal bonding (chapter 13) 289–308
  diffusion-based methods 289
  electro-chemical deposition 291

eutectic bonding 301
  Al/Ge 302–304
  Au/Si 302
  Au/Sn 304
  hermetic sealing applications
     301
  microfabrication 301
  void formation 297–298
metal thermocompression bonding
   298
  grain growth 300
  grain reorientation 299
  interface formation 299
metal wafer bond technologies
   290
solid liquid interdiffusion bonding
   (SLID) 290
  Au/Ga and Cu/Ga 294–297
  Au/In and Cu/In 291–294
  Au/Sn and Cu/Sn 297
  diffusion-based methods 289
  electric and/or thermal
     conductance 289
  Kirkendall voids 298
  3D-integration and high precision
     sensors 289
metal sacrificial 140
metal surface micromachining
   (chapter 6) 99–112
  digital micromirror device (DMD)
     104–110
  amorphous $TiAl_3$-O 108
  anti-stick layer 110
  creep resistance 108
  hinge memory 105–109
  lifetime of DMD 109
  time-sequence control 105
dynamic device 103
high-temperature annealing 99
LIGA like ultraviolet (UV)
   technology 99
MEMS switch 103

metal surface micromachining (chapter 6) (contd.)
    back-end CMOS interconnect process 103
    low resistivity metal and poly Si 103
    single-pole four-throw multiplexer configuration 104
    working frequencies 103
  residual stress 99
  static device 100–101
  static structure fixed after the single movement 101–103
  stress-free film 99
  X-ray lithographie, galvanoformung, abformung (LIGA) 100
micro cantilever 174
micro-electro-mechanical-systems (MEMS) 3, 15, 30
microencapsulation 370
microfabrication 3
microfluidic inclinometer 180
micro holes interetch & sealing 42, 49
micro-hole trenches 49
micromachining 3
micro systems 3
minimally invasive surgery 44, 49
molten solder filling 459
monolithically integrated switched capacitor circuit 379
morphotropic phase boundary (MPB) 246

n
nano-imprinting resist mr-I 9000 series 352
nano-imprint lithography 346
nanosecond laser heating/bonding process 362
narrow-ditch trench 49
Ni electroplating 403, 404
non-cryogenic steady-state etching 452
non-evaporable getter (NEG) 411

o
open circuit potential (OCP) 18
overview (chapter 1) 3–11
  Low-temperature oxide (LTO) 7
  LSI first approach 4
  micro-electro-mechanical-systems (MEMS) 3
  microstructures (MICS) 7
  phosphosilicate glass (PSG) 7
  system on chip (SoC) MEMS 3
  system in package (SiP) MEMS 3
oxide based integrated reactive material systems (oiRMS) 313
oxide dry etching 162
oxide-rich double-ended tuning-fork (DETF) resonator 145
oxygen generation 410

p
packaged resonator 398
packaging, sealing and interconnection (chapter 17) 377–407
  conventional Al metallization 377
  deposition sealing 380–385
  encapsulation and electrical interconnection 377
  interconnection by electroplating 401–404
  lateral feedthrough interconnection 395–401
  metal compression sealing 385–388
  reaction sealing 378–380
  shell packaging 380, 383

vertical feedthrough
    interconnection 388–395
  through glass via interconnection
    388–393
  through Si via interconnection
    393–395
  wafer level packaging (WLP)
    377–378
packaging stress suppressing
    suspension 49
pad module 86
parasitic capacitance 149, 155
parylene 176
periodic pulse reverse (PPR) currents
    458
photolithographic techniques 3
photolithography 3, 17
photosensitive polymer 348
physical sensors 181
physical vapor deposition 313
piezoelectric cantilevers 245
piezoelectric material 8
piezoelectric MEMS (chapter 10)
    243–256
  cantilever and microscanner
    251–254
  composition 246
  orientation 246
  electrode materials and lifetime of
    PZT thin film 250–251
  lead zirconate titanate (PZT) 243
  orientation control 248–249
  poling 254–255
  property as an actuator 244–246
  PZT MEMS fabrication 251–255
  PZT thin films 249–251
  sol-gel 249–250
  sputtering 247–248
  thick film deposition 249–250
  thin film composition and
    orientation 246
piezoresistive accelerometer 15

piezoresistive sensing 21, 23, 25
piezoresistive sensors 183
piezoresistive Si accelerometer 396
piezoresistive Wheatstone-bridge 55
piezoresistors 16, 437
piranha treatment 158
Pirani vacuum gauge 418
planar fabrication technology 131
plasma based etching 19
plasma enhanced chemical vapor
    deposition (PECVD) 71
plastic deformation 103
plastic molding 401
plug-up process 380
poling 254
poly-2 as piezoresistive sensors 152
poly-1 as thermal actuation 152
polycrystalline Si (poly Si) 6
polyethylene glycol (PEG) 458
poly-ethylene2,6-naphthalate 285
polyimide sensing film 171
polymer bonding (chapter 15)
    331–360
  dry-etch BCB 351
  localized polymer wafer bonding
    348–350
  polymer adhesion mechanisms
    332–335
  process parameters 341–348
  properties of polymers 335–337
  root mean square roughness (RMS)
    332
  thermoplastic 338
  UV curing 338
  with thermosetting polymers 352
  in wafer bonding 337–341
  wafer-to-wafer alignment in
    350–351
polymer insulators 455
polymers 170
polypropylene glycol (PPG) 458
polypyrrole film 171

poly-SiGe surface micromachining (chapter 5) 69–98
  Al-Ge bonding for microcaps 87
  atmospheric-pressure chemical vapor deposition (APCVD) 71
  CMEMS* process 78
  coefficient of thermal expansion (CTE) 79
  cost analysis 72
  deposition methods 70–71
  desired SiGe properties for MEMS 70
  electrode module 85–86
  film thickness and Ge content 76–77
  hydrogen peroxide release 86
  interface challenges 79–80
  low-temperature CVD oxide (LTO) 93
  LPCVD polycrystalline SiGe 73
  material properties comparison 71
  nano-electro-mechanical (NEM) switch 92
  pad module 86
  particle control 75
  plasma enhanced chemical vapor deposition (PECVD) 71
  plug module 84
  poly-Ge 70
  poly-SiGe 69
  poly-$Si_{1-x}Ge_x$ 70
  poly-SiGe
    Nano-Electro-Mechanical (NEM) switches 92
  post-CMOS integration 71
  probability density function (PDF) 124
  process monitoring and maintenance 75–76
  process space mapping 77–78
  process flow 80
  pulsed laser deposition (PLD) 71
  reduced-pressure chemical vapor deposition (RPCVD) 71
  resonator for electronic timing 88–92
  SiGe applications in IC and MEMS 70
  slit module 85
  spacer module 83, 85
  structural SiGe module 85
  structure module 85
  top metal module 80–84
  vertical furnace 73–75
  X-ray fluorescence (XRF) spectrometry 76
polysilicon 456
poly-silicon deposition 55
polysilicon piezoresistors 174
poly-silicon thin-film diaphragm 57
pressure sensors 54, 149, 437
printed circuit board (PCB) 58
pulse method 391
Pyrex glass wafer 392

q
quality factor 419

r
radio frequency (RF) magnetron sputtering 313
reaction sealing 378
reactive bonding (chapter 14) 309–330
  electrochemical deposition 315–319
    dual bath technology (DBT) 316–318
    single bath technology (SBT) 318–319
  foil type RMS 313
  integrated reactive material systems (iRMS) 309

material systems   311–312
oxide-based integrated reactive material systems (oiRMS)   313
physical vapor deposition   313–315
planar multilayer stack   310
planar multilayer systems   310
reactive material systems (RMSs)   309
self-propagating exothermic reaction (SER)   309–310
thermite reactions   310
vertical multilayer stack   310
vertical pillar systems   310
vertical reactive material systems   319–323
  dimensioning   320–321
  fabrication of   321–323
wafer bonding application   309
reactive deposition   269
reactive ion etching (RIE)   29, 145, 156, 282
redistribution layer (RDL)   114
release etching   424
resistive heating   365
resistive sensors   183
resonant sensors   190
resonators   150
resorcinol-formaldehyde (RF) aerogel   172

S

sacrificial etching   224
sacrificial layer   3
"scar-free" pressure sensor chip   57
SCREAM process   34
sealing and interconnection methods   10
selective etching   417, 418
self-aligned metallization   22
self-aligned process   22
self-alignment effect   351
self-assembled beam-steering micromirror   102
self-assembly   102
self-propagating exothermic reaction (SER)   310, 370
self-propagating high-temperature synthesis (SHS)   312
self-propagative reaction heating   370
Semiconductor Manufacturing International Corporation (SMIC)   69
sensing films   172
SensorNor   132
sensors hub   132
shape memory alloy actuators   401
silicon beam   439
silicon dioxide insulators   454
silicon dry etching   321
silicon fusion bonding   20
silicon micromirror array   472
silicon-on-insulator (SOI)   20, 279
SOI multiuser MEMS process (SOIMUMPs)   23
silicon on nothing (SON)   40, 49
silicon on sapphire (SOS)   423
silicon optical bench (SiOB) assembly   367
single-crystal silicon (SCS)   147, 456
single-wafer single-sided silicon-on-nothing processes   44
smart cut process   279
Smartview* approach   350
SMERALDO technology   467
solder glass   8
soldering by local heating (chapter 16)   361–376
  inductive heating   368–370
  in MEMS packaging   361
  laser soldering   362–365

soldering by local heating (chapter 16) (contd.)
   resistive heating 365–368
   self-propagative reaction heating 370–371
   ultrasonic frictional heating 371–373
solder jet bumping process 364
solder reflow process 364
sol-gel method 227
sol-gel orientation control 248
solid polysilicon TSVs 460
spacer layer deposition 82
spacer layer etch 83
spacer module 83, 85
spontaneous bonding 279
sputtering 247
state-of-the-art Bosch etchers 451
statistical element selection (SES) technique 124
steady-state etching 452
STMicroelectronics 437
structure module 85
sub-atomspheric chemical vapor deposition (SACVD) 454
superconformal Cu plating 458
superconformal plating 458
super-critical rinsing and drying 233
surface acoustic wave filters 237
surface activated bonding (SAB) 283
surface-micromachined poly Si microstructures 7
surface micromachining 7, 381, 400
surface micromachining like process 21–23
surface sputter etching 283

t
tactile sensors 154, 229–231
temperature-dependent resistor 101
tensile stress 263
test evaluation group (TEG) 266
tetramethyl ammonium hydroxide (TMAH) 146
thermal annealing 7
thermal conduction process 372
thermal expansion coefficient 390, 392
thermal oxidation 55, 66, 378, 380
thermal piezoresistive resonators (TPR) 141, 152
thermoelastic damping (TED) 201
thermoplastic polymers 335, 343
thermosetting polymers 335, 336
thin film deposition 285
through glass via interconnection 388
through-substrate-vias (TSV) (chapter 20) 443–492
  conductor formation 455–460
    conductor materials 459–460
    copper 457–459
    polysilicon 456
    single crystalline silicon 456–457
    tungsten 457
  configuration of TSVs 444–445
    air-gap TSVs 445
    hollow TSVs 445
    solid TSVs 444–445
  deep hole etching 541–454
    deep reactive ion etching 451–452
    laser ablation 452–454
  insulator formation 454–455
    air-gaps 455
    polymer insulators 455
    silicon dioxide insulators 454–455
  metal TSVs 469
    air-gap metal TSVs 480

hollow metal TSVs   474
solid metal TSVs   470
MIDIS technology   467
1,2,3-Benzotriazole (BTA)   458
Periodic pulse reverse (PPR) current   458
polysilicon TSVs   460
  air-gap polysilicon TSVs   464
  solid polysilicon TSVs   460
silicon TSVs   464–469
  air-gap silicon TSVs   467
  solid silicon TSVs   465
3D MEMS integration and WLVP   443
3-mercapto-1-propanesulfonic acid (MPS)   458
through glass vias (TGVs)   449
TSV applications in MEMS   445–450
  CMOS-MEMS 3D integration   446–447
  considerations for   450
  MEMS and CMOS 2.5D integration 447–448
  signal conduction to wafer backside   446
  wafer-level vacuum packaging (WLVP)   448–449
TiN-C free-free beam (FFB) resonator   144, 145
TiN-composite (TiN-C)   143
tire pressure monitoring system (TPMS)   137
traditional bulk micromachining process   25
transversal ultrasonic bonding   372, 373
tri-axis accelerometer   162
tunable SAW filter   240, 241
tungsten   457
two multi-pore dopant injectors   75

u

ultra-miniaturized wireless sensor nodes   473
ultrasonic Al–Al bonding   373
ultrasonication   372
ultrasonic frictional heating   371
ultrasonic soldering technique   372
UV-curable thermosetting polymers   348
UV nanoimprint lithography   321

v

vacuum packaging (chapter 18)   409–421
  by anodic bonding   409–414
  cavity vacuum pressure   412
  high vacuum cavity   411
  non-evaporable getter (NEG)   411–414
  optical deflection sensor   412
  Si diaphragm vacuum gauges   413
  soldering   410
  vacuum encapsulation   413
  with controlled cavity pressure   414–416
  by deposition   417
  by metal bonding   416–417
  hermeticity testing   417–419
  problems of   409
vacuum sealing   388
Venice process for sensors (VENSENS)   41, 437
vertical furnaces   73
vertical-parallel-plate (VPP) sensing capacitor   171
viscoelastic heating   372

w

wafer direct bonding   279, 282
wafer level liquid compression seal   386

wafer level packaging 377–378
wafer-level vacuum packaging 448
wafer transfer (chapter 9) 221–242
    BaSeTiO$_3$ (BST) 237
    BCB (benzocyclobutene) 224
    boron-doped diamond (BDD) 224
    chip level transfer 236–241
    device transfer 226–236
        via-first 231–236
        via-last 226–231
    electrochemical oxidation 232, 233
    film bulk acoustic resonator (FBAR) 223
    film transfer 221, 223–227
    heterogeneous integration 221
    low-temperature co-fired ceramics (LTCC) 221, 271, 389
    massive parallel electron beam write (MPEBW) 232, 395
    PZT (lead zirconate titanate) 228
    PZT MEMS switch 228
    tactile sensor network 228–230
    tunable SAW filter 241
    voltage–controlled oscillator 223
    wafer level packaging (WLP) 222
    wafer level transfer 223
wet etching 62, 143
wire-bonded through-Si vias 10
wire bonding 401